Boundary Element Methods in Engineering

Proceedings of the Fourth International Seminar, Southampton, England, September 1982

Editor: C. A. Brebbia

Seminar sponsored by the International Society for Computational Methods in Engineering

With 291 Figures

Springer-Verlag Berlin Heidelberg GmbH 1982

Dr. CARLOS A. BREBBIA
Computational Mechanics Centre,
Ashurst Lodge,
Ashurst,
New Forest, Hampshire.
England SO4 2AA

Library of Congress Cataloging in Publication Data
Main entry under title:

Boundary element methods in engineering.

(CME publications)
Includes bibliographical references.
1. Boundary value problems--Congresses.
I. Brebbia, C. A. II. International Society
for Computational Methods in Engineering.
III. Series.
TA347.B69B67 1982 620'.001'51535 82-10665

ISBN 978-3-662-11275-5 ISBN 978-3-662-11273-1 (eBook)
DOI 10.1007/978-3-662-11273-1

Originally published by Springer-Verlag Berlin Heidelberg New York in 1982
Softcover reprint of the hardcover 1st edition 1982

2061/3020 - 5 4 3 2 1 0

PREFACE

One of the most interesting developments in engineering analysis during the last few years has been the rapid growth of boundary element methods. The first and second international conferences on this topic held in 1978 and 1980 attracted approximately 30 papers each, most of them from a few well known groups around the world. The third meeting in 1981, produced instead approximately 40 papers, many of them from young investigators working in newly created research groups. They have been attracted to boundary elements by the many advantages of the technique and were able to assimilate rapidly, the new ideas unencumbered by previous conceptions.

That third conference held in 1981 constituted in many ways a turning point for boundary elements and it indicated for the first time a general awareness of the industry to the research being carried out in the new technique. Engineering firms started to appreciate the advantages of the method mainly from the computational aided engineering point of view. The advantages of simple data input and output was rapidly understood by those professional engineers who were forced up to them to use cumbersome finite element codes. Boundary element practitioners in close contacts with the industry started to perceive that the method was gathering a critical momentum of its own. This is now more evident by the diversity and quality of the papers in this volume, which are the edited Proceedings of the 4th International Conference, held at the University of Southampton in September 1982.

Past conferences were characterized by papers reflecting a tentative trust into non-linear and time dependent problems; while a substantial number of contributions on the relationships of finite and boundary element techniques pointed out the difficulty of abandoning a series of preconceptions. Variational techniques for instance were invocated to justify boundary integral treatments, though the same results could be obtained using error minimization concepts. The variational approach is the one frequently associated with finite element techniques, while finite differences for instance are usually related to the vectorial theory of mechanics.

The predominance of one or the other of these two current of thought, has characterized the two main philosophies behind the development of analytical methods in engineering sciences.

The Boundary Element Method, however, defies being classified in any of these categories, while having some characteristics common to variational methods it requires the vectorial approach to find fundamental solutions or influence functions. This intermediate position is one of the strongest advantages of the method as it can profit from advances in the two different mathematical ways of describing a natural phenomena, i.e. the Newtonian vectorial mechanics and the energy approach of Leibnitz and many others.

If the third conference marked the turning point for boundary elements, this 4th meeting has consolidated the newly found confidence in the future of the method. These Proceedings cover the solution of a wide range of new problems and applications, including material and - for the first time - geometrically non-linear problems. Time dependent and vibrations studies are also well represented in this book. More effort is also present in practical applications and industrial uses of boundary element codes. This demonstrates the healthy growth of the techniques which is now becoming an accepted tool of engineering analysis.

In view of the considerable interest shown by Japanese researchers in the new method it is fitting that the next Conference should be held in Japan, and the Organizing Committee has accepted an invitation from the University of Hiroshima to accommodate the Conference there in 1983. The meeting will continue to be sponsored by the International Society for Computational Methods in Engineering, which has become a well established focus for boundary element research.

Carlos Brebbia
Southampton, 1982.

C O N T E N T S

Session I
Basic Principles

THEORETICAL AND EXPERIMENTAL ASYMPTOTIC CONVERGENCE OF THE BOUNDARY INTEGRAL METHOD FOR A PLANE MIXED BOUNDARY VALUE PROBLEM

U. Lamp, T. Schleicher, E. Stephan and W. Wendland
Department of Mathematics, Technische Hochschule Darmstadt, Germany

INTRODUCTION

In this paper we present a numerical procedure via the direct boundary integral method for solving plane mixed boundary value problems in smooth domains. Since the solution behaves at collision points as at crack tips we modify the Galerkin collocation method of [5],[6] by using besides smooth quadratic B splines appropriate singular elements. This improves significantly both the accuracy of computations and the order of convergence.

Our procedure has the following advantages:
(i) the stress intensity factors are computed simultaneously together with the approximate desired boundary charges,
(ii) the numerical results are accurate showing superconvergence and are in agreement with the theoretical orders of convergence in [2],[7],[8] and [11],
(iii) as for most BIE-methods the input of data is very simple and our program handles easily arbitrarily shaped smooth boundary curves and besides the mixed problem also the pure Dirichlet and Neumann problems.

Our method hinges crucially on the decomposition of the solutions into sums of singularity functions and smooth remainders where the latter are approximated up to the collision points. This is due to Galerkin's method and we do not blend the splines with the singularity functions as in the collocation methods [4],[12]. Furthermore our modified Galerkin collocation is more accurate than the usual collocation method [1], [10] and our numerical experiments can be compared with our theoretical error estimates.

FORMULATION OF THE BOUNDARY VALUE PROBLEMS AND THE BOUNDARY INTEGRAL METHOD

Here we apply the boundary integral method to the mixed boundary value problem

$$\begin{aligned} \Delta u &= 0 \quad \text{in} \quad \Omega \subset \mathbb{R}^2 \ (\text{or } \mathbb{R}^2 \setminus \bar{\Omega}) \,, \\ u &= g_1 \quad \text{on} \quad \Gamma_1 \,, \\ \frac{\partial u}{\partial n} &= g_2 \quad \text{on} \quad \Gamma_2 \,. \end{aligned} \tag{1}$$

At infinity for the exterior problem we require the asymptotic behaviour

$$u = b \log |z| - \frac{1}{2}\omega + O\left(\frac{1}{|z|}\right) \tag{2}$$

where b is a given constant and $\frac{1}{2}\omega$ the logarithmic capacity to be determined. The bounded domain Ω is bounded by a smooth Jordan curve $\Gamma = \Gamma_1 \cup \Gamma_2 \cup Z_1 \cup Z_2$ where Γ_1, Γ_2 denote two disjoint open parts of Γ with common end points Z_1 and Z_2. $\frac{\partial}{\partial n}$ denotes the outer normal derivative. We identify $\mathbb{R}^2$ with the complex plane $\mathbb{C}$ by $z = x + iy \simeq (x,y)$. For applications yielding problem (1) see [2],[9],[11].

The boundary integral equations are derived by the direct method from the Green representation formula,

$$\begin{aligned} \lambda\left(u(z) + \frac{1}{2}\omega\right) = \frac{1}{2\pi} \int_\Gamma u(\zeta) \frac{\partial}{\partial n_\zeta} (\log |z-\zeta|) ds_\zeta \\ - \frac{1}{2\pi} \int_\Gamma \frac{\partial u}{\partial n}(\zeta) \log |z-\zeta| ds_\zeta \end{aligned} \tag{3}$$

where $\lambda = +1$ (and $\omega=0$) for the interior problem, $\lambda = -1$ for the exterior problem, s_ζ the arc length at $\zeta \in \Gamma$. In addition we have

$$\int_\Gamma \frac{\partial u}{\partial n} ds = 2\pi b \tag{4}$$

with $b = 0$ for the interior problem and b given by Equation (2) for the exterior problem, respectively. From Equation (3) one obtains a system of boundary integral equations on Γ_2, Γ_1 for the yet unknown layers $u|_{\Gamma_2}$, $\frac{\partial u}{\partial n}|_{\Gamma_1}$ and ω. Note that this system including ω and the side condition (4) is always uniquely solvable whereas the system without ω and condition (4) for the interior problem possesses an eigensolution for Γ having conformal radius 1. Thus we shall use always the full system.

As was already pointed out in [7], the approximations of $u|_{\Gamma_2}$ and $\frac{\partial u}{\partial n}|_{\Gamma_1}$ by splines converge poorly due to the pollution generated by the singularities of grad u at Z_1, Z_2. These singularities are given by the representation of the variational solution in the form [11]

$$u(z) = \sum_{j=1}^{2} \alpha_j \chi_j \tilde{\rho}_j^{1/2} \sin \frac{1}{2} \Theta_j + \tilde{v}(z) \tag{5}$$

where $\tilde{\rho}_j = |z-Z_j|$, Θ_j the respective angle between the tangent vector at the collision point Z_j in the direction of Γ_1 and the ray $z - Z_j$, χ_j a suitable smooth cut-off function, $\alpha_j \in \mathbb{R}$ the respective stress intensity factor and $\tilde{v}(z)$ a continuously differentiable remaining function. The representation (5) is the first order section of a local expansion near corners and collision points [2],[3] and serves as a first regularization via the Fix method. More precisely we write

$$u|_{\Gamma_2} = \sum_{j=1}^{2} \alpha_j \rho_j^{1/2} \chi_j + w \tag{6}$$

and

$$\frac{\partial u}{\partial n}\Big|_{\Gamma_1} = -\frac{1}{2} \sum_{j=1}^{2} \alpha_j \rho_j^{-1/2} \chi_j + \psi \tag{7}$$

with the unknowns $\alpha_j \in \mathbb{R}$, w, and ψ satisfying

$$w(Z_j) = g_1(Z_j) \quad \text{and} \quad \psi(Z_j) = g_2(Z_j) \; , \quad j = 1,2 \; . \tag{8}$$

For our computations, the boundary curve will be given by a smooth 2-periodic parameter representation

$$z = z(t) = x(t) + iy(t) \; , \quad t \in \mathbb{R} \tag{9}$$

where for $0 < t < 1$ Equation (9) gives Γ_1 and for $1 < t < 2$ Equation (9) gives Γ_2. All functions on Γ will be identified with 2-periodic functions of t.

The distance functions ρ_j on Γ_p are chosen as

$$\rho_j(t) = \begin{cases} |t| \dfrac{|\dot{z}(t)|^2}{|\dot{z}(0)|} \; , & j = 1 \; , \; |t| \le 1 \; , \\[2ex] |1-t| \dfrac{|\dot{z}(t)|^2}{|\dot{z}(1)|} \; , & j = 2 \; , |1-t| \le 1 \end{cases} \tag{10}$$

and their 2-periodic extensions. Equation (10) consides on Γ locally near Z_j with the Euclidian distance corresponding to the local decomposition (5) of the exact solution to problem (1). The factors $|\dot{z}(t)|^2 / |\dot{z}(t_j)|$ are chosen to simplify the integrations in the Galerkin procedure. The cut-off functions on Γ are chosen by

$$\chi_j(t) = \begin{cases} 1 & \text{for } |t-(j-1)| \le \frac{\delta}{2} \\ \sqrt{|t-(j-1)|}\; q(t-(j-1)) & \text{for } \frac{\delta}{2} < |t-(j-1)| \le \delta \\ 0 & \text{otherwise} \end{cases} \qquad (11)$$

where

$$q(t) = \frac{3}{2}\left(\frac{2t}{\delta}\right)^3 - \frac{13}{2}\left(\frac{2t}{\delta}\right)^2 + 8\left(\frac{2t}{\delta}\right) - 2 \,. \qquad (12)$$

The parameter $\delta = 0{,}2$ is chosen according to numerical experiments [7],[8] . With Equations (6), (7) the boundary integral equations and side conditions now read as

$$\begin{aligned} \lambda w(z) &- \frac{1}{\pi}\int_{\Gamma_2} w(\zeta)d\Theta_z + \frac{1}{\pi}\int_{\Gamma_1} \psi(\zeta)\log|z-\zeta|ds_\zeta \\ &+ \sum_{j=1}^{2} \alpha_j[\lambda\rho_j^{\frac{1}{2}}\chi_j - \frac{1}{\pi}\int_{\Gamma_2}\rho_j^{\frac{1}{2}}\chi_j d\Theta_z - \frac{1}{2\pi}\int_{\Gamma_1}\rho_j^{-\frac{1}{2}}\chi_j \log|z-\zeta|ds_\zeta] \\ &= \frac{1}{\pi}\int_{\Gamma_1} g_1 d\Theta_z - \frac{1}{\pi}\int_{\Gamma_2} g_2 \log|z-\zeta|ds_\zeta - \lambda\omega \\ &= F_1(z) - \lambda\omega \qquad \text{for } z \in \Gamma_2 \end{aligned} \qquad (13)$$

and

$$\begin{aligned} &- \frac{1}{\pi}\int_{\Gamma_1} \{\psi(\zeta) - \frac{1}{2}\sum_{j=1}^{2}\alpha_j\rho_j^{-\frac{1}{2}}\chi_j(\zeta)\}\log|z-\zeta|ds_\zeta \\ &+ \frac{1}{\pi}\int_{\Gamma_2} w d\Theta_z + \sum_{j=1}^{2}\alpha_j \frac{1}{\pi}\int_{\Gamma_2}\rho_j^{\frac{1}{2}}\chi_j d\Theta_z \\ &= \lambda g_1(z) - \frac{1}{\pi}\int_{\Gamma_1} g_1 d\Theta_z + \frac{1}{\pi}\int_{\Gamma_2} g_2 \log|z-\zeta|ds_\zeta + \lambda\omega \\ &= F_2(z) + \lambda\omega \qquad \text{for } z \in \Gamma_1 \,, \end{aligned} \qquad (14)$$

$$\int_{\Gamma_1} \{\psi - \frac{1}{2}\sum_{j=1}^{2}\alpha_j\rho_j^{-\frac{1}{2}}\chi_j\}ds = -\int_{\Gamma_2} g_2 ds + 2\pi b = B \,, \qquad (15)$$

$$w(Z_j) = g_1(Z_j) \quad \text{and} \quad \psi(Z_j) = g_2(Z_j) \,, \quad j = 1,2 \,. \qquad (16)$$

Here

$$d\Theta_z = \frac{\partial}{\partial n_\zeta}(\log|z-\zeta|)ds_\zeta \,. \qquad (17)$$

For the error analysis and precise formulation of the foregoing boundary integral equations we specify the corresponding function spaces of the data and solutions by using Sobolev spaces on Γ and Γ_p . $H^r(\Gamma)$, $r \in \mathbb{R}$ is given by all distributions f on Γ with Fourier series

$$f(t) = \sum_{k=-\infty}^{+\infty} \hat{f}_k \, e^{ik\pi t} , \tag{18}$$

$$\hat{f}_k = \frac{1}{2} \int_0^2 f(t) \, e^{-ik\pi t} dt \tag{19}$$

satisfying

$$\| f \|_{H^r(\Gamma)} := \{ \sum_{|k|>0} |k|^{2r} |\hat{f}_k|^2 + |\hat{f}_o|^2 \}^{\frac{1}{2}} < \infty . \tag{20}$$

Then

$$H^r(\Gamma_p) = \{ f = F_{|\Gamma_p} \quad \text{with} \quad F \in H^r(\Gamma) \quad \text{and} \quad \| f \|_{H^r(\Gamma_j)} = \inf_F \| F \|_{H^r(\Gamma)} \} . \tag{21}$$

If the boundary data in problem (1) are given with

$$g_1 \in H^{\frac{3}{2} + \sigma}(\Gamma_1) \quad \text{and} \quad g_2 \in H^{\frac{1}{2} + \sigma}(\Gamma_2) , \quad 0 < \sigma < \frac{1}{2} , \tag{22}$$

then the unknowns in Equations (13) - (16) are

$$\omega, \alpha_j \in \mathbb{R} , \; j = 1,2 , \; w \in H^{\frac{3}{2} + \sigma}(\Gamma_2) , \; \psi \in H^{\frac{1}{2} + \sigma}(\Gamma_1)$$

which are determined such that

$$\tilde{u} := \{ g_1 \text{ on } \Gamma_1 \text{ and } w \text{ on } \Gamma_2 \} \in H^1(\Gamma) \quad \text{and}$$
$$\tilde{\phi} := \{ \psi \text{ on } \Gamma_1 \text{ and } g_2 \text{ on } \Gamma_2 \} \in H^{\frac{1}{2} + \sigma}(\Gamma) . \tag{23}$$

(The detailed analysis of the integral equations (13) - (16) with regularity (22), (23) will be presented in [8]).

THE NUMERICAL PROCEDURE

The numerical solution of Equations (13) - (16) is executed by a modification of the Galerkin collocation method, see [5],[7], [8]. For the approximation of w and ψ we use smooth quadratic splines on [0,1] and [1,2], respectively, subject to a uniform grid of the t-variable in the form

$$w_h(t) = \sum_{\ell=-2+M}^{2M-1} w_{h\ell} \, \mu\left(\frac{t}{h} - \ell\right) \quad \text{for} \quad t \in [1,2] , \tag{24}$$

$$\psi_h(t) = \sum_{\ell=-2}^{M-1} \psi_{h\ell} |\dot{z}(t)|^{-1} \mu(\frac{t}{h} - \ell) \quad \text{for} \quad t \in [0,1] \tag{25}$$

where $\mu(\eta)$ is the piecewise quadratic shape function with supp $\mu = [0,3]$, knots at $0,1,2,3$ and symmetric with respect to $3/2$ (see Equation (2.22) in [7]) and $h^{-1} = M$, a sequence of natural numbers. By $H_h(\Gamma_p)$ we denote the splines spanned by functions of the form (24) for $p = 2$ and form (25) for $p = 1$. For the Fix method we augment the space $H_h(\Gamma)$ by the singularity functions $\Xi_j := \rho_j^{-1/2}\chi_j$. The Galerkin equations to Equations (13) - (16) read as

$$\begin{aligned}\int_{\Gamma_2} \{\lambda w_h - \frac{1}{\pi}\int_{\Gamma_2} w_h d\Theta_z + \frac{1}{\pi}\int_{\Gamma_1} \psi_h \log|z-\zeta| ds_\zeta \\ + \sum_{j=1}^{2} \alpha_{jh}[\lambda\rho_j^{1/2}\chi_j - \frac{1}{\pi}\int_{\Gamma_2} \rho_j^{1/2}\chi_j d\Theta_z \\ - \frac{1}{\pi}\int_{\Gamma_1} \rho_j^{-1/2}\chi_j \log|z-\zeta| ds_\zeta]\} \frac{v_h}{|\dot{z}(t)|} ds_z \\ = \int_{\Gamma_2} \{\frac{1}{\pi}\int_{\Gamma_1} g_1 d\Theta_z - \frac{1}{\pi}\int_{\Gamma_2} g_2 \log|z-\zeta| ds_\zeta - \lambda\omega_h\} \frac{v_h}{|\dot{z}(t)|} ds_z\end{aligned} \tag{26}$$

for all $v_h \in H_h(\Gamma_2)$ with $v_h(Z_j) = 0$, $j = 1,2$,

$$\begin{aligned}\int_{\Gamma_1} \{-\frac{1}{\pi}\int_{\Gamma_1} [\psi_h - \frac{1}{2}\sum_{j=1}^{2}\alpha_{jh}\rho_j^{-1/2}\chi_j] \log|z-\zeta| ds_\zeta \\ + \frac{1}{\pi}\int_{\Gamma_2} w_h d\Theta_z + \sum_{j=1}^{2}\alpha_{jh}\frac{1}{\pi}\int_{\Gamma_2}\rho_j^{1/2}\chi_j d\Theta_z\}\Xi\, ds_z \\ = \int_{\Gamma_1} \{\lambda g_1(z) - \frac{1}{\pi}\int_{\Gamma_1} g_1 d\Theta_z + \frac{1}{\pi}\int_{\Gamma_2} g_2 \log|z-\zeta| ds_\zeta + \lambda\omega_h\}\Xi ds_z\end{aligned} \tag{27}$$

for all $\Xi \in H_h(\Gamma_1)$ with $\Xi(Z_j) = 0$ and $\Xi = \Xi_j$, $j = 1,2$,

$$\int_{\Gamma_1} [\psi_h - \frac{1}{2}\sum_{j=1}^{2} \alpha_{jh}\rho_j^{-1/2}\chi_j] ds = -\int_{\Gamma_2} g_2 ds + 2\pi b , \tag{28}$$

$$w_h(Z_j) = g_1(Z_j) , \quad \psi_h(Z_j) = g_2(Z_j) , \quad j = 1,2 . \tag{29}$$

In order to perform all the integrations in the Galerkin equations only with respect to the variable t ($ds = |\dot{z}(t)|dt$) we have chosen the splines in the special form (24) and (25). This choice offers us several advantages: Many of the integrals become simply computable and independent of the shape of Γ, a suitable choice of the parametrization $z(t)$ gives enough freedom for incorporating the shape of the boundary Γ as well as oscillations of the solution. E.g. at sharply curved parts of Γ the parametrization should be chosen such that a dense distribution of grid points corresponds to the equidistant subdivision of the parameter intervals.

The evaluation of the right hand sides has been performed in two steps. First the functions g_p are approximated by g_{ph} via L_2-projection, namely

$$g_{1h}(t) = \sum_{\ell=-2}^{M-1} g_{1h\ell}\,\mu(\tfrac{t}{h} - \ell) \quad \text{for} \quad t \in [0,1]$$

satisfying

$$\int_{\Gamma_1} g_{1h}\, v_h\, ds = \int_{\Gamma_1} g_1\, v_h\, ds \quad \text{for all} \quad v_h \in H_h(\Gamma_1)$$

and

$$g_{2h}(t) = \sum_{\ell=-2+M}^{2M-1} g_{2h\ell}\,|\dot{z}(t)|^{-1}\,\mu(\tfrac{t}{h} - \ell) \quad \text{for} \quad t \in [1,2]$$

satisfying

$$\int_{\Gamma_2} g_{2h}\, v_h\, ds = \int_{\Gamma_2} g_2\, v_h\, ds \quad \text{for all} \quad v_h \in H_h(\Gamma_2)\ .$$

Then g_p everywhere in (26) - (29) is replaced by g_{ph} .

In the second step we obtain for the evaluation of the right hand sides the same double integrals as for the stiffness matrix coefficients on the left hand sides. For the weights involving the logarithmic kernel and the shape function we have used the numerical integrations developped in [5] for Galerkin collocation. For the weights containing $d\Theta$ we used a combination of integration by parts and appropriate numerical integration against the shape function μ (see Equations (3.2) - (3.5) in [7]). For the weights involving the singularity functions we first transformed the integration variable and then used Gaussian quadrature formulas of seventh degree (with and without logarithmic weight) (see Equations (3.11) - (3.13) in [7]).

The smooth remaing parts are also treated with the same Gaussian quadratures in contrary to the collocation formulas in [5] .

ASYMPTOTIC CONVERGENCE

As is pointed out in [7],[8] the consistency estimates of the numerically performed Galerkin method provide the same oders of convergence of the computed solutions as for the Galerkin method, namely:

Theorem [7] : *There exists a meshwidth* $h_o > 0$ *such that the numerically integrated Galerkin equations corresponding to (26) - (29) are uniquely solvable for any* h *with* $0 < h \le h_o$. *For* $h \to 0$ *we have the asymptotic error estimates*

$$\sum_{j=1}^{2} |\alpha_{jh} - \alpha_j| + \| \psi_h - \psi \|_{\tilde{H}^{t-1}(\Gamma_1)} + \| w_h - w \|_{H^t(\Gamma_2)} + |\omega_h - \omega| \le c h^{r-t-\varepsilon} \{ \| g_1 \|_{H^r(\Gamma_1)} + \| g_2 \|_{H^{r-1}(\Gamma_2)} \} \tag{30}$$

for $1 \le t \le r < 2$ *and any* $\varepsilon > 0$ *and*

$$\| \psi_h - \psi - \sum_{j=1}^{2} \frac{1}{2} (\alpha_{jh} - \alpha_j) \rho_j^{-1/2} \chi_j \|_{\tilde{H}^{t-1}(\Gamma_1)} + |\omega_h - \omega| + \| w_h - w + \sum_{j=1}^{2} (\alpha_{jh} - \alpha_j) \rho_j^{1/2} \chi_j \|_{H^t(\Gamma_2)} \le c h^{r-t-\varepsilon} \{ \| g_1 \|_{H^r(\Gamma_1)} + \| g_2 \|_{H^{r-1}(\Gamma_2)} \} \tag{31}$$

for $-1 < t \le r < 2$, $t < \frac{3}{2}$ *and* $\varepsilon > 0$ *if* $0 < t < \frac{3}{2}$ *and* $\varepsilon = 0$ *if* $-1 < t \le 0$. *The constant* c *is independent of* h *and the solutions but may depend on* ε .

Remark: For $t = 0$ Estimate (31) gives for smooth g_j the order of convergence $O(h^{2-\eta})$ with any $\eta > 0$. If one neglects the singularity functions Ξ_j in the numerical procedure then the corresponding order is only $O(h^{1-\eta})$ due to the pollution effect of the singularities (Estimate (2.10) in [7]). Moreover our augmented method computes the stress intensity factors simultaneously without additional time consuming computations, and their order of convergence is $O(h^{1-\eta})$. Further note that the Estimate (31) provides even superapproximation for $t = -1 + \varepsilon$, i.e. an order $O(h^{3-\eta})$. As we shall discuss later this order is revealed by our numerical experiments as a superconvergence effect at the nodal points.

NUMERICAL EXAMPLES

The following numerical experiments have been carried out on the IBM H 370-168 computer at the Technische Hochschule Darmstadt. The computing times were in all 4 examples for 40 grid points 20 sec., for 80 grid points 47 sec. and for 160 grid points 127 sec. In all examples we take $\delta = 0.2$.

For Ω we choose the unit disc with

$$\Gamma_1 = \{z = \sin \pi t - i \cos \pi t \mid 0 < t < 1\},$$

$$\Gamma_2 = \{z = \sin \pi t - i \cos \pi t \mid 1 < t < 2\},$$

$$Z_1^- = -i , \ Z_2 = +i .$$

Example 1:

$$u = \operatorname{Im} \sqrt{z-i} = -\tilde{\rho}_2^{1/2} \sin \frac{1}{2} \Theta_2$$

with $\tilde{\rho}_2 \cos \Theta_2 = x$, $\tilde{\rho}_2 \sin \Theta_2 = 1-y$, $\tilde{\rho}_2^2 = x^2 + (1-y)^2$.

Here $\alpha_1 = 0$, $\alpha_2 = -1$.

The given data are

$$u|_{\Gamma_1} = -\tilde{\rho}_2^{1/2} \sin \frac{1}{2} \Theta_2 , \quad 0 \le \Theta_2 \le \frac{\pi}{2} \quad \text{and}$$

$$\frac{\partial u}{\partial n}\Big|_{\Gamma_2} = \frac{1}{2} \tilde{\rho}_2^{-1/2} (x \sin \frac{1}{2} \Theta_2 + y \cos \frac{1}{2} \Theta_2)|_{\Gamma_2} .$$

Table 1 : Largest absolute errors in Example 1

Number of grid points	40	80	160	Exp.order
error of $w\|_{\Gamma_2}$	$2.6 \cdot 10^{-2}$	$3.4 \cdot 10^{-3}$	$4 \cdot 10^{-4}$	3.08
error of $\psi\|_{\Gamma_1}$	$1.1 \cdot 10^{-2}$	$1.9 \cdot 10^{-3}$	$2.4 \cdot 10^{-4}$	2.85
error of α_j	$5.9 \cdot 10^{-3}$	$3.1 \cdot 10^{-3}$	$1.6 \cdot 10^{-3}$	0.99
values of ω	$1 \cdot 10^{-6}$	$3.7 \cdot 10^{-7}$	$2.1 \cdot 10^{-6}$	——

The experimental asymptotic convergence rates show superconvergence underlining the theoretical order in (31) for w and ψ with $t = -1+\eta$ and $r = 2-\eta$ whereas the order $O(h^{1-\varepsilon})$ of $|\alpha_{jh} - \alpha_j|$ in (30) coincides with the numerical results.

In all examples ω is so small that the rounding effects destroy the convergence order.

Example 2:

$$u = \operatorname{Re} \{\sqrt{2} \frac{\sqrt{(x+iy)^2+1}}{x+iy+1} - \frac{1-x-iy}{1+x+iy}\}^2 .$$

Here $\alpha_1 = \alpha_2 = 2\sqrt{2} = 2.8284 \ldots$

and the given data are

$$u|_{\Gamma_1} = \mathrm{Re}\,\{\sqrt{2}\,\frac{\sqrt{(x+iy)^2+1}}{x+iy+1} - \frac{1-x-iy}{1+x+iy}\}^2\Big|_{\Gamma_1}\,, \quad \frac{\partial u}{\partial n}\Big|_{\Gamma_2} = 0\,.$$

Table 2: Largest absolute errors in Example 2

Number of grid points	40	80	160	Exp.order
error of $w|_{\Gamma_2}$	$7\cdot 10^{-2}$	$9.5\cdot 10^{-3}$	$1.1\cdot 10^{-3}$	2.95
error of $\psi|_{\Gamma_1}$	$5.3\cdot 10^{-2}$	$2.9\cdot 10^{-2}$	$1.9\cdot 10^{-2}$	0.82
error of α_j	$8.3\cdot 10^{-2}$	$3.9\cdot 10^{-2}$	$1.9\cdot 10^{-2}$	1.06
values of ω	$9.8\cdot 10^{-5}$	$2.1\cdot 10^{-6}$	$1.2\cdot 10^{-5}$	——

The experimental order of the error for $\psi(z)$ depends significantly on the distance τ of z to Z_j. The maximal errors above appear at $\tau = 10^{-2}\pi$ whereas at $\tau = 6\cdot 10^{-2}\pi$ the experimental order is 2.29.

Example 3:

$$u = y^2 - x^2\,.$$

Here $\alpha_1 = \alpha_2 = 0$

and the given data are

$$u|_{\Gamma_1} = \cos^2\pi t - \sin^2\pi t \qquad \text{for} \quad 0 < t < 1\,,$$

$$\frac{\partial u}{\partial n}\Big|_{\Gamma_2} = 2(\cos^2\pi t - \sin^2\pi t) \qquad \text{for} \quad 1 < t < 2\,.$$

Table 3: Largest absolute errors in Example 3

Number of grid points	40	80	160	Exp.order
error of $w|_{\Gamma_2}$	$2.5\cdot 10^{-4}$	$3.1\cdot 10^{-5}$	$8.6\cdot 10^{-6}$	2.4
error of $\psi|_{\Gamma_1}$	$5\cdot 10^{-4}$	$6.2\cdot 10^{-5}$	$7.6\cdot 10^{-4}$	(2.93)
error of α_j, j=1,2	$2.2\cdot 10^{-4}$	$1.9\cdot 10^{-5}$	$3.4\cdot 10^{-5}$	——
values of ω	$7\cdot 10^{-10}$	$1.4\cdot 10^{-12}$	$1.4\cdot 10^{-10}$	——

Since u is very smooth, the contamination effect dominates for large numbers of grid points and the above experimental orders in brackets are evaluated only between 40 and 80 grid points. Moreover the errors for ψ for 160 grid points are larger than expected because the integrals involving the singularity functions in Equation (27) are not evaluated accurately enough. In this example the choice of $\delta = 0.01$ improved the results significantly to $4 \cdot 10^{-6}$ for w, $1 \cdot 10^{-5}$ for ψ and 2.10^{-8} for α_j, $j = 1,2$.

Example 4:

$$u = x$$

Here $\alpha_1 = \alpha_2 = 0$ and

$$u|_{\Gamma_1} = \sin \pi t \quad \text{for} \quad 0<t<1 \; ; \quad \frac{\partial u}{\partial n}\Big|_{\Gamma_2} = \sin \pi t \quad \text{for} \quad 1<t<2 \; ;$$

Table 4: Largest absolute errors in Example 4

Number of grid points	40	80	160	Exp.order
error of $w\|_{\Gamma_2}$	$3.3 \cdot 10^{-5}$	$4 \cdot 10^{-6}$	$7.5 \cdot 10^{-7}$	3.1
error of $\psi\|_{\Gamma_1}$	$1.2 \cdot 10^{-4}$	$3.8 \cdot 10^{-5}$	$6.4 \cdot 10^{-5}$	(1.7)
error of α_j, j=1,2	$6 \cdot 10^{-4}$	$5.5 \cdot 10^{-6}$	$4.3 \cdot 10^{-3}$	—
values of ω	$1.9 \cdot 10^{-9}$	$4.3 \cdot 10^{-13}$	$1.8 \cdot 10^{-10}$	—

Here we have contamination effects as in Example 3.

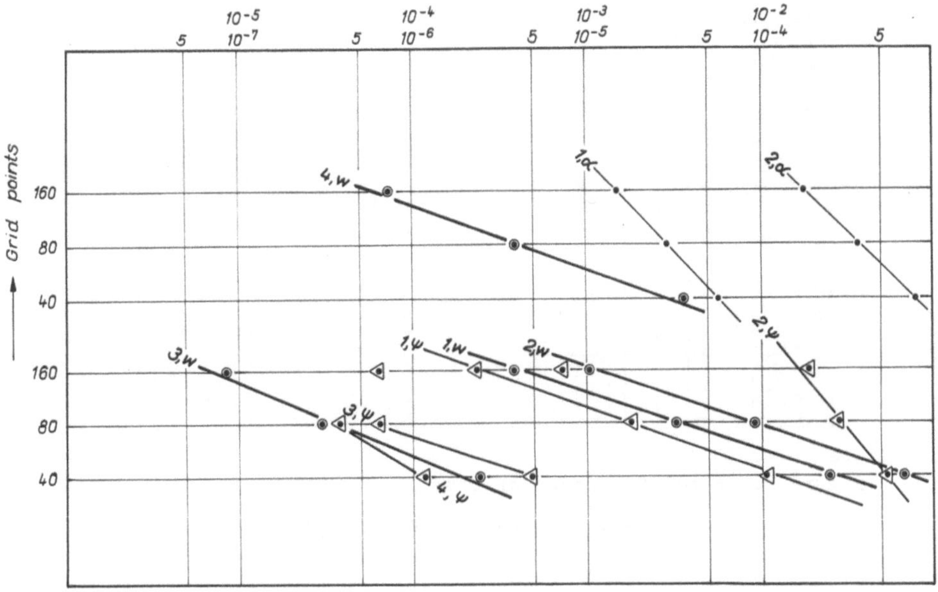

Figure 1: Max. absolute errors of w •, ψ △, α • in Examples 1-4

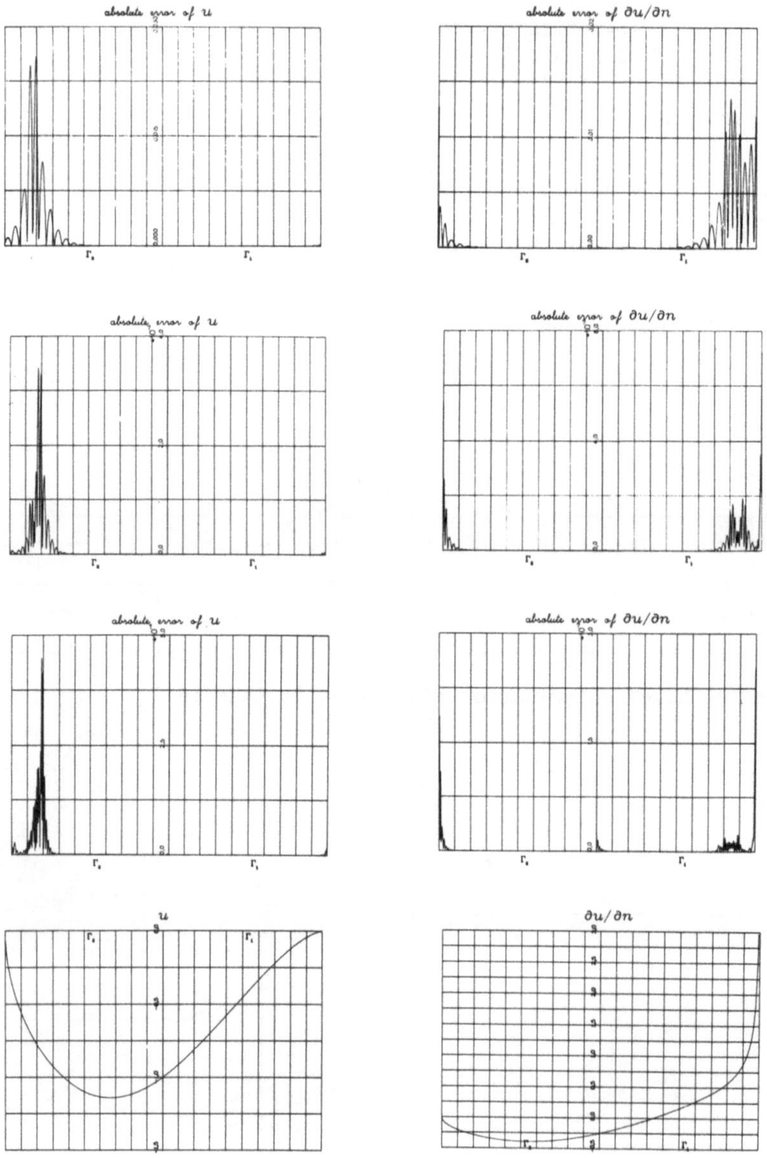

Figure 2: Absolute errors of Example 1 for 40, 80 and 160 grid points and the exact solutions u and $\frac{\partial u}{\partial n}$ on Γ .

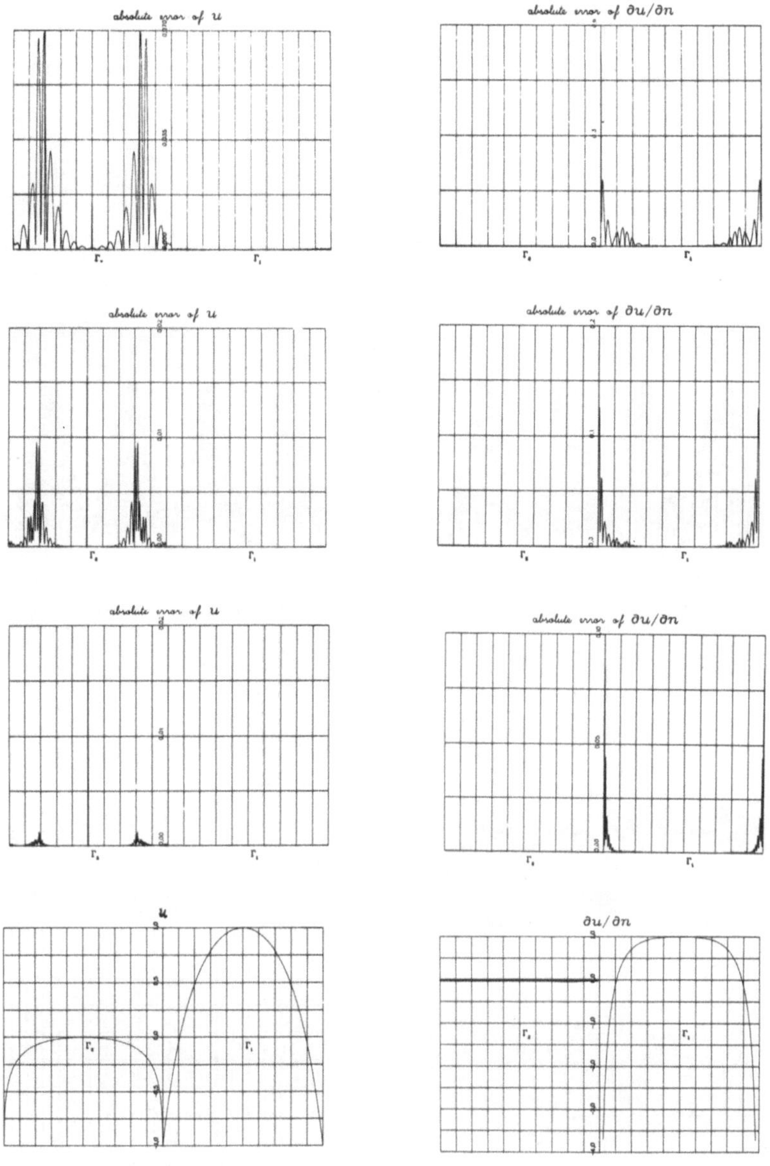

Figure 3: Absolute errors of Example 2 for 40, 80 and 160 grid points and the exact solutions u and $\frac{\partial u}{\partial n}$ on Γ .

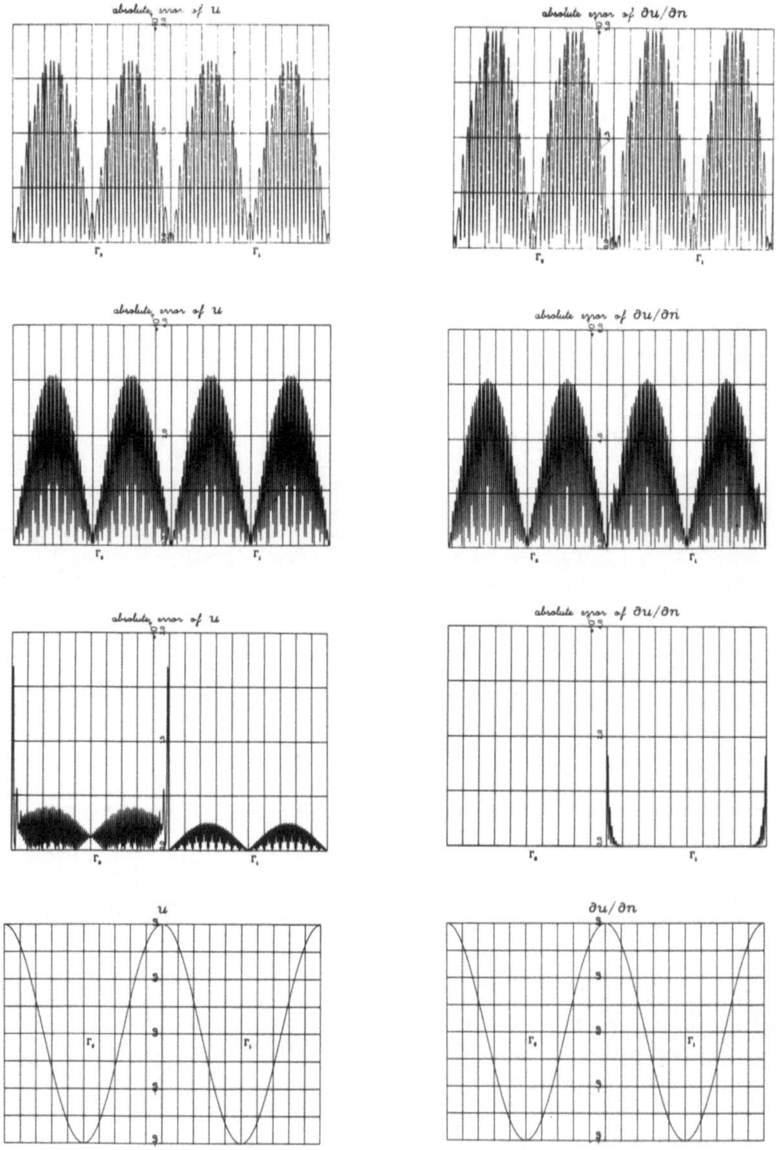

Figure 4: Absolute errors of Example 3 for 40, 80 and 160 grid points and the exact solutions u and $\frac{\partial u}{\partial n}$ on Γ .

All the plots show the influence of the cut-off function χ_j in our numerical procedure. Whereas the singularities are well approximated we see local loss of accuracy produced by the cut-off functions and corresponding oscillations.

REFERENCES:

[1] Blue, J.L. (1978) Boundary integral solutions of Laplace's equation. The Bell System Tech. Journal, 57, 2797-2822.

[2] Costabel, M. and Stephan, E. (1981) Boundary integral equations for mixed boundary value problems in polygonal domains and Galerkin approximation. TH Darmstadt Preprint #593.

[3] Grisvard, P. (1980) Boundary value problems in non-smooth domains, University of Maryland, Dept. Mathematics, College Park, Md. 20742, Lecture Notes #19.

[4] Hough, D.M. and Papamichael, N. (1980) The use of splines and singular functions in an integral equation method for conformal mapping, Brunel Univ. Report 99.

[5] Hsiao, G.C., Kopp, P. and Wendland, W.L. (1980) A Galerkin collocation method for some integral equations of the first kind. Computing, 25, 89-130.

[6] Hsiao, G.C., Kopp, P. and Wendland, W.L. (1980) The synthesis of the collocation and the Galerkin method applied to some integral equations of the first kind, New Developments in Boundary Element Methods (C.A. Brebbia ed., CML Publ., Southampton), 122-136.

[7] Lamp, U., Schleicher, T., Stephan, E. and Wendland, W.L. (1981) The boundary integral method for a plane mixed boundary value problem , Advances in Computer Methods for Partial Differential Equations - IV, IMACS, 223-229.

[8] Lamp, U., Schleicher, T., Stephan, E. and Wendland, W.L. (1982) Galerkin collocation for an improved boundary element method for a plane mixed boundary value problem, to appear.

[9] Stephan, E. and Wendland, W.L. (1982) Boundary element method for membrane and torsion crack problems. Comp. Meth. Appl. Mech. Eng. (to appear).

[10] Symm, G.T. (1973) Treatment of singularities in the solution of Laplace's equation by an integral equation method, NPL Report NAC 31.

[11] Wendland, W.L. Stephan, E. and Hsiao, G.C. (1979) On the integral equation method for the plane mixed boundary value problem of the Laplacian. Math. Meth. in the Appl. Sci., 1, 265-321.

[12] Xanthis, L.S., Bernal, M.J.M. and Atkinson, A. (1981) The treatment of singularities in the calculation of stress intensity factors using the boundary integral equation method. Comp. Meth. Appl. Mech. Eng., 26.

COLLOCATION VERSUS GALERKIN PROCEDURES FOR BOUNDARY INTEGRAL METHODS

D.N. Arnold

Department of Mathematics and
Institute for Physical Science
and Technology
University of Maryland
College Park, Md. 20742 U.S.A.

W.L. Wendland

Fachbereich Mathematik
Technische Hochschule
Darmstadt
Schlossgartenstr. 7
D-6100 Darmstadt, Germany

INTRODUCTION

A comparison of the computational complexities of the finite element domain method with the boundary element method for elliptic interior boundary value problems shows that these are roughly the same in two as well as in three dimensions, respectively [16, p. 444] . Hence a decision for one of these methods depends on further, more specific properties of the problem to be solved. If the solution is also required on the boundary curve or boundary surface Γ , respectively, or if an exterior problem is to be solved, then the boundary element method will usually be preferable. Whereas usually the data administration for the boundary element method is simple and, moreover, a few boundary elements often provide rather accurate solutions, one of the disadvantages of the method is seen in the large amount of computing time for the computation of the stiffness matrix. Here we estimate these times for two-dimensional problems for the Galerkin procedure involving double integrations and for the standard collocation procedure which mostly is used in applications. We first compare the time consuming evalutations of the smooth remaining kernel functions of both methods involving Gaussian quadrature formulas against the order of convergence of the L_2-error terms using splines of the same degree.

Then we compare the number of evaluations if the highest orders of convergence, i.e. superapproximation coincide. It turns out that here the Galerkin method needs much more time than collocation whilst using much lower degree splines. But even for collocation the evaluation of the stiffness matrix with Gaussian quadrature is rather costly. Moreover, a mesh refinement requires the new evaluation of the kernel function values at all nodal points since these change with any change of step size.

Numerical quadrature using the regular grid, however, can be performed much more economically and requires the kernel function values only at the regular grid points. These can be computed at the beginning and then be stored for further use. Mesh refinements require the new function values only at the new additional grid points whereas at the same number of odd grid points these values are still available. Such a numerical integration requires new quadrature formulas for integrals containing the shape function as a factor. (See [8],[15],[17] and Table 3.) For the Galerkin method this is the Galerkin collocation [8],[15], whereas for collocation we use these integrations for the inner integrals on the trial functions. Such procedure will require e.g. only half to one fifth of the computing time for the stiffness matrix in the example of the integral equations with logarithmic principal part in comparison with Gaussian quadrature, i.e. $\alpha = -1/2$ in Table 2.

All these comparisons are made under the assumption that the elements of the stiffness matrix belonging to the principal part are computed exactly and already are known either from analytic integration or via the Toeplitz matrix belonging to a convolutional principal part [15]. The comparisons are based on the asymptotic error analysis for Galerkin's method [14], [10] and the recent results for the collocation method [2]. For strongly elliptic equations the latter yet are known only for odd degree splines. Hence some of our comparisons contain corresponding gaps.

The paper is organized as follows. In Section 1 we formulate the class of problems and refer the corresponding asymptotic error bounds. Section 2 repeates the error estimates including numerical integration. In Section 3 we compare Galerkin's method with collocation using Gaussian quadratures. In Section 4 we compare Galerkin collcoation and collocation with grid point quadratures.

1. ASYMPTOTIC ERROR ESTIMATES

Here we consider systems of boundary integral equations of the form

$$\begin{aligned} Au + B\omega &= f \ , \\ \Lambda u &= \beta \end{aligned} \tag{1}$$

on a sufficiently smooth Jordan curve $\Gamma \subset \mathbb{R}^2$ with a 1-periodic parameter respresentation. $u = (u_1(t),\dots,u_p(t))$ denotes the 1-periodic unknown boundary function and $\omega \in \mathbb{R}^q$ is an unknown vector whereas $f = (f_1(t),\dots,f_p(t))$ and $\beta \in \mathbb{R}^q$ are the given data. A is a given matrix of bounded linear boundary integral operators mapping $H^{\lambda+\alpha} \to H^{\lambda-\alpha}$, B a $p\times q$ matrix of functions in $H^{\lambda-\alpha}$ and Λ a $q\times p$ matrix of bounded linear functionals on $H^{\lambda+\alpha}$. By H^s we denote the Sobolev space of order s of 1-periodic (generalized) functions on Γ equipped with the scalar product

$$\langle f,g\rangle_s = \hat{f}_o\overline{\hat{g}_o} + \sum_{o\neq k\in Z} \hat{f}_k\overline{\hat{g}_k}|2\pi k|^{2s}$$

and $||f||_s^2 = \langle f,f\rangle_s$ where $\hat{f}_k, \hat{g}_k$ denote the k-th Fourier coefficients of f and g, respectively. Note that

$$\langle f,g\rangle_o = \int_0^1 f(t)\overline{g(t)}\, dt .$$

In Equations (1) we further specify A to be a pseudodifferential operator of order 2α on Γ and require strong ellipticity which in turn implies coercivity in form of Gårding's inequality:

There exists a positive constant C and a compact bilinear form $k[u,v]$ on $(H^{j+\alpha})^2$ such that

$$\mathrm{Re}\, \langle Av,v\rangle_\lambda \geq C||v||^2_{\lambda+\alpha} - \mathrm{Re}\, k[v,v] \quad \text{for} \quad v\in H^{\lambda+\alpha} . \tag{2}$$

$\lambda\in\mathbb{R}$ will be specified later on.

Most boundary integral methods for stationary and time harmonic problems in the applications belong to this class of mathematical problems, for short surveys see [2 , §2.3],[15] . The most frequent cases in applications are $\alpha = -1/2$ for Symm's integral equation and related equations of the first kind with logarithmic principal part with applications in conformal mapping, torsion problems, plane elasticity, Stokes flows and electrostatics; $\alpha = 0$ for singular integral equations involving Cauchy's kernel including Fredholm integral equations of the second kind as a special case and with applications in plane elasticity and thermoelasticity, electromagnetic fields, acoustics, classical potential theory, incompressible flows; $\alpha = 1/2$ for the normal derivative of double layer potentials and for the operator of Prandtl's wing theory with applications in acoustics, ideal flows and plane elasticity; $\alpha = 1$ for integrodifferential operators of second order with many applications involving periodic solutions of second order ordinary differential equations.

For the approximations of Equations (1) we select an increasing sequence of mesh points $\Delta = \{t_i\}$, $i\in\mathbb{Z}$ satisfying $t_{i+N} = t_i + 1$ for fixed N and all $i\in\mathbb{Z}$, and denote by $S_m(\Delta)$ the space of all 1-periodic , m-1 times continuously differentiable splines of degree m subordinate to the partition Δ . We also write $S_m(\Delta)$ for $(S_m(\Delta))^p$.

Then the standard Galerkin method for Equations (1) reads as:

Find $u_h \in S_m(\Delta)$, $\omega_h \in \mathbb{R}^q$ such that

$$\langle Au_h, v\rangle_o + \langle B\omega_h, v\rangle_o = \langle f, v\rangle_o \quad \underline{\text{for all}} \quad v \in S_m(\Delta) \; , \qquad (3)$$
$$\Lambda u_h = \beta \; .$$

Clearly, Equations (3) are equivalent to $N\times p+q$ linear equations for the $N\times p$ coefficients of trial functions spanning u_h and the q entries of ω_h.

In the following error analysis we consider a family of meshes Δ and $S_m(\Delta)$ with diminishing meshwidth $h := \max_{i\in\mathbb{Z}} (t_{i+1}-t_i)$.
For the asymptotic error bounds of the difference $u-u_h$, $\omega-\omega_h$ between the exact solution and the Galerkin approximation we now recall well known results from [14],[10],[2].

<u>Theorem 1.1: Assume strong ellipticity of A, uniqueness for Equations (1) and Gårding's inequality (2) with $\lambda = 0$. Then there exist positive constants h_o and c such that the Galerkin equations (3) are uniquely solvable for any $0<h\le h_o$. For $-m-1+2\alpha \le r \le \alpha \le s \le m+1$ and $\alpha < m + 1/2$ we have the asymptotic error estimates</u>

$$\| u - u_h \|_r + |\omega-\omega_h| \le ch^{s-r} \| u \|_s \; . \qquad (4)$$

<u>In case of a quasiuniform family of meshes [4] the Estimate (4) also holds for $\alpha \le r < m + 1/2$ and $r \le s \le m+1$.</u>

At present, for the collocation methods corresponding error estimates are available only for <u>odd degree splines</u>, i.e. assume

$$m = 2j - 1 \quad \text{with} \quad j \in \mathbb{N} \; .$$

Then the collocation method with trial functions in $S_m(\Delta)$ and nodal points Δ reads for Equations (1) as :

<u>Find</u> $u_\Delta \in S_m(\Delta)$, $\omega_\Delta \in \mathbb{R}^q$ <u>such that</u>

$$Au_\Delta(t_i) + B(t_i)\omega_\Delta = f(t_i) \quad \underline{\text{for all}} \quad i = 1,\dots,N \; , \qquad (5)$$
$$\Lambda u_\Delta = \beta \; .$$

For the collocation method we have the asymptotic error bounds from [2].

<u>Theorem 1.2: Assume strong ellipticity of A, uniqueness for Equations (1) and Gårding's inequality (2) with $\lambda = j$ and assume</u>

$$j - \alpha > \frac{1}{2} \; , \quad \text{i.e.} \quad m > 2\alpha \; .$$

<u>Then there exist positive constants h_o and c such that the collocation equations (5) are uniquely solvable for any $0<h\le h_o$. For $2\alpha \le r \le j+\alpha \le s \le m+1$ we have the asymptotic error estimates</u>

$$\| u-u_\Delta \|_r + |\omega-\omega_\Delta| \le ch^{s-r} \| u \|_s . \qquad (6)$$

In case of a quasiuniform family of meshes the Estimate (6) also holds for $j+\alpha \le r < m+1/2$ and $r \le s \le m+1$.

Now we are in the position to compare the two methods. Let us first consider the case of using the same degree splines $S_m(\Delta)$ for both methods. Further let us restrict to quasiuniform families of meshes and smooth solutions. Then the highest rate of convergence achieved by the collocation method is $O(h^{m+1-2\alpha})$ in $H^{2\alpha}$, whilst the Galerkin method converges with rate $O(h^{2m+2-2\alpha})$ in $H^{2\alpha-m-1}$. This situation is summarized in Figure 1 .

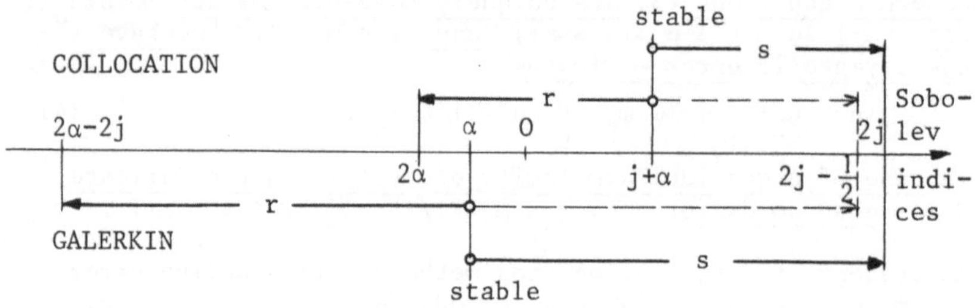

The case $\alpha \le 0$ ($j+\alpha > 0$ pictured, $j+\alpha \le 0$ also possible)

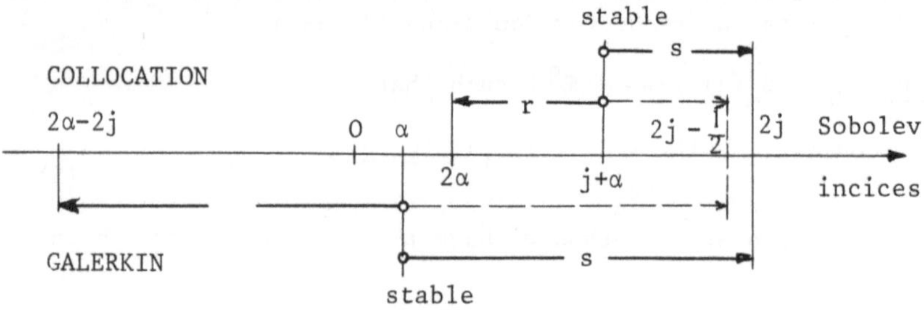

The case $\alpha \ge 0$

Figure 1. The indices $r \le s$ for which $\| u-u_\Delta \|_r \le ch_\Delta^{s-r} \| u \|_s$. Dashed lines indicate estimates requiring a quasiuniform mesh family.

In the boundary integral methods the physical fields at an observation point away from Γ are defined by integrals of u over Γ involving smooth kernels. Hence, there the approximate value of the field will converge at a rate equal to the highest rate of convergence achieved by the approximate solution to u, i.e. the above mentioned rates. Moreover numerical experiments even show superconvergence of the same highest rates at nodal points, for collocation see [1],[6] and for Galerkin's method see [9],[12],[13].

Hence, theoretically the Galerkin method seems to be superior to the collocation as long as the same kind of trial functions is used. To obtain the same order of superapproximation for both methods, one must use splines of different orders m_G for the Galerkin method and m_c for the collocation method which are related by

$$m_c = 2m_G + 1 . \tag{7}$$

Then for both methods we have the same rate $O(h^{2m_G+2-2\alpha}) = O(h^{m_c+1-2\alpha})$ of convergence in $H^{-m_G-1+2\alpha}$ and $H^{2\alpha}$, respectively. On the other hand, the construction of the stiffness matrix for the Galerkin method requires the evaluation of double integrals whilst the collocation method only requires single integrals. Therefore we shall compare the computational expenses of both methods by further estimating the quadrature errors and corresponding consistencies.

2. ESTIMATES INVOLVING NUMERICAL INTEGRATION

For the numerical implementation of both methods the entries of the respective stiffness matrices in the Galerkin equations (3) or the collocation equations (5) must be computed accurately enough to provide high enough consistency such that the asymptotic error estimates in the Theorems 1.1 and 1.2 are prevailed to the corresponding numerically integrated equations. As we pointed out in [18], the further error estimates hinge upon stability estimates for the respective discrete Equations (3) and (5). These are given for Galerkin's method in [10, Corollary 2.1] and for collocation in [2, Theorem 2.1.8]. For their formulation let us introduce the L_2-projection P_h of H^r onto $S_m(\Delta)$ and also the interpolation operator I_h interpolating a given function $f \in H^\sigma$, $\sigma > 1/2$ by a spline $I_h f \in S_m(\Delta)$ at the nodal points t_i.

Theorem 2.1: Let us assume that the family of meshes Δ is quasiuniform and that the assumptions of Theorem 1.1 respectively Theorem 1.2 are fulfilled. Then we have the stability estimates

$$\| u_h \|_o \le c\{h^{2\alpha'} \| P_h f \|_o + |\beta|\} \ , \ |\omega_h| \le c\{\| P_h f \|_o + |\beta|\} \ ,$$

$$\| P_h A u_h \|_o \le c h^{2(\alpha'-\alpha)} \| u_h \|_o \ ,$$

<u>respectively</u>

$$\| u_\Delta \|_o \le c\{h^{2\alpha'} \| I_h f \|_o + |\beta|\} \ , \ |\omega_\Delta| \le c\{\| I_h f \|_o + |\beta|\} \ ,$$

$$\| I_h A u_\Delta \|_o \le c h^{2(\alpha'-\alpha)} \| u_\Delta \|_o \ ,$$

<u>where</u> $\alpha' = \min\{\alpha, 0\}$.

Besides the above stability estimates we also need consistency which corresponds to the numerical accuracy of the elements in the stiffness matrices. To this end let us denote by

$$g_{\ell k} = \langle A\mu_\ell, \mu_k \rangle_o \ , \quad g_k = \langle B\cdot, \mu_k \rangle_o \ , \quad \gamma_\ell = \Lambda\mu_\ell$$

the Galerkin weights and by $\tilde{g}_{\ell k}$, $\tilde{g}_k$, $\tilde{\gamma}_\ell$ their numerically integrated counterparts. The latter define by

$$\langle \tilde{A}_h \mu_\ell, \mu_k \rangle_o = \tilde{g}_{\ell k} \ , \ \langle \tilde{B}_h \omega, \mu_k \rangle_o = \tilde{g}_k \omega \ , \ \tilde{\Lambda}_h \mu_\ell = \tilde{\gamma}_\ell$$

linear mappings $\tilde{A}_h : S_m(\Delta) \to S_m(\Delta)$, $\tilde{B}_h : \mathbb{R}^q \to S_m(\Delta)$ and $\tilde{\Lambda}_h : S_m(\Delta) \to \mathbb{R}^q$ which are approximations to $P_h A P_h, P_h B$ and ΛP_h , respectively.

Correspondingly we denote by

$$c_{\ell k} = A\mu_\ell(t_k) \quad \text{and} \quad c_k = B(t_k) \ , \ \Lambda\mu_\ell = \gamma_\ell$$

the collocation weights, by $\tilde{c}_{\ell k}$, $\tilde{c}_k$, $\tilde{\gamma}_\ell$ their numerically integrated counterparts and by $\tilde{A}_\Delta$, $\tilde{B}_\Delta$, $\tilde{\Lambda}_\Delta$ the corresponding approximations to $I_h A P_h$, $I_h B$ and Λ defined by

$$\tilde{A}_\Delta \mu_\ell(t_k) = \tilde{c}_{\ell k} \quad \text{and} \quad \tilde{B}_\Delta(t_k) = \tilde{c}_k \ , \ \tilde{\Lambda}_\Delta \mu_\ell = \tilde{\gamma}_\ell \ .$$

<u>Theorem 2.2: Let the degree of precision of the numerical integrations of the weights be</u> L [5, p. 49] , <u>i.e. let us assume</u>

$$\begin{aligned} |g_{\ell k} - \tilde{g}_{\ell k}| + h|g_k - \tilde{g}_k| + h|\gamma_\ell - \tilde{\gamma}_\ell| &\le c h^{L+3} \ , \\ |c_{\ell k} - \tilde{c}_{\ell k}| + h|c_k - \tilde{c}_k| + |\gamma_\ell - \tilde{\gamma}_\ell| &\le c h^{L+2} \ . \end{aligned} \tag{8}$$

<u>Then the corresponding operators provide the consistency estimates</u>

$$\| (P_hA-\tilde{A}_h)w_h\|_o + |(\Lambda-\tilde{\Lambda}_h)w_h| \le ch^{L+1}\| w_h\|_o ,$$

$$\| (P_hB-\tilde{B}_h)\omega\|_o \le ch^{L+1}|\omega| \tag{9}$$

and

$$\| (I_hA-\tilde{A}_\Delta)w_h\|_o + |(\Lambda-\tilde{\Lambda}_\Delta)w_h| \le ch^{L+1}\| w_h\|_o ,$$

$$\| (I_hB-\tilde{B}_\Delta)\omega\|_o \le ch^{L+1}|\omega| . \tag{10}$$

Proof: The proof of Inequality (9) in view of Estimate (8) can be obtained in the same way as that of [18, Theorem 6.1] and we omit the details.

For Estimate (10) we use for $w_h = \sum_{j=1}^{N} \alpha_j\mu_j$ the equivalence of weighted discrete norms on $S_m(\Delta)$ to the L_2-norm and the Schwarz inequality,

$$\| (I_hA-\tilde{A}_\Delta)w_h\|_o \le c\sqrt{h}\{\sum_{k=1}^{N}(\sum_{\ell=1}^{N}(A\mu_\ell(t_k)-\tilde{c}_{\ell k})\alpha_\ell)^2\}^{1/2}$$

$$\le c\{\sum_{\ell,k=1}^{N}|c_{\ell k}-\tilde{c}_{\ell k}|^2\}^{1/2}\{h\sum_{\ell=1}^{N}|\alpha_\ell|^2\}^{1/2}$$

$$\le ch^{L+2}N\| w_h\|_o = ch^{L+1}\| w_h\|_o .$$

The remaining estimates in (10) follow in the same manner.

In the following let us denote by $\tilde{u}_h,\tilde{\omega}_h$, respectively $\tilde{u}_\Delta,\tilde{\omega}_\Delta$ the solutions of the Galerkin equations or the collocation equations involving numerical quadratures. Then we have:

Theorem 2.3: Let the family of meshes Δ be quasiuniform, let the assumptions of Theorem 1.1 respectively Theorem 1.2 be fulfilled and let $L+1 + 2\alpha' > 0$. Then we have the stability estimates

$$\| \tilde{u}_h\|_o + |\tilde{\omega}_h| \le c\,\{h^{2\alpha'}\| P_hf\|_o + |\beta|\}$$

respectively

$$\| \tilde{u}_\Delta\|_o + |\tilde{\omega}_\Delta| \le c\,\{h^{2\alpha'}\| I_hf\|_o + |\beta|\}$$

and the error estimates

$$\| \tilde{u}_h-u\|_r + |\tilde{\omega}_h-\omega| \tag{11}$$

$$\le c(h^{L+1+2\alpha'+(-r)'}\{\| f\|_{L+1} + \| f\|_{-2\alpha}\} + h^{s-r}\| f\|_{s-2\alpha})$$

provided $-m-1+2\alpha \le r \le s \le m+1$, $r < m + \frac{1}{2}$,
respectively

$$\|\tilde{u}_\Delta - u\|_r + |\tilde{\omega}_\Delta - \omega| \le c(h^{L+1+2\alpha'+(-r)'}\|f\|_{-2\alpha} + h^{s-r}\|f\|_{s-2\alpha}) \tag{12}$$

provided $2\alpha \le r \le s \le m+1$, $(-r)' = \min\{0,-r\}$, $r < m + \frac{1}{2}$.

Since the proof is very similar to the proofs of [18, Theorems 6.3, 6.4 and 7.2] we omit the details.

3. COMPARISON INVOLVING GAUSSIAN QUADRATURE

Most numerical implementations of the Galerkin or the collocation method are based on Gaussian quadrature formulae on the patches, i.e. on the subintervalls $[t_{i-1},t_i]$. For singularities of the kernels, however, a special treatment is necessary. Therefore we shall require in the following that the operator A has convolutional principal part, i.e.

$$Au = A_1u + A_2u = \int(p_1(t-\tau) + \log|t-\tau|p_2(t-\tau))u(t)dt + \int K(t,\tau)u(t)dt ,$$

where p_1 and p_2 are homogeneous functions of degree $\beta = -2\alpha - 1$ and where K denotes a smooth remaining kernel [15],[18]. Moreover, let the family of meshes Δ be uniform and the spline spaces $S_m(\Delta)$ be generated by one shape function $\mu(\eta)$ as in [18, Chap. 9.5] . Then the weights of the principal part can be computed in terms of a Toeplitz matrix whose entries we consider to be known exactly. For the Galerkin method these are given by

$$(A_1\mu_\ell,\mu_k) = h^{1-2\alpha}\{\iint_{(\mathrm{supp}\,\mu)^2} p(t'-\tau'+(\ell-k))\mu(t')\mu(\tau')dt'd\tau' + \log h \iint_{(\mathrm{supp}\,\mu)^2} p_2(t'-\tau'+(\ell-k))\mu(t')\mu(\tau')dt'd\tau'\} \tag{13}$$

where $p = p_1 + p_2 \log|\cdot|$, whereas for collocation we have

$$A_1\mu_\ell(t_k) = h^{-2\alpha}\{\int_{\mathrm{supp}\,\mu} p(t'-\frac{m+1}{2} + (\ell-k))\mu(t')dt' + \log h \int_{\mathrm{supp}\,\mu} p_2(t'-\frac{m+1}{2} + (\ell-k))\mu(t')dt'\}. \tag{14}$$

(For special equations see [8, Section 3].)
For all remaining integrals on each patch we use a Gaussian quadrature formula with degree L of precision [5]. This

requires the evaluation of the integrand at $[\frac{L}{2}] + 1$ Gaussian nodal points, where $[\frac{L}{2}]$ denotes the largest integer $\leq \frac{L}{2}$. For the double integrals of Galerkin's stiffness matrix we either use $([\frac{L}{2}] + 1)^2$ nodal points and product integration or we use a two-dimensional formula [5, 9.4.2] as was also indicated in [11].

The Gaussian integration has here the disadvantage that for a mesh refinement the function values on the coarser grid cannot be used on the finer grid again since all Gaussian nodal points are placed differently. Hence the evaluation of the stiffness matrix becomes rather time consuming. Without mesh refinement, however, it is easily used since Gaussian quadrature is accessible everywhere.

Comparison for same orders of L_2-errors

Here we choose $t = 0$ and $s = m+1$, i.e. the same splines for both methods in the Estimates (11), (12) and then balance $L + 1 + 2\alpha' \geq m+1$. In Table 1 we compare for various α and m the numbers of evaluations of the kernel function per element of the stiffness matrix denoted by eg for Galerkin's method and ec for collocation. Note that here $eg = (ec)^2$ and $m+1$ = order of convergence. The Table 1 contains only cases satisfying the assumption $\alpha < m+1/2$ for Galerkin's procedure, respectively $2\alpha < m$ for the collocation method. For even m and collocation we put ? into the table since in this case we don't know the rate of convergence yet. In the lower row of eg we note the number of nodal points involving two-dimensional Gaussian quadrature following [5, p. 424].

α	$-\frac{1}{2}$				0				$\frac{1}{2}$				1		
order	1	2	3	4	1	2	3	4	1	2	3	4	2	3	4
L	1	2	3	4	0	1	2	3	0	1	2	3	1	2	3
eg	1	4	4	9 7	1	1	4	4	1	1	4	4	1	4	4
ec	?	2	?	3	-	1	?	2	-	-	?	2	-	-	2

Table 1: Evaluations per element of the stiffness matrix for optimal orders of L_2-errors.

Comparison for same order of superapproximation and different splines

As was pointed out in Section 1, we choose different orders of splines m_c for collocation and m_G for Galerkin's method satisfying Equation (7).

Now we balance the errors in (2.10) finding

$$L_G \geq 2m_G + 1 - 2(\alpha'+\alpha) \quad \text{in case} \quad 2\alpha \leq m_G + 1 \tag{15}$$

and

$$L_G = m_G \qquad \text{in case} \quad m_G + 1 < 2\alpha < 2m_G + 1 \, . \tag{16}$$

Similarly we find for collocation from balancing the errors in (12) ,

$$L_c \geq m_c - 2\alpha' \, . \tag{17}$$

Note that in general the required degrees of precision L_G and L_C for the Galerkin method, respectively, collocation will be different. In Table 2 we compare again the numbers of evaluations of the kernel function depending on the order of superapproximation for various α . For two-dimensional Gaussian quadrature we refer to [5, p. 424 and 427].

α	$-\frac{1}{2}$				0				$\frac{1}{2}$			1		
order	3	5	7	9	2	4	6	8	3	5	7	2	4	6
m_G	0	1	2	3	0	1	2	3	1	2	3	1	2	3
L_G	3	5	7	9	1	3	5	7	2	4	6	1	3	5
eg	4	9 7	16 12	25 18	1	4	9 7	16 12	4	9 7	16 12	1	4	9 7
m_c	1	3	5	7	1	3	5	7	3	5	7	3	5	7
L_c	3	5	7	9	1	3	5	7	3	5	7	3	5	7
e_c	2	3	4	5	1	2	3	4	2	3	4	2	3	4

Table 2: Evaluations per element of the stiffness matrix for same optimal superapproximation orders.

Note that also in this case the evaluation of the stiffness matrix for Galerkin's method requires almost always significantly more time than for the collocation method. But one should be aware of the fact that the latter deals with more than twice higher degree splines yielding more complicated coding and much higher requirements for the smoothness of Γ and the boundary charges u .

The reason for all the trouble is the use of Gaussians quadrature formulas that implies computational expenses of orders $N^2 \cdot (eg) = h^{-2}(eg)$, respectively, $N^2(ec) = h^{-2}(ec)$ for the evaluation of the stiffness matrix. As we shall see in the following section, the consequent use of the regular nodal points t_i also for the numerical quadratures will require a computational expense of order $(\frac{N}{\gamma_G})^2 = h^{-2}\gamma_G^{-2}$ for Galerkin collocation and $\frac{1}{\gamma_c} N^2 = h^{-2}\gamma_c^{-1}$ for collocation only. In our examples we have always $\gamma_c = 1$ whereas $\gamma_G = 1$ or $\frac{1}{2}$. The latter approaches prove also to be advantageous for uniform mesh refinements since the kernel functions are to be computed only at the new meshpoints.

4. GALEKRIN COLLOCATION AND COLLOCATION WITH GRIDPOINT QUADRATURE

In the following we use special integration formulas for integrals of products of smooth functions with the shape functions μ involving only regular nodal points of the regular grid of meshwidth h or γh with $\gamma^{-1} = 2$ (or 4,8,...), i.e. the grid of the following next refinements. As in [7],[8],[9],[15], [18, Section 9.7] we use formulas

$$\int f(t)\mu_k(t)dt = h \sum_{\ell=-M}^{M} b_\ell f(t_{k\ell}) + \text{remainder}$$

where

$$t_k = h(k + \frac{m+1}{2}) \quad \text{and} \quad t_{k\ell} = h(k + \frac{m+1}{2} + \ell\cdot\gamma) ,$$

which are of degree of precision

$$L = 2M + 1 .$$

Hence we again use Equation (13) , respectively Equation (14) for the terms of the stiffness matrix belonging to the principal part. For the smooth remainder we use the quadrature formula

$$(\tilde{A}_{2h}\mu_k,\mu_p) = h^2 \sum_{\ell,i=-M}^{M} b_\ell b_i K(t_{k\ell},t_{pi})$$

in the case of Galerkin's method, i.e. Galerkin collocation, respectively

$$\tilde{A}_{2\Delta}\mu_k(t_i) = h \sum_{\ell=-M}^{M} b_\ell K(t_{k\ell},t_i)$$

for the implementation of the collocation method.

The explicit forms of the shape functions for $m = 0,1$ and 2 can be found in [18, (9.40)] and for arbitrary m in [3, Section 4] . The derivation of the weights $b_\ell = b_{-\ell}$ for various cases is performed in [8, Section 5] and some of the weights we present here in Table 3.

γ	M	m=0		m=1			m=2			
		b_o	b_1	b_o	b_1	b_2	b_o	b_1	b_2	b_3
1	0	1	-	1	-	-	1	-	-	-
	1	$\frac{11}{12}$	$\frac{1}{24}$	$\frac{5}{6}$	$\frac{1}{12}$	-	$\frac{3}{4}$	$\frac{1}{8}$		
$\frac{1}{2}$	2			$\frac{13}{30}$	$\frac{4}{15}$	$\frac{1}{60}$	$\frac{2}{5}$	$\frac{7}{30}$	$\frac{1}{15}$	-
	3						$\frac{358}{945}$	$\frac{157}{630}$	$\frac{19}{315}$	$\frac{1}{945}$

Table 3: Weights of numerical integrations against splines.

If the computation of the values of the kernel functions at the grid points is executed in advance and these values then are stored for further use then the transition of the method to different shape functions can be done in a most efficient way (see [7, Table 4]) .

The computational expense for computing the stiffness matrices now depends only on γ and is proportional to

$(\frac{1}{\gamma_G})^2 N^2 = (\gamma_G \cdot h)^{-2}$ for the Galerkin collocation and to

$(\frac{1}{\gamma_c}) N^2 = \gamma_c^{-1} h^{-2}$ for the collocation with grid point quadrature.

For comparison we consider again the two cases as in Section 3.

Comparison for same orders of L_2-errors

With same splines for both methods we balance

$$2M + 1 + 2\alpha' \geq m = \text{order} - 1 \quad .$$

In Table 4 we collect orders, M and γ for the four cases $\alpha = -\frac{1}{2}, 0, \frac{1}{2}, 1$ and observe that here always $\gamma = 1$, i.e. we need only one function value per stiffness element (and appropriate organization of the code) in contrary to the case using Gaussian quadrature, Table 1.

		$\alpha = -\frac{1}{2}$				$\alpha = 0$				$\alpha = \frac{1}{2}$			$\alpha = 1$		
Galerkin	order	1	2	3	4	1	2	3	4	2	3	4	2	3	4
	M	0	1	1	2	0	0	1	1	0	1	1	0	1	1
Collo-cation	order	-	2	3	4	-	2	3	4	-	3	4	-	-	4
	M		1	?	2		1	?	2		?	2			2

Table 4: Orders and M for optimal L_2-errors.
Note $\gamma = 1$ and order = m+1 .

As we can see, for our four cases both methods are of the same computational expense, the least prossible for boundary element methods.

Comparison for same orders of superapproximation and different splines

As in Section 3 we have Equation (7) and the Restrictions (15) - (17) with $L_G = 2M_G + 1$, $L_c = 2M_c + 1$. In Table 5 we collect m , M and γ for four cases of equations and the two methods for equal orders of superapproximation.

		$\alpha = -\frac{1}{2}$				$\alpha = 0$				$\alpha = \frac{1}{2}$			$\alpha = 1$		
	orders	3	5	7	9	2	4	6	8	3	5	7	2	4	6
Galerkin	m_G	0	1	2	3	0	1	2	3	1	2	3	1	2	3
	M_G	1	2	3	4	0	1	2	3	0	2	3	0	1	2
	γ_G	1	$\frac{1}{2}$	$\frac{1}{2}$	$\frac{1}{2}$	1	1	$\frac{1}{2}$	$\frac{1}{2}$	1	$\frac{1}{2}$	$\frac{1}{2}$	1	1	1
Collo-cation	m_c	1	3	5	7	1	3	5	7	3	5	7	3	5	7
	M_c	1	2	3	4	0	1	2	3	1	2	3	1	2	3
	γ_c	is always 1													

Table 5: Orders and degrees of the splines and precisions for Galerkin collocation and collocation with grid point quadrature for same orders of superapproximation.

For our four cases the collocation with grid point quadrature is alway most efficient and much faster than with Gaussian quadrature (Table 2).

The Galerkin collocation requires in several cases four times as long for computing the stiffness matrix but is still able to compete with the above collocation since the corresponding splines are of less than half the degrees. If we compare the Galerkin collocation with Galerkin's method using Gaussian quadrature (Table 2) then we see again that Galerkin collocation is significantly faster and hence superior.

ACKNOWLEDGEMENTS:

The research of Professor Arnold was supported in part by the National Science Foundation under contract MCS-81-02012 and a NATO Postdoctoral Fellowship. The research of Professor Wendland was partially carried out while he was a visitor to the Depts. of Computer Science and Mathematics at the Chalmers University in Göteborg, Sweden, 1982.

REFERENCES:

[1] Abou El-Seoud, M.S. (1980) Ein Vergleich von Kollokationsmethode und Galerkinverfahren für die numerische Lösung von schwach singulären Integralgleichungen erster Art. ZAMM, T278-T280.

[2] Arnold, D.N. and Wendland, W.L., On the asymptotic convergence of collocation methods. To appear in Math. Comp..

[3] Aubin, J.P. (1972) Approximation of Elliptic Boundary-Value Problems. Wiley-Interscience, New York.

[4] Babuška, I. and Aziz, K.A. (1972) Survey lectures on the mathematical foundations of the finite element method. The Mathematical Foundation of the finite Element Method with Applications to Partial Differential Equations (A.K. Azisz ed., Academic Press, New York) : 3-359.

[5] Engels, H. (1980) Numerical Quadrature and Cubature. Academic Press, London, New York, Toronto, Sydney, San Francisco.

[6] Hoidn, H.-P., Die Kollokationsmethode angewandt auf die Symmsche Integralgleichung. ETH Zürich, Switzerland, in preparation.

[7] Hsiao, G.C., Kopp, P. and Wendland, W.L. (1980) The synthesis of the collocation and the Galerkin method applied to some integral equations of the first kind. New Developments in Boundary Element Methods (C.A. Brebbia ed., CML Publ., Southampton) : 122-136.

[8] Hsiao, G.C., Kopp, P. and Wendland, W.L. (1980) A Galerkin collocation method for some integral equations of the first kind. Computing 25 : 89-130.

[9] Hsiao, G.C., Kopp, P. and Wendland, W.L., Some applications of a Galerkin collocation method for some integral equations of the first kind. In preparation.

[10] Hsiao, G.C. and Wendland, W.L. (1981) The Aubin-Nitsche lemma for integral equations. J. of Integral Equations 3 : 299-315.

[11] Katz, C. (1981) The use of Green's functions in the numerical analysis of potential, elastic and plate bending problems. Boundary Element Methods (C.A. Brebbia ed.), Springer, Berlin, Heidelberg, New York : 609-622.

[12] Lamp, U., Schleicher, T., Stephan, E. and Wendland, W.L. (1982) Theoretical and experimental asymptotic convergence of the boundary integral method for a plane mixed boundary value problem. In Boundary Element Methods in Engineering, 4th Int. Conf., these proceedings.

[13] Lamp, U., Schleicher, T., Stephan, E. and Wendland, W.L., Galerkin collocation for an improved boundary element method for a plane mixed boundary value problem. In preparation.

[14] Stephan, E. and Wendland, W.L. (1976) Remarks to Galerkin and least squares methods with finite elements for general elliptic problems. Springer, Berlin, Lecture Notes Math. 564 : 461-471, Manuscripta Geodaetica 1 : 93-123.

[15] Wendland, W.L. (1980) On Galerkin collocation methods for integral equations of elliptic boundary value problems. Intern. Ser. Num. Math. 53 (J. Albrecht, L. Collatz ed., Numerical Treatment of Integral Equations, Birkhäuser Basel) : 244-275.

[16] Wendland, W.L. (1981) Asymptotic convergence of boundary element methods. Lectures on the Numerical Solution of Partial Differential Equations (I. Babuška, T.-P. Liu, J. Osborn ed. Lecture Notes #20, Univ. Maryland, Dept. Math., College Park) : 435-477.

[17] Wendland, W.L. (1981) On the asymptotic convergence of boundary integral methods. Boundary Element Methods (C.A. Brebbia ed.) Springer-Verlag,Berlin, Heidelberg, New York : 412-430.

[18] Wendland, W.L. (1981) Asymptotic accuracy and convergence. Progress in Boundary Element Methods (C.A. Brebbia ed., Pentech Press, London, Plymouth) 1 : 289-313.

SOLUTION OF FREE BOUNDARY PROBLEMS USING C-COMPLETE SYSTEMS.

Gonzalo Alduncin* and Ismael Herrera**

* División de Estudios de Posgrado de la Facultad de Ingeniería, UNAM, 04510 México, D.F. MEXICO.

** Instituto de Investigaciones en Matemáticas Aplicadas y en Sistemas (IIMAS), UNAM, Apdo. Postal 20-726, Admón No. 20, 01000 México, D.F. MEXICO.

ABSTRACT.

There are two main approaches to the formulation of boundary methods, these are boundary integral equations and approximations by complete systems of solutions. The latter has been the subject of extensive studies by one of the authors oriented to clarifying the foundations of the method and increasing its versatility. The present paper is devoted to explain the application of this procedure to free boundary problems such as Signorini's and contact problem [1-6].

1. INTRODUCTION

There are two main approaches for the formulation of boundary methods; one is based on boundary integral equations and the other one, on the use of complete systems of solutions. One of the authors has given previously extensive descriptions of the latter method [1-11]. Its theoretical foundations and development embrace the following aspects: a) approximating procedures and conditions for their convergence; b) formulation of variational principles; and c) development of complete systems of solutions. It has been shown that a suitable criterion for completeness is c-completeness. A method of considerable generality, for generating such systems is described in [6] and [10]. A general version of separation of variables procedures yields biorthogonal systems which are c-complete [9]. Convenient features of c-complete systems are their simplicity, as in the case of plane waves [11], and the fact that the same system can be applied to large classes of regions and boundary conditions.

It has been shown [3,10] that under general conditions a system which is c-complete for a region has this property for any region which contains the first one. The possibility of using this property to treat problems subjected to floating boundary conditions such as seepage flow was suggested previously [12]. In the present paper we initiate the systematic development of this subject. First, a very simple version of a contact problem is presented as an example and then theoretical results that can be applied to a general class of contact problems are developed. These theoretical developments are based on the theory of variational inequalities [13-16].

2. AN EXAMPLE

Let Ω (Fig. 1) be a bounded and connected set in $\mathbb{R}^n$ with a Lipschitz continuous boundary $\Gamma=\Gamma_1\cup\Gamma_2$ and $\Gamma_1\cap\Gamma_2=\phi$. It will be assumed that meas$(\Gamma_1)\neq 0$.

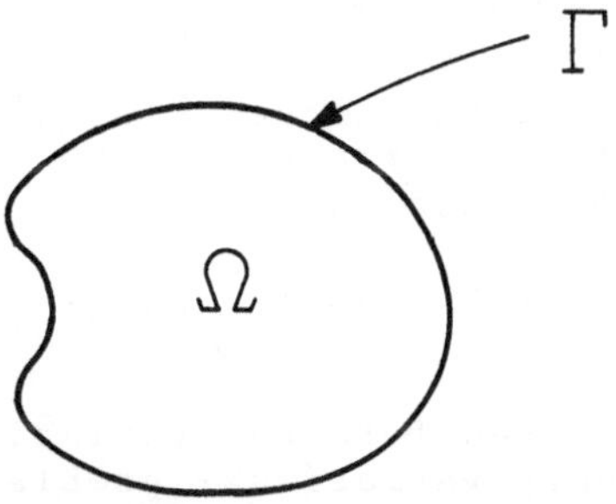

FIGURE 1

Write

$$V = H^1(\Omega) \hookrightarrow H=L^2(\Omega) \tag{1}$$

where the symbol $\hookrightarrow$ means densely contained. Given the functions

$$\begin{aligned} F &\in L^2(\Omega), \\ g &\in H^{1/2}(\Gamma), \\ \hat{u} &\in H^{3/2}(\Gamma), \\ h_o &\in H^{3/2}(\Gamma), \end{aligned} \tag{2}$$

one can formulate the distributional boundary value problem which consists in finding $u\in V$, such that

$$\left.\begin{aligned} -\Delta u &= F \quad \text{in } \Omega, \\ u &= \hat{u} \quad \text{on } \Gamma_1, \\ \left.\begin{aligned} u &\geq h_o \\ \partial u/\partial\nu &\geq g \\ (\partial u/\partial\nu - g)(u-h) &= 0 \end{aligned}\right\} & \text{ on } \Gamma_2 \end{aligned}\right\} \tag{3}$$

Consider the continuous bilinear form

$$a(u,v) = \int_\Omega \nabla u\cdot\nabla v dx, \quad u,v \in H^1(\Omega), \tag{4}$$

As it is usual $\gamma_o : H^1(\Omega)\to H^{1/2}(\Gamma)$ and $\gamma_1 : H^1(\Omega)\to H^{-1/2}(\Gamma)$ are the trace operators. Given $\hat{u}\in H^{3/2}(\Gamma_1)$ and $h_o \in H^{3/2}(\Gamma_2)$, define

$$K = \{v\in H^1(\Omega) : \gamma_o v = \hat{u} \;\text{ on } \Gamma_1,\; \gamma_o v \geq h_o \;\text{ on } \Gamma_2\} \tag{5}$$

and $f\in V'$ by

$$<f,v> = \int_\Omega F v dx + \int_{\Gamma_2} g\gamma_o v ds \;,\; v\in H^1(\Omega), \tag{6}$$

Clearly, K is a non-empty, closed and convex subset of V.

Using standard techniques [13,14] it follows that problem (3) is characterized by the variational problem, find $u\in K$ such that

$$\int_\Omega \nabla u\cdot(\nabla v-\nabla u) d\underset{\sim}{x} \geq \int_\Omega F(v-u) d\underset{\sim}{x} + \int_{\Gamma_2} g(\gamma_o v-\gamma_o u) d\underset{\sim}{x},\; \forall\, v\in K \tag{7}$$

In order to transform the variational problem (7) into a boundary variational problem, let $w\in H^2(\Omega)$ be such that

$$-\Delta w = F \quad , \quad \text{in } \Omega \tag{8}$$

and define the functions

$$\left.\begin{aligned} g_o &= g - \gamma_1 w \in H^{1/2}(\Gamma) \\ \hat{u}_o &= \hat{u} - \gamma_o w \in H^{3/2}(\Gamma) \\ h_o &= h - \gamma_o w \in H^{3/2}(\Gamma) \end{aligned}\right\} \tag{9}$$

Consider the closed convex set

$$K_o = \{v\in V_o \,|\, \gamma_o v = \hat{u}_o \text{ on } \Gamma_1,\; \gamma_o v \geq h_o \;\text{ on } \Gamma_2\} \tag{10}$$

where

$$V_o = \{v\in H^1(\Omega) \,|\, \Delta v = 0 \;\text{ in } \Omega) \tag{11}$$

Then under the transformation

$$u_o = u - w \tag{12}$$

where $w \in H^2(\Omega)$ is a fixed element satisfying (8), problem (7) and therefore (3) is equivalent to the boundary variational problem, find $u_o \in K_o$ such that

$$\int_{\Gamma_2} (v-u_o)\partial u_o/\partial n d\underset{\sim}{x} \geq \int_{\Gamma_1} (v-u_o) g_o d\underset{\sim}{x}, \ \forall \ v \in K_o \tag{13}$$

The method we propose for solving the boundary variational problem (13), is based on the use of a basis $\{\varphi_1, \varphi_2, ...\}$ of V_o. The construction of such basis has been extensively studied by Herrera [6,10] recently for a large class of systems of partial differential equations. For Laplace's equation, for example, it has been shown that a basis of V_o, when Ω is bounded and simply connected, is the system of harmonic polynomials [6]

$$\{\text{Re } z^n, \text{ Im } z^n \ ; \ n = 0,1,2,...\} \tag{14}$$

Given such basis define

$$V_{om} = \text{span}\{\varphi_1, \varphi_2, ..., \varphi_m\}, \ m \geq 1 \tag{15}$$

Let

$$\{p_1, p_2, ...\} \quad \text{be a basis of } H^o(\Gamma_1) \tag{16}$$

$$\{q_1, q_2, ...\} \quad \text{be a basis of } H^o(\Gamma_2) \tag{17}$$

It will be assumed that $q_j \geq 0$, $j = 1,2,...$. Elements v_m of the convex subset K_{om}, will be required to satisfy

$$\int_{\Gamma_1} p_j(v_m - \hat{u}_o) d\underset{\sim}{x} = 0 \ ; \ \int_{\Gamma_2} q_j(\gamma_o v_m - h_o) d\underset{\sim}{x} \geq 0 \ ; \ j=1,...,m. \tag{18}$$

More precisely

$$K_{om} = \{v_m \in V_{om} | v_m \text{ satisfies (18)}\} \tag{19}$$

If K_o is replaced by K_{om} in the definition of the boundary variational problem (13), one obtains a family of variational problems. Let $u_{om} \in K_{om}$ be the solution of such problem, then it can be shown that $u_{om} \to u_o$. The example given here is a particular case of the general theory explained next.

3. NOTATION

$(V, \|\cdot\|)$ is a real Hilbert space with topological dual $(V', \|\cdot\|_*)$; $<\cdot,\cdot>$ denotes the duality pairing on $V' \times V$. $(H, (\cdot,\cdot), |\cdot|)$ is a real Hilbert space identified with its dual, in which V is densely and continously embedded: V H V'. K is a nonempty, closed and convex subset of V.

$a: V\times V\to \mathcal{R}$ is a continuous bilinear form (not necessarily symmetric) which satisfies the condition c: (K-ellipticity) there is a constant $\alpha>0$ such that

$$a(u-v,u-v) \geq \alpha\| u-v\|^2, \ \forall\ u,v\in K.$$

$A\in\mathcal{L}(V,V')$ is the corresponding continuous linear operator:

$$a(u,v) = \langle Au,v\rangle, \quad u,v\in V. \tag{20}$$

$\gamma\in\mathcal{L}(V, B)$ is a linear continuous surjection with kernel V_o dense in H, B being a Hilbert space isomorphic to the quotient space V/V_o, and the quotient map $\hat{\gamma}: V/V_o\to B$ norm-preserving.

$A\in\mathcal{L}(V,V_o')$ is the linear continuous composition ρA, where $\rho: V'\to V_o'$ is the restriction to V_o of functionals on V, called the formal operator determined by $a(\cdot,\cdot)$, V and V_o:

$$a(u,v) = \langle Au,v\rangle, \quad u\in V,\ v\in V_o. \tag{21}$$

Hence (cf. [13]), the following abstract Green's formula holds:

$$\langle Au,v\rangle - (Au,v) = [\partial u,\gamma v], \quad u\in D_o,\ v\in V, \tag{22}$$

where $D_o=\{u\in V: Au\in H\}$, $\partial\in L(D_o,B')$ is the abstract Green's operator and $[\cdot,\cdot]$ is the duality pairing on $B\times B'$.

$j: K\to(-\infty,+\infty]$ is a proper, convex and lower semicontinuous functional. In addition $f\in V'$.

4. ABSTRACT BOUNDARY VALUE PROBLEM

With the above notation in force, we now consider the variational problem (P) corresponding to a kind of abstract boundary value problems. Toward this end, let K be characterized only by boundary constraints:

$$K + V_o \subset K, \tag{23}$$

Clearly, condition (23) is equivalent to

$$K + V_o = K \tag{24}$$

Let $a_1: V\times V\to\mathcal{R}$ and $a_2: B\times B\to\mathcal{R}$ be continuous bilinear forms such that

$$a(u,v) = a_1(u,v)+a_2(\gamma u,\gamma v), \quad u,v \in V, \tag{25}$$

satisfies condition (c). Similarly, let $h: B \to (-\infty,+\infty]$ be a functional such that

$$j(v) = h(\gamma v), \quad v \in K \tag{26}$$

is proper, convex and lower semi-continuous. Let $F \in H$ and $g \in B'$, and define

$$f(v) = (F,v)+[g,\gamma v], \quad v \in V, \tag{27}$$

which belongs to V'. Then the variational problem (P) takes the following form:

$$\left.\begin{array}{l} \text{Find } u \in K \text{ such that} \\ a_1(u,v-u)+a_2(\gamma u,\gamma v-\gamma u)+h(\gamma v)-h(\gamma u) \\ \qquad \geq (F,v-u)+[g,\gamma v-\gamma u], \quad \forall\ v \in K \end{array}\right\} \tag{28}$$

The following theorem determines the abstract boundary value problem to which (28) is equivalent.

Theorem 1. *Let* $A_1 \in \mathcal{L}(V,V')$ *and* $A_2 \in \mathcal{L}(B,B')$ *be the operators corresponding to* $a_1(\cdot,\cdot)$ *and* $a_2(\cdot,\cdot)$, *respectively. Let* $A \in \mathcal{L}(V,V_0')$ *be the formal operator determined by* $a_1(\cdot,\cdot)$ *(or, equivalently,* $a(\cdot,\cdot)$*),* V *and* V_0*, and let* $\partial_1 \in L(D_0,B')$ *be the abstract Green's operator defined by* (22):

$$\langle A_1 u,v\rangle - (Au,v) = [\partial_1 u,\gamma v], \quad u \in D_0, \ v \in V. \tag{29}$$

Then the problem (28) *is equivalent to the problem*

$$\left.\begin{array}{l} \textit{Find } u \in K \textit{ such that} \\ \qquad Au = F \quad \textit{in} \quad H, \\ [\partial_1 u+A_2(\gamma u),\gamma v-\gamma u]+h(\gamma v)-h(\gamma u) \\ \qquad \geq [g,\gamma v-\gamma u], \quad \forall \quad v \in K \end{array}\right\} \tag{30}$$

Proof. Let $u \in K$ be the solution of problem (28). Then, in accordance with (23), we can set $v=u-v_0 \in K$, $v_0 \in V_0$, and obtain $Au=F$ in H, since V_0 is dense in H, and $u \in D_0$. Now, upon using (29), the boundary inequality of (30) follows from (28).

Conversely, let $u \in K$ be a solution of problem (30). Then $u \in D_0$ and because of (29), u is a solution of (28).

5. BOUNDARY VARIATIONAL PROBLEM

In order to transform the abstract boundary value problem (30) into a boundary variational problem, we introduce equation "Au=F in H" in K as an additional constraint. Hence, reconsider problem (30) on the closed convex set

$$\tilde{K} = \{v \in K : Au = F \text{ in } H\}. \tag{31}$$

Then problem (30) transforms into the problem

Find $u \in K$ such that

$$\left.\begin{array}{l}[\partial_1 u + A_2(\gamma u), \gamma v - \gamma u] + h(\gamma v) - h(\gamma u) \\ \qquad \geq [g, \gamma v - \gamma u], \quad \forall\, v \in \tilde{K}\end{array}\right\} \tag{32}$$

Theorem 2. *The abstract boundary value problem* (11) *[or, equivalently, the variational problem* (28)*] is equivalent to the boundary variational problem* (32).

Proof. It is clear that solutions of (30) are also solutions of (32). The converse follows by observing that given $v \in K$, there exist a $\tilde{v} \in \tilde{K}$ such that $\tilde{v} - v \in V_0$ and, consequently, (32) holds for such a v.

6. HOMOGENEIZATION OF THE DOMAIN CONSTRAINT

For convenience in our study on internal approximations of problem (32), we make homogeneous the domain constraint of $\tilde{K}$ and incorporate it into the space V. Toward this end, let

$$w \in D_0 : Aw = F \text{ in } H \tag{33}$$

be a known function, and consider the change of dependent variable

$$u_0 = u - w. \tag{34}$$

Then, according to the definitions

$$\left.\begin{array}{rl} g_0 &= g - \partial_1 w - A_2(\gamma w) \in B' \\ h_w(v) &= h(v + \gamma w), \quad \forall\, v \in B \end{array}\right\} \tag{35}$$

and

$$\left.\begin{array}{rl} V_A &= \{v \in V : Av = 0 \text{ in } H\} \\ K_A &= \{v \in V_A : v + w \in K\} \end{array}\right\} \tag{36}$$

the following result is easily established:

Theorem 3. *Via the relation* (34), *the problem* (32) *is equivalent to the problem*

Find $u_o \in K_A$ such that

$$\left.\begin{array}{r}[\partial_1 u_o + A_2(\gamma_o), \gamma v - \gamma u_o] + h_w(v) - h_w(u_o) \\ \geq [g_o, \gamma v - \gamma u_o], \ \forall\ v \in K_A\end{array}\right\} \quad (37)$$

7. INTERNAL APPROXIMATIONS

Let $\{V_m, K_m\}_{m>1}$ be an internal approximation of $\{V_A, K_A\}$ in the sense of [15]:

$$\left.\begin{array}{rl}
\text{i)} & \{V_m\}_{m\geq 1} \text{ is a family of finite dimensional subspace of } V_A \text{ with parameter } m = \dim V_m; \\
\text{ii)} & \forall\ v \in V_A,\ \exists\ v_m \in V_m : v_m \to v \text{ in } V \text{ as } m \to \infty; \\
\text{iii)} & \text{For each } m \geq 1,\ K_m \text{ is a nonempty, closed and convex subset of } V_m; \\
\text{iv)} & \forall\ v \in K_A,\ \exists\ v_m \in K_m : v_m \to v \text{ in } V \text{ as } m \to \infty; \\
\text{v)} & \text{If } v_m \in K_m \text{ and } v_m \to v \text{ (weakly) in } V \text{ as } m \to \infty, \text{ then } v \in K_A.
\end{array}\right\} \quad (38)$$

For each $m \geq 1$, we associate to problem (37) the discrete problem:

Find $u_m \in K_m$ such that

$$\left.\begin{array}{r}[\partial_1 u_m + A_2 \gamma u_m, \gamma v - \gamma u_m] + h_w(v) - h_w(u_m) \\ \geq [g_o, \gamma v - \gamma u_m], \ \forall\ v \in K_m\end{array}\right\} \quad (39)$$

REFERENCES

1. Herrera, I. (1982) Boundary Methods for Fluids, in Finite Elements in Fluids Vol. IV, Chapter 19, RH Gallagher, et al., eds., John Wiley & Sons.
2. Herrera, I. (1981) An Algebraic Theory of Boundary Value Problems, Kinam, 3, 2:161-230.
3. Herrera, I. (1982) Theoretical Foundations for Numerical Applications of Complete Systems of Solutions, Comunicaciones Técnicas, IIMAS-UNAM (in press).
4. Herrera, I. (1977) General Variational Principles Applicable to the Hybrid Element Method, Proc. Nat'l Acad. Sc. USA, 74, 7:2595-2597.
5. Herrera, I. (1977) Theory of Connectivity for Formally Symmetric Operators, Proc. Nat'l Acad.

Sc. USA, 74, 11:4722-4725.
6. Herrera, I., and Sabina, F.J. (1978) Connectivity as an Alternative to Boundary Integral Equations. Construction of Bases, Proc. Nat'l Acad. Sc. USA, 75, 5:2059-2063.
7. Herrera, I. (1980) Variational Principles for Problems with Linear Constraints. Prescribed Jumps and Continuation Type Restrictions, Jour. Inst. Maths. & its Applics., 25, 1:67-96.
8. Herrera, I. (1980) Boundary Methods. A Criterion for Completeness, Proc. Nat'l Acad. Sc. USA, 77, 8:4395-4398.
9. Herrera, I., and Spence, D.A. (1981) Framework for Biorthogonal Fourier Series. Proc. Nat'l Acad. Sc. USA, 78, 12:7240-7244.
10. Herrera, I. (1982) Boundary Methods. Development of Complete Systems of Solutions. To appear in Proc. Fourth International Symposium on Finite Element Methods in Flow Problems, July 26-29, Tokyo, Japan.
11. Sânchez-Sesma, F.J., Herrera, I., and Avilês, J (1982) A Boundary Method for Elastic Wave Diffraction. Application to Scattering of SH Waves by Surface Irregularities, Bull. Seism. Soc. Am. 72, 3:
12. Herrera, I. (1980) Boundary Methods in Flow Problems. Proc. Third International Conference on Finite Elements in Flow Problems, Banff, Canada, 1, 30-42.
13. Lions, J.L. and Stampacchia, G. (1967) Variational Inequalities, Comm. Pure Applied Math. XX, 493-519.
14. Showalter, R.E. (1977) Hilbert Space Methods for Partial Differential Equations, Pitman, London, San Francisco, Melbourne.
15. Glowinski, R., Lions, J.L., and Trêmolieres, R. (1976) Analyse Numêrique des Inêquations Variationelles, Vol. 1, Dunod, Paris.
16. Brezis, H. (1968) Equations et Inêquations non Linêaires dans les Espaces Vectoriels en Dualitê, Ann. Inst. Fourier, Grenoble, 18, 1:115-175.

ON THE USE OF FUNDAMENTAL SOLUTIONS IN TREFFTZ METHOD FOR POTENTIAL AND ELASTICITY PROBLEMS

C. Patterson and M. A. Sheikh

Dept. of Mechanical Engineering, University of Sheffield, U.K.

SUMMARY

IN this paper a variation of the usual Trefftz Method is presented as applied to harmonic and two dimensional elasticity problems. Here, the fundamental solutions (Brebbia, 1980) of the given problem are employed as the expansion functions (Patterson and Sheikh, 1981-A) and their singularities are located outside the domain (Patterson and Sheikh, 1981-B) to avoid any handling of singular quantities (Patterson and Sheikh, 1982-A). This method has a significant computational advantage in that here no integrations are required. Test problems have been analysed using the proposed method and the results compared with those generated by the Singular Boundary Element Method with constant elements. The convergence behaviour of the approximate solutions given by the Modified Trefftz Method with regard to the mesh size needed to obtain a prescribed precision and the singularity location is appraised.

INTRODUCTION

THE remarkable success of Finite Element Method has led to progressively increased demands being made of it. In particular, there is increasing pressure to use sophisticated three dimensional geometric models which result in large costly numerical models.

Because of confinement of freedoms to the boundary, Boundary Domain Techniques show promise in alleviating the problems of size. While several boundary domain techniques are possible, the Boundary Element Method has received most attention (Brebbia, 1978). Although this method has the advantage of having boundary freedoms only, it has a serious disadvantage in that the singularity of the fundamental solution is taken on the boundary of the problem domain. This requires the evaluation of a series of singular integrals which incurs a sizeable computational overhead.

A number of other boundary domain techniques are available to solve field problems, one being the method of Trefftz (Trefftz, 1926). In this method, no integrations are required; the solution is expressed as a polynomial expansion and expansion co-efficients are determined by satisfying the prescribed boundary conditions. In this paper a variation of this approach is presented.

In the Boundary Element Method, the fundamental solutions of the governing equations are employed to generate the boundary solution. Being solutions of homogeneous equations of the problem, such functions are presumably efficient as expansion functions for the solution of other homogeneous problems, as shown in this paper. Here, the Trefftz Method is applied to potential and two dimensional elasticity problems but the approximate solutions are generated using appropriate fundamental solutions with singularity outside the domain (Patterson and Sheikh, 1981-C), as in the Regular Boundary Element Method (Patterson and Sheikh, 1981-D). The co-efficients of the series are, as usual, determined by imposition of boundary conditions. It is shown that the Modified Trefftz Method is easily implemented and give accurate results cheaply.

Various test cases having regular and singular solutions are examined to demonstrate the practicality of this method and their approximate solutions are compared with those generated by the Boundary Element Method. Problems analysed include: (1) Heat conduction in a rectangular prism resulting in (i) linearly varying temperatures and (ii) a singular temperature gradient, (2) Steady-state, inviscid, laminar fluid flow past a disc in a channel and (3) Stress concentration due to a notched circular hole in a rectangular plate under uniform tension. Also outlined, are the results of a systematic study undertaken to determine: (1) the discretization required to yield a solution of given precision and (2) the best location of fundamental solution singularity outside the domain of the given problem.

THEORY

FOR Potential Problems the governing Laplace's equation for the scalar field function ϕ_0 within the domain of the problem is:

$$\nabla^2\phi_0 = 0 \qquad (1)$$

A Trefftzian trial solution having 'N' freedoms defined on the boundary, can be written as:

$$\phi = \sum_{i=1}^{N} \alpha_i \phi_i \qquad (2)$$

where α_i : unknown co-efficients

ϕ_i : linearly independent functions which in Modified Trefftz Method are taken to be the fundamental solutions which satisfy Equation (1)

In the case of two dimensional potential problems:

$$\phi_i = \ln \frac{1}{r_i}$$

and $$\frac{\partial \phi_i}{\partial n} = \frac{-1}{r_i^2} \hat{\underline{n}} \cdot \underline{r}_i \tag{3}$$

where r_i = distance between source $(\underline{x}_i)$ and field point $(\underline{x})$

$\hat{\underline{n}}$: unit outward normal at $\underline{x}$

$\underline{r}_i$: position vector from $\underline{x}_i$ to $\underline{x}$

The singularity in the solution given by equation (3) is circumvented by locating the source outside the domain (Patterson and Sheikh, 1981-E) as in the Regular Boundary Element Method (Patterson and Sheikh, 1982-B). For potentials and their normal derivatives, equation (2) can be written as:

$$\phi(\underline{x}) = \sum_{i=1}^{N} \phi_i(\underline{x}_i, \underline{x})\underline{\alpha}_i = \underline{\underline{P}}(\underline{x})\underline{C}$$

and $$\frac{\partial \phi(\underline{x})}{\partial n} = \sum_{i=1}^{N} \frac{\partial \phi_i}{\partial n}(\underline{x}_i, \underline{x})\underline{\alpha}_i = \underline{\underline{D}}(\underline{x})\underline{C} \tag{4}$$

where $\underline{C}$ is the column vector of unknown parameters to be determined by satisfaction of the boundary conditions.

For a node 'j' (say) $\phi = \bar{\phi}_j$, so that from equation (4);

$$\bar{\phi}_j = \Sigma \phi_i(\underline{x}_i, \underline{x}_j)\underline{\alpha}_i \tag{5a}$$

Similarly, for another node 'k' where normal derivative boundary conditions are given through $\frac{\partial \phi}{\partial n} = \frac{\overline{\partial \phi_k}}{\partial n}$; equation (4) gives;

$$\frac{\overline{\partial \phi_k}}{\partial n} = \Sigma \frac{\partial \phi_i}{\partial n}(\underline{x}_i, \underline{x}_k)\underline{\alpha}_i \tag{5b}$$

In matrix form equations (5a) and (5b) can be written as:

$$\underline{\psi} = \underline{\underline{A}}\,\underline{C} \tag{6}$$

where $\underline{\psi}$ is the vector of known boundary conditions.

Now since the problem is discretized by defining nodes which ar distinct and the fundamental solutions are linearly independent; det $\underline{\underline{A}} \neq 0$ and thus:

$$\underline{C} = \underline{\underline{A}}^{-1} \underline{\psi} \tag{7}$$

Substituting $\underline{C}$ in equation (4), the solution is obtained as:

$$\Phi(\underline{x}) = \underline{P}(x)\underline{\underline{A}}^{-1} \underline{\psi}$$

and $$\frac{\partial \phi}{\partial n}(\underline{x}) = \underline{D}(x)\underline{\underline{A}}^{-1} \underline{\psi} \tag{8}$$

Similarly, for two dimensional elasticity, the method can be implemented in similar manner; the difference is that now there are two degrees of freedom at each node. The fundamental solutions, which satisfy the homogeneous equation of elasticity, are employed as expansion functions. They are given by:

$$u^*_{ij} = \frac{1}{8\pi G(1-\nu)} \left[(3-4\nu)\ln\left(\frac{1}{r}\right)\delta_{ij} + \frac{\partial r}{\partial x_i} \cdot \frac{\partial r}{\partial x_j} \right] \tag{9}$$

and

$$t^*_{ij} = \frac{1}{4\pi(1-\nu)r} \left\{ \frac{\partial r}{\partial n} \left[(1-2\nu)\delta_{ij} + 2\frac{\partial r}{\partial x_j} \cdot \frac{\partial r}{\partial x_i} \right] - (1-2\nu) \left[\frac{\partial r}{\partial x_i} \cdot n_j - \frac{\partial r}{\partial x_j} \cdot n_i \right] \right\} \tag{10}$$

where 'r' is the distance between the point of application of load to the point under consideration; δ_{ij} is the kronecker delta; G is the rigidity modulus; ν the poisson's ratio; n_j and n_i the direction cosines and n is the normal to the surface so that $\frac{\partial r}{\partial n} = n_s \cdot \frac{\partial r}{\partial x_s}$. For 'N' nodes used in the discretization of the boundary of the domain of the given problem, the displacements and tractions can be approximated by a set of linearly independent functions (fundamental solutions in this case) as:

$$\underline{u} = \sum_{k=1}^{N} \alpha_k u_k$$
$$\underline{t} = \sum_{k=1}^{N} \alpha_k t_k \qquad (11)$$

or by taking into account the components of displacement and traction:

$$u_i(x,y) = \sum_{k=1}^{N} u_{i1}(\underline{x}_k,\underline{y}_k,\underline{x},\underline{y})\alpha_k + u_{i2}(\underline{x}_k,\underline{y}_k,\underline{x},\underline{y})\alpha_{k+N} \qquad (12)$$

and
$$t_i(x,y) = \sum_{k=1}^{N} t_{i1}(\underline{x}_k,\underline{y}_k,\underline{x},\underline{y})\alpha_k + t_{i2}(\underline{x}_k,\underline{y}_k,\underline{x},\underline{y})\alpha_{k+N} \qquad (13)$$

where u_{ij} : fundamental solution for displacement at point (x,y) in direction 'i' due to a unit load at (x_k,y_k) in direction 'j'

t_{ij} : traction at (x,y) in direction 'i' for the source at (x_k,y_k) and load direction 'j'

α_k : expansion coefficients

Expressions (12) and (13) can be written in matrix form as:

$$u(\underline{x}) = \underline{\underline{U}}(x)\underline{C}$$
and
$$t(\underline{x}) = \underline{\underline{T}}(x)\underline{C} \qquad (14)$$

To avoid any singularity in the solution given by equation (9), the source (x_k, y_k) is located outside the domain (Patterson and Sheikh, 1981-F). The expansion coefficients α_k can be determined by equating the given boundary conditions at each node to the appropriate R.H.S. of equation (12) or (13) which gives a system of equations of order 2N;

$$\underline{W} = \underline{\underline{A}}\ \underline{C} \qquad (15)$$

or
$$\underline{C} = \underline{\underline{A}}^{-1}\ \underline{W} \qquad (16)$$

where $\underline{W}$ is the vector of prescribed boundary conditions.

Substituting $\underline{C}$ in equation (14), we get:

$$u(\underline{x}) = \underline{U}(x)\,\underline{\underline{A}}^{-1}\,\underline{W}$$

and

$$t(\underline{x}) = \underline{T}(x)\,\underline{\underline{A}}^{-1}\,\underline{W} \qquad (17)$$

APPLICATIONS

Heat Conduction in a Rectangular Prism Resulting in Linearly Varying Temperatures

The domain of this problem was discretized using 16 nodes with the boundary conditions as shown in Fig.1. The problem was solved using the Modified Trefftz Method for different singularity locations given by 'αL' (L is the length of the segment containing the node) where α=0.5, 1.0, 1.5, 2.0. The functions (temperature or temp. normal grad.) calculated at various points on the boundary and in the interior are given in Table 1, where they are compared against the exact values and the values given by the singular boundary element method for the same grid. The best results for modified trefftz are obtained by the singularity location given by α=2, however the interior solution is not affected much for $\alpha \geq 1$.

Heat Conduction in a Rectangular Prism Resulting in Singular Temp. Normal Gradient

(i) $T = 0$ at $x=a,\ 0 \leq y \leq b$

(ii) $\frac{\partial T}{\partial n} = 0$ at $0 \leq x \leq a,\ y=0$

(iii) $\frac{\partial T}{\partial n} = 0$ at $0 \leq x \leq a,\ y=b$

(iv) $T = 300$ at $x=0,\ {}^{b}/2 \leq y \leq b$

$\frac{\partial T}{\partial n} = 0$ at $x=0,\ 0 \leq y \leq {}^{b}/2$

Condition (iv) implies a discontinuity in temp. normal gradient (t.n.g.) at point 'O'.

The problem was solved for three different singularity locations given by α=1.0, 1.5, 2.0, in case of Modified Trefftz Method. The variation of temperature normal gradient along the edge 'DO' of the prism for all these singularity locations are compared in Fig.3 with the one given by the singular boundary element method for the same discretization. The computed temperatures at 9 internal points are also compared in Table 2.

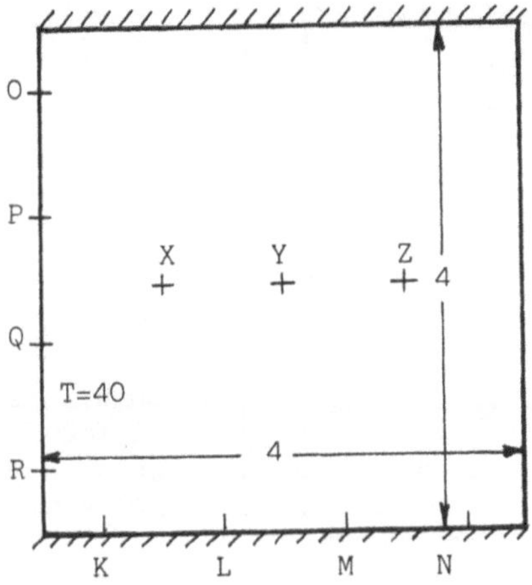

T=0

Fig. 1
Heat Conduction through a prism resulting in linearly varying temperatures

BOUNDARY SOLUTION:

Function	Point	Exact Sol.	Sing. B.E.M.	Modified Trefftz Solutions			
				α =0.5	1.0	1.5	2.0
Temp. (T)	K(0.5,0)	35.0	35.247	34.802	34.861	34.923	34.956
	L(1.5,0)	25.0	25.073	25.312	25.021	25.009	25.040
	M(2.5,0)	15.0	14.930	15.607	15.016	14.996	14.998
	N(3.5,0)	5.0	4.751	6.250	5.314	5.137	5.069
T.n.g. ($\frac{\partial T}{\partial n}$)	O(0,3.5)	10.0	10.556	14.130	10.983	10.415	10.206
	P(0,2.5)	10.0	9.837	10.905	9.933	9.957	9.980
	Q(0,1.5)	10.0	9.837	10.905	9.933	9.957	9.980
	R(0,0.5)	10.0	10.556	14.130	10.983	10.415	0.206
INTERNAL SOLUTION:							
Temp. (T)	X(1.0,2.0)	30.0	30.020	30.116	30.002	29.998	29.999
	Y(2,2)	20.0	20.001	20.264	20.012	20.002	20.000
	Z(3,2)	10.0	9.981	10.285	10.017	10.005	10.003

Table 1. Comparison of Modified Trefftz Solution for different singularity locations with the exact and Singular Boundary Element Solutions

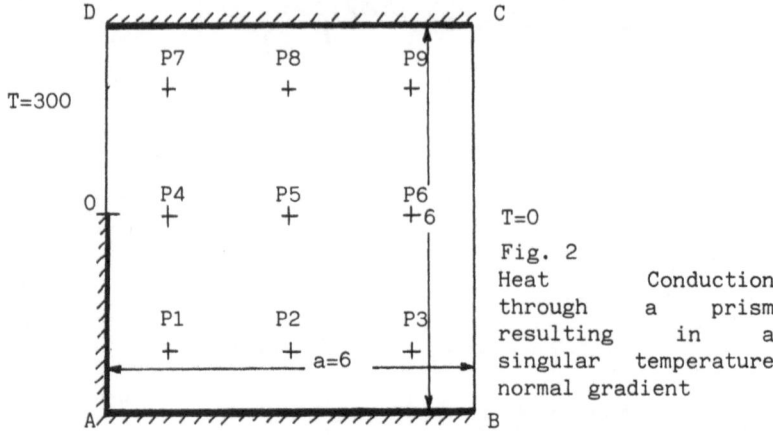

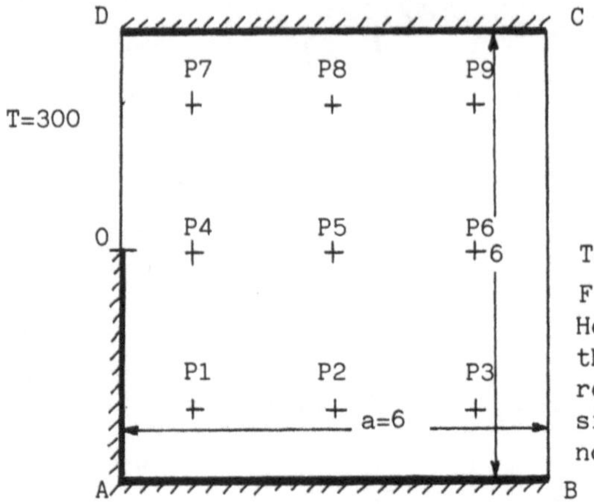

Fig. 2
Heat Conduction through a prism resulting in a singular temperature normal gradient

INTERNAL SOLUTION :

Function	Point	Singular B.E.M.Sol.	Modified Trefftz Solutions		
			$\alpha=1.0$	$\alpha=1.5$	$\alpha=2.0$
Temp. (T)	P1(1,1)	43.649	45.000	44.448	44.640
	P2(3,1)	136.276	136.333	136.171	137.050
	P3(5,1)	241.111	240.644	240.820	241.628
	P4(1,3)	40.697	40.978	41.025	41.590
	P5(3,3)	123.272	123.877	123.876	125.419
	P6(5,3)	211.976	212.518	212.618	214.821
	P7(1,5)	36.826	38.491	38.072	38.652
	P8(3,5)	107.747	109.201	109.240	111.563
	P9(5,5)	160.247	161.812	162.041	167.100

Table 2. Comparison of Modified Trefftz Solution for different singularity locations with the solution generated by the Singular Boundary Element Method

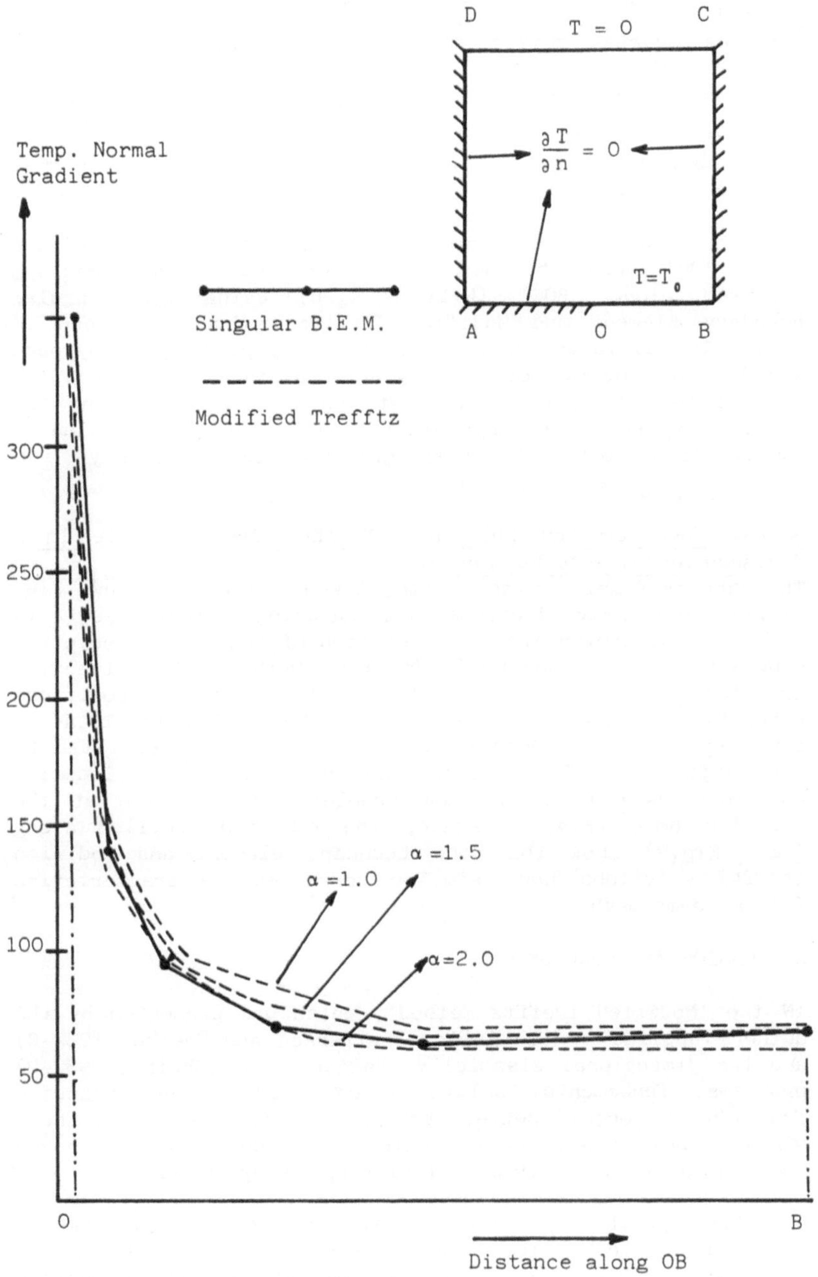

Fig. 3. Comparison of temperature normal gradient along 'OB'

Steady-state, Inviscid, Laminar Fluid Flow Past a Disc in a Channel

For the quarter domain taken for analysis as allowed by symmetry, the boundary conditions are as shown in Fig.4. The solution of this problem (Milne-Thompson, 1968) involves a singularity due to the discontinuity in velocity ($\frac{\partial \psi}{\partial n}$) at point 'O'; the edge of the plate which is kept at normal incidence to the flow.

The flow velocity profiles along the surface of the plate 'BO' were obtained for 4 different mesh gradings employed along 'BOC' (N=11,16,20,26) using the Singular Boundary Element and Modified Trefftz Methods as shown in Fig.5. In the latter case the source corrsponding to any node was located outside the domain at a distance 'αL' from the node where 'L' is the length of the element containing the node. As a result of a systematic study, the optimal value of 'α' was found to be 1.5 for maximum numerical efficiency.

Stress Concentration Due to a Notched Circular Hole in a Rectangular Plate Under Tension

The quarter domain of the plate, taken due to symmetry (Fig. 6) was first analysed without a notch using 34 nodes discretization to determine the best position of fundamental solution singularity for Modified Trefftz Method. The traction profiles along 'DE' were obtained for two such positions given by 'αL' where α=0.25, 1.0. As the singularity location given by =1.0 produced results closer to the ones given by the singular boundary element method (Fig.7), this location was then used to solve the problem with a notch at the circular hole (Fig.8). Again, the traction profiles along 'DE' (Fig.9) show that the boundary element and modified trefftz solutions have similar convergence characteristics for the same mesh.

DISCUSSION AND CONCLUSIONS

IN the 'Modified Trefftz Method', which was presented by the authors applied to fluid flow (Patterson and Sheikh, 1982-C) and two dimensional elasticity (Patterson and Sheikh, 1982-D) problems, fundamental solutions are used as the expansion functions to obtain better results (Brebbia, 1978) and their singular points are located outside the domain of the given problem to avoid any treatment of singular quantities.

Although the theoretical details regarding the implementation of the method to potential and elasticity problems are outlined in this paper, nevertheless, the central objective of this study was to appraise the convergence behaviour of the solutions given by the method with regard to the mesh size required to generate a solution of a prescribed precision and the location of the fundamental solution

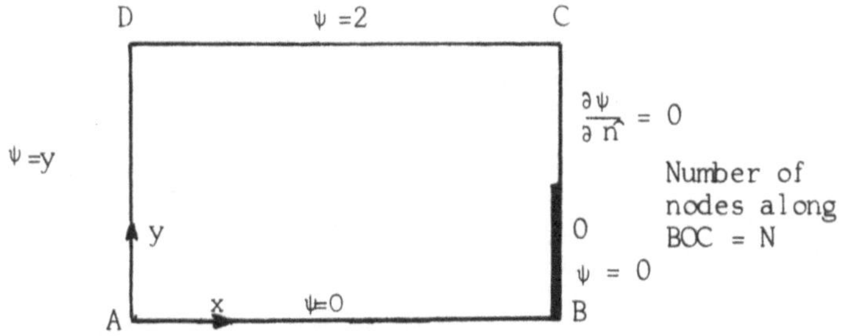

Fig. 4. Boundary Conditions for quarter domain

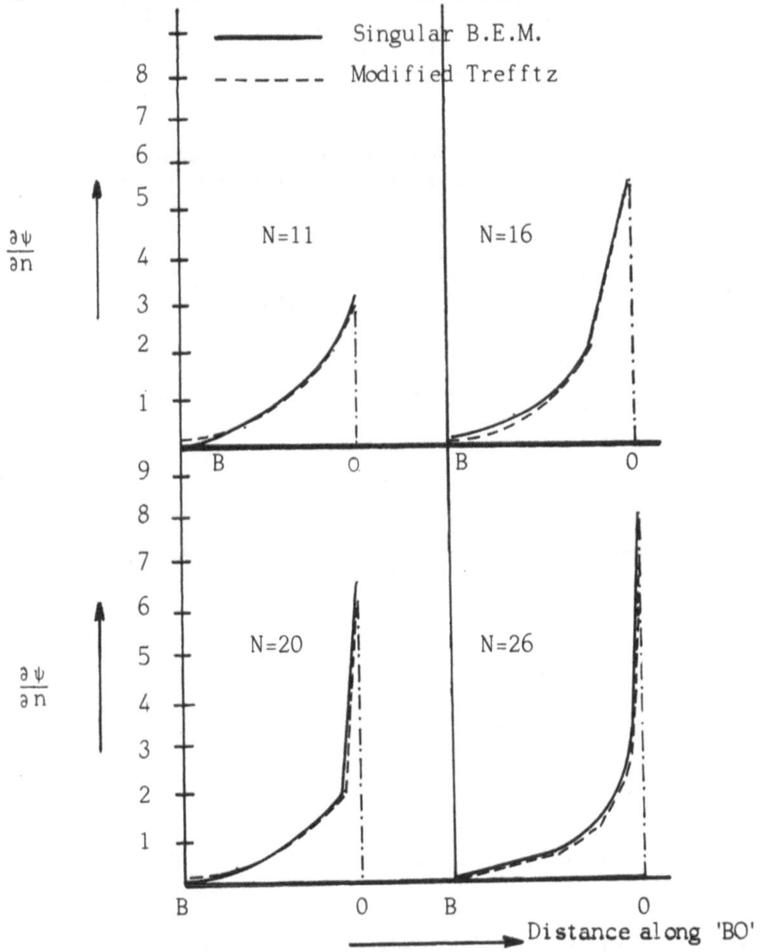

Fig. 5. Velocities along 'BO' given by singular B.E. and Modified Trefftz Methods for different meshes around the singularity

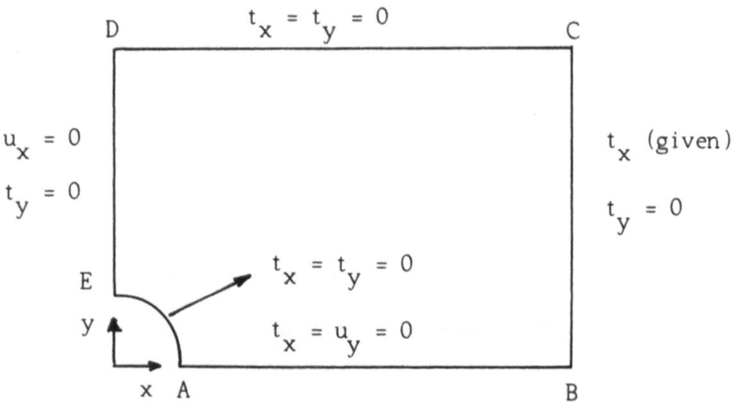

Fig. 6. Boundary Conditions for quarter domain (Plate with a circular hole under tension)

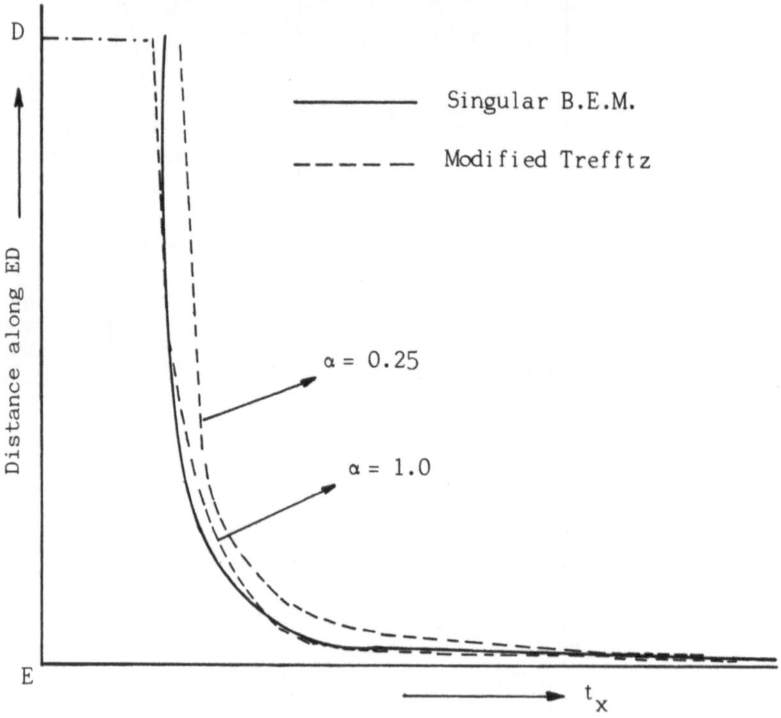

Fig. 7 Traction profiles along 'ED' given by Singular Boundary Element and Modified Trefftz Methods

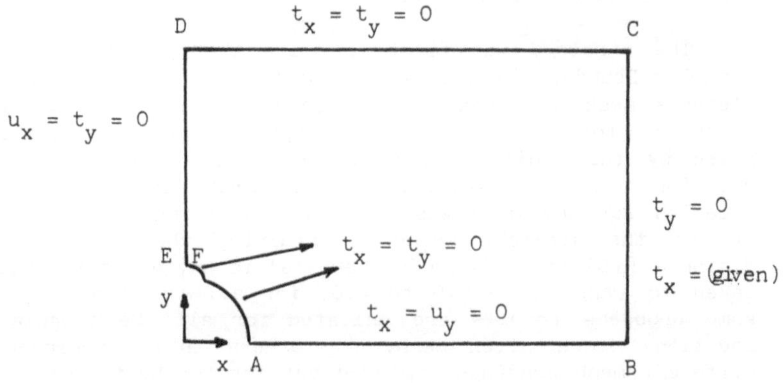

Fig. 8. Boundary Conditions for quarter domain (Plate with a notched circular hole under tension)

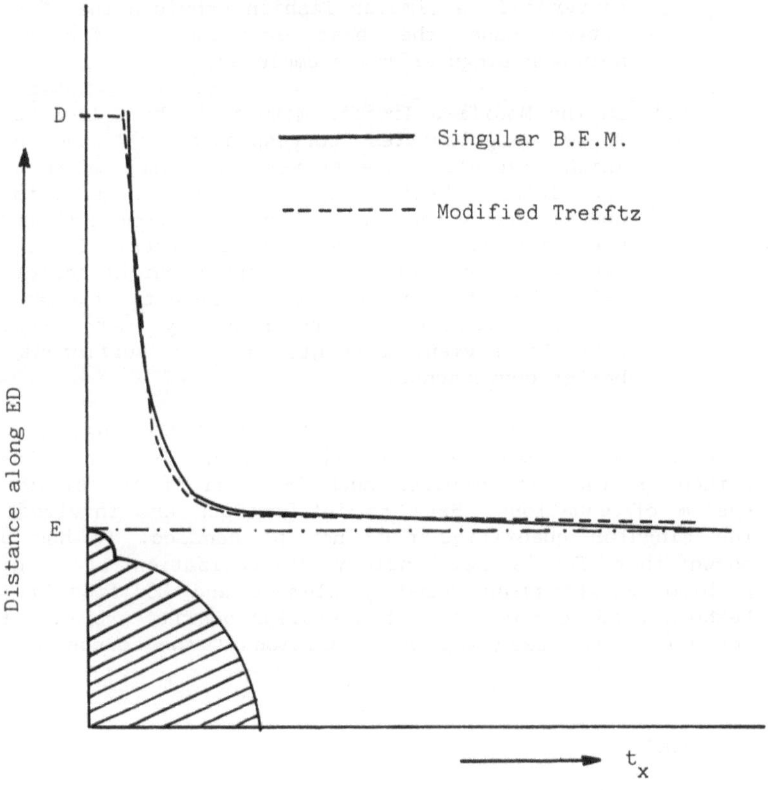

Fig. 9. Traction profiles alond 'ED' given by Singular B.E. and Modified Trefftz Methods

singularity outside the domain.

The strategy was first to solve the problem using the singular Boundary Element Method with a satisfactory constant elements mesh and then use the solution thus obtained as a reference for the purpose of comparison to the solutions given by the Modified Trefftz for a similar discretization but for different singularity locations. This singularity location for any node was expressed in terms of the length (L) of the straight segment containing the node, as the distance (αL) from the node where 'α' is any arbitrary value, taken to range from 0.5 to 2.0, for all the problems. For some problems results are enlisted for all the singularity locations to a given mesh, for others they correspond to different mesh gradings employed but for the best location of the singularity. The following observations are made:

(1) For any given satisfactory mesh size, the Singular Boundary Element and Modified Trefftz solutions converge in a similar fashion provided that for the latter case the best location of fundamental solution singularity is employed.

(2) In the Modified Trefftz Method the best location of these singularities corresponding to the nodes which discretize the system (expressed in terms of the distance 'αL' along the positive normals to the nodes) are given by 'α' ranging from 1.0 to 2.0 depending upon the degree of refinement of the mesh employed. For coarse meshes and straight geometries $\alpha=2$ seems to produce better results whereas for curved geometries represented by more refined straight segment elements $\alpha=1$ is sufficient for better convergence.

In conclusion: The Modified Trefftz Method has a clear computational advantage over the Singular Boundary Element Method in that it requires much less effort to set up the system of equations. Here no integrations are involved and the singular quantities need not be handled. It has been shown that for a satisfactory discretization of a given problem the Singular Boundary Element and Modified Trefftz Methods produce solutions of a similar quality provided that for the latter case the best position of the source outside the domain is employed.

REFERENCES

Brebbia, C.A. (1978) Boundary Element Methods for Engineers, Pentech Press, London.

Brebbia, C.A. (1980) Fundamentals of Boundary Elements. In:

Proc. 2nd. Int. Seminar Boundary Element Methods, Ed. C.A. Brebbia, CML Publ. Ltd., Southampton.

Milne-Thompson, L.M. (1968) Theoretical Hydrodynamics, Macmillan, London.

Patterson, C. and Sheikh, M.A. (1981-A) On the use of Fundamental Solutions in the Trefftz Method for Harmonic Problems, In: Proc. IV MAFELAP Conf., Brunel University, Uxbridge.

Patterson, C. and Sheikh, M.A. (1981-B) A Regular Boundary Method for Coupled Subdomains. In: Proc. Int. Conf. Num. Meth. Coupled Prob., Eds. R.W. Lewis, P. Bettes and E. Hinton, Pineridge Press, Swansea.

Patterson, C. and Sheikh, M.A. (1981-C) Discontinuous Boundary Elements for Heat Conduction. In: Proc. Int. Conf. Num. Meth. Thermal Prob. -II, Eds. R. W. Lewis, K. Morgan and B.A. Schrefler, Pineridge Press, Swansea.

Patterson, C. and Sheikh, M.A. (1981-D) Regular Boundary Integral Equations for Stress Analysis. In: Proc. 2nd. Int. Conf. Boundary Element Methods, Ed. C.A. Brebbia, Springer Verlag, Berlin.

Patterson, C. and Sheikh, M.A. (1981-E) Non-Conforming Boundary Elements for Stress Analysis. In: Proc. 2nd. Int. Conf. Boundary Element Methods, Ed. C.A. Brebbia, Springer Verlag, Berlin.

Patterson, C. and Sheikh, M.A. (1981-F) Regular Boundary Integral Equations for Fluid Flow. In: Proc. Int. Conf. Num. Meth. Laminar and Turb. Flow, Eds. C. Taylor and B.A. Schrefler, Pineridge Press, Swansea.

Patterson, C. and Sheikh, M.A. (1982-A) A Regular Boundary Element Method for Fluid Flow. In: Int. J. Num. Meth. in Fluids, Eds. C. Taylor and P.M. Gresho, John Wiley, London.

Patterson, C. and Sheikh, M.A. (1982-B) A Regular Boundary Integral Method for Linear Elastic Fracture Mechanics. In: Proc. Int. Conf. Finite Element Methods, Shanghai.

Patterson, C. and Sheikh, M.A. (1982-C) A Modified Trefftz Method for Fluid Flow. In: Proc. IV Int. Symposium Finite Elements in Flow Problems, Tokyo.

Patterson, C. and Sheikh, M.A. (1982-D) A Modified Trefftz Method for Sress Analysis. In: Proc. Int. Conf. Finite Element Methods, Shanghai.

Trefftz, E. (1926) Ein Gegenstruck Zum Ritzschem Verfharen. In: Proc. 2nd. Int. Congress. App. Mech., Zurich.

BOUNDARY ELEMENT & DYNAMIC PROGRAMMING METHOD IN OPTIMIZATION OF TRANSIENT PARTIAL DIFFERENTIAL SYSTEMS

Tanehiro FUTAGAMI

Hiroshima Insititute of Technology, Itsukaichi, Hiroshima, Japan

ABSTRACT

By combining the boundary element method with dynamic programming, the BE (Boundary Element) & DP (Dynamic Programming) method is developed and systematized in order to control transient partial differential equation systems with both equality or inequality constraints and a linear/nonlinear objective function. Such systems are frequently encountered in various engineering and scientific problems of control and optimal design. The BE&DP method is applied to optimal control problems in thermal diffusion phenomena. Minimization of total of the controllable loads to meet with temperature requirements is studied. The tractability in the initial and final conditions, the boundary conditions and the equality or inequality constraints makes sure that the BE&DP method becomes a powerful technique for optimization of transient partial differential equation systems.

INTRODUCTION

By combining the finite element method with linear programming, the FE (Finite Element) & LP (Linear Programming) method has been developed in order to solve partial differential equation systems with a linear objective function (Futagami, 1975, 76). The BE (Boundary Element) & LP (Linear Programming) method has been also developed by the combined use of the boundary element method with linear programming (Futagami, 1981). The combined use of finite difference method with linear programming has been also studied in the field of ground-water management (Aguado and Remson, 1974). In order to optimize transient partial differential systems the combined use of the finite element method with dynamic programming has been also studied (Futagami, 1977, 81).

In this paper, by combining the boundary element method with dynamic programming, the BE (Boundary Element) & DP (Dynamic Programming) method is developed and systematized in order to control transient differntial systems with a linear/nonlinear objective function. The BE&DP method utilizes the advantages of established techniques of both the boundary element method and dynamic programming.

The boundary element method, originated at Southampton University from previous work on classical integral equations and the finite element method, is a powerful numerical method for the solution of partial differential equations because of its generality with respect to geometry, its much smaller systems of equations, its simplicity for the input data required, and its numerical accuracy (Brebbia, 1978). Dynamic programming is one of the most useful mathematical methods in optimization of multi-stage decision process (Bellman, 1957 and Futagami, 1970).

In order to show the features and the applicability of the BE&DP method optimal control of thermal diffusion phenomena is also studied.

THE BE (BOUNDARY ELEMENT) & DP (DYNAMIC PROGRAMMING) METHOD

Systems of Basic Differential Equations

The BE (Boundary Element) & DP (Dynamic Programming) method is developed and systematized in order to control the following systems of transient partial differential equations which are frequently encountered in optimal control of field problems (diffusion-convection, heat conduction, seepage flow, electric/magnetic potential, etc.) (see Figs. 1 and 4).

Objective Function (throughout the whole space-time domain $\Omega\times T^e$)

$$Z = \underset{\{\{\theta\}\}}{\text{Opt.}}\ F(\{\{\phi\}\}, \{\{q\}\}, \{\{\theta\}\}) = \underset{\{\{\theta(t)\}\}}{\text{Opt.}} \int_0^{T^e} f(\{\phi(t)\}, \{q(t)\}, \{\theta(t)\})\, dt \tag{1}$$

subject to:

State Equations

Governing Differential Equation (at t=t, in the whole space domain Ω)

$$\text{D.E.}(x, t, \phi, \frac{\partial\phi}{\partial x}, \ldots, \frac{\partial^n\phi}{\partial x^n}, \frac{\partial\phi}{\partial t}, \theta) = 0 \tag{2}$$

Initial Condition (at t=0, in the whole space domain Ω)

$$\text{I.C.}(x, t=0, \phi) = 0 \tag{3}$$

Final Condition (at $t=T^e$, in the whole space domain Ω)

$$\text{F.C.}(x, t=T^e, \phi) \gtreqless 0 \tag{4}$$

Boundary Condtions (at t=t, in the whole space domain Ω)

$$\text{B.C.}(X, t, \phi, \frac{\partial\phi}{\partial n}) = 0 \tag{5}$$

Constraints (at t=t, in the sub-space domain Ω^r)

$$r(x, \phi, q, \theta) \lesseqgtr 0 \tag{6}$$

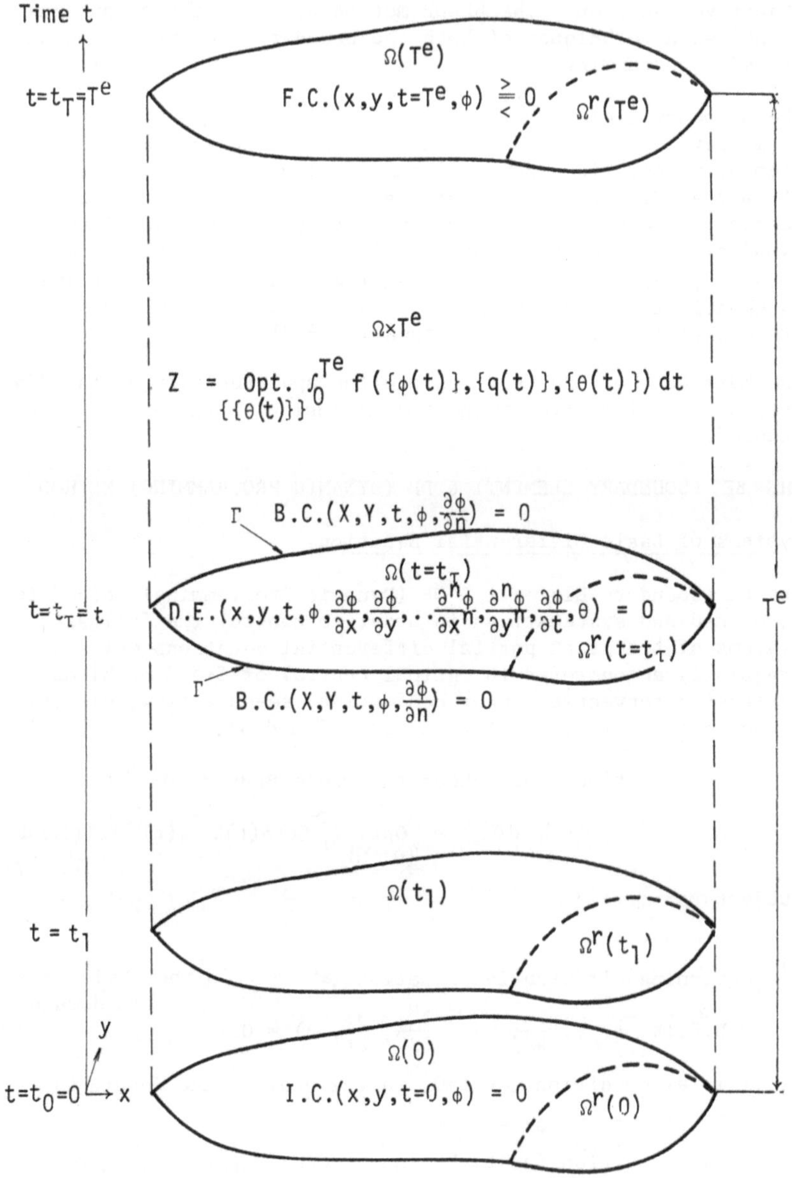

Fig. 1. General Concepts of the BE&DP Method

in which ϕ = the state variable (temperature, concentration, potential, etc.); q = the flux; θ = the decision variable (controllable load, controllable sink/source, etc.); x = Cartesian coordinates (x, y, z); X = Cartesian coordinates of the boundaries (X, Y, Z); t = time; and T^e = the planning period.

In systems governed by two-dimensional unsteady differential equations, for example, Eq. 2 is expressed as follows:

$$\frac{\partial\phi}{\partial t} = \underbrace{c_1\frac{\partial^2\phi}{\partial x^2} + c_2\frac{\partial^2\phi}{\partial x\partial y} + c_3\frac{\partial^2\phi}{\partial y^2} + c_4\frac{\partial\phi}{\partial x} + c_5\frac{\partial\phi}{\partial y} + c_6\phi}_{\phi\text{-tems}} + \underbrace{\theta}_{\theta\text{-term}} + \underbrace{b}_{\text{const.}} \quad (2)'$$

The examples of the initial conditions and the final conditions are as follows:

$$\phi(x,y,t=0) = \Phi^0(x,y), \quad \phi(x,y,t=0) \text{ is free} \quad (3)'$$

$$\phi(x,y,t=T^e) \lesseqgtr \Phi^T(x,y), \quad \phi(x,y,t=T^e) \text{ is free} \quad (4)'$$

The examples of the boundary conditions are as follows:

$$\phi(X,Y,t) = \Phi_b(X,Y,t), \quad \frac{\partial\phi}{\partial n}(X,Y,t) = 0 \quad (5)'$$

As for the constraints, the following simple inequalities are frequently encountered.

$$\underline{\Phi} \leqq \phi \leqq \overline{\Phi} \text{ or } \begin{cases} \phi \geqq \underline{\Phi} \\ \phi \leqq \overline{\Phi} \end{cases}, \quad \underline{\Theta} \leqq \theta \leqq \overline{\Theta} \text{ or } \begin{cases} \theta \geqq \underline{\Theta} \\ \theta \leqq \overline{\Theta} \end{cases} \quad (6)'$$

in which $\underline{\Phi}$ = the lower limit of the state variable; $\overline{\Phi}$ = the upper limit of the state variable; $\underline{\Theta}$ = the lower limit of the decision variable; and $\overline{\Theta}$ = the upper limit of the decision variable.

Formulation of BE&DP Method

Several formulations of the time dependent boundary element method have been studied (Brebbia and Walker, 1980 and Wrobel and Brebbia, 1981). The application of one of these formulations to the discretization of the aforementioned systems (Eqs. 1-6) yields the following matrix-vector forms of the BE&DP method. (As for the details, please see next chapter).

Objective Function (throughout the whole space-time domain $\Omega \times T^e$)

$$Z = \underset{\{\{\theta_i^\tau\}\}}{\text{Opt.}} \sum_{\tau=1}^{T} f^\tau(\{x_j^\tau\}, \{\theta_i^\tau\}) = \underset{\{\{\theta_i^\tau\}\}}{\text{Opt.}} \sum_{\tau=1}^{T} f^\tau(\{\phi_j^\tau\}, \{q_j^\tau\}, \{\theta_i^\tau\}) \quad (7)$$

subject to:

State Transformation Equations ((T×N)-Eqs.)

$$\underset{N\times N}{[H^\tau]}\{\phi_j^\tau\} = \underset{N\times N}{[G^\tau]}\{q_j^\tau\} + \underset{N\times M}{[E^\tau]}\{\phi_m^{\tau-1}\} + \underset{N\times I}{[D^\tau]}\{\theta_i^\tau\} + \{p_n^\tau\}$$

or

$$\underset{N\times N}{[A^\tau]}\{x_j^\tau\} - \underset{N\times N}{[D^\tau]}\{\theta_i^\tau\} = \underset{N\times M}{[E^\tau]}\{\phi_m^{\tau-1}\} + \{P_n^\tau\} \quad (\tau = 1 \sim T) \quad (8\text{-}\tau)$$

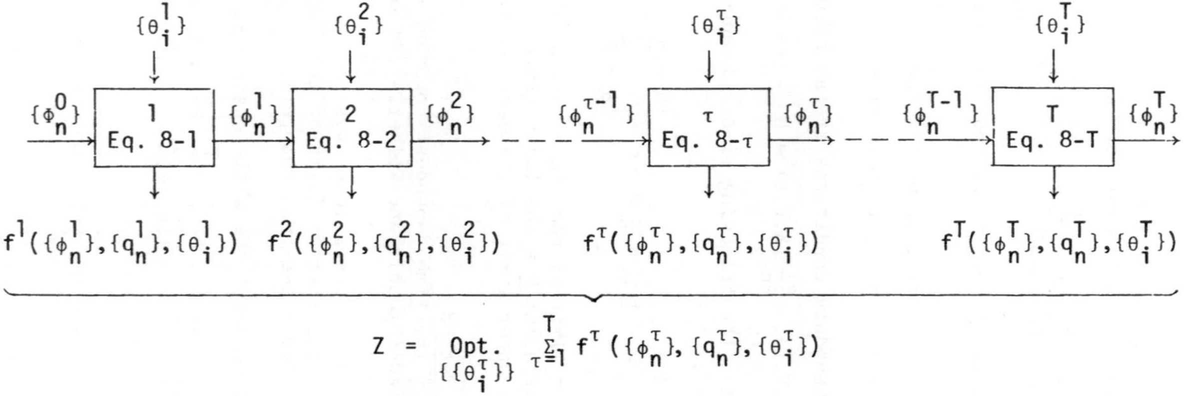

Fig. 2. Multi-Stage Decision Process in the BE&DP Method

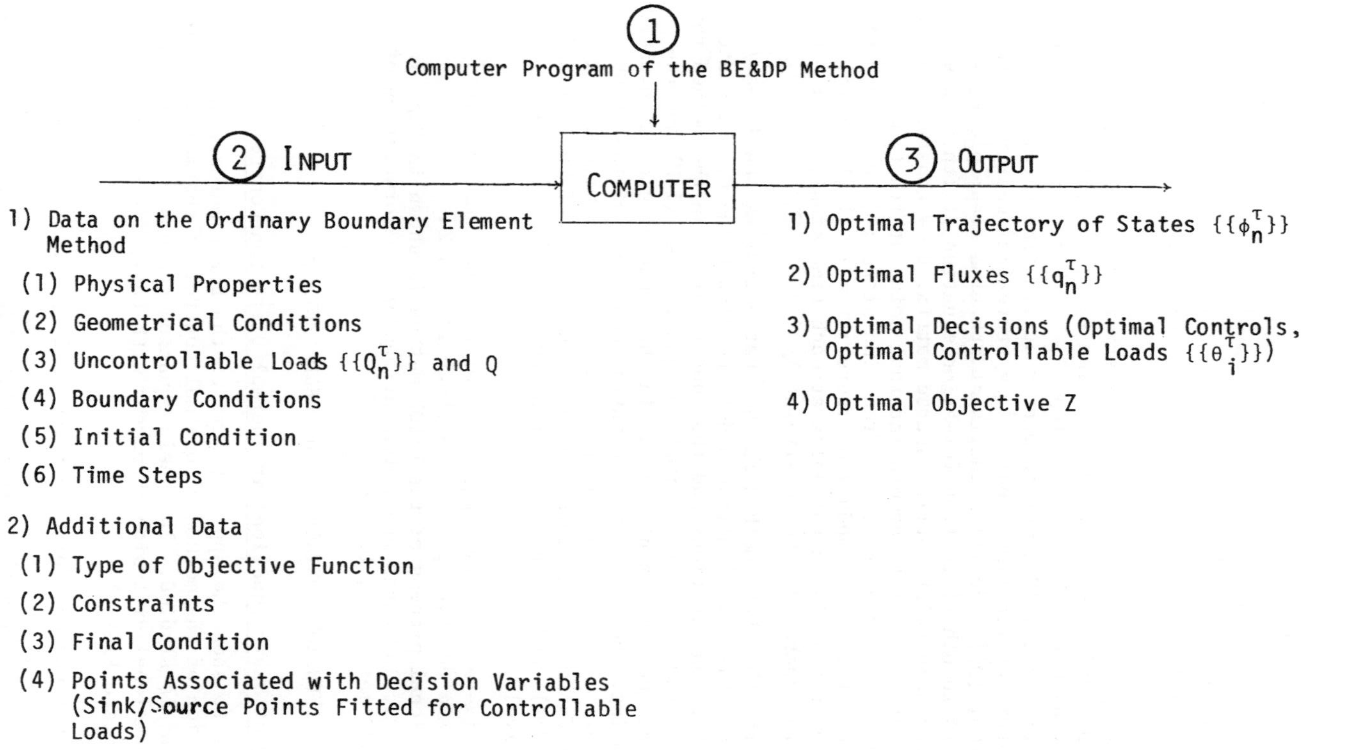

1
Computer Program of the BE&DP Method
2 INPUT
COMPUTER
3 OUTPUT
1) Data on the Ordinary Boundary Element Method
(1) Physical Properties
(2) Geometrical Conditions
(3) Uncontrollable Loads $\{\{Q_n^\tau\}\}$ and Q
(4) Boundary Conditions
(5) Initial Condition
(6) Time Steps
2) Additional Data
(1) Type of Objective Function
(2) Constraints
(3) Final Condition
(4) Points Associated with Decision Variables (Sink/Source Points Fitted for Controllable Loads)
1) Optimal Trajectory of States $\{\{\phi_n^\tau\}\}$
2) Optimal Fluxes $\{\{q_n^\tau\}\}$
3) Optimal Decisions (Optimal Controls, Optimal Controllable Loads $\{\{\theta_i^\tau\}\}$)
4) Optimal Objective Z

Fig. 3. Computational Process of the BE&DP Method

Constraints (L-Eqs.)

$$\underset{L\times(T(N+I))}{[R]}\begin{Bmatrix}\{\{x_j^\tau\}\}\\ \{\{\theta_i^\tau\}\}\end{Bmatrix} \overset{\le}{\underset{>}{=}} \{\{B_l\}\} \quad \text{or} \quad [R]\begin{Bmatrix}\{\{\phi_j^\tau\}\}\\ \{\{q_j^\tau\}\}\\ \{\{\theta_i^\tau\}\}\end{Bmatrix} \overset{\le}{\underset{>}{=}} \{\{B_l\}\} \qquad (9)$$

in which x_n^τ = nth boundary variable at τth time step; ϕ_n^τ = nth state on boundary elements at τth time step; q_n^τ = nth flux on boundary elements at τth time step; θ_i^τ = ith decision variable at τth time step; $[H^\tau]$ = state matrix at τth time step; $[G^\tau]$ = flux matrix at τth time step; $[A^\tau]$ = state-flux matrix at τth time step; $[D^\tau]$ = decision matrix at τth time step; p_n^τ, P_n^τ = constants in nth state transformation equation at τth time step; [R] = the constraint matrix, sparse matrix; B_l = constant in lth constraint; ϕ_m^τ = state at mth internal point at τth time step; $\tau = 1 \sim T$ (T = total number of the time steps); $j, n = 1 \sim N$ (N = total number of the boundary elements); $i = 1 \sim I$ (I = total number of the decision variables at each time step); $l = 1 \sim L$ (L = total number of the constraints).

Therefore, the BE&DP method is one that optimizes the linear/nonlinear objective function under the conditions of the state transformation equations and the constraints. In the method the number of the state transformation equations is (T×N) and the number of the constraints is L. In the sense of general dynamic programming, the state transformation equations of the BE&DP method are also the constraints. Thus the BE&DP method is a kind of dynamic programming in which the number of the variables is (T×(N+I)) and the number of the constraints is (T×N + L), respectively.

The matrix-vector forms (Eqs. 7-9) of the BE&DP method yield a multi-stage decision process as shown in Fig. 2. Computational process of the BE&DP method is shown in Fig. 3. Some particular cases and modifications of the formulation of the BE&DP method may be considered.

THERMAL DIFFUSION CONTROL BY THE BE&DP METHOD

Systems of Basic Equations in Thermal Diffusion Control

In order to show the applicability of the BE&DP method, the application of the method to optimal control of thermal diffusion phenomena is studied. The basic equation systems of transient two-dimensional diffusion phenomena with constraints are as follows (see Fig. 4):

Objective Function (throughout the whole space-time domain $\Omega\times T^e$)

$$Z = \underset{\{\{\theta(t)\}\}}{\text{Opt.}} \int_0^{T^e} f(\{\phi(t)\}, \{q(t)\}, \{\theta(t)\})dt \simeq \underset{\{\{\theta_i(t)\}\}}{\text{Min.}} \int_0^{T^e} \sum_{i=1}^{I} \theta_i(t)\, dt \qquad (10)$$

subject to:

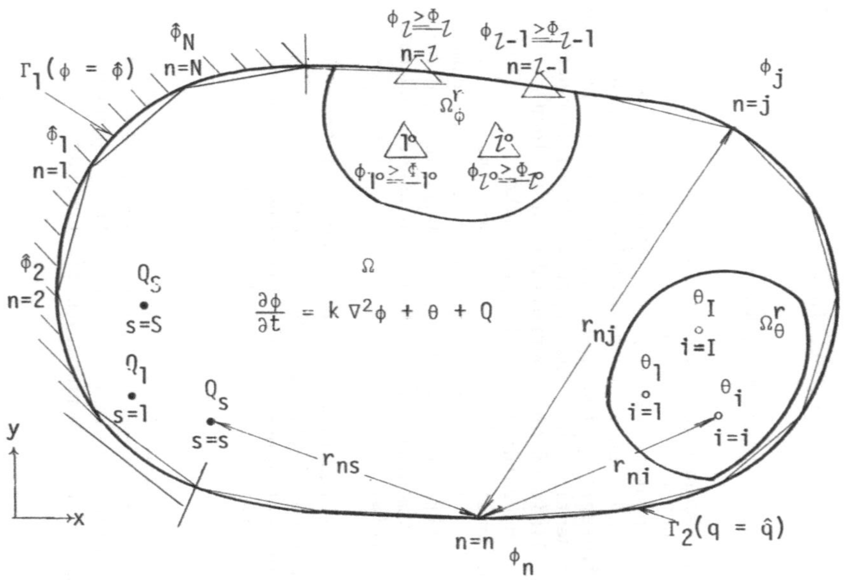

∘ : Source Point Fitted for Controllable Heat Load

• : Source Point Fitted for Uncontrollable Heat Load

△ : Regulated Point in Temperature Requirement

ϕ = Temperature (State Variable)

$\hat{\phi}$ = Prescribed Temperature

$\underline{\Phi}$ = Lower Limit of Temperature

q = Heat Flux

$\hat{q}$ = Prescribed Heat Flux

θ = Controllable Heat Load (Decision Variable, Unknown Heat Load)

Q = Uncontrollable Heat Load (Known Heat Load)

Ω = Whole Space-Domain

Ω^r_ϕ = Sub-domain Associated with Temperature Requirements

Ω^r_θ = Sub-domain Associated with Controllable Heat Loads

$\Gamma = \Gamma_1 + \Gamma_2$ = Boundaries

k = Diffusion Coefficient

r = Distance

Fig. 4. The Boundary Element & Dynamic Programming Method in Thermal Diffusion Control

State Equations

Governing Differential Equation

$$\frac{\partial \phi}{\partial t} = k \nabla^2 \phi + \theta + Q \qquad \text{at } t=t, \text{ in } \Omega \qquad (11)$$

Initial Condition

$$\phi = \phi^0 \qquad \text{at } t=0, \text{ in } \Omega \qquad (12)$$

Final Condition

$$\phi \text{ is free} \qquad \text{at } t=T^e, \text{ in } \Omega \qquad (13)$$

Boundary Conditions

$$\phi = \hat{\phi} \qquad \text{at } t=t, \text{ on } \Gamma_1 \qquad (14)$$

$$q = \frac{\partial \phi}{\partial n} = \hat{q} \qquad \text{at } t=t, \text{ on } \Gamma_2 \qquad (15)$$

Constraints

$$\phi \geqq \underline{\Phi} \qquad \text{at } t=t, \text{ in } \Omega^r_\phi \qquad (16)$$

$$\overline{\Theta} \geqq \theta \geqq 0 \qquad \text{at } t=t, \text{ in } \Omega^r_\theta \qquad (17)$$

in which ϕ = the state variable (temperature); q = heat flux; θ = decision variable = unknown heat load (controllable heat load issuing from variable heat generator); Q = known heat load (uncontrollable heat load issuing from invariable heat generator); k = diffusion coefficient; $\underline{\Phi}$ = the lower limit of the state variable (temperature requirement); $\overline{\Theta}$ = the upper limit of the decision variable (the upper limit of the controllable heat load); $\hat{\phi}$ = the prescribed temperature; $\hat{q}$ = the prescribed heat flux; Ω = the whole space domain; T^e = the whole time domain = the planning period; $\Gamma = \Gamma_1 + \Gamma_2$ = the boundaries; Ω^r_ϕ = sub-domain associated with the state-constraints (sub-domain imposed on the temperature requirements); Ω^r_θ = sub-domain associated with the decision-constraints.

Although, the objective function may be composed of the temperature distribution $\{\{\phi(t)\}\}$, the heat fluxes $\{\{q(t)\}\}$, and the controllable heat loads $\{\{\theta(t)\}\}$ in general, the minimization of the total of the controllable heat loads $\int_0^{T^e} \sum_{i=1}^{I} \theta_i(t)\, dt$ is sought in this study, for simplicity.

As for the constraints, although only the lower limit of the state variable (temperature requirement) $\underline{\Phi}$ and the upper and lower limits of the decision variable are imposed in the above equation systems, we can impose other constraints if necessary.

Formulation of the BE&DP Method in Thermal Diffusion Control

In order to discretize the above-mentioned differential systems, the time dependent boundary element method is used. Several formulations of the time dependent boundary element method have been presented. As for the details of the time dependent boundary element method, one may follow Wrobel and Brebbia (1981). Considering the time dependence of the problem directly in the integration by parts process, the boundary integral equation for the governig differential eauation becomes as follows:

$$c_n\phi_n + k \int_{t_0}^{t_\rho} \int_\Gamma q^* \phi \, d\Gamma \, dt = k \int_{t_0}^{t_\rho} \int_\Gamma \phi^* q \, d\Gamma \, dt + \left[\int_\Omega \phi^* \phi \, d\Omega\right]_{t=t_0}$$

$$+ \int_{t_0}^{t_\rho} \sum_{i=1}^{I} \theta_i \phi^* \, dt + \int_{t_0}^{t_\rho} \sum_{s=1}^{S} Q_s \, \phi^* \, dt + \int_{t_0}^{t_\rho} \int_\Omega Q \, \phi^* \, d\Omega \, dt \qquad (18)$$

in which θ_i = the controllable point load at internal or boundary point 'i'; Q_s = the uncontrollabe point load at internal or boundary point 's'; Q = the distributed uncontrollable load.

The fundamental solution ϕ^* in Eq. 18 is as follows:

$$\phi^*(t_\tau, t) = \frac{1}{4\pi k(t_\tau - t)} \exp\left(-\frac{r^2}{4k(t_\tau - t)}\right) \qquad (19)$$

and

$$q^*(t_\tau, t) = \frac{\partial \phi^*}{\partial n} = -\frac{r}{8\pi k^2 (t_\tau - t)^2} \exp\left(-\frac{r^2}{4k(t_\tau - t)}\right) \frac{\partial r}{\partial n} \qquad (20)$$

in which $r = [(x-x_n)^2 + (y-y_n)^2]^{1/2}$ = the distance from the point to the point under consideration.

Since time variations of ϕ and q are part of the solution, a time stepping scheme (not to be confused with the finite differencing on time) has to be introduced but as the fundamental solution itself is time dependent, large time steps can be adopted. In the following the details of the formulation are described.

In order to obtain the algebraic equation systems for Eq. 18, the boundaries Γ are discretized into elements with space interpolation functions. Let us assume that functions ϕ and q remain constant on time over each time step, for simplicity. Let us also assume that the whole space domain Ω is divided into cells (although it is not necessary).

The following two different discretization procedures could be employed.

1) The First Process

$$\underset{N\times N}{[H^\tau_{nj}]}\{\phi^\tau_j\} = \underset{N\times N}{[G^\tau_{nj}]}\{q^\tau_j\} + \underset{N\times M}{[E^\tau_{nm}]}\{\phi^{\tau-1}_m\} + \underset{N\times I}{[D^\tau_{ni}]}\{\theta^\tau_i\} + \{p^\tau_n\} \qquad (21)$$

$$(\tau = 1 \sim T)$$

with

$$H^{\tau}_{nj} = k \int_{\Gamma_j} \mu^T \int_{t_{\tau-1}}^{t_\tau} q^*(t_\tau,t)\, dt\, d\Gamma + c_n \delta_{nj}$$

$$G^{\tau}_{nj} = k \int_{\Gamma_j} \nu^T \int_{t_{\tau-1}}^{t_\tau} \phi^*(t_\tau,t)\, dt\, d\Gamma$$

$$E^{\tau}_{nm} = \int_{\Omega_m} \phi^*(t_\tau,t_{\tau-1})\, d\Omega \qquad (22)$$

$$D^{\tau}_{ni} = \int_{t_{\tau-1}}^{t_\tau} \phi^*(t_\tau,t)\, dt$$

$$p^{\tau}_{n} = \sum_{s=1}^{S} Q^{\tau}_s \int_{t_{\tau-1}}^{t_\tau} \phi^*(t_\tau,t)\, dt + \int_\Omega Q \int_{t_{\tau-1}}^{t_\tau} \phi^*(t_\tau,t)\, dt\, d\Omega$$

in which δ_{nj} = the Kronecker delta; μ, ν = spacial interpolation functions.

Notice that at the end of each time step, temperatures at a sufficient number (M) of internal points need to be computed in order to be used as initial values for the next time step.

2) The Second Process

$$\sum_{\gamma=1}^{\tau} \underset{N\times N}{[H^{\tau\gamma}_{nj}]}\{\phi^{\gamma}_j\} = \sum_{\gamma=1}^{\tau} \underset{N\times N}{[G^{\tau\gamma}_{nj}]}\{q^{\gamma}_j\} + \underset{N\times M}{[E^{\tau}_{nm}]}\{\Phi^{0}_m\} + \sum_{\gamma=1}^{\tau} \underset{N\times I}{[D^{\tau\gamma}_{ni}]}\{\theta^{\gamma}_i\} + \{p^{\tau}_n\} \qquad (\tau = 1 \sim T) \quad (23)$$

with

$$H^{\tau\gamma}_{nj} = k \int_{\Gamma_j} \mu^T \int_{t_{\gamma-1}}^{t_\gamma} q^*(t_\tau,t)\, dt\, d\Gamma + c_n \delta_{\tau\gamma} \delta_{nj}$$

$$G^{\tau\gamma}_{nj} = k \int_{\Gamma_j} \nu^T \int_{t_{\gamma-1}}^{t_\gamma} \phi^*(t_\tau,t)\, dt\, d\Gamma$$

$$E^{\tau}_{nm} = \int_{\Omega_m} \phi^*(t_\tau,t_0)\, d\Omega \qquad (24)$$

$$D^{\tau\gamma}_{ni} = \int_{t_{\gamma-1}}^{t_\gamma} \phi^*(t_\tau,t)\, dt$$

$$p^{\tau}_{n} = \sum_{\gamma=1}^{\tau} \Big(\sum_{s=1}^{S} Q^{\tau}_s \int_{t_{\gamma-1}}^{t_\gamma} \phi^*(t_\tau,t)\, dt + \int_\Omega Q \int_{t_{\gamma-1}}^{t_\gamma} \phi^*(t_\tau,t)\, dt\, d\Omega \Big)$$

Since in this approach we always start the time integration process at time $t = t_0$, there is no need to compute internal temperatures.

From Eqs. 19 and 20, it follows that the time integrals in Eqs. 22 and 24 are carried out analytically, giving

$$\int_{t_{\gamma-1}}^{t_\gamma} \phi^*(t_\tau,t)\,dt = \frac{1}{4\pi k}[E_1(a_{\gamma-1}) - E_1(a_\gamma)] \qquad (25)$$

$$\int_{t_{\gamma-1}}^{t_\gamma} q^*(t_\tau,t)\,dt = -\frac{1}{2\pi k r}\left(\frac{\partial r}{\partial n}\right)[\exp(-a_{\gamma-1}) - \exp(-a_\gamma)] \qquad (26)$$

with

$$a_\gamma = \frac{r^2}{4k(t_\tau - t_\gamma)} \qquad (27)$$

and $E_1[a_\gamma]$ is the exponential integral function, which can be evaluated by series, i.e.

$$E_1[a_\gamma] = -C - \ln[a_\gamma] + \sum_{n=1}^{\infty}(-1)^{n-1}\frac{a_\gamma^n}{n\cdot n!} \qquad (28)$$

C being the Euler's constant, $C = 0.57721566....$
Notice that $E_1(a_\tau) = \exp(-a_\tau) = 0$.

Then, the matrix-vector forms of the BE&DP method in the thermal diffusion control are obtained below:

Objective Function (throughout the whole space-time domain $\Omega\times T^e$)

$$Z = \underset{\{\{\theta_i^\tau\}\}}{\text{Opt.}} \sum_{\tau=1}^{T} f^\tau(\{\phi_n^\tau\}, \{q_n^\tau\}, \{\theta_i^\tau\}) \simeq \underset{\{\{\theta_i^\tau\}\}}{\text{Min.}} \sum_{\tau=1}^{T}\sum_{i=1}^{I}(t_\tau - t_{\tau-1})\,\theta_i^\tau \qquad (29)$$

subject to:

State Transformation Equations ((T×N)-Eqs.)

$$\underset{N\times N}{[H_{nj}^\tau]}\{\phi_j^\tau\} = \underset{N\times N}{[G_{nj}^\tau]}\{q_j^\tau\} + \underset{N\times M}{[E_{nm}^\tau]}\{\phi_m^{\tau-1}\} + \underset{N\times I}{[D_{ni}^\tau]}\{\theta_i^\tau\} + \{p_n^\tau\} \qquad (30)$$

$(\tau = 1 \sim T)$

,or,

$$\sum_{\gamma=1}^{\tau}\underset{N\times N}{[H_{nj}^{\tau\gamma}]}\{\phi_j^\gamma\} = \sum_{\gamma=1}^{\tau}\underset{N\times N}{[G_{nj}^{\tau\gamma}]}\{q_j^\gamma\} + \underset{N\times M}{[E_{nm}^\tau]}\{\tilde{\Phi}_m^0\} + \sum_{\gamma=1}^{\tau}\underset{N\times I}{[D_{ni}^{\tau\gamma}]}\{\theta_i^\gamma\} + \{p_n^\tau\} \qquad (30)'$$

$(\tau = 1 \sim T)$

Constraints ((T×(L+I))-Eqs.)

$$\{\phi_l^\tau\} \geqq \{\underline{\Phi}_l^\tau\} \qquad (\tau = 1 \sim T) \qquad (31)$$

$$\{\overline{\Theta}_i^\tau\} \geqq \{\theta_i^\tau\} \geqq \{0\} \quad (\tau = 1 \sim T) \qquad (32)$$

The previous assumption of constancy on time over each time step for ϕ and q is not restrictive, and higher order time interpolation functions could be used (Wrobel and Brebbia, 1981).

CONCLUSIONS

The BE (Boundary Element) & DP (Dynamic Programming) method was developed in order to control transient partial differential systems. The application of the method to thermal diffusion control was also studied. The tractability in the initial and final conditions, the boundary conditions and the equality or inequality constraints makes sure that the BE&DP method becomes a powerful technique for optimization of transient partial differential systems with a linear/nonlinear objective function. Efficient computational algorithms of the BE&DP method could be developed.

REFERENCES

1. Aguado, E. and Remson, I. (1974), "Ground-Water Hydraulics in Aquifer Management", Journal of the Hydraulics Division, ASCE, Vol. 100, No. HY1, pp. 103-118.
2. Bellman, R., (1957), "Dynamic Programming", Princeton University Press, Princeton.
3. Brebbia. C.A. (1978), "The Boundary Element Method for Engineers", Pentech Press, London.
4. Brebbia, C.A. and Walker, S. (1980), "Boundary Element Techniques in Engineering", Newnes-Butterworths, London. Boston.
5. Futagami, T. (1970), "Dynamic Programming for a Sewage Treatment Systems", Proceedings, 5th International Water Pollution Research Conference, Jenkins, H.S. ed., Pergamon Press, 1970, pp. II-21/1 - II21/12.
6. Futagami, T. (1975), "Finite Element & Linear Programming Method and Water Polluiton Control", Proceedings, 16th Congress of the International Association for Hydraulics Research, Vol. 3, c7, pp. 54-61.
7. Futagami, T., Tamai, N. and Yatsuzuka, M. (1976), "FEM Coupled with LP for Water Pollution Control", "Journal of Hydraulics Division, ASCE, Vol. 102, HY7, pp. 881-897.
8. Futagami, T., Fukuhara, T. and Tomita, M. (1977), "Transient Finite Element & Linear Programming Method in Environmental Systems Control - The Efficient Computational Algorithm", Proceedings, IFAC Environmental Symposium, Pergamon Press, pp. 143-150.
9. Futagami, T. (1981), "Boundary Element & Linear Programming Method in Optimization of Partial Differentail Systems", Proceedings, 3rd Internatinal Seminar on Boundary Element Methods, Brebbia, C.A., ed., Springer-Verlag, pp. 457-471.
10. Futagami, T. (1981), "The Dynamic Finite Element & Nonlinear Programming Method in Control of Transient Differential Systems", Proceedings, 8th Triennial World Congress, IFAC,

Vol. IV, Sessions 17-20, IV-105 - IV-110.

11. Wrobel, L.C. and Brebbia, C.A. (1981), "Time Dependent Potential Problems", Progress in Boundary Element Methods, Vol. 1, Brebbia, C.A., ed., Pentech Press, London: Plymouth.

SOME IMPROVEMENTS IN 2D BOUNDARY ELEMENTS USING INTEGRATION BY PARTS

C.KATZ MUENCHEN

1. Introduction

Boundary elements or integral equations usually start with some formula relating the solution of a problem in the interior of a domain to the boundary values. Commonly this is done by two integrals one with a singular kernel and the other with the derivatives of this kernel. For example in potential theory the fundamental equation is given by

$$\Phi(z_0) = \oint_{z \in \Gamma} \left(\frac{\partial \Phi(z)}{\partial n} G(z,z_0) - \Phi(z) \frac{\partial G(z,z_0)}{\partial n} \right) d\Gamma \tag{1}$$

z_0 is the point of consideration in the domain, while z is located on the boundary. The kernel G is of logarithmic order while the derivative is of order (1/z). Integrating numerically, the second term will introduce severe errors if the point of consideration approaches the boundary.

Bischoff |1| and Katz |3| used integration by parts for the potential problem to introduce new kernels for the second integral which behave much better. This paper will generalize the approach given in |3| and show how to obtain the so called Green's functions of the second kind for the problem of plain elasticity and plate bending by complex variables.

2. Green's functions of the second kind

Green's functions of the second kind are obtained via integration by parts. Starting from equation (1) the second expression is integrated with respect to the boundary direction s.

$$\int_{\Gamma} \frac{\partial G(z,z_0)}{\partial n} \Phi(z)\, d\Gamma = \int \frac{\partial G(z,z_0)}{\partial n} ds\, \Phi(z) \Big|_{\Gamma}$$

$$- \int_{\Gamma} \frac{\partial \Phi(z)}{\partial s} \left[\int \frac{\partial G(z,z_0)}{\partial n} ds \right] d\Gamma \tag{2}$$

introducing Green's function of the second kind G^* , which is a integral function depending on the boundary shape s.

$$G^*(z,z_0) = \int \frac{\partial G(z,z_0)}{\partial n} ds \tag{3}$$

This function is of logarithmic order for the potential problem and will stabilize the integration process.

If Φ is smooth (that means at least linear shape functions), then the first expression will be some constant value Φ_0 depending on the start and end point of integration. Finally the expression for the potential at an interior location is (1),(2):

$$\Phi(z_0) = \int_{\Gamma} G(z,z_0) \frac{\partial \Phi(z)}{\partial n} d\Gamma + \int_{\Gamma} G^*(z,z_0) \frac{\partial \Phi(z)}{\partial s} d\Gamma + \Phi_0 \tag{4}$$

This definition can be extended easily to other integral equation problems. For elasticity or plate bending problems the fundamental equation corresponding to (1) will be

$$c_{ij} u_i(z_0) + \int_{\Gamma} T_{ij}(z,z_0)\, u_i(z)\, d\Gamma = \int_{\Gamma} U_{ij}(z,z_0)\, t_i(z)\, d\Gamma \tag{5}$$

Where $u_j(z)$ are the displacements and $t_j(z)$ the tractions (i.e. stresses at the boundary). While $U_{ij}(z,z_0)$ is the displacement kernel of a fundamental solution and $T_{ij}(z,z_0)$ is the stress kernel derived from $U_{ij}(z,z_0)$. Assuming a smooth displacement field integration by parts of the second term becomes possible and yields to

$$\int_{\Gamma} T_{ij}(z,z_0)\, u_j(z)\, d\Gamma = T^*_{ij}(z,z_0)\, u_j(z) - \int_{\Gamma} T^*_{ij}(z,z_0) \frac{\partial u_j(z)}{\partial s} d\Gamma \tag{6}$$

where $T^*_{ij}(z,z_0)$ is the Green's function of the second kind defined by

$$T^*_{ij}(z,z_0) = \int T_{ij}(z,z_0)\, ds \tag{7}$$

3. Complex Representation

The evaluation of the desired functions is a very hard job in real arithmetic but it is rather simple to get them in complex variables. We can use the relations for the derivatives

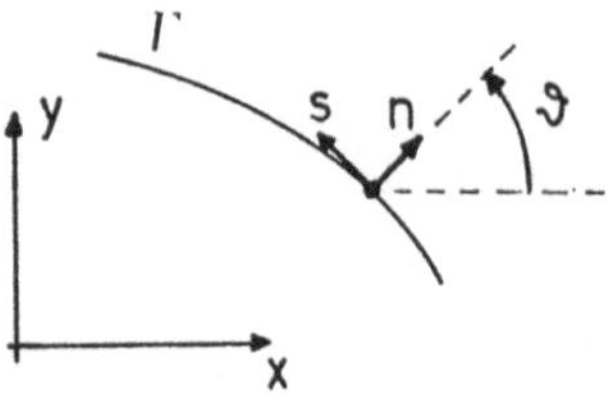

$$\frac{\partial G}{\partial n} = e^{-i\vartheta}\frac{\partial G}{\partial \bar{z}} + e^{i\vartheta}\frac{\partial G}{\partial z} \qquad (8a)$$

$$i\frac{\partial G}{\partial s} = e^{-i\vartheta}\frac{\partial G}{\partial \bar{z}} - e^{i\vartheta}\frac{\partial G}{\partial z} \qquad (8b)$$

ϑ gives the outward direction of the boundary normal vector.

Any solution to the potential problem is given by the real part of an analytical function which can be represented as a sum of a holomorphic function and its conjugate:

$$G(z,z_0) = \varphi(z,z_0) + \overline{\varphi(z,z_0)} \qquad (9)$$

Thus the derivation with respect to the normal n is easily obtained applying equ.(8a). And it is easily verfied that the Green's function of the second kind is given by:

$$G^*(z,z_0) = \frac{1}{i}\left[\varphi(z,z_0) - \overline{\varphi(z,z_0)}\right] \qquad (10)$$

That means that for the potential problem Green's Function and Green's function of the second kind are real and imaginary part of a complex function. Moreover the use of complex differentiation gives the values of the fluxes just by one new expression:

$$\frac{\partial \Phi}{\partial x_0} + i\frac{\partial \Phi}{\partial y_0} = \frac{\partial \Phi}{\partial \bar{z}_0} = \int\left(\frac{\partial G(z,z_0)}{\partial \bar{z}_0}\frac{\partial \Phi}{\partial n} + \frac{\partial G^*(z,z_0)}{\partial \bar{z}_0}\frac{\partial \Phi}{\partial s}\right)d\Gamma \qquad (11)$$

For the free space the complex Green's function is given by its well known real part |2| and is therefore

$$GG = G + iG^* = -\frac{1}{2\pi}\ln(z-z_0) \qquad (12)$$

4. Problem of multivalued logarithm

There is just one problem not discussed yet. The complex logarithm is a multivalued function. Integrating along the boundary we may climb up or down on other Riemann surfaces that means the imaginary part will differ by multiples of 2*PI*i to the main value. However it is rather easy to take care of this during the integration process and adding the multiples to the imaginary part.
The constant term Φ_0 is given by the difference of the function G^* at the start and end point of the integration. If both are identical (i.e. returning to the same point) the real parts will vanish and it remains only the multivalued part

$$\Phi_0 = c\,\Phi_s \tag{13}$$

Where Φ_s is the value of Φ at the start point and c is a constant depending on the kind of problem. If the point of consideration z_0 is inside the domain c will be 1. If it is outside c will be zero. This part of the solution would be lost by the differentiation of the shape functions otherwise. This mode with eigenvalue zero is called rigid body mode in statics and is used very often in boundary element methods to determine diagonal coefficients or to control the matrices.

5. Elasticity Problems

The complex representation of elasticity problems is given by the method of Muskelishvili |5| using the formula of Goursat:

$$\Delta\Delta F = \Delta\Delta \operatorname{Re}\left[\bar{z}\varphi(z) + \chi(z) \right] = 0 \tag{15}$$

Every solution to the plain problem of elasticity can be represented by a stress function F which fullfills the bipotential equation and is given by the two holomorphic functions φ and χ. Setting u for u_1 and v for u_2, the resulting displacements and stresses for plain strain are then given by

$$2G(u+iv) = (3-4\mu)\varphi(z) - z\overline{\varphi'(z)} - \overline{\psi(z)} \tag{16}$$

$$\sigma_x + i\tau_{xy} = \varphi'(z) + \overline{\varphi'(z)} - z\overline{\varphi''(z)} - \overline{\psi'(z)} \tag{17}$$

$$\sigma_y - i\tau_{xy} = \varphi'(z) + \overline{\varphi'(z)} + z\overline{\varphi''(z)} + \overline{\psi'(z)} \tag{18}$$

G is the shear modulus and μ is poisson's ratio

The tractions in the global directions are now given by

$$t_x + i t_y = e^{i\vartheta}\left[\varphi'(z) + \overline{\varphi'(z)}\right] - e^{-i\vartheta}\left[z\,\overline{\varphi''(z)} + \overline{\psi'(z)}\right] \quad (19)$$

The last equation (19) has to be integrated along the boundary. It is easy to verify that the following equation differentiated with respect to s will yield equation (19).

$$it^* = \varphi(z) + z\,\overline{\varphi(z)} + \overline{\psi(z)} \quad (20)$$

5.1. Free space solution

The solution for a point load (X+iY) at the point of consideration z_0 is given by the two functions (Muskhelishvili, |5|):

$$\varphi(z) = -\frac{X+iY}{8\pi(1-\mu)}\ln(z-z_0) \quad (21)$$

$$\psi(z) = (3-4\mu)\frac{X-iY}{8\pi(1-\mu)}\ln(z-z_0) + \bar{z}_0\frac{X+iY}{8\pi(1-\mu)}\frac{1}{z-z_0} \quad (22)$$

Inserting (21) and (22) in (20)

$$t^* = \frac{X+iY}{8\pi i(1-\mu)}\left[(2-4\mu)F_1 - iF_2\right] - \frac{X-iY}{8\pi i(1-\mu)}F_3 \quad (23)$$

with

$$F_1 = \mathrm{Re}\left[\ln(z-z_0)\right] = \ln|z-z_0|$$

$$F_2 = 4(1-\mu)\ \mathrm{Im}\left[\ln(z-z_0)\right] = 4(1-\mu)\arg(z-z_0)$$

$$F_3 = \frac{z-z_0}{\bar{z}-\bar{z}_0}$$

To construct the Kernel $T^*_{ij}(z,z_0)$ we set X=1,Y=0 to get the influence function for the displacement at z_0 in x-direction T^*_{1j} and X=0,Y=1 for the y-direction T^*_{2j}. For convenience these two values are combined to a complex one. Thus we have the following expression for the displacements at z_0 combining equations (5,6,16 and 23)

$$g(z_o) = (u+iv)_o = u_s + iv_s + \tag{24}$$

$$+ \frac{1}{8\pi(1-\mu)} \int_\Gamma \left\{ \left(-(3-4\mu) F_1 + F_3\right) \frac{t_x}{2G} + \left(-(3-4\mu) F_1 - F_3\right) \frac{i t_y}{2G} + \left(-(2-4\mu) F_1 + i F_2 + F_3\right) \frac{i \partial u}{\partial s} + \left((2-4\mu) F_1 - i F_2 + F_3\right) \frac{\partial v}{\partial s} \right\} d\Gamma$$

Indeed the multivalued logarithm will account for the rigid body displacements. The stresses are calculated by complex differentiation and the following expressions.

$$\frac{1}{2}(\sigma_x + \sigma_y) = \frac{2G}{(1-2\mu)} \mathrm{Re}\left[\frac{\partial g}{\partial z_0}\right] \tag{25}$$

$$\frac{1}{2}(\sigma_x - \sigma_y) + i\tau_{xy} = 2G \left[\frac{\partial g}{\partial \bar{z}_0}\right] \tag{26}$$

5.2 Other domains

Neither the shape of the boundary nor the kind of the fundamental solution have any effect on the derived formulae. Equations (16-20) hold for any kernel. It is recommendable to stay with the analytic functions as long as possible. We introduce

$$\varphi(z) = (X+iY)\varphi_2(z) + (X-iY)\varphi_1(z) \tag{27}$$

$$\psi(z) = (X+iY)\psi_2(z) + (X-iY)\psi_1(z) \tag{28}$$

and the combination of x- and y-displacements (X=1,Y=i) will simplify any occurence of the original functions for example by

$$\varphi(z) = 2\varphi_1(z) \tag{29}$$

$$\overline{\varphi(z)} = 2\overline{\varphi_2(z)} \tag{30}$$

The displacements and stresses are then given by

$$
\begin{aligned}
g(z_0) = {} & g_s + \\
\int \Big\{ & \Big((3-4\mu)\, \varphi_1(z) - z\overline{\varphi_2'(z)} - \overline{\psi_2(z)} + (3-4\mu)\overline{\varphi_2(z)} - \bar{z}\varphi_1'(z) - \psi_1(z) \Big) \frac{t_x}{2G} + \\
- & \Big((3-4\mu)\, \varphi_1(z) - z\overline{\varphi_2'(z)} - \overline{\psi_2(z)} - (3-4\mu)\overline{\varphi_2(z)} + \bar{z}\varphi_1'(z) + \psi_1(z) \Big) \frac{i t_y}{2G} + \\
- & \Big(\varphi_1(z) + z\overline{\varphi_2'(z)} + \overline{\psi_2(z)} - \overline{\varphi_2(z)} - \bar{z}\varphi_1'(z) - \psi_1(z) \Big) \frac{i\partial u}{\partial s} + \\
- & \Big(\varphi_1(z) + z\overline{\varphi_2'(z)} + \overline{\psi_2(z)} + \overline{\varphi_2(z)} + \bar{z}\varphi_1'(z) + \psi_1(z) \Big) \frac{\partial v}{\partial s} \Big\}\, d\Gamma
\end{aligned}
\qquad (31)
$$

For the free space problem table I derived from equ. (21),(22) gives the stress functions. The method of Muskelishvili enables us to establish more functions. |4| gives some solutions with rigid circles and lines. Table II gives the solution for the stress-free halfplane.

Table I Stress Functions Free Space

	$\varphi_1(z)$	$\overline{\varphi_2(z)}$	$\bar z\,\varphi_1'(z)+\psi_1(z)$	$z\,\overline{\varphi_2'(z)}+\overline{\psi_2(z)}$
f	0	$-\ln(\bar z-\bar z_0)$	$(3-4\mu)\ln(z-z_0)$	$-\frac{z-z_0}{\bar z-\bar z_0}$
$\frac{\partial f}{\partial z_0}$	0	0	$-\frac{3-4\mu}{z-z_0}$	$\frac{1}{\bar z-\bar z_0}$
$\frac{\partial f}{\partial \bar z_0}$	0	$\frac{1}{\bar z-\bar z_0}$	0	$-\frac{z-z_0}{(\bar z-\bar z_0)^2}$

Table II Stress Functions Halfplane with Stress Free Surface

	$\varphi_1(z)$	$\overline{\varphi_2(z)}$	$\bar z\,\varphi_1'(z)+\psi_1(z)$	$z\,\overline{\varphi_2'(z)}+\overline{\psi_2(z)}$
f	$-\frac{z_0-\bar z_0}{z-\bar z_0}$	$-\ln(\bar z-\bar z_0)$ $-(3-4\mu)\ln(\bar z-z_0)$	$-\frac{z-z_0}{\bar z-\bar z_0}$ $-(3-4\mu)\frac{z-z_0}{\bar z-z_0}$	$\ln(z-\bar z_0)$ $(3-4\mu)\ln(z-z_0)$ $-\frac{(z-\bar z)(z_0-\bar z_0)}{(z-\bar z_0)^2}$
$\frac{\partial f}{\partial z_0}$	$-\frac{1}{z-\bar z_0}$	$\frac{3-4\mu}{\bar z-z_0}$	$\frac{1}{\bar z-\bar z_0}$ $-(3-4\mu)\frac{z-\bar z}{(\bar z-z_0)^2}$	$-(3-4\mu)\frac{1}{z-z_0}$ $-\frac{z-\bar z}{(z-\bar z_0)^2}$
$\frac{\partial f}{\partial \bar z_0}$	$\frac{z-z_0}{(z-\bar z_0)^2}$	$\frac{1}{\bar z-z_0}$	$-\frac{z-z_0}{(\bar z-\bar z_0)^2}$	$-\frac{1}{z-\bar z_0}$ $+(z-\bar z)\left[\frac{1}{(z-\bar z_0)^2}-\frac{2(z_0-\bar z_0)}{(z-\bar z_0)^3}\right]$

All Functions $\times\ \frac{1}{8\pi(1-\mu)}$

6. Plate bending problems

The problem of plate bending is somewhat more complicated due to the existence of higher order derivatives. Complex representation is done via the Goursat formula again. Using the two analytical functions we have for displacements, rotations, moments and shearforces of a Kirchhoff plate (Savin, |6|):

$$w = \mathrm{Re}\left[\bar{z}\varphi(z) + \chi(z)\right] \tag{32}$$

$$\Theta_x + i\Theta_y = \frac{1}{2}\left[\varphi(z) + z\overline{\varphi'(z)} + \overline{\psi(z)}\right] \tag{33}$$

$$m_x + i m_y = -2D(1+\mu)\left[\varphi'(z) + \overline{\varphi'(z)}\right] \tag{34}$$

$$m_x - m_y + i m_{xy} = -2D(1-\mu)\left[z\,\overline{\varphi''(z)} + \overline{\psi'(z)}\right] \tag{35}$$

$$m + i m_t = -D\left[(1+\mu)\left(\varphi'(z) + \overline{\varphi'(z)}\right) + (1-\mu)\,e^{-2i\vartheta}\left(z\,\overline{\varphi''(z)} + \overline{\psi'(z)}\right)\right] \tag{36}$$

$$q_x + i q_y = -4D\,\overline{\varphi''(z)} \tag{37}$$

$$\int q\,ds = 2iD\left[\varphi'(z) - \overline{\varphi'(z)}\right] \tag{38}$$

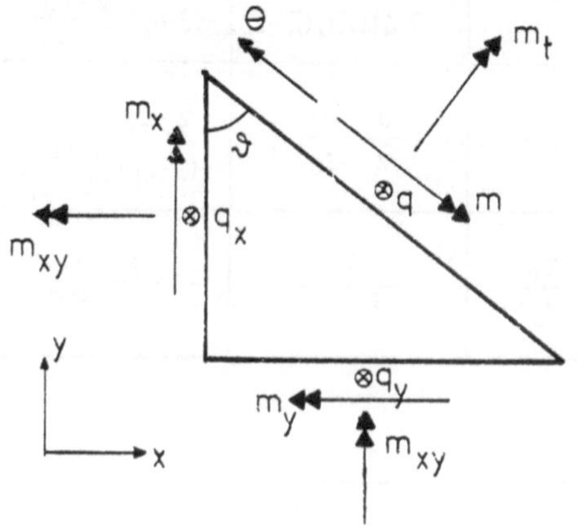

$$D = \frac{E\,t^3}{12(1-\mu^2)}$$

The fundamental equation is given by Tottenham |7|

$$w(z_0) = \int \left(w_a(z,z_0)\, q^*(z) - \Theta_a(z,z_0)\, m(z) \right) d\Gamma$$

$$- \int_{\Gamma} \left(q_a^*(z,z_0)\, w(z) - m_a(z,z_0)\, \Theta(z) \right) d\Gamma \qquad (39)$$

Integration by parts is started with the supporting shear force. This first integration is very simple. Due to the fact that

$$q_a^* = q_a + \frac{\partial m_t}{\partial s} \qquad (40)$$

and that the integral of the shear force is known (38), follows

$$r_a(z,z_0) = \int q_a^* \, ds = 2iD \left[\varphi'(z) - \overline{\varphi'(z)} \right] + m_t \qquad (41)$$

The multivalued functions will introduce the rigid body mode uniform displacement w, lost by the integration otherwise. To proceed we have to consider (41) as a torsional moment. This moment and the bending moment m are combined to a vector of moments with its components in the global directions x and y.

$$\begin{aligned} o_x + i o_y &= e^{i\vartheta} (m_a + i r_a) = \\ &-D(3+\mu)\, e^{i\vartheta} \varphi'(z) \\ &-D(1-\mu) \left[-e^{i\vartheta} \overline{\varphi'(z)} + e^{-i\vartheta} \left(z \overline{\varphi''(z)} + \overline{\psi'(z)} \right) \right] \end{aligned} \qquad (42)$$

Simultaneously the displacements, this kernel has to be multiplied with, change to the global rotations about the x- and y-axis. Now integration becomes possible and we get the Green's function of the second kind.

$$\begin{aligned} s_x + i s_y &= \int o_x + i o_y \, ds \\ &= Di \left[(3+\mu)\, \varphi(z) - (1-\mu) \left(z \overline{\varphi'(z)} + \overline{\psi(z)} \right) \right] \end{aligned} \qquad (43)$$

This time the multivalued function introduces the rigid body rotations. And finally we have the new fundamental equation

$$w(z_0) = w_1 + \Theta_x (y_1 - y_0) - \Theta_y (x_1 - x_0) +$$
$$+ \int \left(w(z,z_0)\, q^*(z) + \Theta_x(z,z_0)\, o_x(z) - \Theta_y(z,z_0)\, o_y(z) \right) d\Gamma$$
$$+ \int \left(s_x(z,z_0) \frac{\partial \Theta_x}{\partial s} - s_y(z,z_0) \frac{\partial \Theta_y}{\partial s} \right) d\Gamma$$
$$+ \text{Loading} + \text{Corners}$$

(44)

Now the requirements for the shape functions are smooth rotations Θ and consistency between displacements and rotations. It is recommendable to use a cubic approach for the displacements along the boundary similar to a normal beam element. The moments and shear forces are derived from (44) via differentiation and:

$$m_x + m_y = -4D(1+\mu) \frac{\partial^2 w}{\partial z_0 \partial \bar{z}_0} \tag{45}$$

$$m_x - m_y + i m_{xy} = -4D(1-\mu) \frac{\partial^2 w}{\partial \bar{z}_0^2} \tag{46}$$

$$q_x + i q_y = -8D \frac{\partial^3 w}{\partial z_0 \partial \bar{z}_0^2} \tag{47}$$

For the infinite plate the following basic functions hold

$$\varphi(z) = \frac{1}{8\pi D} (z - z_0) \ln (z - z_0) \tag{48}$$

$$\chi(z) = \frac{-\bar{z}_0}{8\pi D} (z - z_0) \ln (z - z_0) \tag{49}$$

and displacements and stresses are easily derived herefrom.

7. Numerical Performance for the potential problem

To show the benefits of the new functions a well known constant heat flow example |2| is used. A square with constant temperature at two opposite sides is analysed using linear elements with two nodes at the corners. The exact solution is prescribed at the boundary and the expressions (1) and (4) are used to calculate the potential at interior points. Numerical integration with four Gauss-points is used throughout.

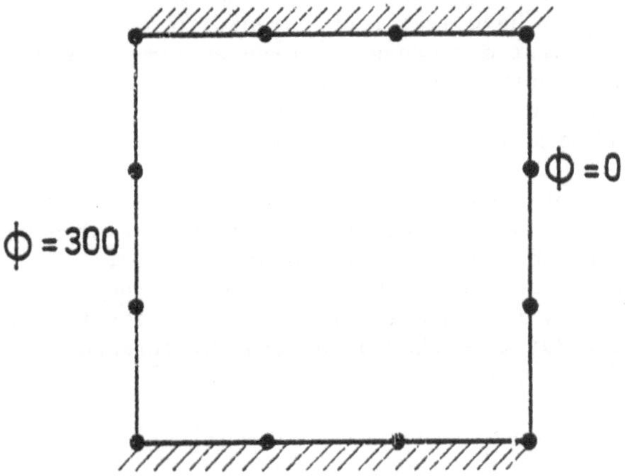

The following table shows the values of the potential at some locations of the square.

coordinates	exact value	Green's first	Green's Second
(-2.50, 0.00)	275.000	271.112	274.916
(-2.75, 0.00)	287.500	258.041	286.900
(-2.95, 0.00)	297.500	178.090	294.720
(-2.90,-2.90)	295.000	387.233	295.414

Table III Numerical Results for Heat Flow Example

As one can see, the stability of the Green's function of the second kind is significantly better. When considering fluxes at interior locations, i.e. the derivatives of the kernels with respect to z_0, the difference between the two approaches becomes greater. And if higher derivatives are needed such as in plate theory to evaluate shear forces, the second kind functions may be the only way to get any reasonable results.

8. Conclusion

So called Green's functions of the second kind are introduced via integration by parts. It is shown how this integration can be performed for the problems of elasticity and plate bending by complex variables. It is proposed that these new functions should replace the derivatives of the Green's functions whenever possible, because

- Second kind functions improve the accuracy
- Generally they have less requirements in computer time
- Complex arithmetic decreases computer time as well as man time in the derivation of code
- The similar structure and degree of Green's function and the second kind function may open the way to new features.

It should be mentioned that this paper seems to provide some arguments for the use of shape functions for the stresses with one degree lower than those for the displacements.
Not dealt herein is the way how to implement the second kind functions not only for interior locations but for the integral equation itself. But this should be straight forward.

9. References

(1) Bischoff,H. (1977) Die Berechnung von Potentialfeldern mit der Randintegralmethode dargestellt am Beispiel der ebenen stationaeren Grundwasserbewegung. Technischer Bericht Nr.18 Institut fuer Hydraulik u. Hydrologie TH Darmstadt.

(2) Brebbia,C.A. (1978) The Boundary Element Method for Engineers Pentech Press,Plymouth.

(3) Katz,C. (1981) The Use of Green's Functions in the numerical Analysis of Potential, Elastic and Plate Bending Problems. C.A.Brebbia(Ed.)Boundary Element Methods Springer Berlin.

(4) Katz,C. (1982) Ein symmetrisches Verfahren zur Berechnung von Problemen der Potential- Scheiben- oder Plattentheorie mit Greenschen Funktionen. Mitteilungen Institut Bauingenieurwesen I TU Muenchen Heft 7.

(5) Muskhelishvili,N.I. (1953) Some Basic Problems of the Mathematical Theory of Elasticity,Groningen.

(6) Savin,G.N. (1961) Stress concentration around holes Pergamon Press, New York

(7) Tottenham,H. (1979) The Boundary Element Methods for Plates and Shells. Banerjee/Butterfield, Developments in Boundary Element Methods I Applied Science Publishers Ltd. London.

Session II
Potential Problems

THE ROBIN PROBLEM IN A MULTIPLY-CONNECTED DOMAIN

George T. Symm

National Physical Laboratory
Teddington, Middlesex, U.K.

ABSTRACT

A Fortran program for the solution of the Robin problem for Laplace's equation in two dimensions has recently been extended from a form applicable only to simply-connected domains to one applicable to multiply-connected domains also. The original program had a modular structure based on three subroutines and advantage has been taken of this structure, and of certain features of the Fortran language, in developing the new program. Like the original, this program uses the direct boundary element method based on Green's third identity with piecewise constant approximations to boundary functions. The program is applied here to a simple test problem with a known solution and to a practical heat conduction problem for which the results are compared with a finite element solution.

INTRODUCTION

In recent years, the boundary element method has been applied to an increasing number and variety of practical problems (Brebbia, 1978a, 1980, 1981). Also, the method has formed the basis of several library programs for the solution of some of the more common problems. In particular, for the solution of Laplace's equation in two dimensions, fully documented programs have been written by Harrington et al. (1969), Hayes (1970), Symm and Pitfield (1974), Brebbia (1978b), Symm (1980) and Danson et al. (1981). One of these programs (Symm, 1980) was developed to solve the general Robin problem, to which earlier programs were not applicable, in a simply-connected domain. In the present paper it is shown how this program is easily extended to solve the corresponding problem in a multiply-connected domain.

Formulation of the Robin problem in terms of boundary integral equations, derived from Green's third identity, is followed by a description of the boundary element discretisation used to obtain a solution. The modular structure of the program for a simply-connected domain is then discussed and the modifications necessary for a multiply-connected domain are described in detail. Finally, two numerical examples are presented - one a simple test problem with a known solution and the other a practical problem of heat conduction in a turbine blade. In the latter example, the results are compared with an existing finite element solution.

THE ROBIN PROBLEM

The problem considered here is the solution of Laplace's equation

$$\nabla^2\phi \equiv \frac{\partial^2\phi}{\partial x^2} + \frac{\partial^2\phi}{\partial y^2} = 0 \qquad (1)$$

in a finite plane domain D, bounded externally by a closed contour C_0 and internally by m closed contours C_1, C_2, ..., C_m, when each contour C_i is subject to a Robin boundary condition, i.e. to a boundary condition of the form

$$a\phi + b\phi' = c, \qquad (2)$$

where ϕ' denotes the derivative of ϕ along the normal to the contour C_i directed into the domain D. In this boundary condition, a, b and c are functions of x and y in general and may be piecewise continuous on C_i. Whilst a and b must never vanish simultaneously, the separate cases a = 0 and b = 0 correspond to Neumann and Dirichlet boundary conditions respectively.

Necessary conditions on a and b for the existence and uniqueness of a solution to this problem are discussed elsewhere (Kellogg, 1929; Tsuji, 1959; Symm, 1980). Let it suffice here to note only two particular cases:-

(i) When a=0 everywhere on the complete boundary

$$C = C_0 + C_1 + \dots + C_m, \qquad (3)$$

which is the case of a pure Neumann problem, a solution exists only if the prescribed values of ϕ', viz. c/b, have zero integral around C. In this case, an arbitrary constant may be added to ϕ in general but a unique solution may be defined by placing an appropriate restriction on ϕ.

(ii) If ab is positive anywhere on C, a solution to the problem may not exist or, if it does exist, it may not be unique.

INTEGRAL EQUATION FORMULATION

In two-dimensional potential theory, Green's third identity may be written in the form

$$\int_C \phi'(q)\log|q-p|dq - \int_C \phi(q)\log'|q-p|dq = \theta(p)\phi(p), \qquad (4)$$

where p and q are vector variables specifying points of the plane and points on C respectively and dq is the differential increment of C at q. When $p \in D$, the parameter $\theta(p)$ has the value 2π, whilst when $p \in C$, θ has the value π if C is smooth or, more generally, the value of the "internal" angle of C at the point p, i.e. the angle in D between the two tangents to C which meet at p (e.g. $3\pi/2$ at a corner of an internal rectangular boundary). With this notation, where the prime denotes differentiation along the normal to the boundary directed into the domain D, equation (4) is exactly the same for a multiply-connected domain as for one which is simply-connected.

In the Robin problem, either ϕ or ϕ' may be eliminated from equation (4) at each point $q \in C$ by means of the boundary condition (2). Then, for $p \in C$, equation (4) becomes an integral equation (or a system of coupled integral equations) for those boundary values of ϕ and ϕ' which are not eliminated.

If a does not vanish identically and ab is never positive on C, this integral equation generally has a unique solution. The only exception arises when C_0 has the particular form of a Γ-contour (Jaswon, 1963) and the boundary conditions on C_0 are of Dirichlet or mixed type, in which case the solution (of the integral equation) is not unique. In the present work, this situation is avoided completely by making a simple change of scale - a method adopted also by Hayes and Kellner (1972). Alternatively, in such a case, a unique solution may be obtained by adding an appropriate auxiliary condition (Christiansen, 1975; Wendland, 1980).

In the Neumann problem, case (i) above, when a is zero everywhere on C and b and c are such that

$$\int_C \phi'(q)dq = 0, \qquad (5)$$

the condition

$$\int_{C_0} \phi(q)dq = 0 \qquad (6)$$

may be imposed on ϕ to make the solution unique. In all other cases ((ii) above) the integral equation, like the differential equation, must be treated with caution since a unique solution may not exist.

In general, by solving the integral equation (4), coupled

with equation (6) if necessary, and substituting the solution, complemented by the boundary condition (2), back into formula (4), we may obtain the value of ϕ at any point $p \in D + C$.

DISCRETISATION

The method outlined above is implemented numerically, in the manner of the earlier NPL programs (Symm and Pitfield, 1974; Symm, 1980), by dividing the boundary C into N intervals in each of which ϕ and ϕ' are approximated by constants. We denote these constants by

$$\phi_i \text{ and } \phi_i', \quad i = 1,2,\ldots,N, \tag{7}$$

and apply equation (4) at one "nodal" point q_i in each interval of C to obtain

$$\sum_{j=1}^{N} \phi_j' \int^{(j)} \log|q-q_i|\,dq - \sum_{j=1}^{N} \phi_j \int^{(j)} \log'|q-q_i|\,dq - \theta(q_i)\phi_i = 0,$$
$$i = 1,2,\ldots,N, \tag{8}$$

where $\int^{(j)}$ denotes integration over the jth interval of C. Eliminating one of the constants (7) from each interval, by applying the boundary condition (2) at the corresponding nodal point, we thus obtain a system of N simultaneous linear algebraic equations (in N unknowns) whose solution provides the approximation

$$\Big[\sum_{j=1}^{N} \Big(\phi_j' \int^{(j)} \log|q-p|\,dq - \phi_j \int^{(j)} \log'|q-p|\,dq\Big)\Big]/\theta(p) \tag{9}$$

to $\phi(p)$.

In order to evaluate the coefficients in equations (8) and (9), we approximate each interval of C by the two chords which join its end points to the nodal point within it. Then all the integrations can be carried out analytically and, in (8), θ becomes the "internal" angle, as defined above, at the nodal point q_i, of an approximation to the boundary C by m + 1 polygons.

In the Neumann problem, equations (8) are supplemented by a discrete form of equation (6), viz.

$$\sum_{j=1}^{N_0} \phi_j h_j = 0, \tag{10}$$

where N_0 is the number of intervals on C_0 and h denotes interval length measured on the corresponding approximating polygon. In this case, the resulting N + 1 equations are solved in the least-squares sense.

PROGRAM STRUCTURE

The NPL program for the Robin problem in a simply-connected domain (Symm, 1980) was written in modular form as a series of three Fortran subroutines. Here we consider each of these routines in turn and describe those program modifications which are necessary when the domain D is multiply-connected.

I) The first routine, called ROBINA, is a discretisation routine designed to simplify the data input for the later routines whenever, as frequently occurs in practice, the boundary is made up of straight lines and circular arcs. If each such boundary segment is divided by "interval" points into a prescribed number of intervals, in each of which one "nodal" point is located, then ROBINA obtains the coordinates of interval and nodal points alternately around the boundary, given the first point of each boundary segment and, in the case of a circular arc, the coordinates of its centre and the direction in which it is to be described.

For a simply-connected domain, the boundary as a whole is described in an anti-clockwise direction, i.e. keeping the domain on its left, and the coordinates are stored in this order in an array X of dimensions 2 by M, where $M = 2N + 1$ and the first point coincides with the last on the closed boundary contour. As above, N is the total number of intervals into which the boundary C is divided,

For a multiply-connected domain, ROBINA is applicable without modification to each boundary contour in turn. The outer boundary C_0 is described in an anti-clockwise direction whilst the internal boundaries C_i, $i = 1,2,\ldots,m$, are described in a clockwise direction (so that the domain is always on the left of the boundary). In this case, the coordinates of the points on successive contours are compiled into a single array X of dimensions 2 by M, where now $M = 2N + m +1$. This is achieved by using two features of the Fortran language:

(i) the fact that arrays are stored by columns and

(ii) the facility whereby an array may be addressed by its first element.

Thus, if the contour C_i is divided into N_i intervals, then the corresponding $M_i = 2N_i + 1$ points on it are added into X by calling ROBINA with X addressed by its element X(1,K) where

$$K = 2(N_0 + \ldots + N_{i-1}) + i + 1. \tag{11}$$

We note, in passing, that the intervals on each boundary segment need not be equal in length and that the nodal points need not be at the centres of the intervals. The routine ROBINA therefore permits grading of intervals, which is particularly useful when solving problems involving domains with corners or boundary singularities. This facility is described more fully and illustrated by examples in the report on the earlier program (Symm, 1980).

Having obtained the coordinates of the boundary points, the routine ROBINA also evaluates the functions a, b and c, in the boundary condition (2), at each nodal point, given the necessary information through a user-supplied subroutine BCI. The computed values are stored in an array ABC of dimensions 3 by N and, for a multiply-connected domain, this again involves use of the Fortran features described above. In this case, when ROBINA is called for the contour C_i, the array ABC is addressed by its element ABC(1,L) where

$$L = N_0 + \dots + N_{i-1} + 1. \tag{12}$$

Finally we note that the routine ROBINA includes a parameter IW which warns the user if the product ab is positive at any nodal point, in which case, as observed above, a unique solution of the problem may not exist.

II) The second routine in the program for a simply-connected domain is called ROBINB. This routine assembles and solves the simultaneous linear algebraic equations which approximate the continuous problem, given the boundary discretisation as derived by ROBINA.

For a multiply-connected domain, this routine is modified only slightly to take account of the jumps in the coordinates from one boundary contour to the next. The new version is called ROBIND. In it, as in ROBINB, the coefficients in the equations are obtained by means of an auxiliary integration routine SLINTA and the elimination of ϕ or ϕ' from each boundary interval is achieved by setting

$$\phi = \frac{c}{a} - \frac{b}{a}\phi', \tag{13}$$

when a is greater in magnitude than b, at the nodal point, and setting

$$\phi' = \frac{c}{b} - \frac{a}{b}\phi \tag{14}$$

otherwise. (The routine checks that a and b are not zero simultaneously.) The solution of the equations is obtained by means of library routines which compute the unique solution of the square non-singular system, in the general case, or the unique least-squares solution of the rectangular system, in the Neumann case, by the method of Peters and Wilkinson (1970). Note that within the routine ROBIND the domain is scaled down so that its maximum diameter is not greater than unity, this being sufficient to ensure that there is no possibility of non-uniqueness due to C_0 being a Γ-contour. The boundary solution is completed by a further application, at each nodal point, of equation (13) or (14) as appropriate.

III) The third routine in the program for a simply-connected domain is called GREENA. This routine evaluates the function ϕ at any selected point p, from the boundary solution, by means of formula (9).

For a multiply-connected domain, this routine again requires only slight modification of the coordinate indices to allow for the jumps between boundary contours. The new version is called GREENC. This routine, which is called as often as required, again uses the auxiliary routine SLINTA for the necessary straight line integrations. These integrations are carried out as described in detail by Symm and Pitfield (1974) and by Jaswon and Symm (1977). SLINTA requires no modification for a multiply-connected domain.

EXAMPLES

There now follow two examples of the use of the routines described above - one a simple test problem with a known solution and the other a practical problem of heat conduction.

Problem 1

To solve Laplace's equation (1) in the domain D bounded externally by the circle

$$C_0: \quad x^2 + y^2 = 4, \tag{15}$$

and internally by the circle

$$C_1: \quad x^2 + y^2 = 1, \tag{16}$$

given the boundary conditions

$$\phi - 2\phi' = 0 \quad \text{on } C_0 \tag{17}$$

and

$$\phi - \phi' = x \quad \text{on } C_1. \tag{18}$$

This problem has the analytic solution

$$\phi = \frac{x}{2(x^2+y^2)}. \tag{19}$$

For this example, the data input, by means of the routine ROBINA, is particularly simple, since each boundary is a single circular arc defined by its starting point, (2,0) or (1,0) as the case may be, and its centre (0,0). Each of these arcs is divided into $N_0 = N_1 = N/2$ equal intervals with nodal points at their mid-points and ϕ is computed at selected points in D. Typical results are compared with the analytic solution in Table 1.

Table 1. Results of Problem 1
Values of ϕ at selected points

x	y	N=16	N=32	N=64	N=128	Anal.
1.0	0.0	0.481	0.497	0.499	0.500	0.500
1.2	0.0	0.396	0.412	0.416	0.416	0.417
1.4	0.0	0.341	0.354	0.356	0.357	0.357
1.6	0.0	0.298	0.309	0.312	0.312	0.313
1.8	0.0	0.264	0.275	0.277	0.278	0.278
2.0	0.0	0.251	0.250	0.250	0.250	0.250
0.0	1.0	0.000	0.000	0.000	0.000	0.000
0.2	1.0	0.141	0.095	0.096	0.096	0.096
0.4	1.0	0.171	0.168	0.171	0.172	0.172
0.6	1.0	0.199	0.220	0.220	0.220	0.221
0.8	1.0	0.226	0.241	0.243	0.244	0.244
1.0	1.0	0.239	0.248	0.249	0.250	0.250
1.2	1.0	0.238	0.244	0.245	0.246	0.246
1.4	1.0	0.233	0.235	0.236	0.236	0.236
1.6	1.0	0.224	0.222	0.225	0.225	0.225

Not unnaturally, the worst results occur (near the inner boundary) when N is small.

Problem 2

To determine the temperature ϕ in a (hypothetical) turbine blade of uniform cross-section, as illustrated in Figure 1, subject to convection boundary conditions of the form

$$K\phi' = h(\phi - \phi_g), \tag{20}$$

where

K is the thermal conductivity of the blade,
h is a heat transfer coefficient dependent upon the fluid conditions, e.g. Reynold's number, etc., and
ϕ_g is the gas adiabatic wall temperature governed by the surrounding cooling flow.

In this example, the smooth boundary is defined by a set of discrete points with x-coordinates ranging from -2.4927 to 0.0750 and y-coordinates ranging from -0.0583 to 2.3487. These points have been simply joined by straight lines, of which there are 268 in total, in the diagram. This diagram, produced on a graph plotter, provides a useful check on the input data. The thermal conductivity K has the constant value 3000 in units appropriate to the prescribed coordinates. Around the outer boundary C_0, the heat transfer coefficient h varies between 668.513 and 11532.988 while ϕ_g varies relatively little - between 1514.8 and 1552.8. On each internal boundary, h has a constant value in the range from 4368.609 to 5998.453 while ϕ_g has a constant value in the range from 948.0 to 1001.2. In this case, the routine ROBINA is not used, though it could have simplified the input of the internal boundaries had the data been presented in a different form.

Here, the temperature ϕ is computed, at selected points within the domain, for two subdivisions of the boundary. In

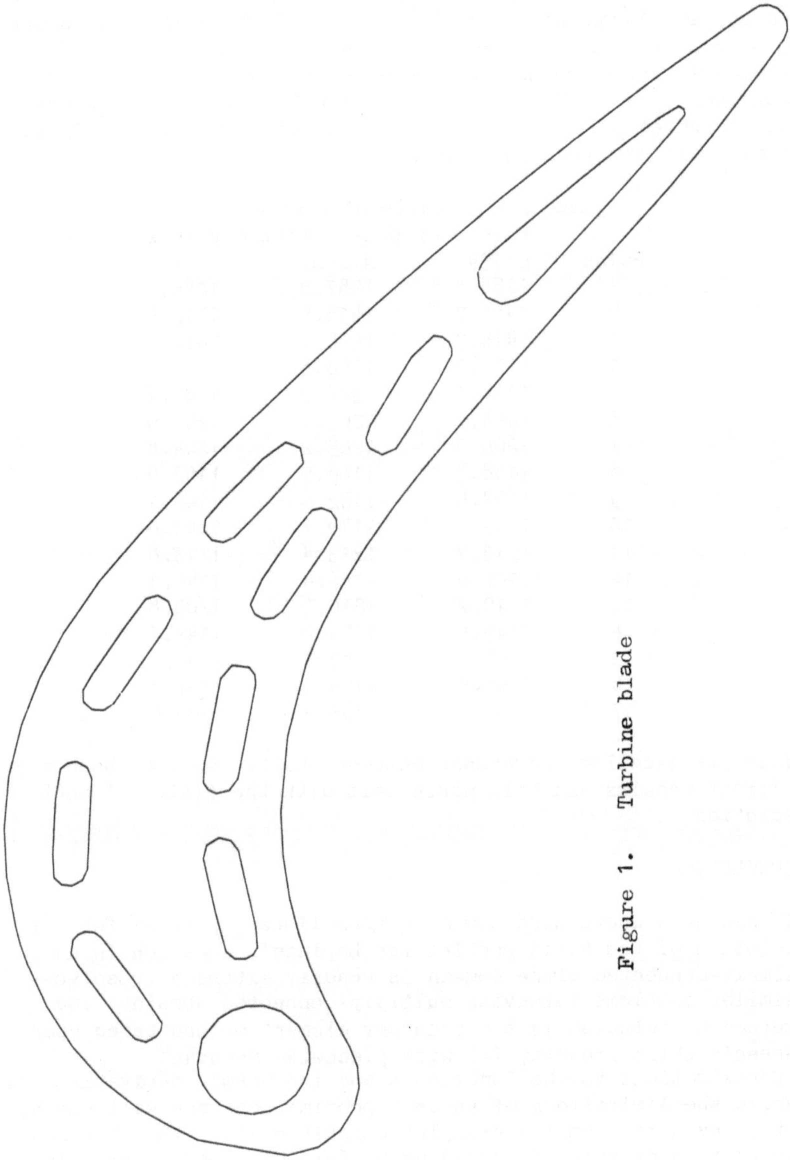

Figure 1. Turbine blade

the first, the prescribed boundary points are taken as interval and nodal points alternately on each contour, so that there are 134 boundary elements. In the second, the prescribed points are taken as interval points, and the nodal points are placed mid-way between these, so that there are 268 boundary elements. The corresponding results are compared with those obtained by a finite element method (FEM) in Table 2, where the selected points are, in fact, the internal points of the finite element quadrangulation of the domain.

Table 2. Results of Problem 2

	Values of ϕ at selected points		
Point	N=134	N=268	FEM
1	1288.4	1287.8	1286.5
2	1416.7	1416.5	1412.6
3	1418.3	1418.1	1414.1
4	1391.3	1390.6	1388.9
5	1388.1	1387.3	1385.7
6	1204.2	1205.1	1205.0
7	1206.0	1205.2	1204.8
8	1168.3	1168.5	1162.9
9	1183.5	1182.6	1180.1
10	1149.1	1149.7	1144.0
11	1213.3	1213.4	1213.8
12	1146.0	1146.8	1140.1
13	1210.2	1210.3	1209.8
14	1149.4	1150.2	1143.2
15	1183.0	1182.7	1180.2
16	1198.6	1199.2	1197.2
17	1153.6	1153.9	1146.6

There is excellent agreement between the two sets of boundary element results and both agree well with the finite element solution.

CONCLUSION

It has been shown here how a Fortran library program for the solution of the Robin problem for Laplace's equation in a simply-connected plane domain is readily extended to solve similar problems involving multiply-connected domains. The method of solution is the boundary element method based upon Green's third identity (4) with piecewise constant approximations to the function ϕ and its normal derivative ϕ'. While the limitations of these approximations are well known, it is evident from the examples described above that the new program is capable of providing useful solutions to practical problems in multiply-connected domains. The simplicity of the boundary element discretisation makes this a very attractive method for such problems.

ACKNOWLEDGMENT

The author wishes to thank Rolls-Royce Limited for permission to include the turbine blade example in this paper and for providing the finite element solution of this problem for comparison.

REFERENCES

Brebbia, C. A., Ed., (1978a) Recent Advances in Boundary Element Methods. Pentech Press, London.

Brebbia, C. A. (1978b) The Boundary Element Method for Engineers. Pentech Press, London: 58-81.

Brebbia, C. A., Ed., (1980) New Developments in Boundary Element Methods. CML Publications, Southampton.

Brebbia, C. A., Ed., (1981) Boundary Element Methods. Springer-Verlag, Berlin.

Christiansen, S. (1975) Integral Equations Without a Unique Solution Can Be Made Useful for Solving Some Plane Harmonic Problems. J. Inst. Maths. Applics., **16**: 143-159.

Danson,D., Brebbia, C.A. and Adey, R.A. (1981) The BEASY System. In "A Handbook of Finite Element Systems", edited by C.A.Brebbia, CML Publications, Southampton: 77-91.

Harrington, R. F., Pontoppidan, K., Abrahamsen, P. and Albertsen, N. C. (1969) Computation of Laplacian Potentials by an Equivalent-Source Method. Proc. IEE, **116**, 10: 1715-1720.

Hayes, J. K. (1970) Four Computer Programs Using Green's Third Formula to Numerically Solve Laplace's Equation in Inhomogeneous Media. Los Alamos Scientific Laboratory Report LA-4423.

Hayes, J. and Kellner, R. (1972) The Eigenvalue Problem for a Pair of Coupled Integral Equations Arising in the Numerical Solution of Laplace's Equation. SIAM J. Appl. Math., **22**, 3: 503-513.

Jaswon, M. A. (1963) Integral Equation Methods in Potential Theory. I. Proc. Roy. Soc. (A), **275**: 23-32.

Jaswon, M. A. and Symm, G. T. (1977) Integral Equation Methods in Potential Theory and Elastostatics. Academic Press, London.

Kellogg, O. D. (1929) Foundations of Potential Theory. Springer-Verlag, Berlin: 315.

Peters, G. and Wilkinson, J. H. (1970) The Least Squares Problem and Pseudo-Inverses. Computer J., 13, 3: 309-316.

Symm, G.T. (1980) The Robin Problem for Laplace's Equation. NPL Report DNACS 32/80.

Symm, G. T. and Pitfield, R. A. (1974) Solution of Laplace's Equation in Two Dimensions. NPL Report NAC 44.

Tsuji, M. (1959) Potential Theory in Modern Function Theory. Maruzen, Tokyo.

Wendland, W.L. (1980) On Galerkin Collocation Methods for Integral Equations of Elliptic Boundary Value Problems. In "Numerical Treatment of Integral Equations", edited by J.Albrecht and L.Collatz, Birkhäuser Verlag, Basel: 244-275.

BOUNDARY ELEMENTS IN AXISYMMETRIC POTENTIAL PROBLEMS

F. Yoshikawa
Power Department
Mizushima Works
Kawasaki Steel Corporation
Kurashiki, Japan

M. Tanaka
Department of Mechanical Engineering
Osaka University
Suita, Japan

INTRODUCTION

The boundary element method has been applied to a wide variety of practical problems in engineering, and its potentiality has been revealed in most cases of the problems analyzed (Brebbia 1978, Brebbia and Walker 1980). Although at present a number of the solution procedures based on the boundary element method are available at least for linear problems, there are still to be put forward further developments of the method for more efficient numerical computations.

Several attempts have been made to present a direct formulation of axisymmetric problems in elastostatics by using some analytical expressions of the fundamental solution to ring sources (Cruse 1977, Mayer et al. 1980). Wrobel and Brebbia (1980) also proposed a boundary element solution procedure for the axisymmetric potential problem, in which the fundamental solution is expressed in terms of a Legendre function and hence is somewhat analytical.

In this paper, a relatively simple boundary element formulation is proposed for the axisymmetric potential problem. The fundamental solution to a ring source is derived not in an analytical way, but in a numerical manner. Namely, in the proposed method, we need only the fundamental solution in the three-dimensional state and its transformation to the axisymmetric state is made in a computational procedure. A numerical implementation of the proposed formulation is made by using a series of constant boundary elements and also linear boundary elements. A mixed boundary condition of the convection type can also be treated. Several sample problems of the steady-state heat conduction are computed to demonstrate the effectiveness of the proposed method.

FORMULATION

The governing equation for the potential problem in a homogeneous isotropic domain Ω and the corresponding boundary condition over the boundary $\Gamma = \Gamma_1 + \Gamma_2 + \Gamma_3$ are,

$$k\nabla^2 u=0 \qquad \text{in } \Omega \qquad (1)$$

B.C. $$u=u_0 \qquad \text{on } \Gamma_1 \qquad (2)$$

$$q=k(\partial u/\partial n)=q_0 \qquad \text{on } \Gamma_2 \qquad (3)$$

$$q=h(u_a-u) \qquad \text{on } \Gamma_3 \qquad (4)$$

where u denotes the potential, e.g. temperature, and k and h the thermal conductivity of the medium and the heat transfer coefficient, respectively. u_0 and q_0 stand for the given values of potential and flux and u_a the potential of an ambient medium, respectively. Equation (4) prescribes the mixed boundary condition of the convection type.

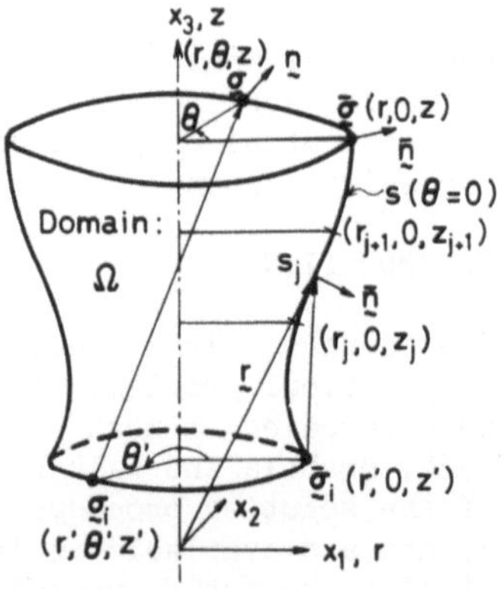

Fig.1 Axisymmertic domain Ω and boundary surface Γ

The boundary integral equation for the three-dimensional problem relating the potential and the normal flux on the boundary is given by (Brebbia 1978)

$$c_i u_i + \iint_\Gamma u q^* d\Gamma = \iint_\Gamma q u^* d\Gamma \qquad (5)$$

where u_i denotes the value of u at a point i and c_i is equal to 0.5 for a smooth boundary. The value of c_i can be determined for a non-smooth boundary including also the smooth one by applying the equipotential condition over the whole boundary (Brebbia 1978). The weighting function u^* satisfies the following equation which governs the effect of a concentrated unit potential at the point i and is called the fundamental solution (Brebbia 1978):

$$k\nabla^2 u^* + \delta_i = 0 \qquad (6)$$

where δ_i is the Dirac delta function. The fundamental solution u^* of equation (6) for a three-dimensional homogeneous isotropic medium and its normal derivative q^* are

$$u^* = 1/4\pi kR \qquad (7)$$

$$q^* = k(\partial u^*/\partial n) = (-1/4\pi R^2)(\partial R/\partial n) \qquad (8)$$

where R is the distance between the point i of application of the unit potential and an observation point under consideration.

Equations (5), (7) and (8) can be also expressed in cylindrical coordinates (r, θ, z). For this purpose, we shall express R and its normal derivative in cylindrical coordinates. We then obtain the following relations(see Fig.1):

$$R=\left[r^2 + r'^2 - 2rr'\cos\theta + (z-z')^2\right]^{\frac{1}{2}} \tag{9}$$

$$\partial R/\partial n=(r-r'\cos\theta)\bar{n}_1/R + (z-z')\bar{n}_3/R \tag{10}$$

where the cylindrical coordinates of the observation point are denoted by (r, θ, z) and those of the source point i in the plane $\theta=0$ by $(r', 0, z')$. We also denote by $\bar{n}_1$, $\bar{n}_2$ and $\bar{n}_3$ the direction cosines of the normal vector $\bar{\underset{\sim}{n}}$ at the observation point to the meridional boundary curve $\tilde{s}$ in the plane $\theta=0$. Since the circumferential component $\bar{n}_2$ of $\bar{\underset{\sim}{n}}$ is equal to 0, the direction cosines of $\underset{\sim}{n}$ are given by those of $\bar{\underset{\sim}{n}}$ as

$$n=(n_1, n_2, n_3)=(\bar{n}_1\cos\theta, \bar{n}_1\sin\theta, \bar{n}_3) \tag{11}$$

In the axisymmetric domain we have the follwing relations:

$$u=\bar{u}, \qquad u_i=\bar{u}_i \tag{12}$$

$$q=\bar{q}, \qquad q_i=\bar{q}_i \tag{13}$$

where $\bar{u}$ and $\bar{q}$ stand for the values of u and q, respectively, on the meridional boundary curve s in the plane $\theta=0$. The subscript i denotes the values at the point i. The statement of equation (5) in cylindrical coordinates becomes

$$c_i\bar{u}_i + \iint_{\Gamma c}\bar{u}\tilde{q}^*|J|\,d\Gamma_c = \iint_{\Gamma c}\bar{q}\tilde{u}^*|J|\,d\Gamma_c \tag{14}$$

where $\tilde{u}^*$ and $\tilde{q}^*$ are u^* and q^* expressed in cylindrical coordinates, respectively. Γ_c is the domain in the new coordinate system corresponding to Γ and $|J|$ is the Jacobian appearing in coordinate transformation.

NUMERICAL FORMULATION

Equation (14) is the integral equation relating potentials $\bar{u}$ and fluxes $\bar{q}$ on the meridional boundary curves s. Equation (14) can be discretized by dividing the meridional boundary curve s into a series of elements in the same way as in the two-dimensional case. The integrals in equation (14), however, become double integrals which should be evaluated over the whole boundary surface Γ_c . Potential $\bar{u}$ and flux $\bar{q}$

over each boundary element are assumed to vary according to certain interpolation functions $\underset{\sim}{\phi}$ as follows :

$$\bar{u}=\underset{\sim}{\phi}\underset{\sim}{\bar{u}}, \qquad \bar{q}=\underset{\sim}{\phi}\underset{\sim}{\bar{q}} \tag{15}$$

where $\underset{\sim}{\bar{u}}$ and $\underset{\sim}{\bar{q}}$ are nodal potentials and nodal fluxes, respectively. Substituting equation (15) into equation (14), we obtain the discretized equation as follows :

$$c_i\bar{u}_i + \sum_{j=1}^{N} H_{ij}\bar{u}_j = \sum_{j=1}^{N} G_{ij}\bar{q}_j \qquad (i=1,2,\cdots N) \tag{16}$$

where the coefficient matrices $\hat{H}_{ij}$ and G_{ij} can be given, for example, in the case of constant boundary elements as

$$\hat{H}_{ij} = \iint_{\Gamma c,j} \tilde{q}^* |J|\, d\Gamma c \text{ and } G_{ij} = \iint_{\Gamma c,j} \tilde{u}^* |J|\, d\Gamma c \tag{17}$$

In order to evaluate these integrals the following method can be applied. That is, r is assumed to be expressed as a linear function of z in the meridional curve segment s_j of the jth boundary element $\Gamma_{c,j}$ in the plane $\theta=0$ as follows:

$$r= az + b = a\Delta z_j t + az_j + b \qquad (t=0 \text{ to } 1) \tag{18}$$

where $\Delta z_j=z_{j+1}-z_j$, and $(r_j, 0, z_j)$ and $(r_{j+1}, 0, z_{j+1})$ are the coordinates of both extreme points of the segment s_j. Using equation (18) we can derive the following parametric expression for the jth boundary element $\Gamma_{c,j}$:

$$\underset{\sim}{r}=(a\Delta z_j t+az_j+b)\cos\theta\underset{\sim}{i} + (a\Delta z_j t+az_j+b)\sin\theta\underset{\sim}{j} + (\Delta z_j t+z_j)\underset{\sim}{k} \tag{19}$$

Where $\underset{\sim}{i}$, $\underset{\sim}{j}$ and $\underset{\sim}{k}$ are the unit vectors in each direction of rectangular coordinates (x_1, x_2, x_3). After some manipulation by means of equation (19), the following relations can be derived (Kreyszig 1967):

$$\underset{\sim}{\bar{n}}=(\Delta z_j/l_j)(\underset{\sim}{i}-a\underset{\sim}{k}) \tag{20}$$

$$|J|=l_j(a\Delta z_j t+az_j+b) \tag{21}$$

where $l_j=\left[(r_{j+1}-r_j)^2+(z_{j+1}-z_j)^2\right]^{\frac{1}{2}}$ is the length of the straigth line segment s_j and $d\Gamma_c=d\theta dt$. The use of these relations makes it possible to evaluate the integrals in equation(16). For constant boundary elements, for example, the integrals become as follows:

$$\hat{H}_{ij} = \left(-\frac{\Delta z_j}{4\pi}\right)\int_0^{2\pi}(b+az'-r'\cos\theta)\,d\theta\int_0^1\frac{pt+q}{R^3}\,dt \tag{22}$$

$$G_{ij} = \left(\frac{l_j}{4\pi k}\right)\int_0^{2}d\theta\int_0^1\frac{pt+q}{R}\,dt \tag{23}$$

where $R=\left[l_j^2t^2 + 2\Delta z_j(z_j(a^2+1)+ab-z'-r'a\cos\theta\ t\right.$ (24)

$\left. + (a^2+1)z_j^2+2z_j(ab-z')+(b^2+r'^2+z'^2)-2r'(az_j+b)\cos\theta\right]^{\frac{1}{2}}$

$p=a\Delta z_j$ and $q=az_j+b$ (25)

Similar expressions to the above can be obtained for linear boundary elements. For boundary elements on a plane perpendicular to the axis of revolution, the same procedure as mentioned above can be employed.

For computational purposes, analytical solutions are used for integration with respect to t, and the Gauss-Chebyshev quadrature (Stroud and Secrest 1966) with respect to ξ by letting $\xi=\cos\theta$. The integrals become singular and must be evaluated in the sense of its Cauchy principal values when the node under consideration is contained by the element over which integration is performed. In the same manner as the other integrals, however, the singular integrals can be evaluated by applying the method described above.

After the evaluation of the integrals, equation(16) may be rewritten as the following system of equations:

$$\underset{\sim}{H}\underset{\sim}{U} = \underset{\sim}{G}\underset{\sim}{Q} \tag{26}$$

The diagonal terms of the matrix $\underset{\sim}{H}$ including the coefficient c_i can be calculated by taking into account the fact that when a uniform potential is applied over the whole boundary, the normal fluxes must be zero (Brebbia 1978). That is, the values of the diagonal coefficients can be calculated by the use of the off-diagonal components in the following manner:

$$H_{ii} = - \sum_{j=1(i\neq j)}^{N} H_{ij} \tag{27}$$

Since potential values on Γ_1 and flux values on Γ_2 are prescribed and the boundary condition (4) holds on Γ_3, these are a set of N unknowns in the system (26) which can be reordered as

$$\underset{\sim}{A}\underset{\sim}{X} = \underset{\sim}{F} \tag{28}$$

where $\underset{\sim}{X}$ is the column vector of the nodal unknowns on the boundary. If equation (28) is solved for $\underset{\sim}{X}$, then all the values of potential and flux are determined on the boundary Γ. Using the nodal values on the boundary thus obtained, we can calculate the potential and flux values at any internal point through equation (14) with the coefficient c_i equal to unity.

RESULTS OBTAINED AND DISCUSSION

Some example problems of steady-state heat conduction in which the convection-type boundary condition is also taken into account, are computed by employing constant and also linear boundary elements. The 15-point and the 50-point formulas of the Gauss-Chebyshev quadrature are used for integration with respect to the circumferential direction. Results obtained are compared with analytical solutions or other numerical solutions.

Fig.2 shows temperature and heat-flux profiles in a hollow cylinder subject to the following boundary conditions:

$u=0$ at $r=6$, $z=0$ and 10

$q=0$ at $r=2$, $0<z<3$ and $7<r<10$

$q=1$ at $r=2$, $3<z<7$

Because of the symmetry with respect to the plane $z=5$, half a cross section of the cylinder below the plane $z=5$ is discertized into 18 equal elements in the case of constant boundary elements. In the case of linear boundary elements two nodal points are placed near the corners of the domain and also the point at which the boundary conditions are changed, on the inner surface $r=2$. The results obtained are compared with an analytical solution (Carslaw and Jaeger 1959) in Fig.2.

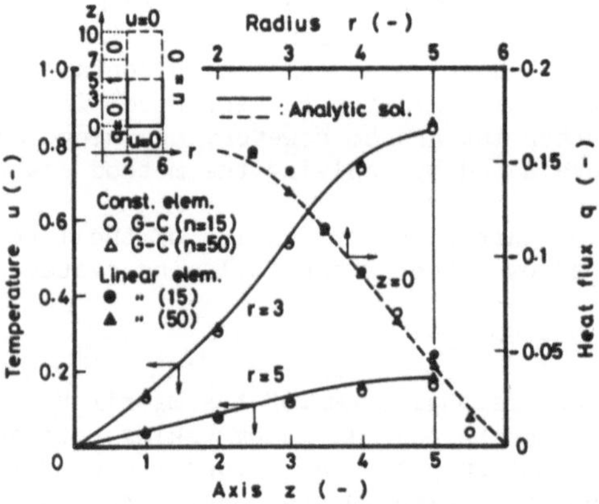

Fig.2 Temperature and heat-flux profiles in a hollow cylinder

Next, let us study a finite solid cylinder subject to the following boundary conditions:

$u=0$ at $z=3$

$u=1$ at $z=0$

$q=0.1(0-u)$ at $r=1$

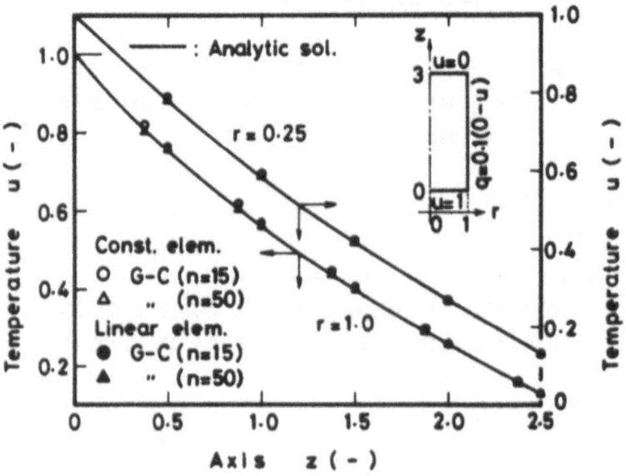

Fig.3 Temperature profiles in a solid cylinder

20 equal elements are used for analysis in the case of constant boundary elements, while two very small elements supplemented in the case of linear boundary elements. No boundary elements are necessary over the axis of revolution. In Fig.3, comparision is made with temperature distributions computed from an analytical solution (Carslaw and Jaeger 1959).

Now we consider a hollow sphere subject to the following boundary conditions:

$u=1 \quad$ at $R_1=1$

$u=0 \quad$ at $R_2=2$

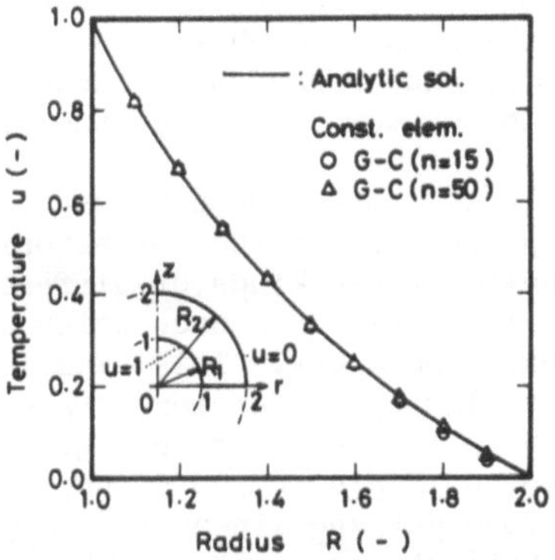

Fig.4 Temperature profile in a hollow sphere

Only one-fourth of the hollow sphere is discretized into 21 elements in a similar manner to that mentioned above. Fig.4 shows a temperature profile obtained together with an analytical solution.

The computed results are in good agreement with analytical solutions. It should be noted that 50 integration points were necessary in the numerical integration to yield the required accuracy in the present examples.

we now apply the proposed method of solution to a practical problem. Here we consider a heat conduction problem in the blast furnace hearth in practical use. A schematic diagram of the blast furnace hearth is represented in Fig.5.

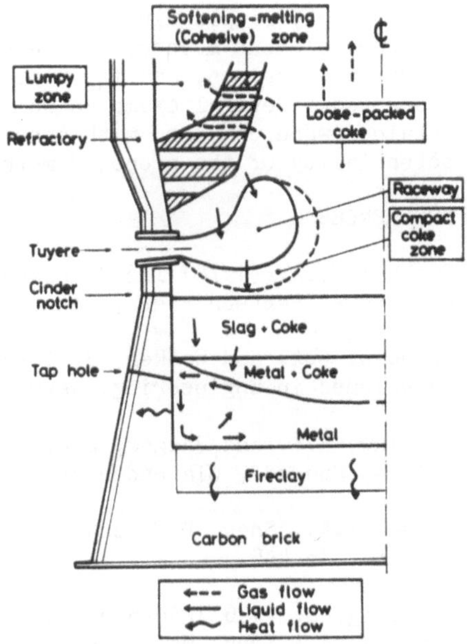

Fig.5 Schematic diagram of the blast furnace hearth

Iron-bearing materials charged into the blast furnace, get heated and reduced by hot air(blast) from the tuyeres of the furnace as they descend through the furnace stack. They are finally converted into molten iron and molten slag. Molten iron and molten slag collect in the hearth from which they are discharged periodically or continuously through the tap holes and the cinder notches. The blast furnace hearth is made of ceramic and/or carbon refractories and is equipped with a cooling system to protect the refractories.

Fig.6 shows the dimensions, the boundary conditions and the properties of the refractories of a hearth. The hearth comprises three kinds of different refractories and pertains to a large blast furnace which can produce 10 000 metric tons of pig iron a day. This example is computed by using 364 triangular finite elements (Zienkiewicz and Cheung 1967) and the results are shown in Fig.7, together with the grid employed. Fig's. 8 and 9 show the results obtained by applying the constant and linear boundary elements, along with the discretizations adopted. 106 and 125 elements are used for the respective cases. It can be seen that similar isotherms are obtained from both the finite element and boundary element methods.

CONCLUSIONS

A relatively simple direct formulation by means of the boundary element method, has been proposed for axisymmetric potential problems, and its numerical implementation has been presented. Several examples were computed and the results obtained were compared with other solutions, whereby the potentiality of the proposed method was revealed.

REFERENCES

Brebbia, C.A. (1978) The Boundary Element Method for Engineers, Pentech Press, London.

Brebbia, C.A. and Walker, S. (1980) Boundary Element Techniques in Engineering, Newnes-Butterworths, London.

Carslaw, H.S. and Jaeger, J.C. (1959) Conduction of Heat in Solids, 2nd ed., Clarendon Press, Oxford.

Cruse, T.A., Snow, D.W. and Wilson, R.S. (1977) Compt. Struct., 7: 445.

Kreyszig, E. (1967) Advanced Engineering Mathematics, 2nd ed., John Wiley & Sons, New York.

Mayer, M., Drexler, W. and Kuhn, G. (1980) Int. J. Solids & Struct., 16: 863.

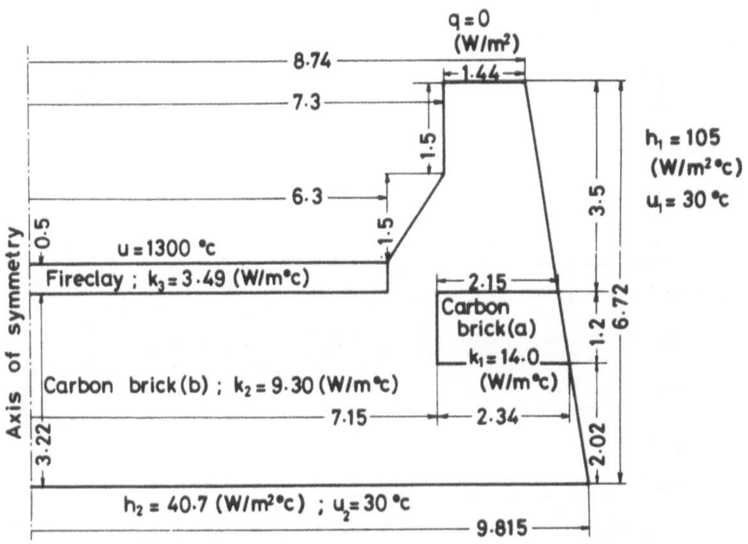

Fig.6 Blast furnace hearth ; dimensions(m), boundary conditions and properties of refractories

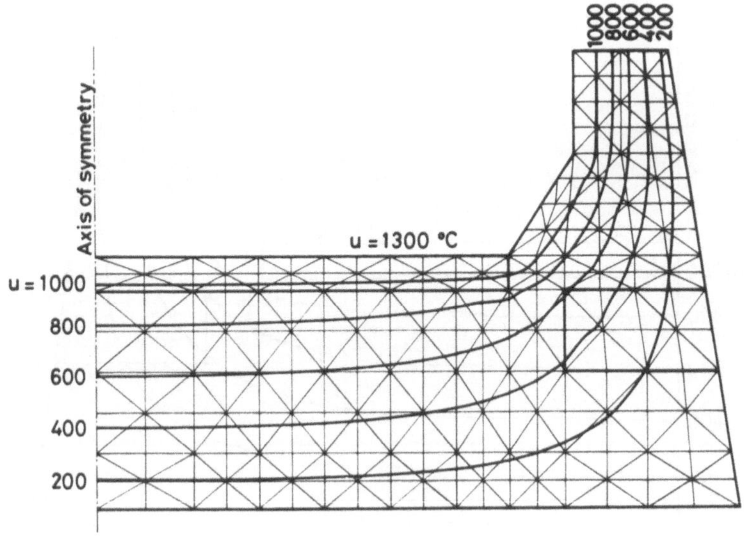

Fig.7 Blast furnace hearth ; FEM grid and isotherms

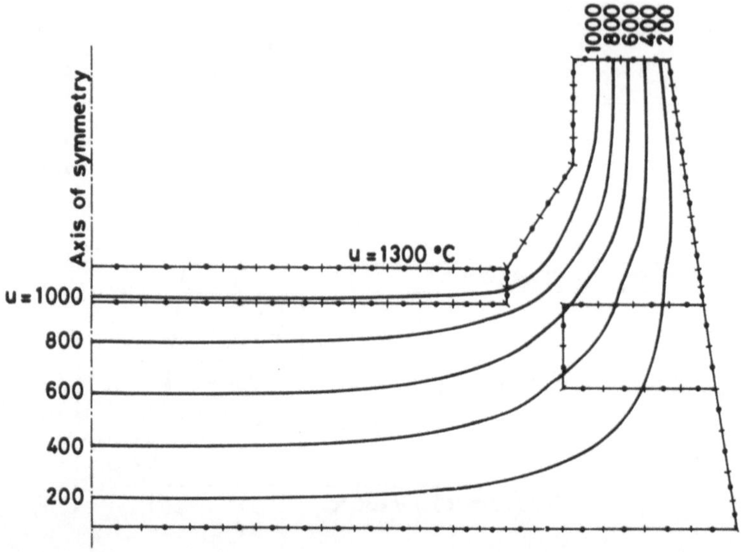

Fig.8 Blast furnace hearth ; constant BEM discretization and isotherms

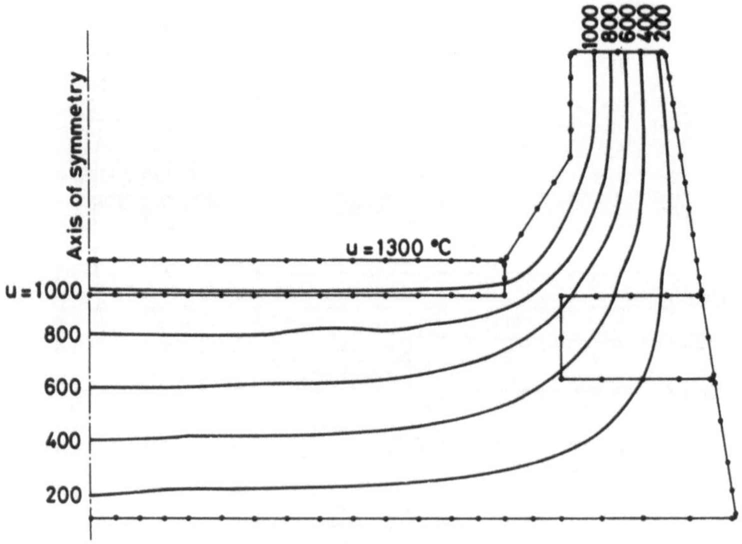

Fig.9 Blast furnace hearth ; linear BEM discretization and isotherms

Stroud, A.H. and Secrest, D. (1966) Gaussian Quadrature Formulas, Prentice-Hall, New York.

Wrobel, L.C. and Brebbia, C.A. (1980) Axisymmertic Potential Problems, in New Developments in BEM's (ed. Brebbia, C.A.), CML, 77-89.

Zienkiewicz, O.C. and Cheung, Y.K. (1967) Finite Element Method in Structural and Continuum Mechanics, McGraw-Hill, New York.

A REGULAR BOUNDARY METHOD USING NON-CONFORMING ELEMENTS FOR POTENTIAL PROBLEMS IN THREE DIMENSIONS

C. Patterson and N. A. S. Elsebai,

Dept. of Mech. Engineering,
University of Sheffield, U.K.

INTRODUCTION

The very success of the Finite Element Method, a domain method for the solution of engineering field problems, has led to ever increasing demands for geometrically complex design models. At present, the majority of design analyses are two dimensional but there is substantial and mounting demand for three dimensional modelling. Because of the square-cube relation of degrees of freedom in comparable two and three dimensional models and because solution times vary as the cube of total degrees of freedom, three dimensional finite element models are very much more expensive than their two dimensional counterparts. The pressure to use three dimensional models coupled with the alarming cost of full domain analyses has given renewed impetus to the examination of Boundary Domain Methods in general, and the Boundary Element Method in particular, in the hope of finding a more cost efficient approach than the finite element method.

In this paper a boundary element method, for three dimensional Laplacian problems, based on Regular Boundary Integral equations, is presented. The regular boundary integral equations are obtained by the simple device of taking the singular point of the fundamental solution kernel function outside the domain of the problem. The application of this method to two dimensional potential problems was shown in(Patterson, C, and Sheikh, M. A., 1982). It has been traditional, but not necessary, to use the conventional (singular) boundary element method in conjunction with continuous elements, where possible. It was shown in(Patterson, C, and Sheikh, M. A., 1981) that this continuity is not necessary and results for two dimensional harmonic problems were presented. In this paper non-conforming boundary elements are again used. This has advantages at geometric singularities and at edges where the given boundary data changes type, e.g. from temperature to temperature normal gradient. Additionally, while the elements must describe the surface geometry adequately, there is no requirement for the topological interelement continuity used in finite element meshes, so that mesh grading is easily accomplished.

Results are presented relating to three three dimensional test problems presenting both regular and singular solutions.

THEORY

Consider a scalar field ϕ, defined over a domain Ω and on its boundary Γ, which satisfies Laplace's equation within the domain,

$$\nabla^2 \phi = 0 \tag{1}$$

and boundary conditions

$$\begin{aligned} \phi &= \bar{\phi} \quad \text{on} \quad \Gamma_1 \\ \frac{\partial \phi}{\partial n} &= \frac{\partial \bar{\phi}}{\partial n} \quad \text{on} \quad \Gamma_2 \end{aligned} \tag{2}$$

where Γ_1 and Γ_2 partition the boundary Γ, n is the outward normal, and $\bar{\phi}$ and $\frac{\partial \bar{\phi}}{\partial n}$ are given functions.

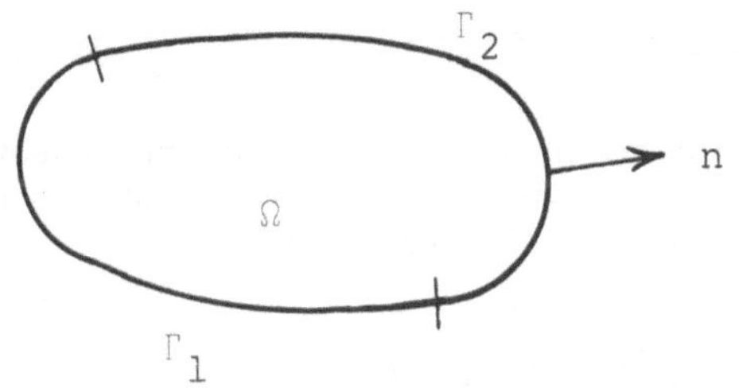

Figure 1 Problem domain

Let $\phi^*(x,y)$ be the fundamental solution of Laplace's equation with singular point at y. Then, taking $\phi^* \nabla^2 \phi$ integrated over Ω with y located on Γ, and integrating by parts twice it follows that

$$c\,\phi(y) + \int_\Gamma \phi(x) \frac{\partial \phi^*}{\partial n}(x,y)\, d\Gamma(x) = \int_\Gamma \frac{\partial \phi}{\partial n}(x)\, \phi^*(x,y)\, d\Gamma(x) \tag{3}$$

where c is a known constant resulting from the Cauchy Principal Value of the surface integral. Equation (3) is the starting point of the singular boundary element method.

If the point y is taken outside the domain of the problem, the integrations by parts may again be taken with a similar result to equation (3) except that the integral is regular and, in consequence, c vanishes. If the boundary Γ be partitioned into n elements Γ_1 .... Γ_n the resulting regular boundary integral equation is

$$\sum_{j=1}^{n} \int_{\Gamma_j} \phi(x) \frac{\partial\phi^*}{\partial n}(x,y)\, d\Gamma(x) = \sum_{j=1}^{n} \int_{\Gamma_j} \frac{\partial\phi}{\partial n}(x)\; \phi^*(x,y)\, d\Gamma\,(x) \qquad (4)$$

Equation (4) is the basic equation of the Regular Boundary Element method. If a boundary element discretization is introduced, in the familiar manner, for ϕ on each element Γ_j a single non homogeneous linear algebraic equation results from equation (4) among the n unknown parameters ϕ_j, $\frac{\partial\phi_k}{\partial n}$. If, in turn, the singular point, y, of the fundamental solution is located at some point along the outward normal at each freedom node, n such equations are generated yielding a determinate linear algebraic system.

Suppose further, that the domain Ω is partitioned into subdomains $\Omega_1 \ldots\ldots \Omega_p$ with subdomain boundaries $\Gamma_1 \ldots\ldots \Gamma_p$ and that each subdomain boundary Γ_j is further partitioned into elements $\Gamma_{j1} \ldots\ldots \Gamma_{jn_j}$; then for each subdomain Ω_j application of equation (4) gives

$$\sum_{k=1}^{n_j} \int_{\Gamma_{jk}} \phi(x) \frac{\partial\phi^*}{\partial n}(x,y)\, d\Gamma_j(x) = \sum_{k=1}^{n_j} \int_{\Gamma_{jk}} \frac{\partial\phi}{\partial n}(x)\; \phi^*(x,y)\, d\Gamma_j(x) \qquad (5)$$

where y is exterior to Ω_j.

In addition to these boundary integral equations, at intersubdomain boundaries the field function and its normal derivative must satisfy continuity conditions

$$\begin{aligned} \phi^{(j)}(x) &= \phi^{(k)}(x) \\ \frac{\partial\phi^{(j)}}{\partial n}(x) &= -\frac{\partial\phi^{(k)}}{\partial n}(x) \end{aligned} \qquad (6)$$

where j and k adjacent subdomains and x is any point on the subdomain interface. On introducing a boundary element discretization of the field function, generating non homogeneous linear algebraic equations for each subdomain by choice of the singular point y, as before, and imposing the boundary conditions (6) at all intersubdomain freedom nodes, a determinate non homogeneous linear algebraic system is obtained. A treatment of the corresponding subdomain model using the singular method is given in (Lachat, J. C., 1975).

For three dimensional potential problems the fundamental solution is given by

$$\phi^*(x,y) = \frac{1}{4\pi r(x,y)} \qquad (7)$$

where r is the distance between the points x,y.

Non-conforming boundary elements

The elements used in the investigation all employed a conventional 8-node isoparametric quadrilateral element to represent the geometry. Here geometric nodes are taken at the corners and 'midside' points and the geometry is represented by

$$x_i(\xi) = \sum_{a=1}^{8} M^a(\xi)\, x_i^a \;;\; i = 1,\, 2,\, 3 \tag{8}$$

where: $\xi \in (\xi_1, \xi_2)$, $\xi_i \in [-1, +1]$

x_i^a are nodal coordinates

$M^a(\xi)$ are interpolation functions

Typical interpolation functions consistent with the assignment in figure 2 are:

corner function $M^1(\xi) = \frac{1}{4}(1+\xi_1)\,(1+\xi_2)\,(\xi_1+\xi_2-1)$

midside function $M^5(\xi) = \frac{1}{2}(1-\xi_1^2)\,(1+\xi_2)$

The rest are given by induction

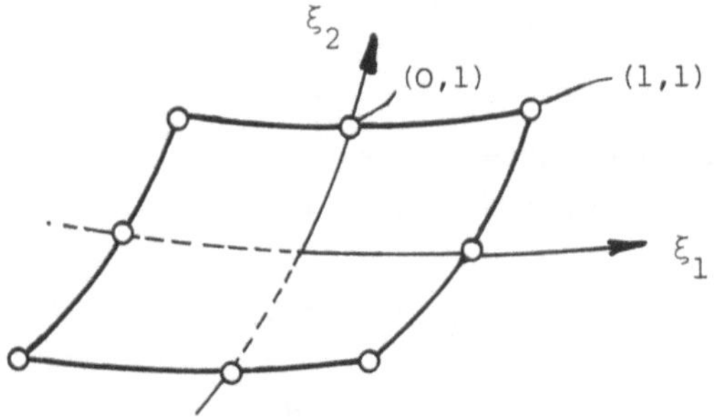

Figure 2 Element geometry

Upon the isoparametric quadrilateral geometry, linear, quadratic and cubic non-conforming elements, for the field function and its normal derivative, were taken. In each case the field quantities are given by

$$\phi(\xi) = \sum_{a=1}^{q} F^a(\xi)\, \phi^a$$

$$\frac{\partial\phi}{\partial n}(\xi) = \sum_{a=1}^{q} F^a(\xi)\, \frac{\partial\phi^a}{\partial n} \tag{9}$$

where: the element has q freedom nodes, that is 4, 8 or 12 for the linear, quadratic or cubic elements respectively,

$\xi = (\xi_1, \xi_2)$, $\xi_i \in [-1,+1]$,

$F^a(\xi)$ are the interpolation functions defined below.

The freedom nodes are geometrically distinct from the geometrical nodes and are taken within the element (and thus implying non-conformity). The location of the freedom nodes, which are symmetrically arranged in the ξ plane, and interpolation functions are as follows:

<u>linear elements</u> (see figure 3)

$$F^a(\xi) = (\tfrac{1}{2} \pm \xi_1)\ (\tfrac{1}{2} \pm \xi_2)$$

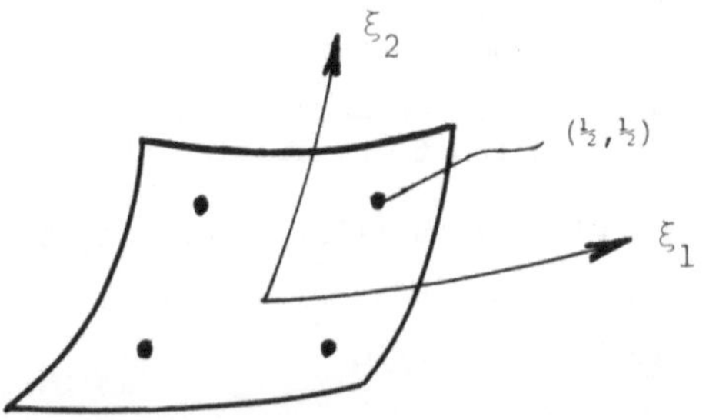

Figure 3 Freedom nodes - linear elements

<u>quadratic elements</u> (see figure 4)

corner function $F^1(\xi) = \frac{27}{32}\,(2/3 + \xi_1)(2/3 + \xi_2)(\xi_1 + \xi_2 - 2/3)$

midside function $F^5(\xi) = \frac{27}{16}\,(4/9 - \xi_1^2)(2/3 + \xi_2)$

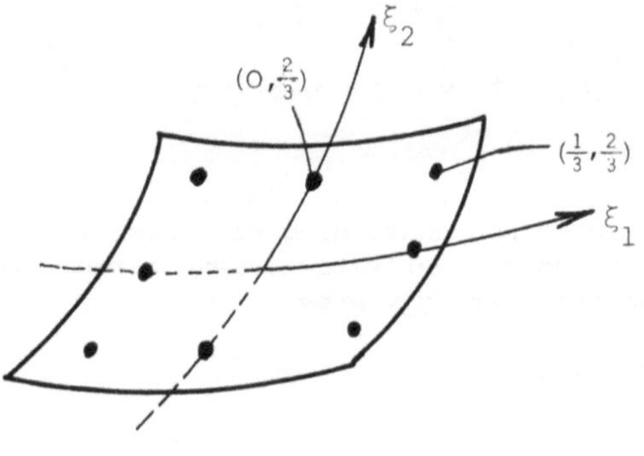

Figure 4 Freedom nodes - quadratic elements

<u>cubic elements</u> (see figure 5)

corner function $F^1(\xi) = 1/9(\tfrac{3}{4} + \xi_1)(\tfrac{3}{4} + \xi_2)(8(\xi_1^2 + \xi_2^2) - 5)$

midside function $F^5(\xi) = 3/2(\tfrac{1}{4} + \xi_1)(1 - 16/9\ \xi_1^2)(\tfrac{3}{4} + \xi_2)$

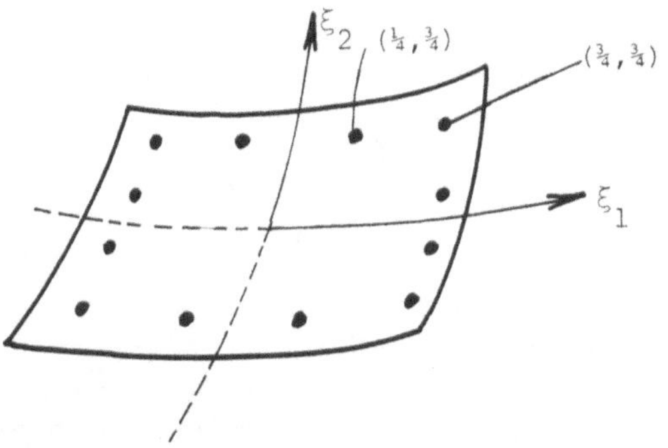

Figure 5 Freedom nodes - cubic elements

The remaining interpolation functions can be deduced by induction.

Location of singular point

As stated earlier, in the Regular Boundary Element method the singular point of the fundamental solution kernel function is taken outside the domain of the problem, in contrast with the conventional method where it is taken on the boundary. In exact arithmetic if there are q freedom nodes in a model, in order to obtain a determinate algebra it is merely necessary to derive q linearly independent kernel functions. This can be achieved by choosing arbitrarily q distinct locations outside the domain of the problem, at which to locate the singularity. In finite arithmetic, it is necessary to locate the singular points reasonably close to the domain in order to avoid ill-conditioning problems; on the other hand if the singular points are brought unduly close to the surface a high integration order must be taken near the singularity in order to maintain accurate integration. Clearly, this implies increased computer cost. In order to obtain a systematic approach to the assignment of singular point location it was decided that it should be located on the outward normal from a freedom location, figure 6, thereby guaranteeing q linearly independent kernel functions. A systematic study was then made in order to determine the 'best' location of the singular point. This involves a compromise between the need for nicely well conditioned algebraic equations and the need for moderate computer effort. It was found that if a moderate aspect ratio was taken (no greater than 3:1) optimal results were obtained if for each element the singularity was located at a distance from the element, along the outward normal, equal to the minimum distance between freedom nodes for that element.

Once the algebraic equations have been solved, and thereby the approximate field function and its normal derivative determined over the bounding surface, Γ_1 the interior solution is determined as usual (see for instance Lachat and Watson,1976).

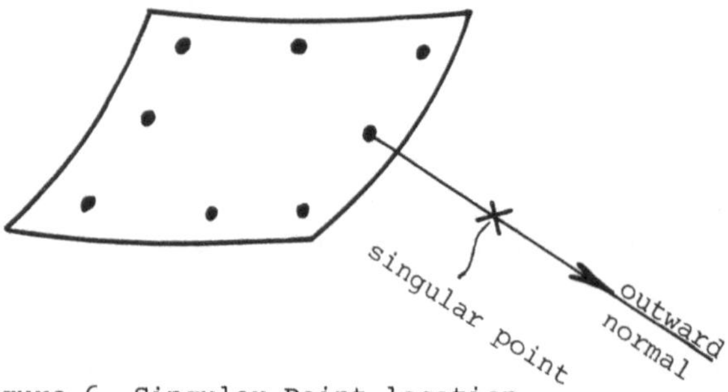

Figure 6 Singular Point location

APPLICATIONS

Three applications to three dimensional potential problems are presented. In the first inviscid laminar flow around a cylinder is considered. Whilst this is a planar problem, it is analysed in 3-space and is interesting as a validity test on the coding and as an application of mismatched non-conforming elements. The second and third problems are of steady state heat flow in a bar and a cylinder. Fourier series solutions are available for each and both problems present singular heat flux behaviour at the boundary. The second problem is analysed using subdomains.

Inviscid laminar flow around a cylinder

The geometry of the problem is depicted in figure 7a only a quarter of which need be analysed due to symmetry. The boundary element discretization employed is given in figure 7b. Twenty-five 8 freedom node elements, giving 200 nodes, are taken using a mismatched mesh. The assumed boundary values of stream function and its normal derivative are shown in figure 7c. The numerical model possesses a plane of symmetry which was observed in the numerical results to a high degree. Again the planar nature of the problem was evidenced to the fourth significant figure of significant quantities. Computed $\partial\phi/\partial n$ at the surface along lines ED and ABC are given in figure 8a together with known values, see for instance (Brebbia, 1978). The mutual agreement is around ± .002 units. In figure 8b the computed trajectory of the streamline ϕ=1 is presented.

Steady state heat conduction along a rectangular bar

The geometry of the bar is given in figure 9. The face at AD is held at temperature 300., and the rest of the surface at temperature 0.. The solution is clearly singular at the edges of the heated end. A boundary element model was made by partitioning the bar, along its length, into two identical halves. Ten 8 freedom node elements were taken on each subregion giving 160 freedom nodes. This problem also has a symmetry plane which was well presented in the computed results. Computed surface values of temperature normal gradient are given in figure 10 and temperature profiles along three internal lines along with the corresponding values given by the fourier series solution are given in figures 11 a,b,c, see for instance (Aypaci, 1966). The mutual agreement of the two sets of values is within ± 0.5 units.

Steady state heat conduction in a short cylinder

The problem analysed is depicted in figure 12a. Assumed dimensions are R=3, L=4 and the upper surface of the cylinder, $(-\pi/2, \leq \theta \leq +\pi/2)$, is at temperature 300. whilst the remaining surface is at 0.. The solution to this problem is singular along the edge of the surface at ϕ=300.. θ=0, π defines a symmetry plane so that only the half geometry need be analysed. The boundary element discretization used is depicted in figure 12b. There are 20 mismatched 8 freedom node elements. Computed surface values of temperature and its normal derivative along the edges ABCD, and EFGH are presented in figures 13a and 13b. There is good evidence of the expected singularities. Computed values of internal temperature along three lines are given in figures 14a,b,c and compared with the corresponding values given by the fourier series solution, see for instance (Aypaci, 1966). The agreement is within ± 0.3 units.

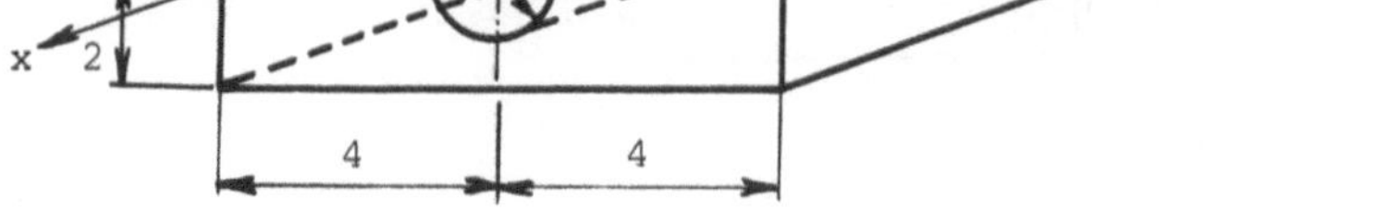

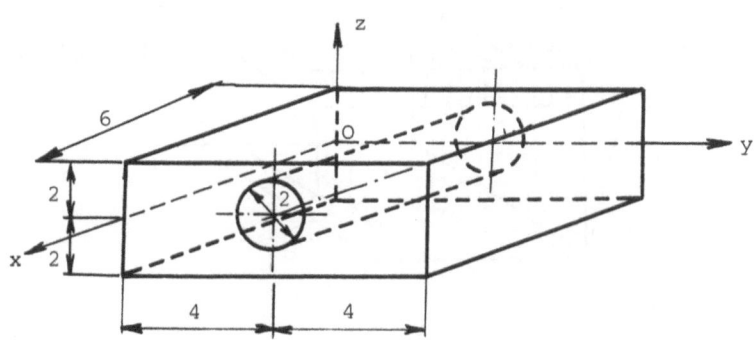

Figure 7a Flow problem - geometry

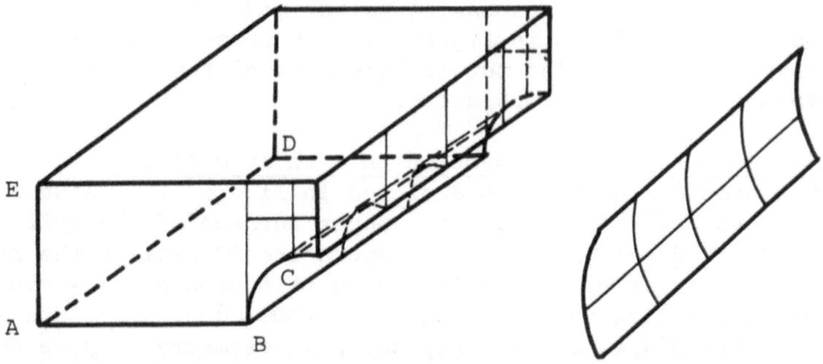

Figure 7b Flow problem - discretization

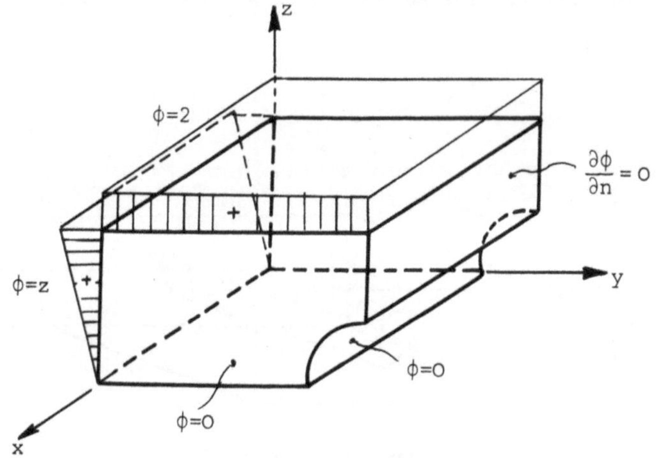

Figure 7c Flow problem - boundary conditions

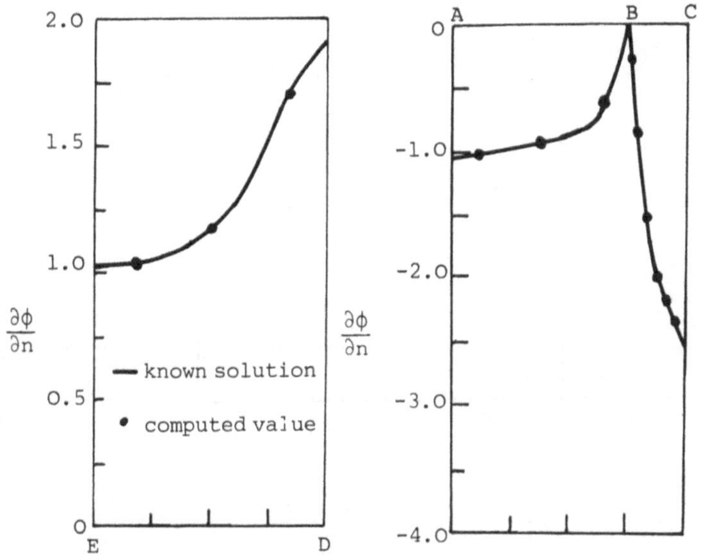

Figure 8a Computed surface values

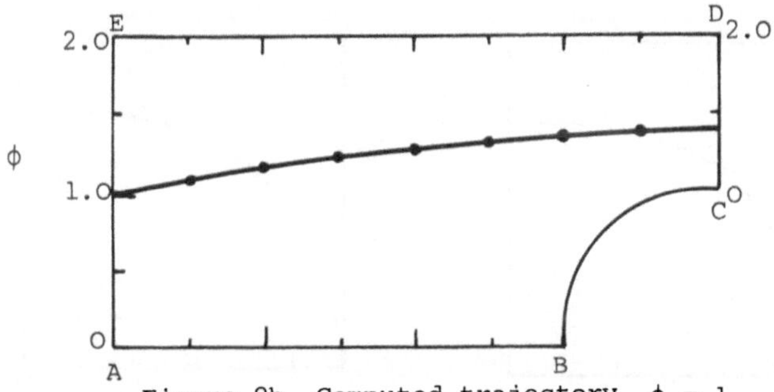

Figure 8b Computed trajectory, $\phi = 1$

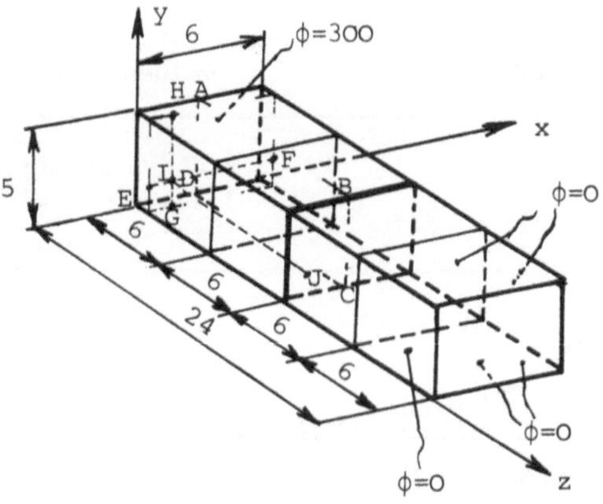

Figure 9 Bar problem - geometry

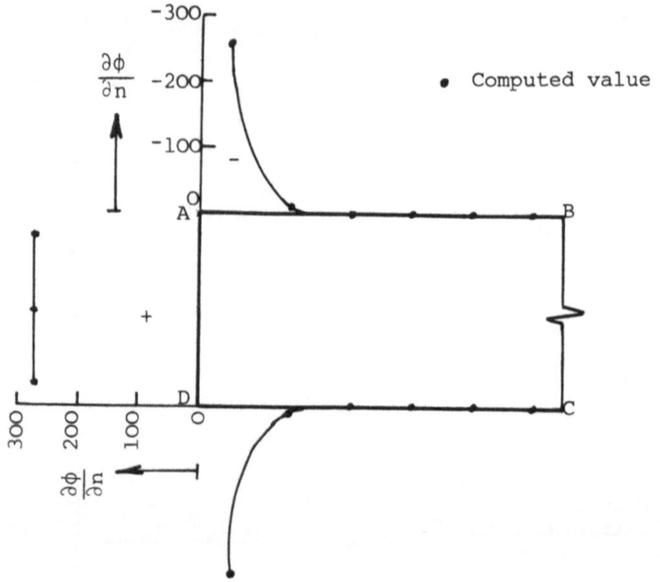

Figure 10 Computed surface values

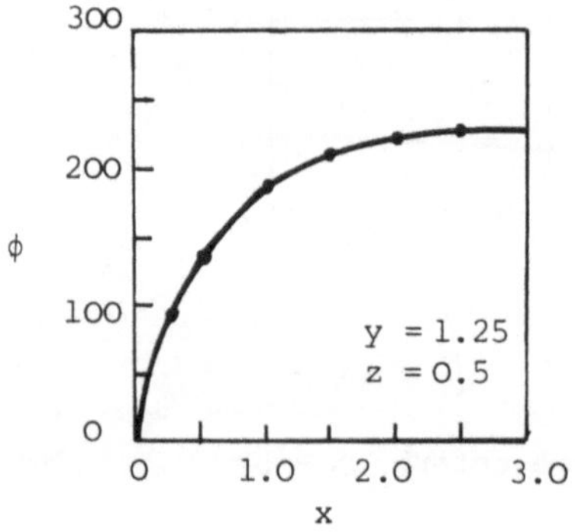

Figure 11a. Temperature values along EF

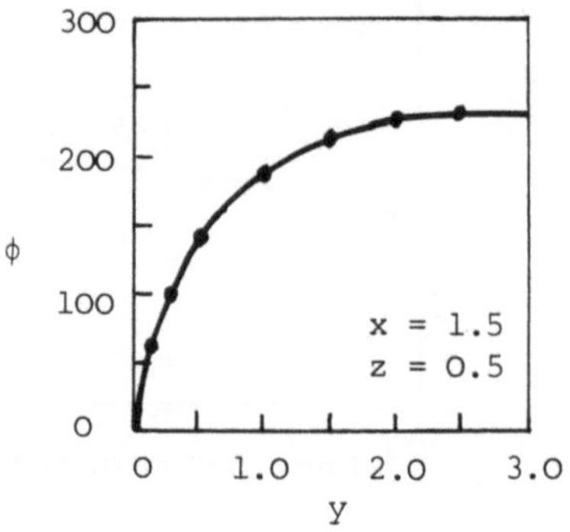

Figure 11b. Temperature values along GH

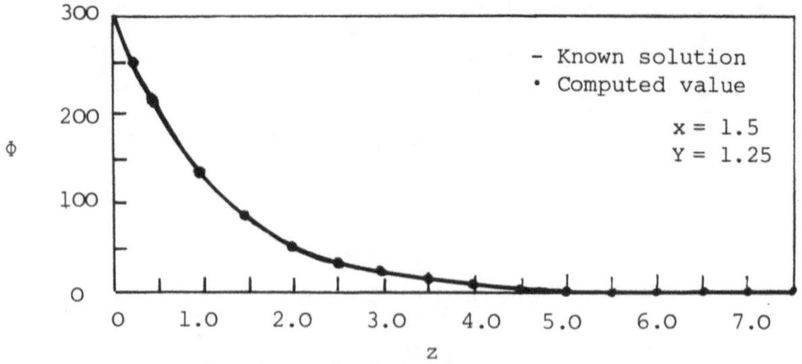

Figure 11c. Temperature values along IJ

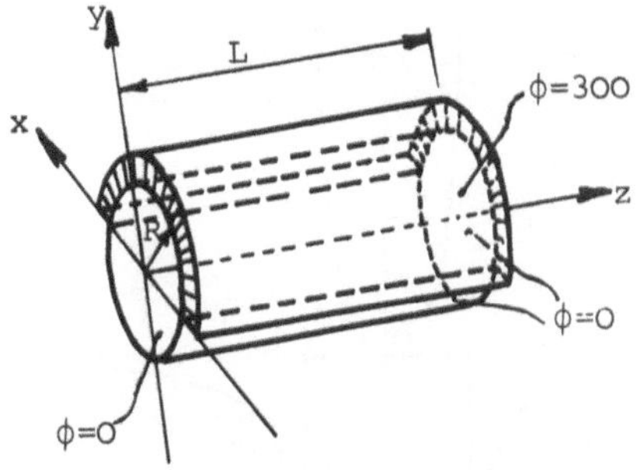

Figure 12a Cylinder problem

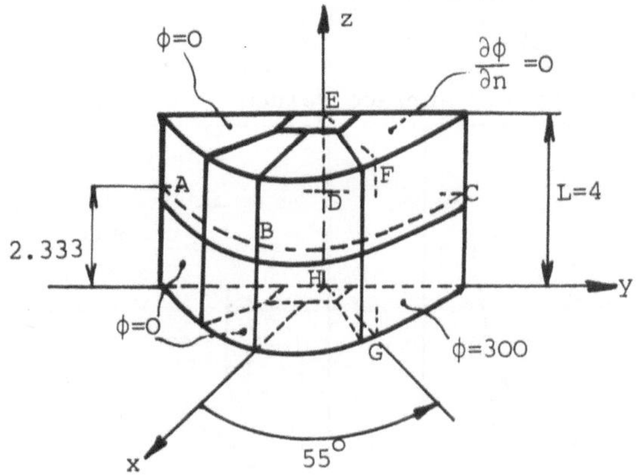

Figure 12b Cylinder problem - discretization

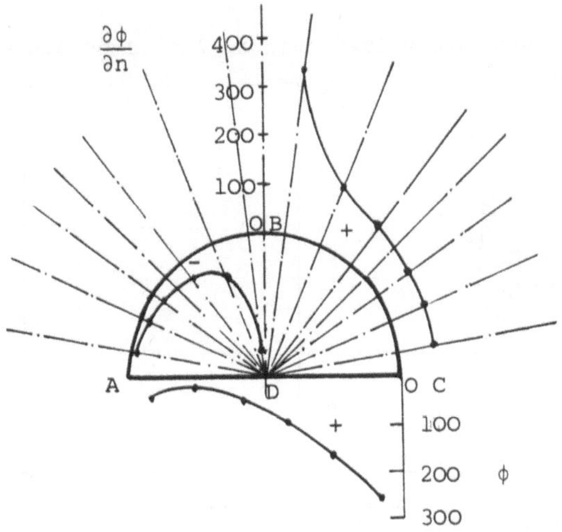

Figure 13a. Computed surface values along ABCD

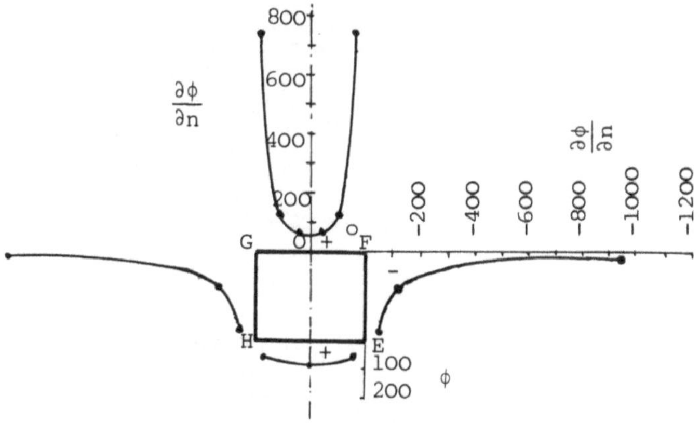

Figure 13b Computed surface values along EFGH

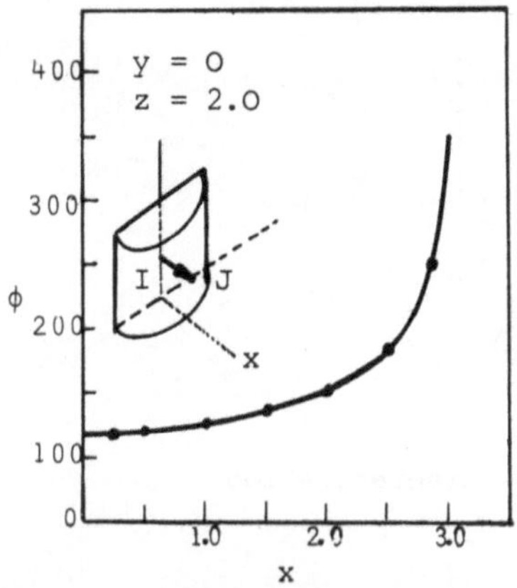

Figure 14a Temperature values along IJ

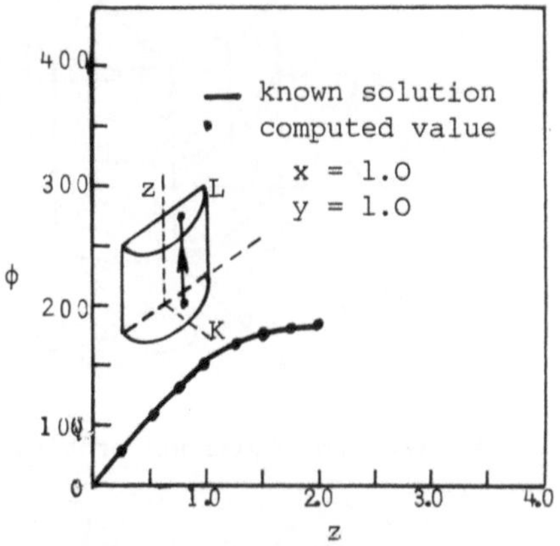

Figure 14b Temperature values along KL

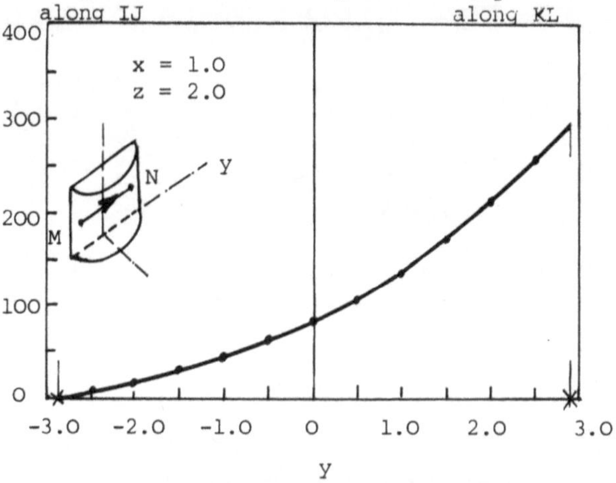

Figure 14c. Temperature values along MN

DISCUSSIONS AND CONCLUSIONS

A regular boundary element method for use with three dimensional potential problems has been presented. This method has been applied using non-conforming boundary elements having linear, quadratic or cubic variation of the field quantity supported by an 8-node isoparametric quadrilateral geometric element. Additionally, the method has been extended to cope with numerical models defined on a set of subdomains obtained by partitioning the original problem domain. A subsidiary investigation showed that the singularity of the fundamental solution kernel function was 'best' located along the outward normal from a freedom node at a distance approximately equal to the shortest internodal distance within the element considered.

Results obtained, using the quadratic variant of boundary element, for three test problems have been presented. The first, inviscid laminar flow around a cylinder, was chosen because the problem is planar and its solution is well known. Principally, it was chosen as a validity test on the coding. The numerical problem analysed had a plane of symmetry which was found, to a high degree, in the computed values. Also, the planar nature of the problem was well evidenced, and the computed numerical values were in good agreement with the known two dimensional values. It is noteworthy that a coarse mesh of mismatched elements was used.

The second and third problems were chosen as truly three dimensional tests of the method. Whilst the geometries are simple, a rod and a bar, the solutions are fully three dimensional and have edge flux singularities. Fourier solutions are available for both problems. The rectangular bar problem was analysed using subregions and a coarse mesh, especially near the heated end of the bar. The computed boundary values showed good evidence of the flux singularity despite the incompleteness of the mesh. Comparison of the interior solution with that of the Fourier expansion proved very favourable showing that the presence of the singularity and lack of completeness of mesh at the boundary could be well tolerated. In the heated cylinder problem the exact solution shows more extensive singular behaviour. Nevertheless, with the coarse mesh employed, the computed boundary results represented the singularities as well as could be expected and again the interior solution was in good agreement with the Fourier series solution.

In conclusion: the regular boundary element method presented here has given satisfactory results for the problems examined; the presence of a singularity in the exact solution can be tolerated without the necessity of highly refined meshes near the singularity in that a good solution away from the singularity can be obtained, non-conforming elements may validly be used and, additionally, there is no requirement for topological element continuity so that freely graded meshes may be used; finally, the subdomain approach may still be applied.

REFERENCES

Aypaci, V. S. (1966) Conduction Heat Transfer, Addison-Welsey, U.S.A.

Brebbia, C. A. (1978) Recent Advances in Boundary Element Methods, Pentech Press, London.

Lachat, J. C. (1975) A Further Development of the Boundary Integral Technique for Elastostatics, Ph.D. Thesis, University of Southampton, U.K.

Lachat, J. C. and Watson, J. O. (1976) Effective Numerical Treatment of Boundary Integral Equations, Int. J. Num. Meth. Eng., 10:991-1005.

Patterson, C. and Sheikh, M. A. (1982) A Regular Boundary Element Method for Fluid Flow, Int. Journal of Num. Meth. in Fluid Flow, (edited by C. Taylor), at press.

Patterson, C. and Sheikh, M. A. (1981) Discontinuous Boundary Elements for Heat Conduction, Numerical Methods for Thermal Problems, (edited by R.W. Lewis, K. Morgan and B.A. Schrefler), 25-35.

A BOUNDARY INTEGRAL EQUATION APPROACH TO THE ONE DIMENSIONAL ABLATION PROBLEM

R.P. Shaw
State University of New York at Buffalo

ABSTRACT

The ablation of a semi-infinite solid under general surface heat flux and initial temperature conditions is examined as a boundary integral equation. The particular case of a constant initial temperature and a constant surface heat flux is briefly described first and the extension to the general case is made directly in terms of specific integrals which have arisen in this formulation. The premelting phase may be solved analytically for a wide variety of surface heat flux and initial temperature forms; these analytical expressions carry over to the postmelting phase thereby simplifying much of the calculation for such cases.

INTRODUCTION

The moving boundary problem associated with melting or ablation has been of interest to both the theoretical and the practical scientific community for some time. The inherent nonlinearity of such problems poses significant mathematical difficulties, e.g. Rubenstein (1971), while their practical applications to a wide variety of physical situations, e.g. Ockendon and Hodgkins (1975) or Wilson, Solomon and Boggs (1978), require some method of solution, albeit approximate or numerical.

One such method is the boundary integral/element approach; this paper is an outgrowth of a review article, Banerjee and Shaw (1981), in a recent text on this area and is designed to provide an illustration of such an approach.

In this study, a specific surface heat flux q(t) is, beginning at t = 0, applied to the exposed boundary of a semi-infinite domain at a specific initial temperature. Once the surface reaches the melting point, the molten material is immediately removed, i.e. ablated. The use of a semi-infinite domain is not particularly restrictive since "other" boundaries, if sufficiently far from the domain of interest, would not be expected to appreciably alter the solution. It must be stated, however, that this "ploy" is used to simplify the arithmetic and is not necessary to the boundary integral/element approach. This is of course a Landau type problem, Landau (1950).

One final point to be made is to emphasize the distinction between this approach and others which used the word "integral" in their title, e.g. Goodman's heat balance integral method, Goodman (1964) and related methods, e.g. Zien (1978). Those methods contain an inherent physical approximation to the original problem which is then solved in some approximate manner, while the present approach is an exact (boundary integral) reformulation which is then solved numerically (the boundary element method), e.g. Shaw (1970), Rizzo and Shippy (1970). These methods are totally unrelated and are not to be confused with each other.

There is of course a strong connection to the work of Chuang and Szekeley (1971) who first applied the BIE method to melting/solidification problems. The major difference between their work and that presented here is in the physical problem considered, i.e. ablation vs melting, and in the present formulation which emphasizes the free boundary location as the primary dependent variable and compartmentalizes the various initial condition and boundary condition effects.

ONE DIMENSIONAL EQUATION FORMULATION

Consider the one dimensional Landau problem of an ablating semi-infinite slab, $0 < x < \infty$, with an initial temperature $T(x,0) = T_I(x)$ and a prescribed heat flux on the exposed face, $q(x^*,t) = q(t)$, where x^* is the location of the exposed face and is zero until melting begins and the moving boundary, $x = x^*(t)$, afterwards. For simplicity, no internal sources are taken.

The governing equations are, e.g. Landau (1950)

$$\begin{aligned} &a^2\partial^2 T(x,t)/\partial x^2 = \partial T(x,t)/\partial t; \; x^* < x < \infty, \; t > o \\ &T(x,o) = T_I(x); \; 0 < x < \infty, \; t = o \\ &-k\partial T(o,t)/\partial x = q(t); \; x = o, \; o < t \lesssim t_m \\ &-k\partial T(x^*,t)/\partial x + \rho\lambda dx^*(t)/dt = q(t); \; x = x^*(t), \; t > t_m \end{aligned} \tag{1}$$

where k is the conductivity, c the specific heat, ρ the density, $a = (k/c\rho)^{1/2}$ the diffusivity, and λ is the latent heat, all of which shall be taken constant in this discussion.

Up to the time t_m, there is no ablation, and the problem has a widely known solution which shall be briefly repeated here as a boundary integral equation solution. The melting temperature T_m is taken to be constant and used as a reference temperature, i.e. $T_m = 0$. The equivalent boundary integral form to equations (1) is

$$\varepsilon T(x,t) = a^2 \int_0^t \{G(x,t;o,t_o)\, q(t_o)/k + T(0,t_o)\, \partial G(x,t;x_o,t_o)/\partial x_o\}dt_o ; x_o = 0 + \int_0^\infty T_I(x_o)\, G(x,t;x_o,o)dx_o \tag{2}$$

This form is well known, e.g. Shaw (1970); here $\alpha = 0, 1/2, 1$ depending on whether $x < 0$, $x = 0$ or $x > 0$ respectively, and G is a fundamental solution to

$$a^2\partial^2 G(x,t;x_o,t_o)/\partial X_o^{\,2} + \partial G(x,t;x_o,t_o)/\partial t_o = -\delta(x-x_o)\;\delta(t-t_o) \tag{3}$$

as discussed for example in section (7.4) of Morse and Feshbach (1953).

For the particular problem at hand, placing x at the face $x = 0$ yields an integral equation on the boundary value of T, i.e. $T(0,t)$. Once this is found, the "field point" x may be placed in the interior of the slab yielding $T(x,t)$ as a quadrature of known boundary and initial values. However, it is clear that the choice of a "double" fundamental solution, symmetric to the surface $x_o = 0$ will give "the" Green's function for this problem, i.e. eliminate the unknown term under the integral, i.e.

$$G_{(2)}\;(x,t;x_o,t_o) = [2a(\pi(t-t_o))^{\frac{1}{2}}]^{-1} \{\exp[-(x-x_o)^2/4a^2(t-t_o)] + \exp[-(x+x_o)^2/4a^2(t-t_o)]\} \tag{4}$$

such that $\partial G_{(2)}/\partial x_o$ is identically zero on $x = 0$ for all x leaving (with ε equal to 1/2 times 2 for the double source)

$$T(o,t) = \int_0^\infty T_I\;(x_o)\; G_{(2)}(o,t;x_o,o)dx_o + (a^2/k)\, G_{(2)}(o,t;o,t_o)q(t_o)dt_o = I(t) + II(t) \tag{5}$$

and (with ε equal to 1 times 1 for one of the sources at $x_o = x$)

$$T(x,t) = \int_0^\infty T_I(x_o)G_{(2)}(x,t;x_o,o)dx_o + (a^2/k) \int_0^\infty G_{(2)}(x,t;o,t_o)q(t_o)dt_o \tag{6}$$

The initiation of melting, or in this case ablation, occurs at $t = t_m$ when $T(0,t_m)$ first reaches $T_m = 0$ thereby defining t_m implicitly through

$$t_m^{\frac{1}{2}} = \frac{-\int_0^\infty T_I(x_o)\exp[-x_o^2/4a^2t_m]dx_o}{(a^2/k)\int_0^{t_m}(t_m-t_o)^{-\frac{1}{2}}q(t_o)dt_o} \tag{7}$$

This presumes that the initial temperature distribution and surface heat flux input are such that melting does not occur at interior points.

For values of t greater than t_m, i.e. post-melting, the latent heat must be included. The standard integral equation formulation now yields

$$\begin{aligned}\varepsilon T(x,t) = {} & (a^2/k)\int_{t_m}^{t}\{G(x,t;x_o^*,t_o)[q(t_o) - \rho\lambda dx_o^*/dt_o] \\ & + kT(x_o^*,t_o)[\partial G(x,t;x_o,t_o)/\partial x_o]\}\,dt_o \Big|_{x_o = x_o^*} \\ & + (a^2/k)\int_0^{t_m}\{G(x,t;x_o,t_o)q(t_o) + kT(x_o,t_o) \\ & \cdot[\partial G(x,t;x_o,t_o)/\partial x_o]\}\,dt_o \Big|_{x_o = o} \\ & + \int_0^\infty T_I(x_o)G(x,t;x_o,o)\,dx_o \\ & + \int_{x^*(t)}^\infty [T(x_o,t_o^*)\,G(x,t;x_o,t_o^*)]\,dx_o \Big|_{t_o^* = t_o(x_o^*)}\end{aligned} \tag{8}$$

The last integral arises from the moving boundary term, e.g. see Chuang and Szekeley (1971) or Banerjee and Shaw (1981); in the present case, it vanishes since $T(x_o,t_o^*) = T_m = 0$, i.e. the temperature is evaluated at the moving boundary which is always at the melting point. The double source Green's function, $G_{(2)}$, is used again for convenience to eliminate the $T(0,t_o)$ term in the second integral. The term involving $T(x_o^*, t_o)$ also vanishes since this again is at the melting point which is zero. Placing the field point at $x = 0$, i.e. outside of the domain of integration, $0 < x^* < x < \infty$, requires that ε be zero and thus

$$0 = \frac{1}{a(\pi t)^{\frac{1}{2}}} \int_o^\infty T_I\ (x_o)\exp(-x_o{}^2/4a^2t)dx_o$$

$$+ (a/k\pi^{\frac{1}{2}}) \int_o^{t_m} q(t_o)\ (t-t_o)^{-\frac{1}{2}}\ dt_o$$

$$+ (a/k\pi^{\frac{1}{2}}) \int_{t_m}^{t} [q(t_o) - \rho\lambda\ dx_o^*/dt_o]\exp[-x_o^{*2}/4a^2(t-t_o)] \cdot (t-t_o)^{-\frac{1}{2}}dt_o \qquad (9)$$

$$= I_m(t) + II_m(t) + III_m(t)$$

CONSTANT SURFACE HEAT FLUX AND CONSTANT INITIAL TEMPERATURE

Consider as the simplest example the case of a constant surface heat flux, q_o, and a constant initial temperature, T_o. The integrals in eqs. (5) and (6) for the premelting solution may be evaluated analytically to give the well-known solutions

$$T(o,t) = I(t) + II(t) = T_o + (2aq_o/k)\ (t/\pi)^{\frac{1}{2}} \qquad (10)$$

$$T(x,t) = T_o + (2aq_o/k)\ (t/\pi)^{\frac{1}{2}} \exp[-x^2/4a^2t] -(q_ox/k)\ \mathrm{erfc}[x/2at^{\frac{1}{2}}] \qquad (11)$$

which, apart from a misprint in the second solution, agrees with Landau's results. The time of initiation of melting is

$$t_m = \pi(kT_o/2aq_o)^2 \qquad (12)$$

For t greater than t_m, eq. (9) yields

$$I_m(t) = T_o \qquad (13\text{-a})$$

$$II_m(t) = (2aq_o/k\pi^{\frac{1}{2}})\ (-(t-t_m)^{\frac{1}{2}} + t^{\frac{1}{2}}) \qquad (13\text{-b})$$

$$III_m(t) = (aq_o/k\pi^{\frac{1}{2}}) \int_{t_m}^{t} [1-(\rho\lambda/q_o)dx_o^*/dt_o] \cdot (t-t_o)^{-\frac{1}{2}} \exp[-x_o^{*2}/4a^2(t-t_o)]dt_o \qquad (13\text{-c})$$

which may be conveniently rewritten, using $\tau = t - t_m$, as

$$0 = T_o k\pi^{\frac{1}{2}}/aq_o + 2\,[(\tau + t_m)^{\frac{1}{2}} - \tau^{\frac{1}{2}}] + \int_o^{\tau} [1-(\rho\lambda/q_o)dx_o^*/d\tau_o](\tau-\tau_o)^{-\frac{1}{2}} \exp\,[-x_o^{*2}/4a^2(\tau-\tau_o)]d\tau_o \tag{14}$$

At this point, it would seem appropriate to nondimensionalize this equation prior to attempting a numerical solution. Since x and t are relative (both extend over an infinite domain), they may be scaled with one parameter, ℓ .

Temperature is scaled by T_O, the initial temperature. If the initial temperature were zero, i.e. the half space is at the melting point, this would of course cause some difficulty, but that special case reduces simply to t_m equal zero (no premelting phase) and dx_o^*/dt equal to $q(t)/\rho\lambda$, i.e. all of the incident heat flux goes into a plation and no thermal energy is conducted away.

$$\bar{x} = x/\ell \qquad \bar{\tau} = 4a^2\tau/\ell^2 \qquad \bar{T} = T/T_o \tag{15}$$

such that eqs. (10, 11, and 14) become

$$\bar{T}\,(o,\bar{t}) = 1 + 2(\bar{t})^{\frac{1}{2}}/\alpha \tag{16}$$

$$\bar{T}\,(\bar{x},\bar{t}) = 1 + (2(\bar{t})^{\frac{1}{2}}/\alpha)\,\exp[-\bar{x}^2/\bar{t}] - (2\bar{x}\pi^{\frac{1}{2}}/\alpha)\text{erfc}\,[\bar{x}/(\bar{t})^{\frac{1}{2}}] \tag{17}$$

and

$$o = \alpha + 2[(\bar{\tau} + \bar{t}_m)^{\frac{1}{2}} - (\bar{\tau})^{\frac{1}{2}}] + \int_o^{\bar{\tau}} [1-\beta\; d\bar{x}_o^*/d\bar{\tau}_o]\;(\bar{\tau}-\bar{\tau}_o)^{-\frac{1}{2}} \exp\,[-\bar{x}_o^{*2}/(\bar{\tau}-\bar{\tau}_o)]\;d\bar{\tau}_o \tag{18}$$

There are two parameters that define such a problem then; $\alpha = 2\,\{k\; T_o\; \pi^{1/2}/\ell q_o\}$ and $\beta = (4a^2\rho\lambda/q_o\ell)$ and $\bar{x}_o^*\;(\tau_o)$ may be expressed in terms of α and β . However, there is also one 'free' parameter, ℓ, introduced in the nondimensionalization which could be used to eliminate either α or β leaving only one true parameter. Note that T_o is negative in this problem and thus α is also negative and $\overline{T}$ is opposite insign to T. The initial melting time in this nondimensional form is $\alpha^2/4$. The simplest level of numerical approximation for this equation is to assume $\overline{x}_o^*$ to be constant over specified, although not necessarily uniform, intervals in τ, i.e.

$$\overline{X}_o^* = X_{JA} = (X_J + X_{J-1})/2 \ ; \ \tau_{J-1} < \overline{\tau}_o < \tau_J \tag{19-a}$$

$$d\overline{X}_o^*/d\overline{\tau}_o = (X_J - X_{J-1})/(\tau_J - \tau_{J-1}) \ ; \ \tau_{J-1} < \tau_o < \overline{\tau}_J \tag{19-b}$$

Clearly higher order approximations for $\overline{x}_o^*$, e.g. linear in $\overline{\tau}_o$, could be used at the expense of additional arithmetic. The integral is then replaced by a finite sum up to J = N where $\overline{\tau} = \tau_N$ represents the current time and $\overline{x}_o^* = X_N$ the current location of the ablating boundary

$$S_N = S(\tau_N) = -\alpha - 2\left[(\tau_N + \overline{t}_m)^{\frac{1}{2}} - \tau_N^{\frac{1}{2}}\right] = \sum_{J=1}^{N} \left[1 - \beta(X_J - X_{J-1})/\Delta\tau_J\right] A_{NJ} \tag{20}$$

with $\Delta\tau_J = \tau_J - \tau_{J-1}$ and $\xi_{NJ} = X_{JA}/(\tau_N - \tau_J)$, $\xi_{NJ1} = X_{JA}/(\tau_N - \tau_{J-1})^{\frac{1}{2}}$

$$\begin{aligned} A_{NJ} &= \int_{\tau_{J-1}}^{\tau_J} (\tau_N - \overline{\tau}_o)^{-\frac{1}{2}} \exp[-X_{JA}{}^2/(\tau_N - \overline{\tau}_o)]\, d\overline{\tau}_o \\ &= 2\left[-(\tau_N - \tau_{J-1})^{\frac{1}{2}}[\pi^{\frac{1}{2}}\xi_{NJ1}\operatorname{erfc}(\xi_{NJ1}) - \exp(-\xi_{NJ1}{}^2)]\right. \\ &\quad \left. + (\tau_N - \tau_J)^{\frac{1}{2}}[\pi^{\frac{1}{2}}\xi_{NJ}\operatorname{erfc}(\xi_{NJ}) - \exp(-\xi_{NJ}{}^2)]\right] \end{aligned} \tag{21}$$

The current unknown appears only in the term A_{NN};

$$A_{NN} = -2(\Delta\tau_N)^{\frac{1}{2}}\{\pi^{1/2}\xi_N\operatorname{erfc}(\xi_N) - \exp(-\xi_N{}^2)\}; \quad \xi_N = X_{NA}/\Delta\tau_N^{\frac{1}{2}} \tag{22}$$

It is therefore appropriate to write eq. (20) in the form

$$X_N = X_{N-1} + \frac{\Delta\tau_N}{\beta} - \frac{\Delta\tau_N}{\beta A_{NN}}\{S_N - \sum_{J=1}^{N-1}[1 - \beta(X_J - X_{J-1})/\Delta\tau_J]A_{NJ}\} \tag{23a}$$

which is ready for iteration using for example as an initial guess for $x_N^{(1)}$ the previous value $x_N - 1$.

This form is not valid for β equal to zero and requires a small time step for small β . Care must be taken in evaluating the coefficients A_{NJ} since, as the argument $X_{JA}/(\Delta\tau_J)^{1/2}$ becomes small, they represent small differences, i.e.

$$\sqrt{\pi}\, z \operatorname{erfc} z - \exp(-z^2) \simeq \exp(-z^2) \sum_{m=1}^{\infty} (-1)^m \frac{1.3\ldots(2m-1)}{(2z^2)^m}$$

EXTENSIONS TO VARIABLE INITIAL TEMPERATURE

The assumption of a constant initial temperature is not particularly restrictive since many "half space" problems would be expected to begin from such an equilibrium state. However, the extension to a variable initial temperature field may be readily accomplished either numerically or by direct integration for a number of specific cases. The integral in question,

$$I(t) = I_m(t) = [a^2\pi t]^{-\frac{1}{2}} \int_0^\infty T_I\,(x_o)\exp[-x_o{}^2/4a^2t]dX_o \tag{24}$$

is the same form for pre- and for post-melting. It is appropriate to put this integral into nondimensional form

$$\bar{I}(\bar{t}) = I(t)/T_o = 2(\pi\bar{t})^{-\frac{1}{2}} \int_0^\infty \bar{T}_I(\bar{x}_o)\, \exp[-\bar{x}_o^2/\bar{t}_o]d\bar{X}_o \tag{25}$$

There are a number of forms for $\bar{T}_I$ for which this integral may be found analytically; furthermore it may be expressed as a Laplace transform, and finally, when all else fails, it may be evaluated numerically. This is not meant to denigrate the efficient numerical integration routines available, but to point out that general programs written for all possible inputs, e.g. $\bar{T}_I$, may overlook some useful savings in computation. For example,

$$\bar{T}_I\,(\bar{x}) = 1\ ;\ \bar{I}(\bar{t}) = 1 \tag{26-a}$$

$$\bar{T}_I\,(\bar{x}) = \bar{T}_1 + \bar{T}_2\,\exp(-\nu\bar{x});\ \bar{I}(\bar{t}) = \bar{T}_1 + \bar{T}_2\,\exp\,(\nu^2\bar{t}/4)\ \operatorname{erfc}\,(\nu\bar{t}^{\frac{1}{2}}/2) \tag{26-b}$$

$$\bar{T}_I\,(\bar{x}) = \bar{T}_1 + \bar{T}_2\,\exp\,(-\nu\bar{x}^2);\ \bar{I}(\bar{t}) = \bar{T}_1 + \bar{T}_2(1+\nu\bar{t})^{-\frac{1}{2}} \tag{26-c}$$

For other forms of $\bar{T}_I$, it may be possible to use Laplace transform tables to determine $\bar{I}(\bar{t})$, i.e.

$$\bar{I}(\bar{t}) = (\pi\bar{t})^{-\frac{1}{2}}\, L\,\{\eta_o{}^{-\frac{1}{2}}\,\bar{T}_I\,(\eta_o{}^{+\frac{1}{2}})\} \tag{27}$$

where L { } represents a Laplace transform with respect to the variable $\eta_o = x_o^2$ using a transform parameter s equal to $(\bar{t})^{-1}$.

Finally, if these approaches are inappropriate, the numerical evaluation of this integral must be included in the solution program for x*(t).

EXTENSION TO VARIABLE SURFACE HEAT FLUX

Consider the surface heat flux to be q(t) equal to $q_o f(t)$ where q_o is an appropriate measure of heat flux and f(t) is nondimensional. Although this might appear to present some mathematical difficulties, the modification to the original solution procedure is actually quite straightforward. For premelting, the surface flux integral in nondimensional form is

$$\overline{II}(\bar{t}) = (\alpha/T_o)II(\bar{t}) = \int_o^{\bar{t}} f(\bar{t}_o)(\bar{t}-\bar{t}_o)^{-\frac{1}{2}}d\bar{t}_o \tag{28}$$

which is integrable for a wide variety of forms for f(t). For example,

$$f(\bar{t}) = 1; \quad \overline{II}(t) = 2(\bar{t})^{\frac{1}{2}} \tag{29-a}$$

$$f(\bar{t}) = 1 + \mu\bar{t}; \quad \overline{II}\ (\bar{t}) = 2\ (\bar{t})^{\frac{1}{2}} + (4\mu/3)\ (\bar{t})^{3/2} \tag{29-b}$$

$$f(\bar{t}) = \exp[\mu\bar{t}]; \quad \overline{II}\ (\bar{t}) = (\pi/\mu)^{\frac{1}{2}} \exp[\mu\bar{t}]\ \mathrm{erf}\ [(\mu t)^{-\frac{1}{2}}] \tag{29-c}$$

Other forms are also integrable analytically as well. For the rest, a numerical integration routine will evaluate $\overline{II}$(E) readily.

The time of initial melting is given, implicitly at least, by eq. (7) or, in nondimensional form, by

$$\alpha\ \bar{I}\ (\bar{t}_m) + \overline{II}\ (\bar{t}_m) = 0 \tag{30}$$

Finally the postmelting surface heat flux integrals become, in nondimensional form,

$$\overline{II}_m(\bar{t}) = (\alpha/T_o)II(t) = \int_o^{\bar{t}_m} f(\bar{t}_o)(\bar{t}-\bar{t}_o)^{-\frac{1}{2}}d\bar{t}_o \tag{31}$$

$$\overline{III}_m(\bar{t}) = (\alpha/T_o)III(t) = \int_{\bar{t}_m}^{\bar{t}} [f(\bar{t}_o)-\beta d\bar{X}_o^*/d\bar{t}_o](\bar{t}-\bar{t}_o)^{-\frac{1}{2}} \cdot \exp[-\bar{X}_o^{*\,2}/(\bar{t}-\bar{t}_o)]d\bar{t}_o \tag{32}$$

The first of these may be treated much as $\overline{II}(\bar{t})$ with a modified upper limit, e.g.

$$f(\bar{t}) = 1;\ \overline{II}_m(\bar{t}) = 2(\bar{t})^{\frac{1}{2}} - 2(\bar{t}-\bar{t}_m)^{\frac{1}{2}} \tag{33-a}$$

$$\begin{aligned} f(\bar{t}) = 1 + \mu\bar{t};\ \overline{II}_m(\bar{t})^{\frac{1}{2}} - 2(\bar{t}-\bar{t}_m)^{\frac{1}{2}} \\ + (4\mu/3)\ (\bar{t})^{3/2} - 2\mu\bar{t}\ (\bar{t}-\bar{t}_m)^{\frac{1}{2}} + (2\mu/3)(\bar{t}-\bar{t}_m)^{3/2} \end{aligned} \tag{33-b}$$

$$f(\bar{t}) = \exp[\mu\bar{t}];\ \overline{II}_m(\bar{t}) = 2\mu^{\frac{1}{2}} \exp[\mu\bar{t}] \tag{33-c}$$

etc.

The final integral must of course be done numerically as in the constant surface heat flux and initial temperature case discussed in section III. However, if the same approximation is made for f(t) as was made for x(t), e.g.

$$f(\bar{t}) = F_{JA} = (F_J + F_{J-1})/2\ ;\ \tau_{J-1} < \bar{\tau} < \dot{\tau}_J \tag{34}$$

there is only a slight change in the form of $\overline{IIIm}(\bar{t})$, i.e.

$$\overline{III}_m(\bar{t}) = \sum_{J=1}^{N} [F_{JA} - \beta(X_J - X_{J-1})/\Delta\tau_J]\ A_{NJ} \tag{35}$$

SOLUTION ROUTINE FOR GENERAL $T_I(x)$ AND q(t)

The premelting phase up to (and defining $\bar{t}_m$) is governed by a single parameter α and the two shape factors, $\bar{T}_I(\bar{x})$ and $f(\bar{t})$. The premelting surface temperature is

$$\bar{T}\ (o,\bar{t}) = \bar{I}(\bar{t}) + \overline{II}(\bar{t})/\alpha \tag{36}$$

and the time of initial melting is, implicitly, given by

$$0 = \bar{I}(\bar{t}_m) + \overline{II}(\bar{t}_m)/\alpha \tag{37}$$

The post-melting solution requires

$$0 = \alpha\bar{I}_m(\bar{t}) + \overline{II}_m(\bar{t}) + \overline{III}_m(\bar{t}) = S(\bar{t}) + III_m(\bar{t}) \tag{38}$$

where $S(\bar{t})$ is independent of $\bar{x}_o^*$. Using the previous piecewise constant approximation for $\bar{x}_o{}^*$ (and f) leads to

$$X_N = X_{N-1} + F_{NA}\,\Delta\tau_N/\beta - (\Delta\tau_N/(\beta A_{NN})).$$

$$\{S_N - \sum_{J=1}^{N-1} [F_{JA} - \beta\,(X_J - X_{J-1})/\Delta\tau_J]\,A_{NJ}\} \qquad (39)$$

The program used to solve for X_n has subprograms to calculate $\bar{t}_m$, $T(o,\bar{t})$, S_N and A_{NJ} for any given form for $\bar{T}_I(\bar{x})$ and $f(\bar{t})$ and then solve eq. (39) by iteration at each new time step. It is convenient to choose ℓ to have $\bar{t}_m$ equal to 1 in the constant flux case, i.e. α equal to -2 or ℓ equal to $-kT_o\Pi^{1/2}/q_o$. Then β becomes $(4/\Pi^{1/2})\,(-\lambda/cT_o)$ which is $+2/m$ as m is defined by Landau (1950), eq. 3.6. This parameter represents a ratio of heat content change due to ablation to that due to a change of temperature from T_o to melting. Alternatively, ℓ could be chosen as $-2\,k\,T_o/q_o$ such that α is $-\pi^{1/2}$, t_m is $\pi/4$ and β is $2\,\lambda/c\,T_o$ which is 2ν as ν is defined by Zien (1978).

ASYMPTOTIC SOLUTION

As time becomes large, the ablation front for a uniform surface heat flux and uniform initial temperature is expected to move at a steady rate which, on physical grounds, e.g. Landau (1950) would be given by

$$(dx^*/dt)_{ASY} = qo/[\rho\lambda + \rho c\,(-T_o)] \qquad (40)$$

The present formulation for large $\bar{\tau}$ should then lead to an asymptotic solution $\bar{x}_o^* \sim A\bar{\tau}_o$

where A is given by

$$0 \sim \alpha + \bar{t}_m/\bar{\tau}^{1/2} + .. + (1-\beta A) \int_{\lambda=o}^{\lambda=\bar{\tau}^{1/2}} (1-\lambda/\bar{\tau}^{1/2})^{-1/2} \qquad (41)$$

$$.\exp\,[-A^2\lambda^2/(1-\lambda/\bar{\tau}^{1/2})]d\lambda$$

As $\bar{\tau}$ approaches ∞, this equation yields $o \simeq \alpha + (1-\beta A).\sqrt{\pi}/2A$

or

$$A = (\beta - 2\alpha/\sqrt{\pi})^{-1} \qquad (42)$$

which is the nondimensional form of the asymptotic rate given above. The same form will hold for variable initial temperatures that go to a constant asymptotic value, T_1, in space provided that this value is used in place of T_o in the forms for α and β, i.e.

$$(dx^*/dt)_{ASY} = qo/[p\lambda + pc(-T_1)] \tag{43}$$

NUMERICAL RESULTS

Values for the location of and the speed of the ablating front are given in Fig. 1 for the simple case of a uniform initial temperature and a uniform surface heat flux with $\alpha = -2$, $\beta = +2$ and $\beta = 1$ to correspond to the solution of Landau (1950) with m = 1 and 2 respectively. A comparison of solutions indicates very good agreement; the variables are $u = (p\lambda/qo)dxo^*/d\tau$ (a front speed) and $y = \tau/t_m$. Similarly, the approximate solutions of Zien (1978) are in good agreement with these as well as Landau's results for these cases, but do not extend far enough in time to adequately represent the asymptotic behavior of the speed of the front.

Some care must be taken when calculating values near the asymptote; as mentioned in section III, the coefficients A_{NJ} become small - especially the critical coefficient A_{NN} which changes values during the iterative procedure used to find x_o^*. The overall behavior of the solution is best seen on a semi-logarithmic plot, indicating that a reasonable solution procedure would allow the time step to increase as time increases, e.g. taking $\Delta\tau_N$ equal to some constant multiplied by the previous time, τ_{N-1}. Results in Fig. 1 used a multiplier of .15 with an initial time step of $\Delta_1 = 0.02$, and an instability arose for $y = \tau/tm$ about 10.0, due primarily to a small value for A_{NN}.

CONCLUSION

The value of this work is not meant to be in the actual numerical results given nor is it meant to merely show that boundary integral/element methods may be used in ablation problems since that has already been established by Chuang and Szekeley (1971) for the analogous melting problem. This work is rather to place emphasis on the "boundary" part of this problem since the location of the free boundary is after all of paramount importance and to illustrate the very straightforward in which this boundary may be found. Such an approach is readily extendable to multi-dimensional ablation problems as well as the corresponding melting/solidification situations. Furthermore, the inclusion of a variable surface heat influx input and/or a variable initial temperature field can be readily written in the solution routine as a separate, clearly defined integral term, thereby making the addition of these effects a very straightforward step. In summary, the point of this paper is to present the ease with which the BIE method may be used to solve this general class of problems and to encourage its use for these solutions.

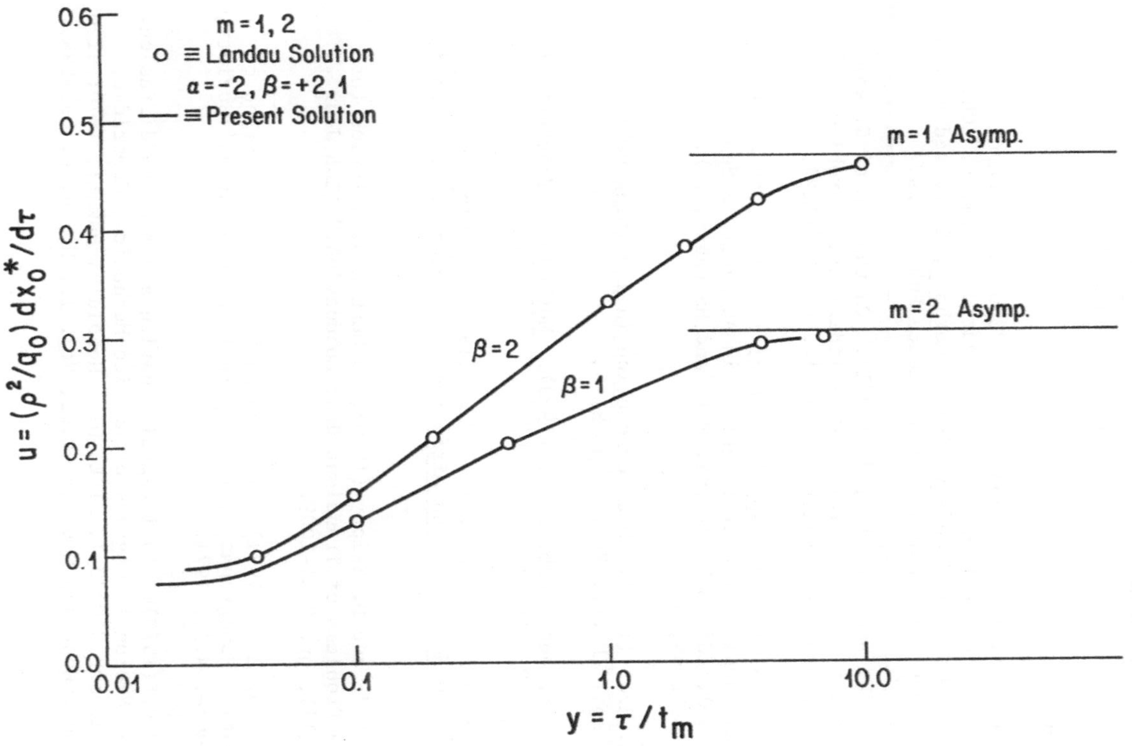

Fig. 1: Speed of Ablation Front for Uniform Initial Temperature and Surface Heat Flux

ACKNOWLEDGMENT

This research has been supported by the Office of Naval Research - Structural Mechanics - under contract N 00014-75-C-0302 and distribution is unlimited.

REFERENCES

Banerjee, P.K. and R.P. Shaw (1981), "Boundary Element Formulation for Melting and Solidification Problems," Chap. 1 in Developments in Boundary Element Methods, 2, edited by P.K. Banerjee and R.P. Shaw, Applied Science Publishers.

Chuang, Y.K. and Szekeley, J. (1971). On the use of Greens function for solving melting and solidification problems. Int. Jour. Heat and Mass Transfer, Vol. 14, p. 1285-1294.

Goodman, T.R. (1964). Application of integral methods to transient nonlinear heat transfer, in Advances in Heat Transfer-I, Edited by Irvine, T.F. and Hartnett, J.P., Academic Press, N.Y.

Landau, H.G. (1950). Heat conduction in a melting solid. Quart. Appl. Math., Vol. 8, p. 81-94.

Morse, P.M. and Feshback, H. (1953). Methods of Theoretical Physics, McGraw Hill, New York.

Ockendon, J.R. and Hodgekins, W.R. (1975). Moving Boundary Problems in Heat Flow and Diffusion. Claredon Press, Oxford, England.

Rizzo, F.J. and D. Skippy (1970), "A Method of Solution for Certain Problems of Transient Heat Conduction," AIAA Jour., Vol. 8(11), pp. 2004-2009.

Rubinstein, L.I. (1971). The Stefan Problem, Translation of Mathematics Monographs, Vol. 27, American Mathematical Society, Providence, R.I., 1971.

Shaw, R.P. (1970). An integral equation approach to diffusion. 6th U.S. National Congress of Applied Mechanics, Cambridge, Mass.; Report HIG-70-I, JTRE, HIG, Honolulu, Hawaii, 1970; also Int. Jour. Heat and Mass Transfer, Vol. 17, No. 6, p. 693-699, 1974.

Wilson, D.G., Solomon, A.D. and Boggs, P.T. (Editors) (1978). Moving boundary problems, Academic Press, New York.

Zien, T.F. (1978), "Integral Solutions of Ablation Problems with Time-Dependent Heat Flux," AIAA Jour., Vol. 16(12), pp. 1287-1295.

BOUNDARY ELEMENT METHOD IN SINGULAR AND NONLINEAR HEAT TRANSFER

K. Onishi and T. Kuroki

Applied Mathematics Dept., and Civil Engineering Dept., Fukuoka University, Fukuoka 814-01, Japan.

INTRODUCTION

This paper is concerned with three advanced topics: (1) treatment of singularities in BE solution, (2) heat equation subject to nonlinear boundary conditions, and (3) temperature dependent nonlinear heat conduction. Specific examples are presented in applications using the method of subregions and the double point technique developed for BEM. Iterative BE solution to nonlinear problems are discussed.

Various types of singularities are encountered in the solution of transient heat conduction problems. Boundary singularities often arise due to ill-behaved initial and boundary conditions imposed or due to irregular geometry of the material under consideration. Difficulties involving corners are called corner problems by Brebbia[1978]. We shall consider an example involving two re-entrant corners.

Heggs et al.[1981] considered BIEM for steady heat transfer in an L-shaped domain. Their discussion was essentially similar to the method of subregions. They treated singularity at a re-entrant corner.

The BE solution to the heat equation subject to Newton's superficial cooling was obtained by Wrobel and Brebbia [1981]. In practice, the surface is subjected to the convection and radiation simultaneously. We shall consider an example with simultaneous forced convection and Stefan-Boltzmann's radiation boundary conditions.

Heat conduction with high temperatures becomes nonlinear because of the temperature dependent thermal coefficients. The simple iterative scheme by Butterfield[1978] is extended to the solution of nonlinear steady problems. We shall show the convergence of the scheme by an example of concentration dependent steady diffusion equation.

Banerjee[1979] considered BE formulation to nonlinear potential flow problems. He split a linear part from material nonlinearities. Khader and Hanna [1981] used Kirchhoff's transformation for the iterative BE solution to nonlinear steady heat conduction.

The method of subregions introduced in BE formulation by Brebbia[1978] was called by other names, e.g., zoned body procedure by Butterfield[1978], or multidomain boundary element formulation by Blandford[1981].

The double points were defined due to Brebbia[1978] as two adjacent nodes having the same spatial coordinates with different numberings.

As the weighting function, time dependent fundamental solution is used for transient problems. Constant boundary elements are used for steady-state problems, and linear boundary elements are used for transient problems. Convergence and accuracy of the numerical solution are discussed for two-dimensional problems.

BOUNDARY ELEMENT APPROXIMATION --- PRELIMINARIES

Let Ω be a two-dimensional domain enclosed by the boundary Γ . The domain is supposed to be a heat conducting medium. We denote spatial variables by x_i (i =1,2, in meters) and a time variable by t (in seconds).

The conservation of thermal energy without heat generation in the body is expressed by

$$\rho c \frac{\partial u}{\partial t} + \frac{\partial q_i}{\partial x_i} = 0 \qquad (1)$$

where ρ is the mass density (in kilograms per cubic meter), c is the specific heat capacity (in joules per kilogram, degrees Kelvin), u is the temperature (in degrees Celsius), and q_i is the heat flux component (in joules per square meter, second) in the i-th direction.

Assume that the heat flux components obey Fourier's law of heat conduction written by

$$q_i = - k \frac{\partial u}{\partial x_i} \tag{2}$$

For the sake of simplicity, assume that ρ , c, k are constant in this section. Consider the fundamental solution v* satisfying

$$- \rho c \frac{\partial v^*}{\partial t} + \frac{\partial q_i^*}{\partial x_i} = \rho c \delta(P_i) \delta(\tau) \tag{3}$$

with the instantaneous point heat charge of the intensity ρc at an arbitrary spatial point P_i at any time τ . The heat flux component associated with v* is defined by $q_i^* = - k \, \partial v^*/\partial x_i$.

From Gauss-Green's identity and following the usual limiting process, we have the boundary integral equation

$$w(P_i) \rho c u(P_i, t_k) - \int_\Gamma d\Gamma \int_{t_{k-1}}^{t_k} u q^* dt = - \int_\Gamma d\Gamma \int_{t_{k-1}}^{t_k} q v^* dt + \int_\Omega \rho c u^{k-1} v^* d\Omega \tag{4}$$

where t_k = break points in time (k=1,2,...) for which $\Delta t = t_k - t_{k-1}$ is the time increment. The weight $w(P_i)$ is determined according to the geometry of the domain in the vicinity of the point P_i. Here q denotes the boundary flux in the outward normal direction with its components n_i and we define $q^* = q_i n_i$.

The boundary Γ is divided into small line segments called boundary elements. The base domain Ω is divided into finite elements or cells. The cell division is required only for convenience to perform numerical integrations over the domain.

The cell division need not be consistent with the boundary element subdivision.

By interpolating the temperature u and the flux q along the boundary in terms of corresponding nodal values using linear shape functions, we can obtain the linear system of boundary element equations in the matrix form

$$[\, H \,]\, \{\, u^k \,\}_\Gamma = [\, G \,]\, \{\, q^k \,\}_\Gamma + \{\, b(u_\Omega^{k-1}) \,\} \quad (5)$$

in which $\{u^k\}$ and $\{q^k\}$ are column vectors with components u_j^k and q_j^k on the boundary. These are the approximations to the exact temperature $u(P_j, t_k)$ and to the exact flux $q(P_j, t_k)$, respectively. The b-column vector on the right side can be evaluated from temperatures at internal node points. If the time increment is kept constant, the coefficient matrices are independent on the time step indicator k.

Equation (5) is a time marching scheme. At the end of each time step, temperatures at internal points must be calculated in order to continue the time stepping procedure. The approximate temperatures at internal points can be calculated directly from Equation (4). Heat flux components at any internal points can be calculated also from an integral equation by using known boundary temperatures and fluxes.

SINGULARITIES

Consider an example having two re-entrant corners. Figure 1 shows boundary element and cell meshes. Mixed boundary conditions are indicated together with the initial condition.

Since the domain is not convex, we must divide it into convex subregions in order to retain the accuracy. The convex subdivision is not unique. The domain was subdivided into 5 rectangules. The mesh has 136 boundary nodes, 115 internal nodes.

Bell and Crank [1973] solved this problem by incorporating the analytical expression of singularities into FDM. Bruch and Zyvoloski [1974] solved the same problem using FEM. They obtained fairly good agreement between FDM and FEM results. Bruch and Lewis [1975] reconsidered the problem by the least squares.

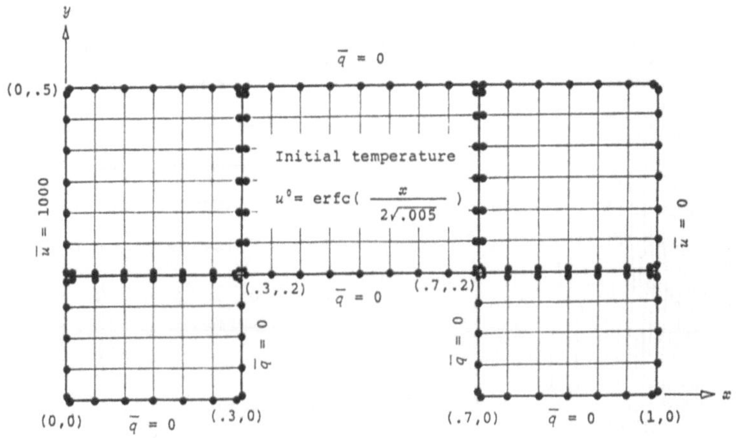

Figure 1. Pi-shaped domain and boundary conditions imposed. Element meshes and subregions are shown.

Bell and Crank took very small time increment of order $\Delta t = 0.0005$ for stability, because they used the explicit FD scheme. Bruch and Zyvoloski took the same time increment. Here we take $\Delta t = 0.005$, ten times larger than that, on account of the fact that the present BE scheme is unconditionally stable.

Figure 2 shows calculated isotherms (a) and heat fluxes (b) for a small time. In spite of the steep gradient near the left side, the calculated temperature is not oscillatory. The temperature is smooth, and it remains positive.

Figure 3 shows calculated results at the Fourier's number t = 0.1 . Due to the antisymmetry of the problem, the exact isotherm of u = 500 must run on the vertical center of the domain in the ultimate steady-state. The calculated isotherm of u = 500 is approaching to the vertical center line. Isotherms intersect the adiabatic boundaries with right angles. This suggests the high accuracy of the calculated temperatures.

Calculated boundary fluxes in outward normal directions are depicted along with fluxes on interfaces between subregions. We can observe very large fluxes at re-entrant corners.

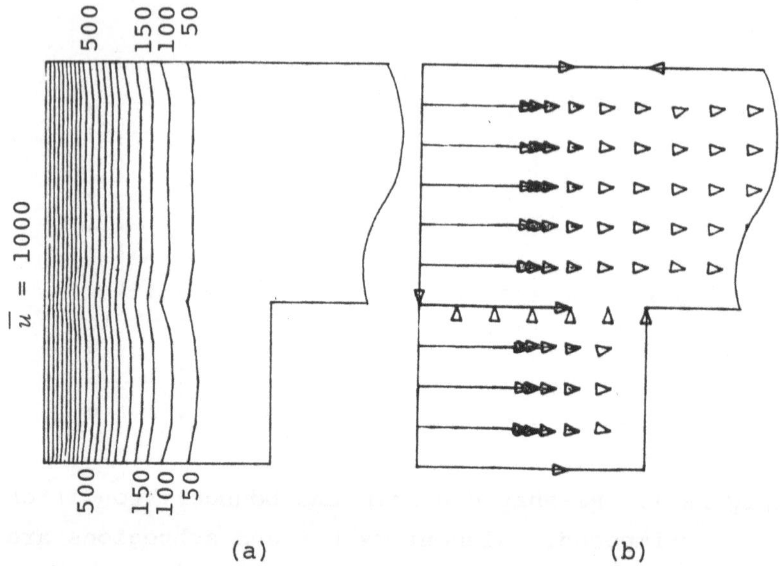

Figure 2. Calculated isotherms (a) and heat fluxes (b) at the time $t = \Delta t = 0.005$. The method of subregions is used.

If we tentatively calculate the solution of the same problem not using the method of subregions, we will no longer hold the positivity of the temperature, and we will not obtain large fluxes at re-entrant corners.

NONLINEAR BOUNDARY CONDITIONS

Let us consider the linear heat Equations (1) and (2) in an infinite slab with its thickness 2L as shown in Figure 4.

If both convection and radiaton are the same order of magnitude, we have to take boundary convection and radiation simultaneously into account in the form

$$- k \frac{\partial u}{\partial n} = h (u - \bar{u}_C) + \sigma E (u^4 - \bar{u}_R^4) \qquad (6)$$

where h is the heat transfer coefficient (in joules per square meter, second, degrees Kelvin), σ is the Stefan-Boltzmann constant, and E is the emissivity. Here $\bar{u}_C$ denotes the ambient fluid temperature in forced convection, and $\bar{u}_R$ denotes the temperature of the radiation source.

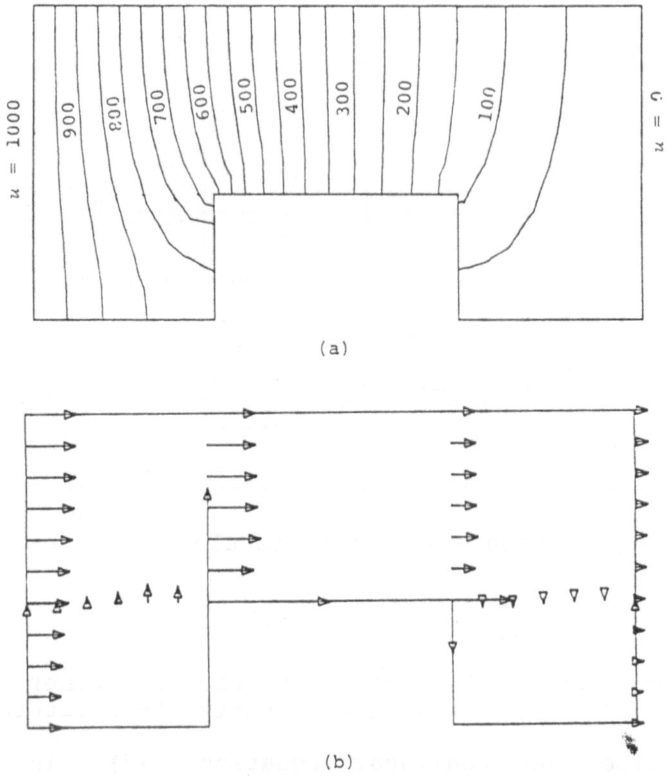

Figure 3. Calculated isotherms (a) and heat fluxes (b) at the time $t = 0.1$ ($\Delta t = 0.005$). The method of subregions is used.

We assume that $\bar{u}_C = \bar{u}_R = 0$ for the sake of simplicity. The dimensionless form associated with Equation (6) is

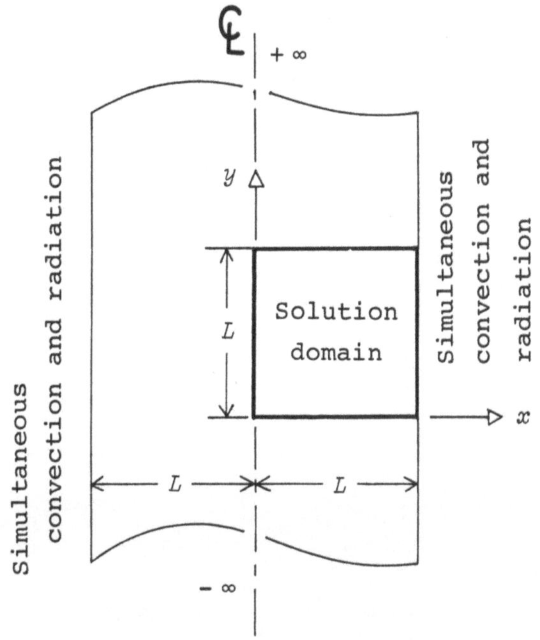

Figure 4. Infinite slab

$$-\frac{\partial u}{\partial n} = B_i \cdot u + SB \cdot u^4 \tag{7}$$

where $B_i = hL/k$ represents Biot's number, and $SB = \sigma E(u^0)^3 L/k$ for a representative temperature u^0.

We write the nonlinear Equation (7) in the pseudo-linear form

$$-\frac{\partial u}{\partial n} = H(u) \cdot u \tag{8}$$

having the temperature dependent heat transfer coefficient H defined by $H(u) = B_i + SB \cdot u^3$.

By substituting Equation (8) into Equation (5), we obtain the nonlinear system of boundary element equations with temperature dependent coefficients. The system can be written formally by

$$[H(u_\Gamma^k)]\{u^k\}_\Gamma = [G(u_\Gamma^k)]\{q^k\}_\Gamma + \{b(u_\Omega^{k-1})\} \tag{9}$$

In order to solve this nonlinear system, the method of iterative refinement is resorted to. Let us denote the m-th refinement by ${}^{m}u$. For m=1,2,3,...., the linear system of equations

$$[H({}^{m-1}u_{\Gamma}^{k})]\{{}^{m}u^{k}\}_{\Gamma} = [G({}^{m-1}u_{\Gamma}^{k})]\{{}^{m}q^{k}\}_{\Gamma} + \{b(u_{\Omega}^{k-1})\} \qquad (10)$$

is iterated until the convergence is attained. This process is called the inner iteration. Once the approximate limit is obtained at m = M, we then set $\{u^{k}\}_{\Gamma} = \{{}^{M}u^{k}\}_{\Gamma}$ and $\{q^{k}\}_{\Gamma} = \{{}^{M}q^{k}\}_{\Gamma}$ at the k-th time step.

We calculated the transient temperatures over the period, 0<t<2, with the time increment Δt = 1/128. The uniform initial temperature u^{0} = 1000 was assumed. Eight times inner iteration was necessary only for the initial step. No iterations were used thereafter. Boundary element and cell meshes are shown in Figure 5. The dimensionless constants were B_i = 4 and SB = 4.

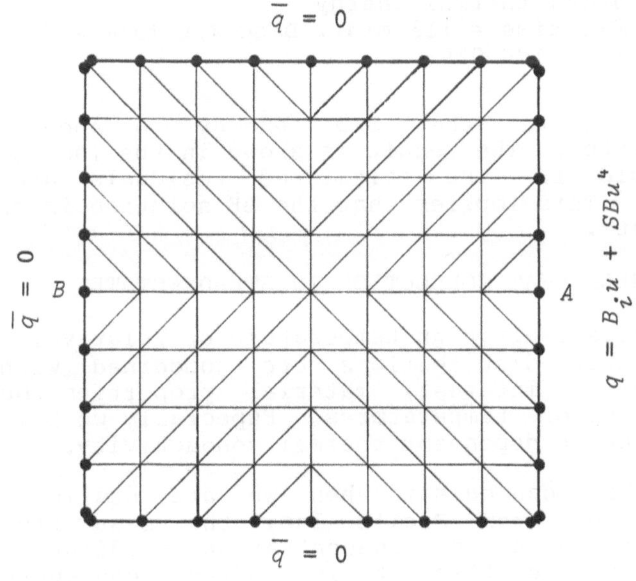

Figure 5. Boundary elements and cell mesh

Calculated results are summarized in Table 1. Figure 6 shows the calculated temperature plots against the dimensionless time. The exact solution is not available. The BE solution is compared with FE ADINAT solution by Bathe[1978].

Table 1 Calculated temperatures for simultaneous boundary convection and radiation.

Step No.	Time	Node A	Node B	T.E.*	Residual
1	.0078125	661.4	1000.0	965.5	.213E-01
2	.015625	549.4	1000.0	938.4	.156E-01
4	.03125	468.0	1000.0	896.3	.338E-02
8	.0625	384.5	993.3	829.3	.289E-02
16	.125	304.5	930.9	724.2	.267E-02
32	.25	227.5	747.3	564.0	.260E-02
64	.5	139.6	459.8	346.1	.258E-02
128	1.0	52.8	173.6	130.7	.258E-02
256	2.0	7.5	24.8	18.7	.258E-02

* Total thermal energy

** CPU time = 113 min., Disc I/O time = 13 min. on PRIME 250.

Residuals in the I/O energy balance were calculated. The result is shown in the last column of Table 1. The calculated residuals are very small. This implies that the BE solution is highly accurate.

NONLINEAR HEAT CONDUCTION IN STEADY-STATE

If a relatively high temperature or relatively low temperature distributions are concerned with the problem of interest, material properties become functions of temperatures. Especially we have the temperature dependent thermal conductivity.

Same thing can be said when we are dealing with diffusion. From Fick's law, the mass flux is proportional to the concentration gradient. The diffusion coefficient is often concentration dependent.

Let u denote the temperature as before. We consider the Fourier's law

$$q_i = - k(u) \frac{\partial u}{\partial x_i} \tag{11}$$

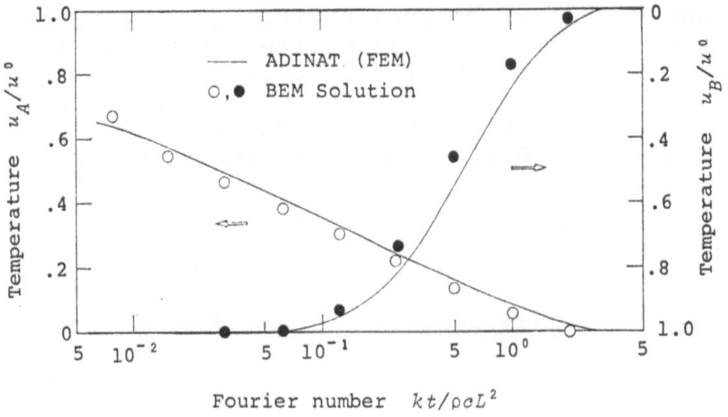

Figure 6. Calculated temperatures for simultaneous convection and radiation (B_i = 4.0 , SB = 4.0)

with the temperature dependent thermal conductivity. Here we assumed that the conductivity is independent on the orientation and spatial positions explicitely.

From the usual procedure, we can derive the boundary integral equation

$$(cku)_i + \int_\Gamma ku \frac{\partial v^*}{\partial n} d\Gamma = - \int_\Gamma qv^* d\Gamma + \int_\Omega \frac{\partial k}{\partial x_i} u \frac{\partial v^*}{\partial x_i} d\Omega \tag{12}$$

By the BE discretization, we can obtain the system of boundary integral equations in the form

$$[\ H\ ]\ \{\ k(u)u\ \}_\Gamma = [\ G\ ]\ \{\ q\ \}_\Gamma + \{\ b(u_\Omega)\ \} \tag{13}$$

Note that the product k(u)u was chosen as the primitive unknown.

Consider an infinite plane sheet shown in Figure 7. In the context of diffusion, the variable u denotes the concentration, and is written by the letter C. The concentration on the left side of the sheet is kept constant as $C = C_0$. The concentration on the other side is kept C = 0. Under fully developed condition, the diffusion is in steady-state.

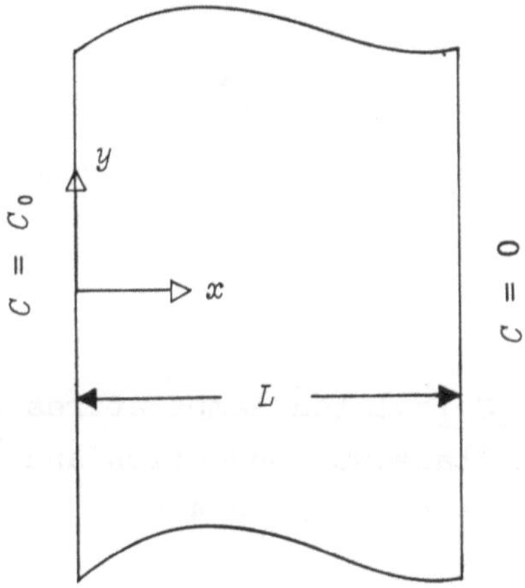

Figure 7. Infinite membrane

For the computational purpose, we considered concentration dependent diffusivities in two forms

$$k(u) = k_0 (1 + 100 \cdot u) \tag{14}$$

and

$$k(u) = k_0 (1 - u) \quad \text{for which} \quad 0 < u < 1 \tag{15}$$

for some constant diffusivity k_0.

A square domain was subdivided into boundary elements and cells as shown in Figure 8. There are 16 boundary nodes in the coarse mesh, and 32 bounary nodes in the fine mesh.

A simple iteration technique is employed for the solution of Equation (13). The solution to linear equation was used as the initial estimate. Four iterations were required for the convergence. Calculated concentration profiles along the x-axis are ploted in Figure 9. Exact solutions found in Crank[1975] are represented by solid curves. We can see that the BE solutions are highly accurate

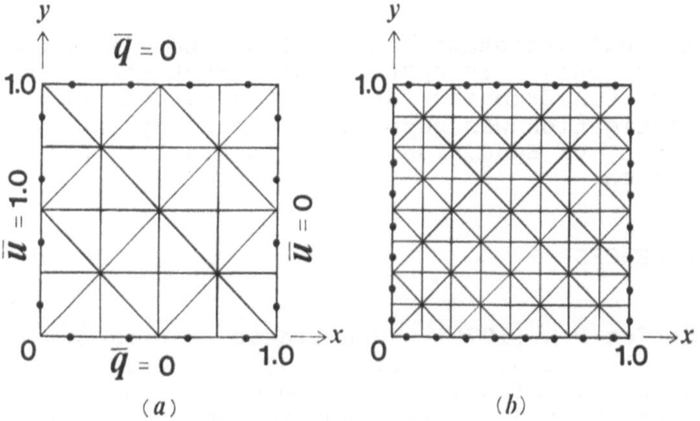

Figure 8. Boundary element meshes and cells.
(a) Coarse mesh , (b) Fine mesh

k / k_0	Meshes Coarse	Fine
$1 + 100\,u$	●	○
$1 - u$	▲	△

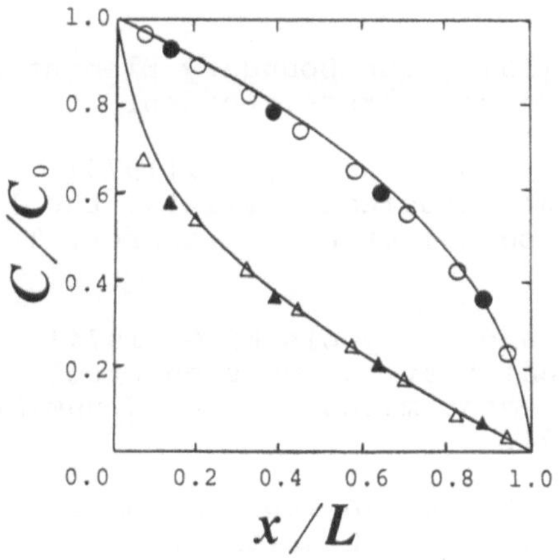

Figure 9. Calculated concentration profiles.

even for the coarse mesh.

ACKNOWLEDGMENT

Technical assistance of T.Ito and K.Obata in Fukuoka Univ. is appreciated. Much thanks are due to Dr. C.A.Brebbia for his encouragement. This investigation was supported in part by The Japanese Ministry of Education, Grant in Aid for Scientific Research.

REFERENCES

Banerjee,P.K.[1979] Non-linear problems of potential flow. Ch.2: 21-30, in P.K.Banerjee and R.Butterfield(Eds.) Developments in Boundary Element Methods -- 1. Applied Science Publishers. London.

Bathe,K.J.[1978] ADINAT, A Finite Element Program for Automatic Dynamic Incremental Nonlinear Analysis of Temperatures. Rev. Report 82448-5, Mechanical Engineering Dept., MIT.

Bell,G.E. and Crank,J.[1973] A method of treating boundary singularities in time-dependent problems. J.Inst.Maths and its Applics., 12: 37-48.

Blandford,G.E.[1981] Two-dimensional stress intensity factor computations using the boundary element methods. Int.J.num.Meth.Engng., 17, 3: 387-404.

Brebbia,C.A.[1978] The Boundary Element Method for Engineers. Pentech Press, London.

Bruch,J.C. and Lewis,R.W.[1975] Transient two-dimensional problems utilizing the least square algorithm. Journal of Heat Transfer, Trans. ASME., 8: 467-469.

Bruch,J.C. and Zyvoloski,G.[1974] Transient two-dimensional heat conduction problems solved by the finite element method. Int.J.num.Meth.Engng., 8, 3: 481-494.

Butterfield,R.[1978] An application of the boundary element method to potential flow problems in generally inhomogeneous bodies: 123-135, in C.A.Brebbia(Ed.) Recent Advances in Boundary Element Methods. Pentech Press, London.

Crank,J.[1975] The Mathematics of Diffusion. Second ed.: 161, Clarendon Press, Oxford.

Heggs,P.J., Ingham,D.B., and Manzoor,M.[1981] Boundary integral equation analysis of fin assembly heat transfer. Numerical Heat Transfer, 4: 285-301.

Khader,M.S. and Hanna,M.C.[1981] An iterative boundary integral numerical solution for general steady heat conduction problems. Journal of Heat Transfer, Trans. ASME., 103, 2: 26-31.

Wrobel,L.C. and Brebbia,C.A.[1981] Boundary elements in thermal problems. Ch. 5: 91-113, in R.W.Lewis, K.Morgan, and O.C.Zienkiewicz (Eds.) Numerical Methods in Heat Transfer. John Wiley & Sons, Chichester.

Wrobel,L.C. and Brebbia.C.A.[1981] A formulation of the boundary element method for axisymmetric transient heat conduction. Int.J.Heat Mass Transfer, 24, 5: 843-850.

UNSTEADY HEAT CONDUCTION USING THE BOUNDARY ELEMENT METHOD

J.L.M. Fernandes and H.L.G. Pina,
Instituto Superior Técnico, Lisbon.

ABSTRACT

The unsteady heat equation is solved by the boundary element method using piecewise constant approximation in time. The influence of different numerical evaluation of the resulting integrals is assessed. We also, point out the necessity of a consistent choice of the time step and the spatial discretization.

1. FORMULATION OF HEAT CONDUCTION PROBLEMS

This paper deals with the solution of heat conduction problems. These problems can be stated as follows [CARSLAW and JAEGER, 1959]: to find a function u $(\underset{\sim}{x},t)$ satisfying the partial differential equation

$$\frac{\partial u}{\partial t} = \kappa \nabla^2 u \qquad \text{in } \Omega , \tag{1}$$

and the following boundary conditions,

$$u = \alpha \qquad \text{on } \Gamma_1 \text{ and } t>t_0 , \tag{2a}$$

$$\kappa \frac{\partial u}{\partial n} + \beta u = \gamma \qquad \text{on } \Gamma_2 \text{ and } t>t_0 , \tag{2b}$$

and the initial condition

$$u = u_0 \qquad \text{in } \Omega \text{ and } t=t_0 . \tag{2c}$$

Here Ω denotes a domain of the euclidean plane R^2, Γ is its boundary and Γ_1 and Γ_2 are complementary parts of Γ, i.e., $\Gamma_1 \cup \Gamma_2 = \Gamma$. The symbol $\frac{\partial}{\partial n}$ stands for the derivative along the exterior unit normal to Γ. Quantities present in equation (1-2) are identified in the context of heat conduction as:

u - temperature

κ - termal diffusivity (a positive constant)

$\alpha = \alpha(\underset{\sim}{x},t)$, $\beta = \beta(\underset{\sim}{x})$, $\gamma = \gamma(\underset{\sim}{x},t)$ are fields specifying the boundary conditions, with the restriction that $\beta \geq 0$. The function $u_0 = u_0(\underset{\sim}{x})$ delivers the value of the temperature field at (initial) instant $t=t_0$.

2. ASSOCIATED INTEGRAL EQUATIONS

It is already well established that the solution of the previous problem possesses the following integral representation, which has been used by several authors [BREBBIA and WALKER, 1980, CARSLAW and JAEGER, 1959, CHANG et.al., 1973, WROBEL (1981), etc.]

$$c(\underset{\sim}{x})u(\underset{\sim}{x},T)=\int_{\Omega} u_0(\underset{\sim}{y})g(\underset{\sim}{y},\underset{\sim}{x},T,t_0)\, d\,\Omega\,(\underset{\sim}{y})$$

$$+\int_{\Gamma}\int_{t_0}^{T} \kappa\left[g(\underset{\sim}{y},\underset{\sim}{x},T,\tau)\frac{\partial u}{\partial n}(\underset{\sim}{y},\tau)\right]d\tau\ \ d\Gamma(\underset{\sim}{y})$$

$$-\int_{\Gamma}\int_{t_0}^{T} \kappa\ \left[\frac{\partial g}{\partial n}(\underset{\sim}{y},\underset{\sim}{x},T,\tau)u(\underset{\sim}{y},\tau)\right]d\tau\ d\Gamma(\underset{\sim}{y}). \qquad (4)$$

In this expression g is the fundamental solution

$$g(\underset{\sim}{y},\underset{\sim}{x},t,\tau)=\frac{1}{4\pi\kappa(t-\tau)}\ \exp\left[-\frac{r^2}{4\kappa(t-\tau)}\right],\ t>\tau, \qquad (5)$$

Where r is the euclidean distance between points $\underset{\sim}{x}$ and $\underset{\sim}{y}$, i.e., $r=|\underset{\sim}{x}-\underset{\sim}{y}|$. Making use of well known properties of the fundamental solution (5) it can be proved that $c(\underset{\sim}{x})=1$ for interior points and for boundary points $\underset{\sim}{x}\in\Gamma$, $c(\underset{\sim}{x})$ equals the angle subtended by Γ at $\underset{\sim}{x}$, measured in units of 2π. If $\underset{\sim}{x}$ is a regular point of Γ then $c(\underset{\sim}{x})=1/2$.

The integral representation (4) can be used to compute the value of u at any interior point x and at any instant $t>t_0$, provided we know the functions $u(\underset{\sim}{y},\tau)$ and $\frac{\partial u}{\partial n}(\underset{\sim}{y},\tau)$ on the boundary. To determine this functions we set $\underset{\sim}{x}$ on the boundary, thus obtaining integral equations to be satisfied by u on Γ_2 and $\frac{\partial u}{\partial n}$ on Γ_1. Next we show how to solve these integral equations by appropriate discretization of the space-time domain.

3. DISCRETIZATION

The discretization of equation (4) involves the discretization of the space domain Ω together with its boundary Γ and the discretization of the time domain, $]t_0, T]$.

Discretization of the domain Ω and its boundary is done as usual by finite elements. For the time domain we use piecewise constant variations for the unknowns. Specifically, we make

$$\left.\begin{aligned} u(\underset{\sim}{x},t) &\simeq \sum_i u_{im}\,\psi_i(\underset{\sim}{x}) \\ \kappa\frac{\partial u}{\partial n}(\underset{\sim}{x},t) &\simeq \sum_i q_{im}\,\psi_i(\underset{\sim}{x}) \end{aligned}\right\}\ \text{for } t_{m-1}<t\leq t_m. \qquad (6)$$

The parameters u_{im} represent the nodal value of the temperature at node i and time t_m, and the parameter q_{im} represent the nodal values of the flux also at node i and time t_m. The functions $\psi_i(x)$ are the interpolation functions of the finite element mesh, therefore satisfying the relation

$$\psi_i(x_j) = \delta_{ij} \tag{7}$$

Because of its major computational advantages, which we shall make apparent below, we have chosen a constant time step, that is,

$$t_m - t_{m-1} = \Delta t = \text{constant and } t_m = t_0 + m\Delta t. \tag{8}$$

Introducing the approximations (6) in the integral representation (4) and collocating at the boundary nodes and at time t_1 (for instance) we obtain for each node the following relation,

$$\begin{aligned} c_i\, u_{i1} = & \int_\Omega u_0(\underset{\sim}{y})\, g(\underset{\sim}{y},\underset{\sim}{x}_i,t_1,t_0)\, d\Omega(\underset{\sim}{y}) \\ & + \sum_j \left[\int_\Gamma \left(\int_{t_0}^{t_1} g(\underset{\sim}{y},\underset{\sim}{x}_i,t_1,\tau)\, d\tau \right) \psi_j(\underset{\sim}{y})\, d\Gamma(\underset{\sim}{y}) \right] q_{j1} \\ & + \sum_j \left[\int_\Gamma \left(\int_{t_0}^{t_1} \kappa \frac{\partial g}{\partial n}(\underset{\sim}{y},\underset{\sim}{x}_i,t_1,\tau)\, d\tau \right) \psi_j(\underset{\sim}{y})\, d\Gamma(\underset{\sim}{y}) \right] u_{j1} \end{aligned} \tag{9}$$

The time integrals can be evaluated analytically (WROBEL, 1981), yielding

$$\int_{t_0}^{t_1} g(\underset{\sim}{y},\underset{\sim}{x}_i,t_1,\tau)\, d\tau = \frac{1}{4\pi\kappa}\, Ei\left(\frac{r^2}{4\kappa\Delta t}\right), \tag{10a}$$

Where Ei is the exponential integral. Also

$$\int_{t_0}^{t_1} \kappa \frac{\partial g}{\partial n}(\underset{\sim}{y},\underset{\sim}{x}_i t_1,\tau)\, d\tau = \frac{1}{2\pi} \exp\left(-\frac{r^2}{4\kappa\Delta t}\right) \frac{\underset{\sim}{r}\cdot\underset{\sim}{n}}{r^2}, \tag{10b}$$

and

$$g(\underset{\sim}{y},\underset{\sim}{x}_i,t_1,t_0) = \frac{1}{4\pi\kappa\Delta t} \exp\left(-\frac{r^2}{4\kappa\Delta t}\right). \tag{10c}$$

Now we have to evaluate the space integrals. The integrals over Γ can be evaluated in the same way as for the stationary case (BREBBIA et al., 1981). For elements not containing the collocation node, the integrands are regular so we employed Gauss-Legendre quadrature rules with 2 points for linear elements and 3 points for parabolic elements. When the collocation node coincides with a node of the element, the exponential integral exhibits a logarithmic type singularity. In this case we separate the singular and regular parts (ABRAMOWITZ et al., 1968). The singular part integral was evaluated by a Berthod--Zaborowiski (MINEUR, 1966) quatrature rule and the regular one by the Gauss-Legendre rule described above.

The exponential integral function was computed by a subroutine of (SSP, 1970) which makes use of two polynomial approximations according to the range of the respective argument (LUKE et al., 1963).

The evaluation of the integrals when collocating at interior nodes was perfomed by the selective integration scheme described in (BREBBIA et al., 1981), using up to 5 points.

To evaluate the first integral in the right hand side of equation (9) we divide the domain Ω in finite elements. Since our method proceeds in a step by step fashion, i.e., the temperature field calculated at one time step acts as the initial condition for the next time step, we need a suitable representation for the temperature field in the domain Ω to perform the above integral. We experienced with the two methods employed in (WROBEL, 1981).

In the first method (Method A) the initial temperature field is computed at integration points by formula (4). In the second method (Method B) the initial temperature field is evaluated at the nodal points of the finite element mesh of the domain Ω. In Method A the finite elements perform merely as integration cells, but not so in Method B, where they are used to give an approximate representation for u_0. In the examples shown below we have employed several integration formulas, namely (STROUD, 1971):

- Triangles (linear and quadratic) - seven point quintic rule of Hammer.
- Linear quadrilaterals - 2 x 2 Gauss rule.
- Quadratic quadrilaterals - seven point quintic rule of Radon, Albrecht and Collatz.

After performing all the integrations the final form of the system of equations can be written in matricial notation as

$$[H]\{u_1\} + [G]\{q_1\} = [B]\{v_1\} + [C]\{u_0\} . \qquad (11)$$

In this equation $\{u_1\}$ is the vector of the unknown nodal temperatures on Γ_2 at time t_1 and $\{q_1\}$ is the vector of the unknown nodal fluxes on Γ_1 at time t_1. The vector $\{v\}$ represents the known boundary conditions at time t_1 and the vector $\{u_0\}$ represents the nodal values of temperature at time t_0.

This equation can be reorganized in the following way

$$[H \mid G] \begin{Bmatrix} u_1 \\ \hline q_1 \end{Bmatrix} = [B]\{v_1\} + [C]\{u_0\} \quad , \qquad (12)$$

or, putting,

$$[A] = [H \mid G] \qquad (13a)$$

$$\{W\} = \left\{ \begin{array}{c} u_1 \\ \hline q_1 \end{array} \right\} \qquad (13b)$$

We have that

$$\{w\} = [A]^{-1} [B] \{v_1\} + [A]^{-1} [C] \{u_0\} \qquad (14)$$

The matrices $|H|$, $|G|$, $|B|$, and $|C|$ represent the influence of the several variables. These equations are constant for all instants t_m, $m=1,2,...$ if we use a constant time step Δt, so they need be calculated once at the first time step and saved for subsequent time steps. In each time step the operations to be performed are the computations of vector $\{v_m\}$ and $\{u_m\}$ and the matrix multiplications on the right hand side of equation (14).

EXAMPLES

We present in the sequel, several examples demonstrating the capability of the boundary element method and exhibit the influence of the choice of different elements, integration schemes and time steps.

Example 1.

This example was taken from (WROBEL, 1981). It consists in the study of the transient solution on a rectangle $-5<x<5$, $-4<y<4$ with unit diffusivity and initial ($t=0$) zero temperature and a Dirichlet boundary condition of unit temperature for $t>0$.

The discretization employed is shown in Figure 1. Table 1 presents our results and compares them with those of (WROBEL, 1981).

For the case of time step $\Delta t=1.0$ and linear elements Method A performs better then Method B. However if we increase the time step to $\Delta t=2.0$ Method B is significantly better than Method A.

In both cases quadratic elements perform best, both in accuracy and required computing time.

Example 2.

This example studies a 0.1 x 0.1 m plate initially at 0 ^{0}C and put in contact with a fluid at 100^{0}C (KÖHLER et al., 1974, WROBEL, 1981). The values of the thermal conductivity, specific heat and heat transfer coefficients are respectively 18 Kcal/hm^{0}C, 912 Kcal/m^3 ^{0}C and 5000 Kcal/hm^2 ^{0}C.

The mesh employed is depicted in Figure 2.

TABLE 1

Results for central point .($\Delta t = 1.0$)

TIME	WROBEL Method-A	WROBEL Method-B	Method-A	Method-B	Method-B	ANALYTIC
2	0.093	0.112	0.104	0.112	0.104	0.114
4	0.390	0.450	0.421	0.450	0.424	0.420
6	0.623	0.686	0.659	0.685	0.659	0.646
8	0.770	0.822	0.804	0.822	0.800	0.786
10	0.860	0.899	0.889	0.900	0.883	0.871
12	0.915	0.943	0.939	0.943	0.932	0.922
14	0.948	0.968	0.959	0.968	0.960	0.953
16	0.968	0.982	0.978	0.982	0.977	0.972
18	0.981	0.990	0.989	0.990	0.986	0.983
20	0.988	0.994	1.001	0.994	0.992	0.990

TABLE 2

Results for central point ($\Delta t = 2.0$)

TIME	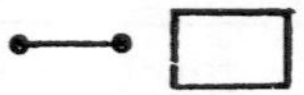 Method-A	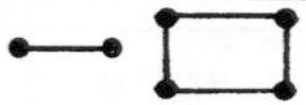 Method-B	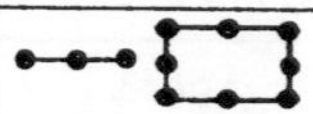 Method-B	ANALYTIC
2	0.0811	0.0819	0.0882	0.1136
4	0.3815	0.3848	0.3907	0.4202
6	0.6269	0.6279	0.6310	0.6456
8	0.7811	0.7790	0.7803	0.7858
10	0.8731	0.8692	0.8697	0.8708
12	0.9273	0.9227	0.9228	0.9221
14	0.9591	0.9543	0.9542	0.9530
16	0.9778	0.9730	0.9729	0.9717
18	0.9888	0.9840	0.9840	0.9829
20	0.9952	0.9905	0.9906	0.9897

Time steps from 0.005 to 0.02 hours were tried. In all cases Method B performed better thant Method A. Figure 3 shows typical results at 3 interior points for Method B with time step $\Delta t=0.005$.

Example 3.

This is one-dimensional test problem used in (CURRAN et al., 1980). The region Ω is the interval $0<x<1$ subject to the initial condition $u_0(x)=\sin \pi x$ and to the Dirichlet boundary conditions $u(0,t)=u(1,t)=0$ for all $t>0$. The analytical solution is $u(x,t) = \exp(-\pi^2 t) \sin \pi x$.

The meshes employed for this example are shown in Figure 8. Methods A and B yield similar accuracy, Method A proving to be marginally better.

With linear elements the best accuracy was obtained with a time step $\Delta t \simeq 0.01$ and with quadratic elements with $\Delta t \simeq 0.005$. The accuracy of Method A remains almost unchanged for a large range of time steps. Method B is more sensitive to the choice of time step. Typical results are presented in Figures 4 and 5.

Example 4.

In this example we study the distribution of temperature in a half-plane of unit diffusivity subjected to a periodic surface temperature of unit amplitude and frequency $\omega=3\pi$. The analytical solution is $u(x,t)= \exp(-\sqrt{\frac{\omega}{2}}\, x) \cos(\omega t-\sqrt{\frac{\omega}{2}}\, x)$. The half-plane was discretized as shown in Figure 9, the boundary condition at infinity was replaced by the approximate boundary condition $u(\Delta,t)=0$.

Quadratic elements with Method B and a time step $\Delta t=0.05$ gave the results shown in Figures 6 and 7 which are quite good. Decreasing the time step to $\Delta t=0.025$ allowed a even greater accuracy.

An attempt to check these results with Method A gave poorly accurate solutions. We tried the 7-point quadrature rule of Radon et al., (STROUD, 1971) to verify the influence of a better integration formula in the domain but the results did not show any significant improvement and by decreasing the time step we faced instability in the results.

CONCLUSIONS

We have shown that the boundary element method can provide very accurate solutions to the unsteady heat equation. In particular quadratic elements prove to be superior in accuracy and also in computational speed.

This is more marked as the number of nodes of the mesh increases.

Regarding the integration in the domain Method B seems to be preferable if there are limitations in memory size since it requires smaller matrices. It proved also more rapid than Method A is less sensitive to the choice of time step Δt. It requires some care in the choice of the size of the cells, since too small cells induce integration points very near the boundary Γ where the fundamental solution changes rapidily, with the correspondent increase in integration errors.

Also the space discretization and the time-step should be chosen consistently. The determination of the optimum time step for a given mesh is an open problem deserving some consideration in the future.

ACKNOWLEDGEMENT

The work reported in this paper was partially supported by CTAMFUTL-INIC, Portugal.

REFERENCES

1. ABRAMOWITZ, M. and STEGUN, I.A. (1968) Handbook of Matherma-tical Functions, Dover (1968)

2. BREBBIA, C.A., PINA, H.L.G. and FERNANDES, J.L.M. (1981) The effect of Mesh Refinement in the Boundary Element Solution of Laplace's Equation with Singularities, Proceed. of the 3rd Int. Sem. on Boundary Element Methods, Springer-Verlag.

3. BREBBIA, C.A. and WALKER, S. (1980) Boundary Element Techniques in Engineering, Newnes-Butterworths.

4. CARLSLAW, H.S. and JAEGER, J.C. (1959) Conduction of Heat in Solids, 2nd Edn, Clarendon Press, Oxford.

5. CHANG, Y.P., KANG, C.S. and CHENG, D.J. (1973) The Use of Fundamental Green's Functions for the Solution of Problems of Heat Conduction in Anisotropic Media, Int. Journal of Heat and Mass Transfer, 16, pp-1905-1918.

6. CURRAN, D., CROSS, M. and LEWIS, B.A. (1980) A Preliminary Analysis of Boundary Element Methods Applied to Parabolic Partial Differential Equations, Proceed. of the 2nd Int. Sem. on Recent Advances in Boundary Element Methods, Southampton.

7. KÖHLER, W. and PITTR, J. (1974) Calculation of Temperature Fields with Finite Elements in Space and Time Dimensions, Int. Journal of Numerical Methods in Eng., 8 pp 625-631.

8. LUKE and WIMP (1963) Jacobi Polynomial Expansions of a Generalized Hipergeometric Function over a Semi-Infinite Range, Mathematical Tables and Other Aids to Computation, 17, issue 84, pp 395-404.

9. MINEUR, H. (1966) Techniques de Calcul Numérique, Dunod.

10. STROUD, A.H. (1971) Approximate Calculation of Multiple Integrals, Prentice Hall, N.J.

11. SSP, System/360 Scientific Package, version III, Programmer's Manual, IBM (1970).

12. WROBEL, L.C. (1981) PhD Thesis, University of Southampton.

13. ZIENKIEWICZ, D.C. and PAREKH, C.J. (1970) Transient Field Problems: Two-Dimensional and Three-Dimensional Analysis by Isoparametric Finite Elements, Int. Journal of Numerical Methods in Engineering, 2 pp 61-71.

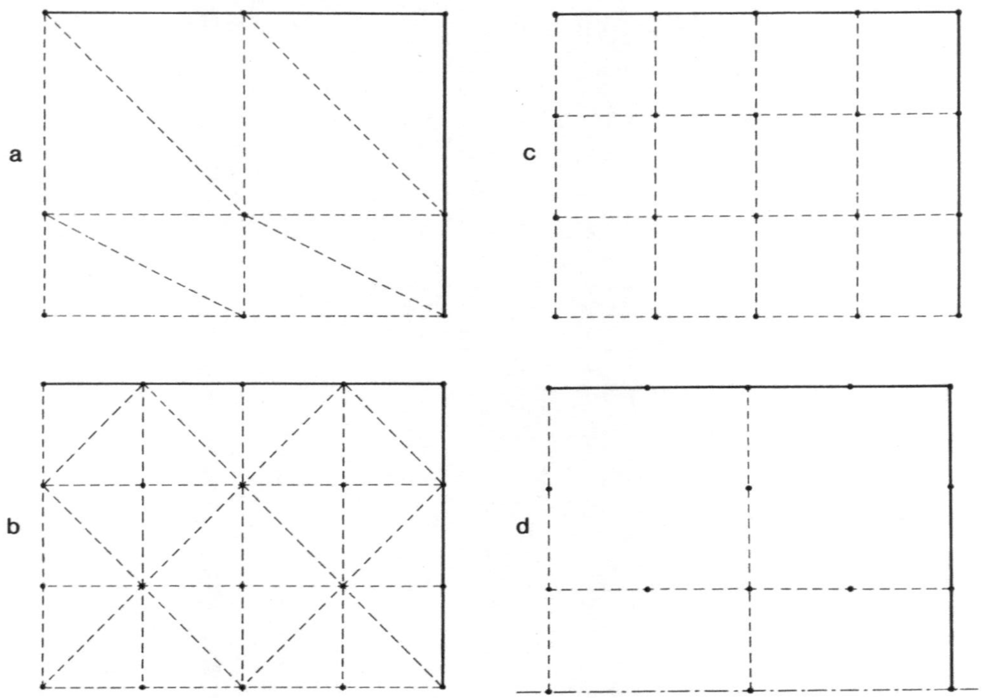

Figure 1. Discretizations: Wrobel a)method-A b) method-B; This paper c)linear elements, method A and B d) parabolic elements, method-B

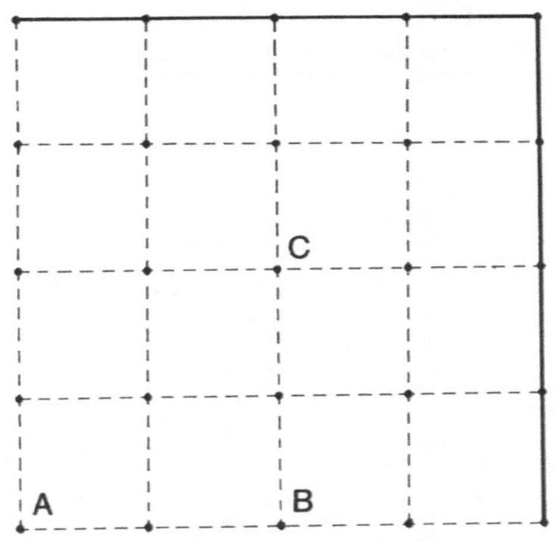

Figure 2. Discretization of one quarter of square region, Example-2

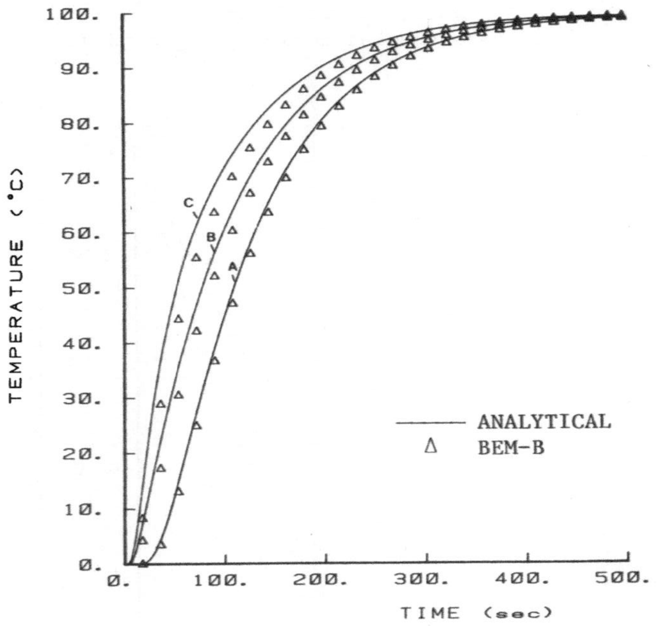

Figure 3 - Temperature at points (0;0), (0.025;0) and (0.025;0.025) for Δt = 0.005, Example - 2

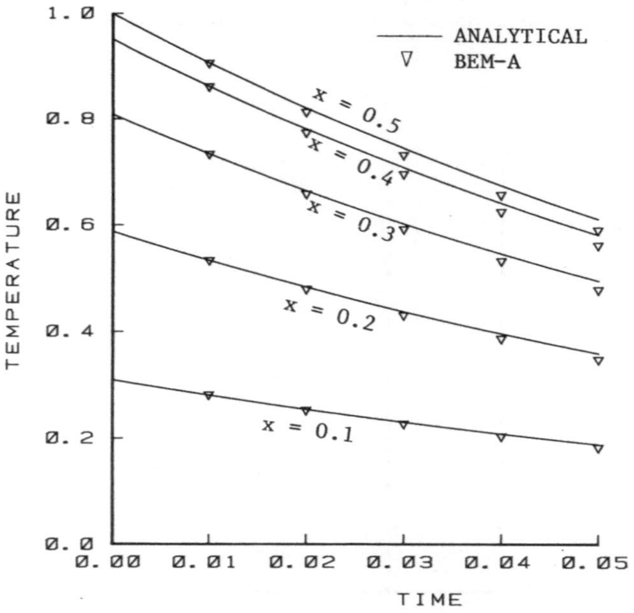

Figure 4 - Results for Example-3 at some internal points with $\Delta t=0.01$

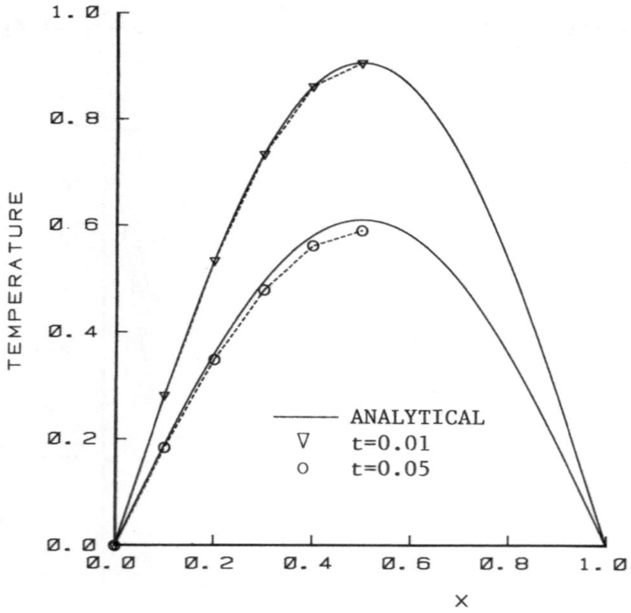

Figure 5 - Interior temperatures for Example-3 (method-A)

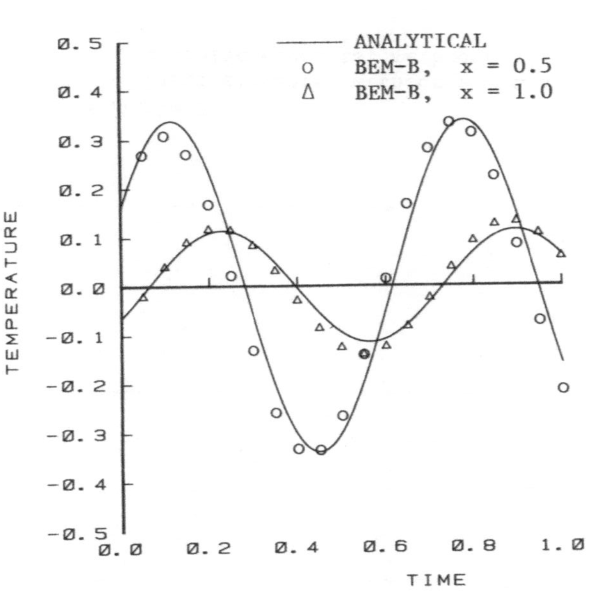

Figure 6 – Temperature variation with time at two internal points Example-4, Δt=0.05

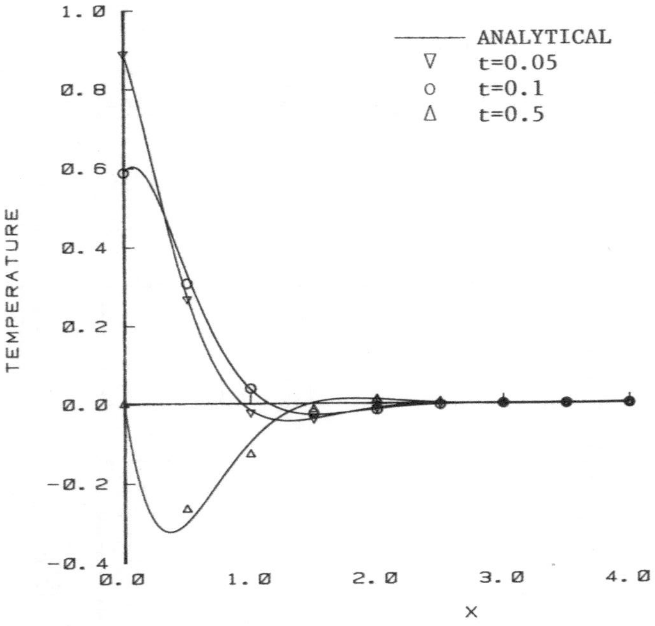

Figure 7 – Spatial (parabolic) representation of internal temperature Example-4, (method-B)

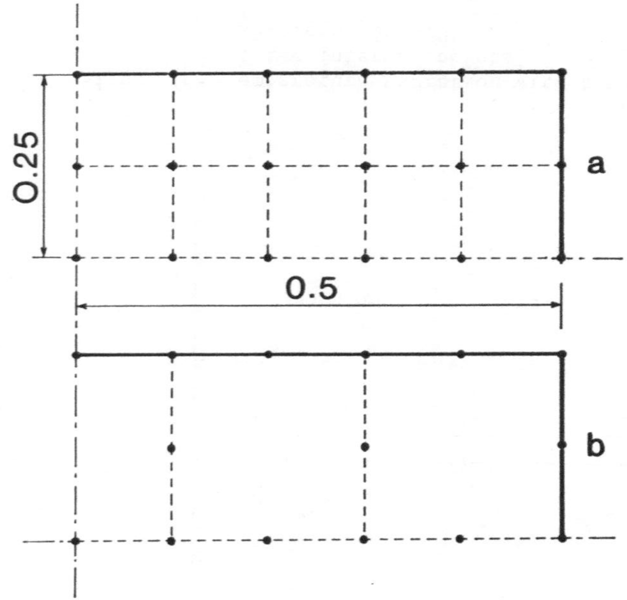

Figure 8. Example-3:
a) linear elements, methods A and B
b) parabolic elements, method-B

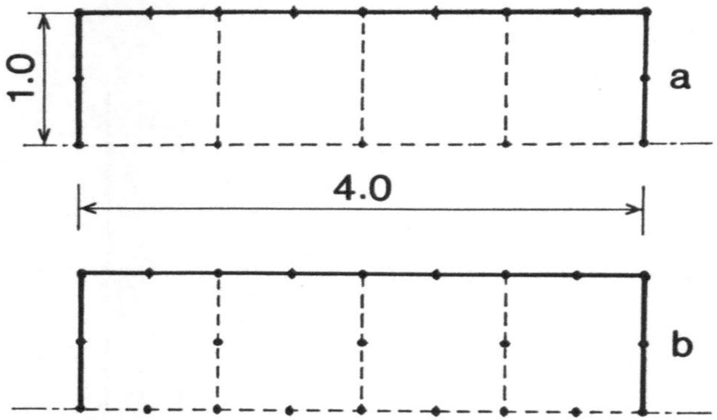

Figure 9. Example-4, parabolic elements
a) method A b) method B

SOLUTION OF WAVE PROPAGATION PROBLEMS BY BOUNDARY ELEMENTS

W.J. Mansur and C.A. Brebbia

Federal University of Rio de Janeiro (Brazil) presently at Southampton University and Southampton University

1. INTRODUCTION

Despite the growing number of applications of the boundary element method [1][2] much remains to be done in the field of wave propagation. The first work in this area can be traced to Rizzo and Cruse [3],[4] who solved the transient elastodynamic problem in the Laplace transform domain and used a numerical algorithm to obtain time domain solutions. Unfortunately the numerical results that they presented were accurate only for early times.

As an extension of Cruse's work, Manolis et al. [5] improved the results by using a better numerical algorithm for the inversion of the Laplace's Transform. They also studied the steady state elastodynamic problem and pointed out that the governing integral equations for this case are similar to those employed by Cruse provided that the Laplace parameter 'k' is replaced by '$i\omega$', where $i\omega$ is the exciting frequency. Alarcon et al. [6] used the same idea later on to find the dynamic stiffness of foundations.

Time stepping techniques have also been used to analyse transient wave propagation problems. Groenenboom [7] studied unsteady potential flow in three dimensions and used a time retarded formulation for some applications. Cole et al. [8] analysed two dimensional transient elastodynamic antiplane motion but restricted their formulation to problems for which the boundary integral involving the potential disappears.

In the present paper the governing equations for the Boundary Element Method are formulated to analyse two dimensional transient scalar wave propagation problems. Time and space interpolation functions are used for the potential and its derivative and the system of equations obtained is solved in a stepwise fashion. Initial conditions are taken into account by integrating over internal cells.

The paper starts by reviewing the integral formulation of the scalar wave equation and then deals with its numerical implementation. Some problems were studied to illustrate how the time stepping technique was implemented and to show the accuracy of the numerical solution. Some important conclusions regarding the size of the time step versus accuracy and stability can be inferred from the applications.

2. BOUNDARY INTEGRAL FORMULATION

The equation under study is the inhomogeneous scalar wave equation, i.e.,

$$\nabla^2 u - \frac{1}{c^2}\frac{\partial^2 u}{\partial t^2} = -\gamma \tag{1}$$

where u is the potential, γ is the source density, t is time and c is the velocity of wave propagation. On the Γ boundary the following conditions are prescribed,

$$\begin{aligned} &\text{Essential conditions} \qquad u = \bar{u} \text{ on } \Gamma_1 \\ &\text{Natural conditions} \qquad q = \frac{\partial u}{\partial n} = \bar{q} \text{ on } \Gamma_2 \end{aligned} \tag{2}$$

where $\Gamma = \Gamma_1 + \Gamma_2$, q is the normal derivative of u and $\bar{q}$ and $\bar{u}$ indicate respectively known values of q and u on Γ. q can also be written as $q = \vec{n}.\nabla u$, where ∇u is the gradient of u, $\vec{n}$ is the unit outward vector normal to Γ and '.' stands for dot product.

In addition, the problem presents the following initial conditions in the domain,

$$\begin{aligned} u_o &= \bar{u}_o \\ v_o &= \left(\frac{\partial u}{\partial t}\right)_o = \bar{v}_o \end{aligned} \tag{3}$$

$\bar{u}_o$ and $\bar{v}_o$ are prescribed values of u and $\frac{\partial u}{\partial t}$ at t = 0.

The Green's function for the problem above described is the solution of equation (1) for an unbounded domain [9] and a particular concentrated source, i.e. $\gamma = 4\pi\delta(r-\lambda)\delta(t-\tau)$ where δ is the Dirac delta function with the property,

$$\int_{-\infty}^{+\infty} \delta(x-a)\, f(x)dx = f(a) \tag{4}$$

Equation (1) can now be written as,

$$\nabla^2 u^* - \frac{1}{c^2}\frac{\partial^2 u^*}{\partial t^2} = -4\pi\delta(r-\lambda)\delta(t-\tau) \tag{5}$$

Notice that this Green's function has been called u^* to indicate its unbounded character. Thus u^* is the effect of a source represented by an impulse at $t = \tau$ located at $r = \lambda$. r and λ are referred to in the literature as observation - or field - and source points respectively.

Notice that t can now be replaced by τ in equation (1) and the spatial dependency of u can be indicated by r. Hence equation (1) can be written as,

$$\nabla^2 u(r,\tau) - \frac{1}{c^2}\frac{\partial^2 u(r,\tau)}{\partial \tau^2} = -\gamma(r,\tau) \tag{6}$$

The Green's function represented in equation (5) has the following properties [9]

i) $u^*(r,t/\lambda,\tau) = 0$ whenever $c(t-\tau) < |r-\lambda|$ causality

ii) $u^*(r,t/\lambda,\tau) = u^*(\lambda,-\tau/r,-t)$ reciprocity

iii) $u^*(r,t+t_1/\lambda,\tau+t_1) = u^*(r,t/\lambda,\tau)$ time translation

From the reciprocity property, equation (5) can be written as [9]

$$\nabla^2 u^*(r,t/\lambda,\tau) - \frac{1}{c^2}\frac{\partial^2 u^*(r,t/\lambda,\tau)}{\partial \tau^2} = -4\pi\delta(r-\lambda)\delta(t-\tau) \tag{7}$$

Now the following weighted residual statement can be written (for detailed description, see reference [1])

$$\int_o^{t^+}\int_\Omega \left(\nabla^2 u - \frac{1}{c^2}\frac{\partial^2 u}{\partial \tau^2} + \gamma\right)u^*\, d\Omega d\tau = \int_o^{t^+}\int_{\Gamma_2} (q-\bar{q})u^*\, d\Gamma d\tau - \int_o^{t^+}\int_{\Gamma_1} (u-\bar{u})q^*\, d\Gamma d\tau \tag{8}$$

By t^* it is meant $t+\varepsilon$ where ε is arbitrarily small. With this procedure one avoids ending the integration exactly at the peak of a Dirac delta function. Integrating by parts twice the Laplacian ($\nabla^2 u$) with respect to space and once the derivative ($\partial^2 u/\partial\tau^2$) with respect to time one obtains [10]

$$4\pi c_\lambda u(\lambda,t) = \int_o^{t^+}\int_\Gamma q\, u^*\, d\Gamma\, d\tau + \int_o^{t^+}\int_\Gamma \frac{\partial R}{\partial n}\left(u_o B + \frac{v}{c}u^*\right) d\Gamma\, d\tau +$$

$$+ \frac{1}{c^2}\int_\Omega v_o u_o^*\, d\Omega + \frac{1}{c}\int_\Omega \left(-u_o B_o + \frac{\partial u_o}{\partial R} u_o^* + u_o \frac{u_o^*}{R}\right) d\Omega\ -$$

$$+ \int_o^{t^+}\int_\Omega \gamma u^*\, d\Omega\, d\tau \qquad (9)$$

where

$$u^* = u^*(r,t/\lambda,\tau) = \frac{2c}{\sqrt{c^2(t-\tau)^2 - R^2}} H[c(t-\tau)-R]$$

$$B = B(r,t/\lambda,\tau) = \frac{2c[c(t-\tau)-R]}{\sqrt{[c^2(t-\tau)^2-R^2]^3}} H[c(t-\tau)-R] \qquad (10)$$

In the expression above H(....) stands for the Heaviside function.

In expression (9), the terms with the 'o' subscript refer to initial state ($\tau = 0$), i.e.

$$u_o^* = u^*(r,t/\lambda,0)$$

$$B_o = B(r,t/\lambda,0) \qquad (11)$$

and v indicates velocity as given by

$$v = \frac{\partial u(r,t)}{\partial t} \qquad (12)$$

The space integrals in equation (9) must be performed with respect to the field (observation) point r and must be understood in the Cauchy principal value sense. Notice that R is the distance between r and the source point λ, i.e. $R = |r-\lambda|$, and c_λ is taken to be equal to 0, 1 or 1/2 for λ respectively inside, outside or on a smooth part of the Γ boundary.

The source intensity term γ, can in some cases be represented by a Dirac delta function δ, such that

$$\gamma(r,t) = f_s(t)\ \delta(r-r_s) \tag{13}$$

where r_s gives the position of the source. The last term in equation (9) then becomes,

$$\int_0^{t^+} f_s(\tau)\ \frac{2c}{\sqrt{c^2(t-\tau)^2-R_s^2}}\ H[c(t-\tau)-R_s]\ d\tau \tag{14}$$

where $R_s = |r_s-\lambda|$.

A detailed study of the problem described in this section and of how the integral equation (9) is obtained can be found in reference [10].

3. NUMERICAL FORMULATION

Consider a set of discrete points r_j (nodes) on the Γ boundary, $j = 1,\ldots,J$ and a set of values of time t_n, $n = 1,\ldots,N$. $u(r,t)$ and $q(r,t)$ $(r \in \Gamma)$ can be approximated by using a set of interpolation functions as indicated below,

$$\begin{aligned} u(r,t) &= \sum_{j=1}^{J}\sum_{m=1}^{N} \phi^m(t)\ \eta_j(r)\ u_j^m \\ q(r,t) &= \sum_{j=1}^{J}\sum_{m=1}^{N} \theta^m(t)\ \nu_j(r)\ q_j^m \end{aligned} \tag{15}$$

where m and j refer to time and space respectively. $\phi^m(t)$, $\eta_j(r)$, $\theta^m(t)$ and $\nu_j(r)$ are chosen such that,

$$\begin{aligned} \eta_j(r_i) &= \delta_{ij}\ , \quad \phi^m(t_n) = \delta_{mn} \\ \nu_j(r_i) &= \delta_{ij}\ , \quad \theta^m(t_n) = \delta_{mn} \end{aligned} \tag{16}$$

where δ_{ij} is the Kronecker delta. Therefore,

$$\begin{aligned} u_j^m &= u(r_j,t_m) \\ q_j^m &= q(r_j,t_m) \end{aligned} \tag{17}$$

Next, one can write equation (9) for every node 'i' and for every value of time t_n, and replace the values of u and q by their approximations given by equation (15). The following system of algebraic equations is then obtained,

$$\sum_{m=1}^{N} \sum_{j=1}^{J} H_{ij}^{nm} u_j^m = \sum_{m=1}^{N} \sum_{j=1}^{J} G_{ij}^{nm} q_j^m + F_i^n + S_i^n \tag{18}$$

where

$$H_{ij}^{nm} = 4\pi c_i \delta_{ij} \delta_{mn} - \int_\Gamma \frac{\partial R}{\partial n} \eta_j(r) \int_0^{t_n} \left[\phi^m(\tau) B(r, t_n/r_i, \tau) + \right.$$

$$\left. + \frac{\partial \phi^m_{(\tau)}}{\partial \tau} \frac{u^*(r, t_n/r_i, \tau)}{c} \right] d\tau \, d\Gamma \tag{19}$$

$$G_{ij}^{nm} = \int_\Gamma \nu_j(r) \int_0^{t_n} \theta^m(\tau) \, u^*(r, t_n/r_i, \tau) \, d\tau \, d\Gamma \tag{20}$$

$$F_i^n = \frac{1}{c^2} \int_\Omega v_o \, u^*(r, t_n/r_i, 0) \, d\Omega + \frac{1}{c} \int_\Omega \frac{\partial u_o}{\partial R} u^*(r, t_n/r_i, 0) \, d\Omega +$$

$$+ t_n \int_\Omega u_o \frac{B(r, t_n/r_i, 0)}{R} \, d\Omega \tag{21}$$

$$S_i^n = \int_0^{t_n} \int_\Omega \gamma u^*(r, t_n/r_i, \tau) d\Omega \, d\tau \tag{22}$$

Notice that the third term on the right hand side of equation (21) is the sum of the first and third terms of the fourth integral on the right hand side of equation (9).

When the boundary is of Kellog type [11] it can be demonstrated that $c_i (c_i = c_{r_i})$ in expression (19) is given by,

$$c_i = \frac{\alpha_i}{2\pi} \tag{23}$$

where α_i is the internal angle indicated in figure 1.

A time stepping scheme in which equation (18) is successively solved for n = 1,...,N can now be used to calculate unknowns u_j^N and q_j^N at time t_N.

The boundary discretization adopted in this work is linear that is, Γ is represented by a series of straight line segments (elements) each one of them joining two consecutive nodes of Γ.

In the numerical applications presented in this paper it was decided to use $\phi^m(t)$ linear and $\theta^m(t)$ constant. The spatial interpolation functions $\eta_j(r)$ and $\nu_j(r)$ were assumed linear.

Substitution of linear $\phi^m(t)$ and constant $\theta^m(t)$ into equations (19) and (20) leads to expressions which can be integrated analytically on time. The spatial integrations were performed analytically for the case of elements with singularities and numerically for all of the others, using standard Gauss quadrature formulae.

In order to carry out integrations involving initial conditions (see expression (21)) the domain is discretized into triangular cells,over which initial displacements and initial velocities are linearly interpolated. The procedure used in this work is similar to that described in reference [12], where a system of polar coordinates (R,θ) based at source point is used, and analytical integration with respect to the coordinate R is performed. The integrations with respect to θ are computed numerically for each cell, using standard one-dimensional Gauss quadrature formulae.

Source contribution such as shown in equation (14) can be easily considered. Notice that for them no space integration is required and that time integration is similar to the one shown in expression (20).

4. APPLICATIONS

In order to demonstrate the range of applications of the numerical procedures developed in this paper, the two following problems were studied,

(i) One-dimensional rod under a Heaviside type forcing function.

(ii) Square membrane under prescribed initial velocity.

(i) One-dimensional rod under a Heaviside type forcing function

Results obtained using the two-dimensional boundary element program were compared with analytical results for a one dimensional rod under a Heaviside type forcing function. The boundary element solution considered a rectangular domain with sides of length a and b ($b = a/2$) as indicated in figure 2. The u displacements are assumed to be zero at $x = a$ and their normal derivative q are also taken as null at y = 0 and y = b, for any time 't'. At x = 0 and t = 0 a load **Eq.** is suddenly applied and kept constant until the end of the analysis (E is the Young's modulus). Due to the topology and boundary conditions the problem is actually one-dimensional and its analytical solution can be found in reference [13].

The boundary was discretized into 24 linear elements (figure 3) and double nodes were used at the corners.

The time was subdivided into equal intervals such that

$$\beta = \frac{c(t_m - t_{m-1})}{\ell_j} = \text{constant} \tag{24}$$

where ℓ is the length of an element. For this problem .6 was found to be the optimum value of β.

Figures 4 to 6 show B.E.M. and analytical displacement results at internal and boundary points. The order of accuracy of B.E.M. results is quite good. In figure 7 the normal derivative of u displacements at point (a, b/2) versus ct is presented. Except for the presence of a comparatively small amount of noise, boundary elements and analytical solutions are in good agreement.

Care must be taken on the choice of β in order to avoid noise, which although usually not critical for displacements, can sometimes be excessive for tractions. A study of the effect of varying β on the level of noise can be found in reference [14].

ii) <u>Square membrane under prescribed initial velocity</u>
The subject of this investigation is a square membrane with an initial velocity $v_o = c$ prescribed over the domain Ω_o shown in figure 8 and zero displacements prescribed all over the boundary.

The boundary was discretized into 32 elements and Ω_o was divided into four cells (see figure 9). Analytical (see reference [15]) and B.E.M. results for displacements at point (a/2, a/2) and the normal derivative of displacements at point (a, a/2) were compared.

The values of u and q for β = .6 are respectively plotted in figures 10 and 11. Although the agreement for displacements is reasonable, it was found that a more refined time division was needed to represent q more accurately. Another boundary element analysis was carried out with β = .2 and the results obtained for q are plotted in figure 12, and they show a better agreement.

5. CONCLUSIONS

This paper has presented the formulation for the analysis of two-dimensional transient scalar wave propagation problems using the boundary element method. The numerical implementation has been carried out by using interpolation functions in time and space, and two representative examples have been presented. Care should be taken in the analysis on the choice of time intervals and boundary discretization in order to avoid contradicting too much much the causality property, that is, in each time step waves should not be allowed to travel between nodes far from each other.

The examples demonstrate the remarkably good accuracy of the results and validate the application of the boundary element method to solve transient two-dimensional problems governed by the scalar wave equation.

REFERENCES

1. BREBBIA, C.A. and WALKER, S. "Boundary Element Techniques in Engineering" Butterworths, London,(1979).

2. BREBBIA, C.A. (ed.) " Boundary Element Methods " Springer Verlag, Berlin (1981).

3. CRUSE, T.A. AND RIZZO, F.J. "A Direct Formulation and Numerical Solution of the General Elastodynamic Problem I" J. Math. Analysis and Applic. 22, 244-259 (1968).

4. CRUSE, T.A. "A Direct Formulation and Numerical Solution of the General Elastodynamic Problem II", J. Math. Analysis and Applic. 22, 341-355 (1968).

5. MANOLIS, O.D. AND BESKOS, D.E. "Dynamic Stress Concentration Studies by Boundary Integrals and Laplace Transform", Int. J. for Num. Meth. in Eng., 17, 573-599 (1981).

6. ALARCON, E. and DOMINGUEZ, J. "Elastodynamics" Chapter in "Progress in Boundary Elements" Vol. I, Pentech Press, London and Halstead Press, N.Y. 1981.

7. GROENENBOOM, P.H.L. "The Application of Boundary Elements to Steady and Unsteady Potential Flow Problems in Two and Three Dimensional Problems" in "Boundary Element Methods" by C. Brebbia (Editor), Springer Verlag, Berlin, 1981.

8. COLE, D.M., KOSLOFF, D.D. and MINSTER, J.B. "A Numerical Boundary Integral Equation Method for Elastodynamics I" Bull. Seism. Soc. Amer. 68, 1331-1357, 1978.

9. MORSE, P.M. and H. FESHBACH "Methods of Theoretical Physics", McGraw-Hill, 1953.

10. MANSUR, W.J. and BREBBIA, C.A. "Formulation of the Boundary Element Method for Transient Problems Governed by the Scalar Wave Equation", Applied Mathematical Modelling, in press, 1981.

11. KELLOGG, O.D. "Foundations of Potential Theory", Dover, New York, 1953.

12. TELLES, J.C.F. and BREBBIA, C.A. "The Boundary Element Method in Plasticity", Applied Mathematical Modelling, 1981, 275-281.

13. MILES, J.W. "Modern Mathematics for the Engineer". Edited by Beckenbach, E.F., pp. 82-84 - McGraw Hill Book Co. Inc. 1961.

14. MANSUR, W.J. and BREBBIA, C.A. "Numerical Implementation of The Boundary Element Method for Two Dimensional Transient Scalar Wave Propagation Problems" Applied Mathematical Modelling, in press, 1982.

15. MORSE, P.M. and INGARD, K.V. "Theoretical Acoustics", McGraw-Hill Book Company, 1968.

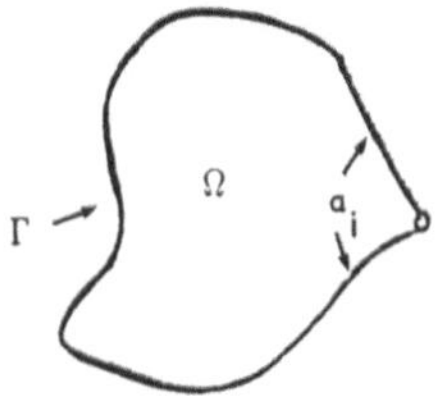

Figure 1 Internal angle used to compute c_i

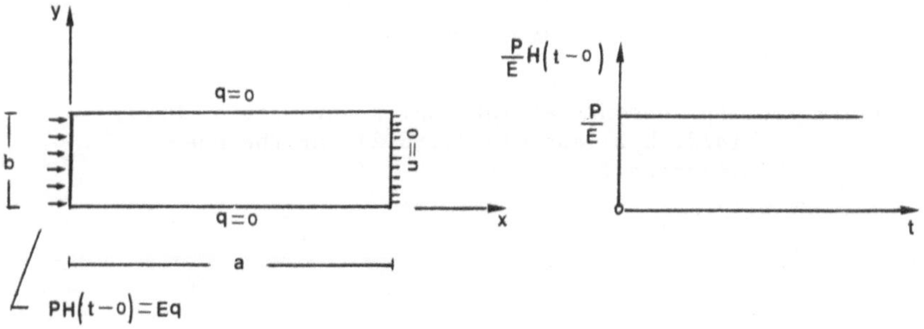

Figure 2 Boundary conditions and geometry definitions for the one-dimensional rod

Figure 3 Boundary discretization for the one-dimensional rod

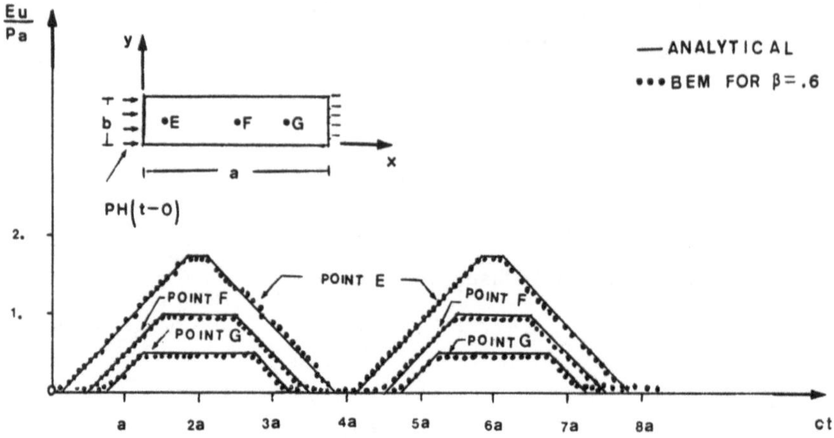

Figure 4 Displacements at internal points E(a/8, b/2), F(a/2, b/2) and G(3a/4, b/2) for the one-dimensional rod

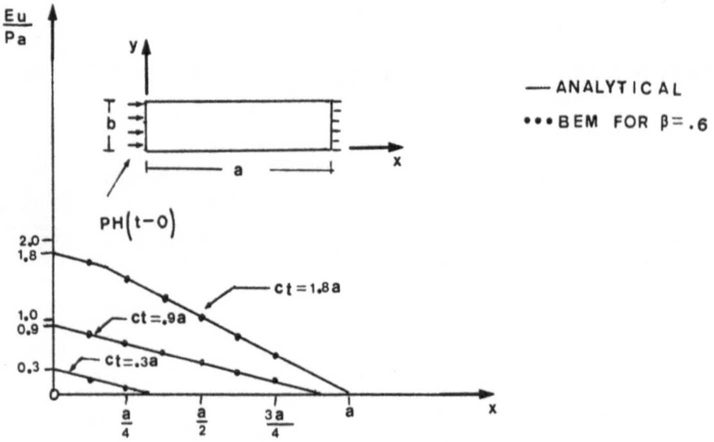

Figure 5 Displacements along the boundary $y = 0$ at times $t = .3a/c$, $t = .9a/c$, $t = 1.8a/c$ for the one-dimensional rod

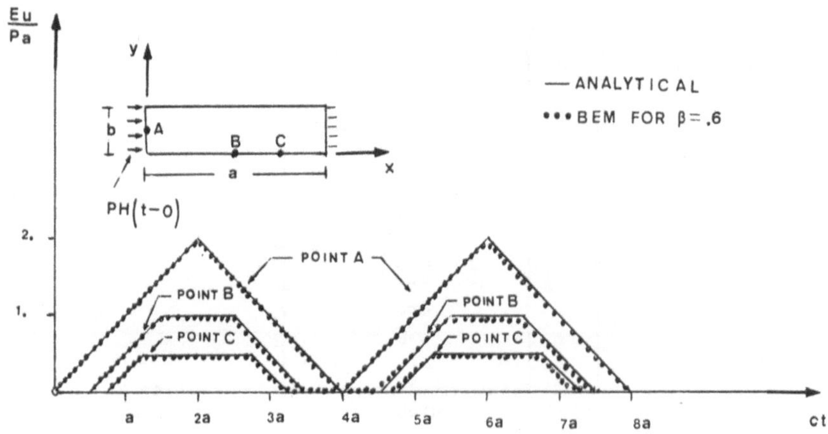

Figure 6 Displacements at boundary points A(0, b/2), B(a/2, 0) and C(3a/4, 0) for the one-dimensional rod

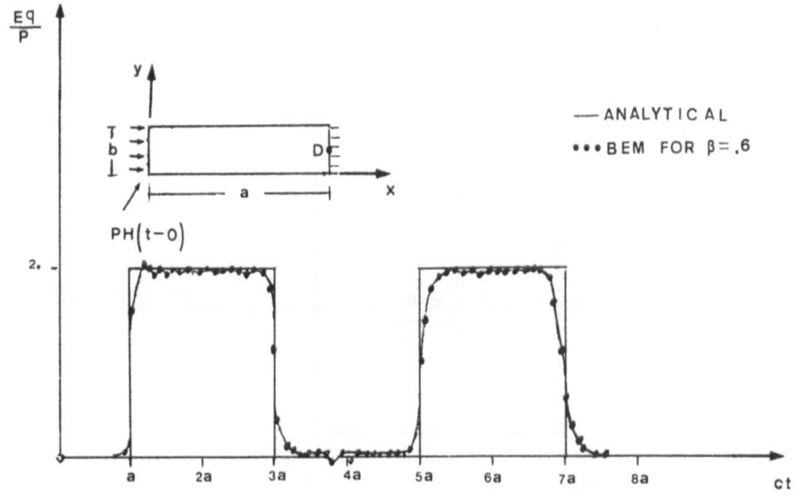

Figure 7 Normal derivative of the displacement at point D(a, b/2) for the one-dimensional rod

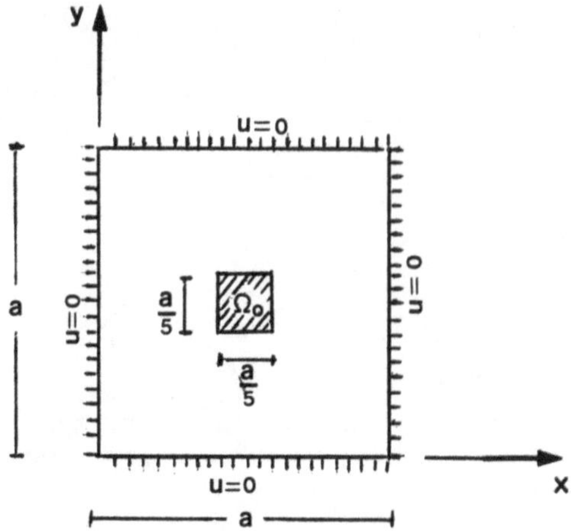

Figure 8 Geometry definition, boundary and initial conditions for the membrane analysis

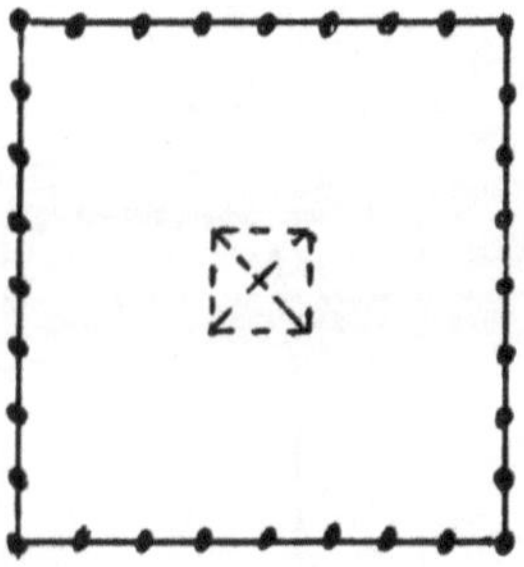

Figure 9 Membrane discretized into 32 elements and four cells

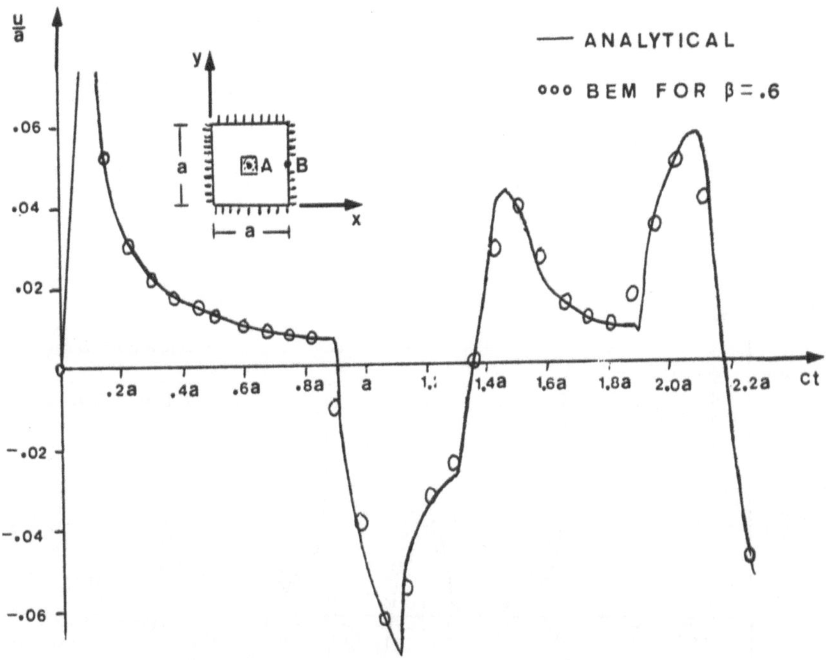

Figure 10 Displacement at point A(a/2, a/2) for the membrane

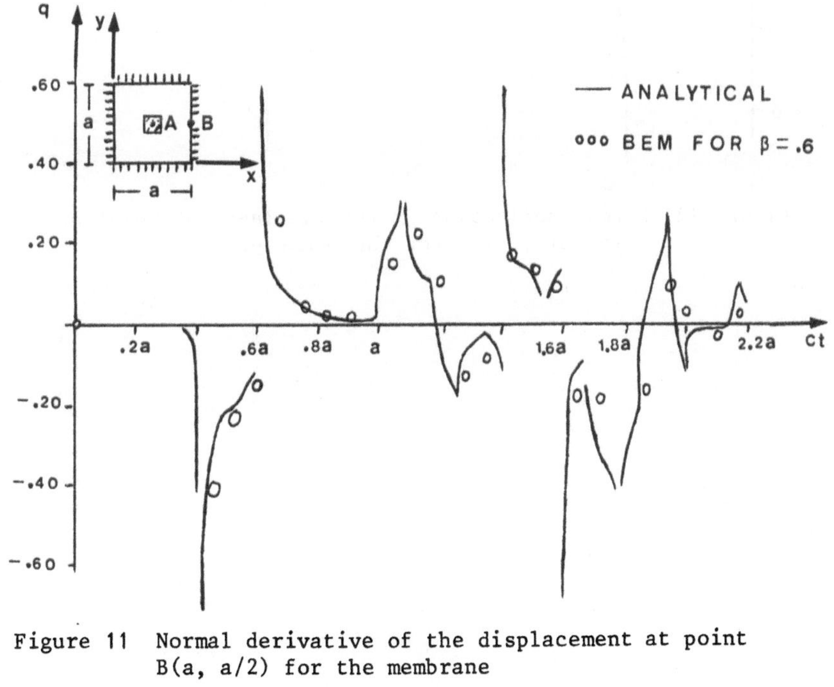

Figure 11 Normal derivative of the displacement at point B(a, a/2) for the membrane

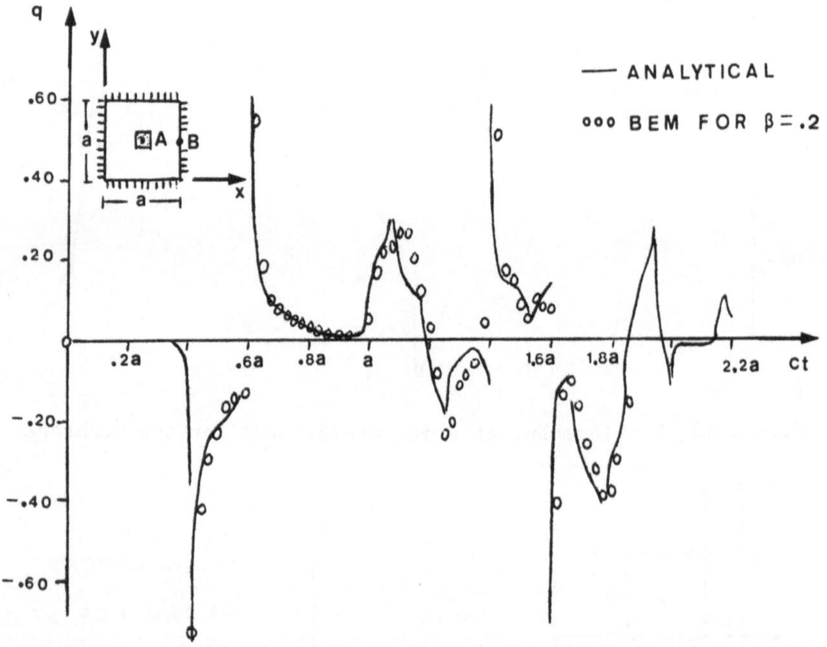

Figure 12 Normal derivative of the displacement point at point B(a, a/2) for the membrane

SOLUTION OF THE WAVE EQUATION WITH STIFFNESS BOUNDARY CONDITIONS

Djuro M. Misljenovic

Mathematical Institute, Belgrade, Yugoslavia

ABSTRACT

In this paper the boundary integral expression for a one-dimensional wave equation is developed for the important case of the stiffness boundary conditions, i.e. a beam with a left-hand side end free and a right-hand side end on elastic support is considered. The integral expression obtained, or its differential counterpart, contains a retarded argument which can be solved using time marching procedure as suggested in the paper.

The case of propagation of a shock waves along a homogeneous beam is analysed in detail.

A way in which boundary elements linear in space can be used for the transformation of the starting problem into a simpler one, with homogeneous initial condition is presented.

The accuracy of the results is assesed by comparison to the shape of initial data. Higher accuracy is obtained for the results at the right-hand side end point than for the other points along the beam.

INTRODUCTION

The boundary integral element method has become very popular for solving problems described by partial differential equations /1/. The method is based on the numerical solution of the boundary integral expressions, some of which have been known from the beginning of the century. A large number of continuum mechanics problems have been solved /2/-/4/, including some time dependent problems /5/-/7/. The theory and application for hyperbolic partial differential equation are not yet well developed. Some important references were recently published /8/-/13/, including two papers by the author /11/, /13/. The present paper extends the work to analyse the case of the beam with stiffness boundary conditions - which has not yet been solved - and demonstrates the treatment and applicability of the BIEM in this area of interest.

FORMULATION OF THE PROBLEM

Let us consider the one-dimensional hyperbolic partial differential equation

$$\frac{\partial^2 u}{\partial t^2} = c^2 \frac{\partial^2 u}{\partial x^2} \qquad x \in \Omega, t \in R \qquad (1)$$

together with the following non-homogeneous initial conditions

$$u(x,o) = \Phi(x) \qquad \text{on } \Omega\times\{o\} \qquad (2)$$

$$\frac{\partial u(x,o)}{\partial t} = \Psi(x) \qquad \text{on } \Omega\times\{o\} \qquad (3)$$

and with non-homogeneous boundary conditions

$$\alpha \frac{\partial u(L,t)}{\partial x} = \beta\, u(L,t) \qquad \text{on } \Omega\times R \qquad (4)$$

$$\frac{\partial u(o,t)}{\partial x} = F(t) \qquad \text{on } \partial\Omega\times R \qquad (5)$$

Functions F, Φ, Ψ and constants α, β are prescribed. $\Omega = (0,L)$ is given as a finite interval. Without loss of generality one can assume that the time is defined by interval (0,T). c is a speed of propagation of the disturbance in the medium under consideration.
Equation (1) is known as the wave equation and governs many problems in continuum mechanics and other fields.

THE PROBLEM UNDER CONSIDERATION

We are solving the problem of propagation of the shock waves along the beam of length L, with a free end at the left-hand side and the other end at the elastic support. At the free end a disturbance is produced by applying a force of magnitude F given by equation (5). The other end is designed to allow damping of the incoming waves and this occurrence is described by(4). We assume that the shape of the beam at the initial moment is given by equations (2) and (3).
This mathematical model is closer to the real life situation than the model which has been discussed in reference /13/.

BOUNDARY INTEGRAL EXPRESSION

Using the standard weighted residual procedure, as given in /1/,/12/ or /13/, we can transform the system of equations (1) to (5) into

$$2u(x_o,t_o) = u(x_o-ct_o,0) + u(x_o+ct_o,0) + $$
$$+ u(0,t_o-x_o/c) + u(L,t_o-(L-x_o)/c) + \qquad (6)$$

$$+ \frac{1}{c}\int_{x_o-ct_o}^{x_o+ct_o} \frac{\partial u(x,o)}{\partial t}\, dx - c\int_{o}^{t_o-x_o/c} \frac{\partial u(o,t)}{\partial x} dt + $$

$$+ c\int_{o}^{t_o-(L-x_o)/c} \frac{\partial u(L,t)}{\partial x} dt$$

which is the starting integral expression. Note that this formula is valid in the generalised sense i.e. the terms are valid only when all arguments are physically meaningful. Hence, time and space coordinates must be for the domain of interest and the integrals are valid for cases for which the lower bound is smaller than upper one; otherwise, they are equal to zero.

Introducing all initial and boundary conditions we obtain

$$2u(x_o,t_o) = A(x_o,t_o) + u(0,t_o-x_o/c) + u(L,t_o-(L-x_o)/c) + c\gamma\int_{0}^{t_o-(L-x_o)/c} u(L,t)\, dt \tag{7}$$

($\gamma=\beta/\alpha$), where the following notation has been used

$$A(x_o,t_o) = \Phi(x_o-ct_o) + \Phi(x_o+ct_o) + \frac{1}{c}\int_{x_o-ct_o}^{x_o+ct_o}\Psi(x)\, dx - c\int_{0}^{t_o-x_o/c} F(t)\, dt \tag{8}$$

a is the time for which a wave advances along the whole beam, $a=L/c$. At the free end $x_o = 0$ one finds

$$u(0,t_o) = A(0,t_o) + u(L,t_o-a) + c\,\gamma\int_{0}^{t_o-a} u(L,t)\, dt \tag{9}$$

and at the other end of the beam, $x_o = L$, we have

$$u(L,t_o) = A(L,t_o) + u(0,t_o-a) + c\gamma\int_{0}^{t_o} u(L,t)\, dt \tag{10}$$

Hence, substituting (9) and (10) into equation (7) we obtain

$$2u(x_o,t_o) = A(x_o,t_o) + A(0,t_o-x_o/c) + A(L,t_o-(L-x_o)/c) +$$

$$+ u(L,t_o-(L+x_o)/c) + u(0,t_o-(2L-x_o)/c) + \quad (11)$$

$$+ c\gamma \left(\int_0^{t_o-(L+x_o)/c} + 2 \int_o^{t_o-(L-x_o)/c} \right) u(L,t)\, dt$$

For $x_o = L$, after using equation (9), formula (11) gives

$$u(L,t_o) = A(L,t_o) + A(0,t_o-a) + u(L,t_o-2a) + \qquad (12)$$

$$+ c\gamma \left(\int_0^{t_o} + \int_0^{t_o-2a} \right) u(L,t)\, dt$$

This is the integral equation we are interested in.
Let us denote by $\underline{v}$ the resulting motion of the right-hand side end at $x_o = \overline{L}$, i.e.

$$v(t) = u(L,t) \equiv u(L,t_o) \qquad (13)$$

Differentiating with respect to time one obtains the following linear non-homogeneous ordinary differential equation of the first order

$$v'(t) - c\,\gamma\, v(t) = A'(1,t) + A'(0,t-a) + (1+c\gamma)v(t-2a) \qquad (14)$$

This equation can be solved if we know the resulting motion at the right-hand end point of the beam in the previous moment $t-2a$. A time marching procedure with time step equals to $2a$ allows us to solve differential equation (14) which has a retarded argument.

Let us denote by G the right-hand side of the expression (14) which consists of the known quantities. The solution of that equation can now be written as

$$v(t) = \mathrm{EXP}(\, c\,\gamma\, t) \int_0^t \mathrm{EXP}(\, -c\,\gamma\,\tau)\, G(\tau)\, d\tau \qquad (15)$$

This equation describes the motion at the right-hand side end of the beam (at which the stiffness boundary conditions are prescribed).

IMPORTANT PARTICULAR CASE

Let us assume that the beam is at rest at the initial moment i.e. the initial conditions are homogeneous. Thus the right-hand side of (14) is equal to zero. The function G

for this case is found to be

$$G(t) = -2c\, H(t-a)\, F(t-a) \tag{16}$$

where H is the Heaviside's Unit Step function.

Before the wave front reaches the right-hand side end there is no movement there and the corresponding differential equation is homogeneous. Later on, the wave produces some effects which must be taken into account and the equation (14) becomes non-homogeneous.

Equation (16) is suitable for substitution in formula (15) which gives us the desired function $\underline{v}$ at the initial time-interval $0< t< 2a$. Afterwards, using (14), we can calculate the movement at the end point $x_o = L$ at any time.

In the more special case of a shock waves the applied external force is described by

$$F(t) = F\,\delta\,(t) \tag{17}$$

i.e. we are concerned with an instant force of magnitude F, acting at the initial moment $t = 0$. Substituting (17) into (16) one finds the solution for the first time-interval

$$v(t) = -2cF\, H(t-a)\, EXP(\, c\gamma(t-a)) \tag{18}$$

Repeating the use of (15) gives us all desired values at the end point $x_o = L$.

The knowledge of the movements at the $x_o = L$ end allows us to find the values of $\underline{u}$ at the other end of the beam, using equation (9).

Hence, the displacement field in the interior points can be found using equation (11), with all the quantities on the right-hand side of this equation known.

We are also interested in the space and the time derivatives of the solution and using (11) one finds these derivatives in the interior points of the beam directly, rather than by applying some numerical procedure.

Equation (15) can be used to find the asymptotic value of $\underline{v}$ when $t\to\infty$. It is easy task to show, using L'Hospital rule, that

$$\lim_{t\to\infty} v(t) = -\frac{1}{c\gamma}\lim_{t\to\infty} G(t) \tag{19}$$

This expression is valid for $\gamma \neq 0$ i.e. there exists damping (or, equivalently, amplification). Hence, in the limiting case the movement of the right-hand end of the beam is proportional to the corresponding value of the function G. Note that G is formed in a cumulative manner and partly consists of the function $\underline{v}$ with retarded argument.

GENERALISATION

We have seen that our original problem starts with homogeneous initial conditions and as time advances the conditions become non-homogeneous. Here, we discuss the possibility of using some kind the inverse process i.e. problem with non-zero initial conditions can be transformed to one with zero initial conditions. It is specially useful when using linear boundary elements in space as the equation we are concerned with will not change its form.

Let assume that we want to solve the problem described by the following one-dimensional hyperbolic partial differential equation

$$\frac{\partial^2 u}{\partial t^2} = c^2 \frac{\partial^2 u}{\partial x^2} \qquad x \in D,\ t \in R \tag{20}$$

with non-homogeneous initial conditions

$$u(x,0) = \Phi(x) \qquad \text{on } D\times\{o\} \tag{21}$$

$$\frac{\partial u(x,0)}{\partial t} = \Psi(x) \qquad \text{on } D\times\{o\} \tag{22}$$

and with non-homogeneous boundary conditions

$$u(0,t) = \xi(t) \qquad \text{on } \partial D\times R \tag{23}$$

$$u(1,t) = \eta(t) \qquad \text{on } \partial D\times R \tag{24}$$

We restrict ourselves to the finite spatial domain $D=(0,1)$.

Introducing a new dependent variable $\underline{w}$ such that

$$u(x,t) = w(x,t) + \Phi(x) + t\Psi(x) \tag{25}$$

it is easy to see that the non-homogeneous initial conditions (21) and (22) are transformed into homogeneous ones

$$w(x,0) = 0 \qquad \text{on } D\times\{o\} \tag{26}$$

$$\frac{\partial w(x,0)}{\partial t} = 0 \qquad \text{on } D\times\{o\} \tag{27}$$

The starting equation (20) becomes more complicated one

$$\frac{\partial^2 w}{\partial t^2} = c^2 \frac{\partial^2 w}{\partial x^2} + c^2 \left[\Phi''(x) + t\Psi''(x) \right] \tag{28}$$

In order to preserve the simplicity of transformed equation (28) one may use only linear elements in spatial domain; so the derivatives in the last bracket disappear and we obtain the same type of equation as (20) but in terms of the new dependent variable $\underline{w}$

$$\frac{\partial^2 w}{\partial t^2} = c^2 \frac{\partial^2 w}{\partial x^2} \tag{29}$$

Let the functions and be given by

$$\Phi(x) = Ax+B \tag{30}$$
$$\Psi(x) = Cx+D \tag{31}$$

in spatial domain D. In practical applications these expressions are valid only locally. The non-homogeneous boundary conditions (23) and (24) are transformed into

$$w(0,t) = \xi(t) - Dt - B \tag{32}$$
$$w(1,t) = \eta(t) -(C+D)t - (A+B) \tag{33}$$

Thus, the new boundary conditions differ from the old ones by a linear in time.
Hence we have shown that the non-homogeneous system of equations represented by (20) to (24) can be transformed into a system with homogeneous initial conditions. This valuable property is valid if the shape functions are linear functions in space domain.

CONCLUDING REMARK

Notice that the method presented here does not introduce any truncation error in the resulting expression at the right hand side point of the beam provided that the G function were represented by polinomial of the fixed degree, equation (15). For the other points of the beam the truncation error will appear and the results will be less accurate.

ACKNOWLEDGEMENT

The author is grateful to Dr. C.A. Brebbia, Department of Civil Engineering, University of Southampton, U.K. for his helpful comments and support in developing these results. The support of the Mathematical Institute, Belgrade, Yugoslavia to undertake this research is higly appreciated.

REFERENCES

(1) BREBBIA C.A., The Boundary Element Method for Engineers, Penctech Press, London, 1978.
(2) BREBBIA C.A. and WROBEL L.C., Steady and Unsteady Potential Problems using the Boundary Element Method, Recent Advances in Numerical Methods in Fluids, Volume 1, Chapter 1, edited by C. Taylor and K. Morgan, University College of Swansea, Swansea, U.K.,1980.
(3) WROBEL L.C. and BREBBIA C.A., Boundary Elements in Water Resources, NATO Instituto at Vimerio, Portugal,1981.
(4) BREBBIA C.A., WROBEL L., Applications of Boundary Elements in Water Resources, 3rd International Conference on Finite Elements in Water Resources, May 1980, University of Mississippi, Oxford, USA, pp 2.3-2.14.
(5) SHAW R.P., An Integral Approach to diffusion, Int.J. Heat Mass Transfer, Vol 17, pp 693-699, Pergamon Press,1974.

(6) LIGGETT J.A. and LIU P.I.F., Unsteady Flow in Confined Aquifers: A Comparison of Two Boundary Integral Methods, Water Resources Research, Vol 15, No.4, August 1979,pp.861-866.

(7) BREBBIA C.A. and WALKER S., Boundary Element Technique in Engineering, Newnes-Butterworth, 1980.

(8) BREBBIA C.A.,(ed.), Progress In Boundary Elements, Vol 1, Pentech Press, London, 1981.

(9) BREBBIA C.A.,(ed.), Boundary Element Methods, Springer Verlag, Heidelberg, 1981.

(10) BREBBIA C.A., Private Communications.

(11) MISLJENOVIC M. DJURO, Boundary Element Method and Wave Equation, Applied Mathematical Modelling, Vol 6, June 1982, pp. 205-208.

(12) GROENENBON H.L. PAUL, The Application of Boundary Elements to the Steady and Unsteady Potential Fluid Problems in Two and Three Dimension, Boundary Element Methods, Proceedings of the Third International Seminar, Irvine, California, July 1981,pp.37-52.

(13) MISLJENOVIC M. DJURO, The Boundary Element Method in Shock Waves Analysis, Applied Mathematical Modelling (to appear).

COMPUTATION OF WAVE FORCES ON THREE DIMENSIONAL OFFSHORE STRUCTURES

M.C. Au and C.A. Brebbia

Southampton University and Computational Mechanics Centre, Southampton.

ABSTRACT

This paper deals with the application of the boundary element method to determine the wave forces on large offshore structures. The simple fundamental solution for the Laplace's equation is used together with the radiation rather than complex fundamental solutions which implicitly satisfy radiation. The free surface and sea bottom have to be discretized using an element mesh which resolution is appropriate for the frequency under consideration. Then the appropriate kinematic and zero velocity boundary conditions are introduced in the free surface and sea bottom respectively.

An original treatment of symmetry conditions is proposed and its implementation is explained in detail. Such treatment considerably reduces the computation time required for large three dimensional structures. Full use of symmetry is made and no elements are required on the symmetry planes.

Several typical examples are presented to illustrate the range of applicability of the technique.

1. INTRODUCTION

Boundary integral equations have been frequently used to solve wave diffraction problems, as illustrated for instance in the work of Garrison [1], Hogben and Standing [2], Faltinsen and Michelson [3] and Taylor [4]. All these authors used a special Green's function which is similar to the one applied by John [5] and by Wehausen and Laitone [6]. This kind of function satisfies all the boundary conditions except those on the structural surface, which needs to be discretized. The main disadvantage of these Green's functions is that they have a complex frequency dependent integral form which requires special integration techniques around the critical frequencies. These techniques have been described by Faltinsen and

Michelsen [3] and Garrison [7]. Furthermore, as the fundamental solution is frequency dependent, completely new computations are required for each different exciting wave, which may be very uneconomical for three dimensional structures.

An alternative to the above integral equation solution is to use finite elements. Here, the whole fluid volume needs to be divided into elements up to a distance far away from the structure where a radiation condition can be applied. This has been done by Berhhoff [8] using a coupling between integral equations and finite elements. Yue, Chen and Mei [9] proposed a hybrid element method. Bettess and Zienkiewicz used infinite elements [10], and Brebbia and Walker [11] deduced a radiation condition for two dimensional problems.

In this paper the boundary element method is used to describe the fluid, and the simple 1/R frequency independent fundamental solution is applied to determine the wave forces on the structure. The fluid surrounding the structure is discretized using boundary elements, and the sea floor, free surface and structure's boundary conditions are applied. In addition the radiation condition is used on the outer elements to simulate the energy losses. In this way the governing boundary element matrices need to be computed only once and can be used for all exciting frequencies.

Symmetry of the structure is taken into consideration to reduce the size of the matrices and the length of the computations. A technique to analyse structures with symmetry is presented and explained in detail. Some applications illustrate the validity of the procedure and the order of economies achieved.

2. FORMULATION OF THE PROBLEM

The problem to be analysed is the prediction of wave forces on three dimensional offshore structures. The assumptions are that the fluid is inviscid, incompressible and irrotational, and that the waves are governed by the linear wave theory. Under these conditions the incident wave potential can be written as,

$$U_o = u_o\, e^{-i\omega t}$$

with
$$u_o = -\frac{iga_o}{\omega}\,\frac{\cosh k(z+h)}{\cosh kh}\,.\,e^{i(k_1x+k_2y)}$$

$$k_1 = k\cos\alpha \qquad \text{and } k_2 = k\sin\alpha \tag{1}$$

where a_o is the incident wave amplitude
k^o is the wave number
ω is the wave frequency

α is the angle of incidence
h is the depth of the water and
g is the acceleration due to gravity.

In the presence of a large structure, this wave potential will be altered and the resulting total potential can be written as consisting of two components,

$$U_t = U_o + U \tag{2}$$

where U is the diffracted wave potential given by

$$U = u\, e^{-i\omega t} \tag{3}$$

The incident potential satisfies Laplace's equation and the total potential will obey the same equation with the appropriate boundary conditions. Under these assumptions, the boundary value problem for the diffracted potential can be expressed as

$$\nabla^2 u = 0 \qquad \text{in the water domain } \Omega(x,y,z) \tag{4}$$

with the following boundary conditions.

(a) On the structural surface (Γ_1)

$$\frac{\partial u}{\partial n} + \frac{\partial u_o}{\partial n} = 0 \tag{5a}$$

(b) On the ocean floor (Γ_2)

$$\frac{\partial u}{\partial n} = 0 \tag{5b}$$

(c) The radiation condition at infinity (Γ_3)

$$\frac{\partial u}{\partial n} - iku = 0 \tag{5c}$$

(d) and finally, on the free water surface (Γ_4)

$$\frac{\partial u}{\partial n} - \frac{\omega^2}{g}\, u = 0 \tag{5d}$$

For simplicity, the above (a) to (d) conditions can be written as

$$\frac{\partial u}{\partial n} + pu - q = 0 \qquad \text{on } \Gamma \tag{6}$$

where $\Gamma = \Gamma_1 + \Gamma_2 + \Gamma_3 + \Gamma_4$

p and q are values which depend on the type of condition.

3. BOUNDARY ELEMENT METHOD

The boundary element method is based on integral equations which can be obtained using weighted residual considerations. For the case of equation (4) and the boundary condition (6) one obtains [11]

$$\int_{\Omega} u^* \nabla^2 u \, d\Omega = \int_{\Gamma} u^* \left(\frac{\partial u}{\partial n} + pu - q\right) d\Gamma \tag{7}$$

where u^* is the weighting function. Integrating left hand side of (7) by parts twice and rearranging the terms, one finds,

$$-\int_{\Omega} u \nabla^2 u^* \, d\Omega + \int_{\Gamma} \left(\frac{\partial u^*}{\partial n} + pu^*\right) u \, d\Gamma$$

$$= \int_{\Gamma} u^* q \, d\Gamma \tag{8}$$

We can now take the simple fundamental solution of $\nabla^2 u^* + \Delta_i = 0$ as the weighting function. That is

$$u^* = \frac{1}{4\pi r} \tag{9}$$

where $r = [(x-x_i)^2 + (y-y_i)^2 + (z-z_i)^2]^{\frac{1}{2}}$

The resulting boundary integral relationships for a point 'i' on the boundary Γ can be written as,

$$c_i u_i + \int_{\Gamma} \left(\frac{\partial u^*}{\partial n} + pu^*\right) u \, d\Gamma = \int_{\Gamma} u^* q \, d\Gamma \tag{10}$$

where $c_i = \frac{1}{2}$ in case the boundary at 'i' is smooth. $c_i = 1$ for any internal point.

As usual, the boundary Γ is discretized into constant, linear or quadratic elements as shown in figure (2) so that the potential within the element 'e' can be written as

$$u = [\Phi] \{u_e^n\} \tag{11}$$

where $[\Phi]$ is the interpolation function. The boundary integral (10) can then be written as

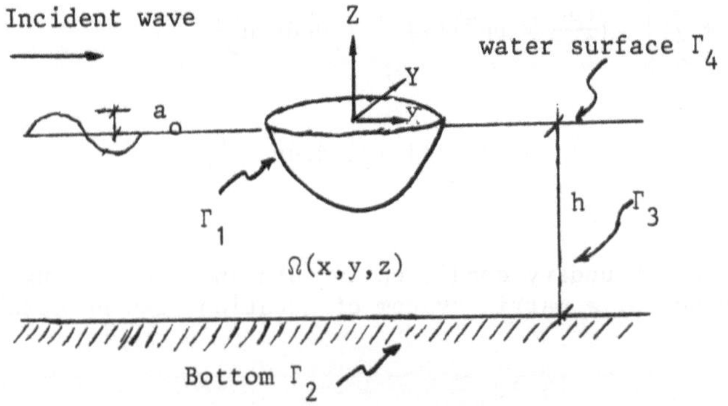

Fig. 1 The wave diffraction of the structure

constant element

$[\Phi] = 1$

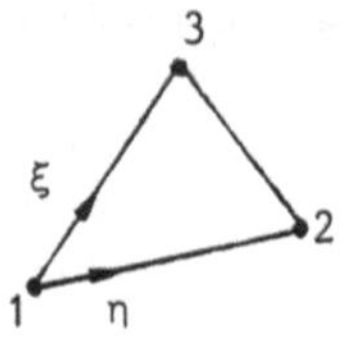

linear element

$[\Phi] = [(1-\xi-\eta),\ \eta,\ \xi]$

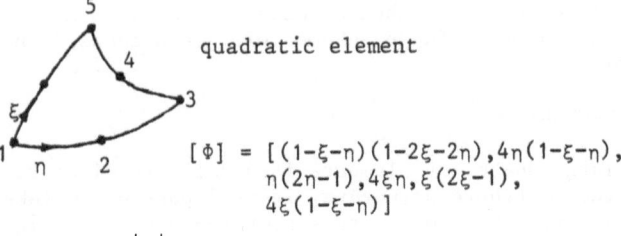

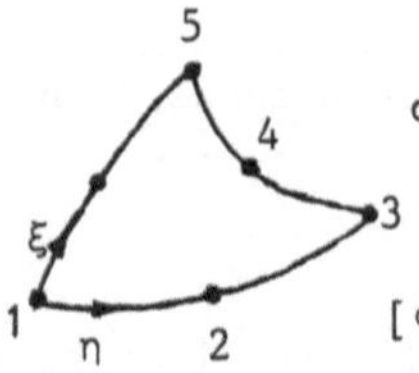

quadratic element

$$[\Phi] = [(1-\xi-\eta)(1-2\xi-2\eta), 4\eta(1-\xi-\eta), \eta(2\eta-1), 4\xi\eta, \xi(2\xi-1), 4\xi(1-\xi-\eta)]$$

with $d\Gamma = |J|\ d\xi d\eta$

where $|J| = \left|\frac{\partial \vec{r}}{\partial \xi} \times \frac{\partial \vec{r}}{\partial \eta}\right|$

Fig. 2 Constant, linear and quadratic elements

$$c_i u_i + \sum_e \int_{\Gamma_e} \left(\frac{\partial u^*}{\partial n} + pu^*\right)[\Phi]\ |J|\ d\xi d\eta \{u_e^n\}$$

$$= \sum_e \int_{\Gamma_e} u^*\ [\Phi]\ |J|\ d\xi d\eta\ \{q_e^n\} \qquad (12)$$

Taking every boundary condition in turn and considering all boundary nodes, a matrix system of equations can be obtained such that,

$$[H_1,\ H_2,\ H_3 - ikG_3,\ H_4 - \frac{\omega^2}{g} G_4] \begin{Bmatrix} u_1 \\ u_2 \\ u_3 \\ u_4 \end{Bmatrix} = [G_1]\ \{q_1\} \qquad (13)$$

where the subscripts 1 to 4 indicate that those matrices are valid for the part of the boundary with conditions (a) to (d) in equation (5). $[H_j]$ and $[G_j]$ ($j = 1,2,3,4$) are the boundary element influence matrices whose elements are produced by integrating the following expressions.

$$\int_{\Gamma_j \cap \Gamma_e} \frac{\partial u^*}{\partial n} [\Phi]\ |J|\ d\xi d\eta \qquad \text{and} \qquad \int_{\Gamma_j \cap \Gamma_e} u^*[\phi]\ |J|\ d\xi d\eta$$

where Γ_j ($j = 1,2,3,4$) indicates the different types of boundaries and Γ_e is the surface of the element 'e'. The analytical solution of these integrals for curved element is generally impossible and it is then necessary to apply numerical integration schemes. These schemes will be discussed in the examples.

4. SYMMETRY

As a large number of elements is usually required to analyse an actual offshore structure, it is important to take advantage of symmetry. Many of these structures present a high degree of symmetry which if taken into consideration can reduce the amount of time required to run an analysis problem. The transformation techniques proposed here have proved to reduce by a considerable amount the computer time required for the solution. This reduction can be up to 80% for some two dimensional wave diffraction applications, as shown in reference [12]. The economies are more marked for three dimensional applications for which computer time becomes very important.

A more compact version of the boundary element matrix equation (13) can be written as

$$[H]\ \{u\} = \lambda\ [G]\ \{q\} \tag{14}$$

where λ is a parameter similar to p defined in (6). Let us consider a transformation applied to the variables {u} and {q}, then

$$\{u\} = [R]\ \{\bar{u}\}$$

and $$\{q\} = [R]\ \{\bar{q}\} \tag{15}$$

Formula (14) can be written as

$$[H]\ [R]\ \{\bar{u}\} = \lambda\ [G]\ [R]\ \{\bar{q}\} \tag{16}$$

Pre-multiplying both sides of (16) by $\frac{1}{\beta}\ [R]^T$, one obtains

$$\frac{1}{\beta}\ [R]^T\ [H][R]\ \{\bar{u}\} = \lambda\ \frac{1}{\beta}\ [R]^T\ [G][R]\{\bar{q}\} \tag{17}$$

or we may write

$$[\bar{H}]\ \{\bar{u}\} = \lambda\ [\bar{G}]\ \{\bar{q}\} \tag{18}$$

where $$[\bar{H}] = \frac{1}{\beta}\ [R]^T\ [H]\ [R] \tag{18a}$$

and $$[\bar{G}] = \frac{1}{\beta}\ [R]^T\ [G]\ [R] \tag{18b}$$

with β as a normalizing factor.

For a symmetrical structure, the elements of the matrices [H] and [G] will have some special arrangements (for instance, one row could be produced by rearranging another) so that in certain cases a matrix [R] can be chosen such that the matrices [H] and [G] will be transformed into a diagonal form as indicated in figure (3). The $[\bar{H}]$ and $[\bar{G}]$ can be written as

$$[\bar{H}] = \begin{bmatrix} \bar{H}_1 & 0 & 0 & 0 & 0 \\ 0 & \bar{H}_2 & \cdot & 0 & 0 \\ 0 & 0 & \cdot & & 0 \\ 0 & 0 & & \cdot & \\ 0 & 0 & & \cdot & \bar{H}_m \end{bmatrix}$$ and

$$[\bar{G}] = \begin{bmatrix} \bar{G}_1 & 0 & 0 & 0 \\ 0 & \bar{G}_2 & 0 & 0 \\ 0 & 0 & \cdot & 0 \\ 0 & 0 & 0 & \bar{G}_m \end{bmatrix} \quad \text{respectively} \tag{19}$$

where m is the total number of diagonal sub-matrices. Then, the solution of the problem can be found by considering the individual sub-matrices in the corresponding diagonal position, i.e.

$$[\bar{H}_r]\{\bar{u}_r\} = \lambda\,[\bar{G}_r]\{\bar{q}_r\}\ ,\ (r = 1,2,\ldots,m) \tag{20}$$

Once these values are obtained the original variables {u} and {q} can be computed by using equation (15).

By applying such transformation, a banded matrix can be used to store $[\bar{H}]$ and $[\bar{G}]$ as indicated in figure (3). In the actual computation, it is not necessary to use the equations (18a) and (18b) to produce $[\bar{H}]$ and $[\bar{G}]$. If $[H^*]$ and $[G^*]$ are the banded matrices of $[\bar{H}]$ and $[\bar{G}]$ respectively, they can be simply written as,

$$[H^*] = [R^*]\{h\}$$

and

$$[G^*] = [R^*]\{g\} \tag{21}$$

where {h} and {g} consists of columns from the [H] and [G] matrices. They are selected depending on the type of symmetry under consideration (see Appendix). $[R^*]$ is a matrix with the coefficients required to produce $[H^*]$ and $[G^*]$. For some of the examples shown later, the type of symmetry and corresponding [R] and $[R^*]$ matrices are given in the Appendix.

5. WAVE FORCES

After the solution in terms of the potential, the dynamic pressure can be obtained by using Bernoulli's equation

$$P = -\rho\,\frac{\partial}{\partial t}\{\mathrm{Re}(U_t)\} \tag{22}$$

where ρ is the density of the water and
$\mathrm{Re}(U_t)$ is the real part of the total potential.

The wave forces on the structure can be obtained by integrating the pressure over the boundary Γ_1. This is given by

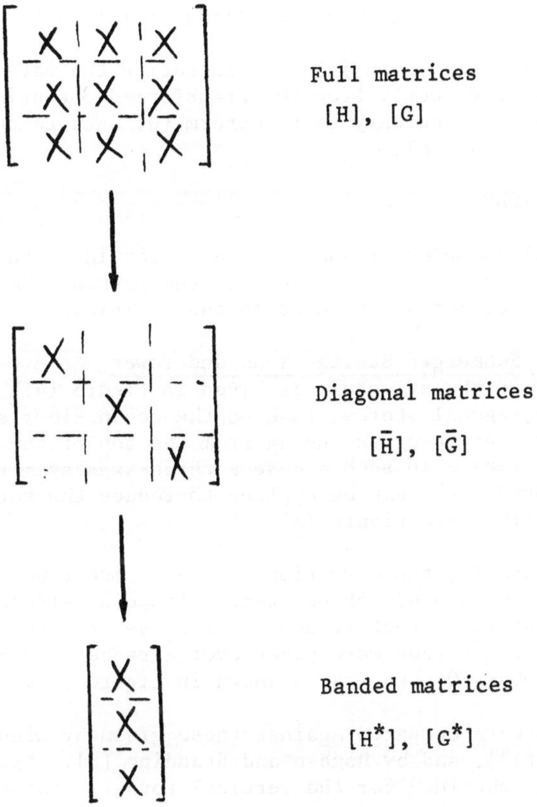

Fig. 3 The transformation of [H] and [G]

$$\{F\} = \int_{\Gamma_1} \begin{Bmatrix} n_x \\ n_y \\ n_z \end{Bmatrix} P \, d\Gamma \tag{23}$$

where $(n_x, n_y, n_z)^T$ is the unit outwards normal on Γ_1.

If symmetry is taken into consideration the wave forces can be obtained directly from the transformed potential $\{\bar{u}\}$. Notice it is not necessary to transform $\{\bar{u}\}$ back to the original potential $\{u\}$.

6. APPLICATIONS

Three typical examples of submerged and floating structures were used to show the applications of the method. Constant and quadratic elements were used in the examples.

Example (i) Submerged Storage Tank and Tower

The geometry of the structure is given in figure (4). It consists of a hexagonal storage tank on the ocean floor and a vertical circular tower extending from the top of the tank to the water surface. In such a case a three-axes symmetry (as shown in appendix ii) can be applied to reduce the total boundary by 1/6 - see figure (6).

As an example, the direction of the incident wave was considered to be $\alpha = 0$. 56 constant triangular elements were used on 1/6 of the structure and the results for the horizontal and the vertical forces were given over a range of dimensionless wave numbers $0 < kh < 4$ as shown in figure (5).

Results were compared against those given by Zienkiewicz and Bettess [13], and by Hogben and Standing [2]. Excellent agreement was obtained for the vertical forces. For the horizontal forces, the results agreed with the 3-D finite element results obtained by Zienkiewicz and Bettess and with Hogben and Standing's results for lower kh values.

Example (ii) Floating Vertical Cylinder

The second example is a vertical cylinder floating on the free surface as shown in figure (7). The cylinder has radius (a) which is two times its draught and is floating vertically in water of depth (1.5a).

Symmetry along four axes was considered and 40 constant elements were used to discretize the boundary of 1/8 domain as shown in figure (9). The results obtained were compared to those due to Garrett [14]. Agreement was found for the horizontal forces as shown in figure (8). However, more elements were needed to determine the correct vertical forces.

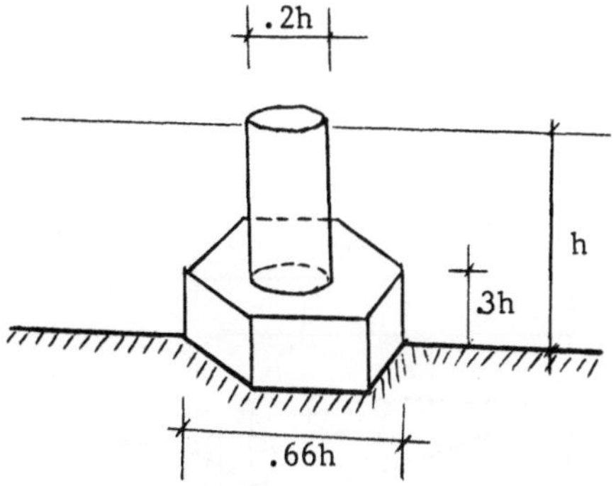

Fig. 4 Submerged storage tank and tower

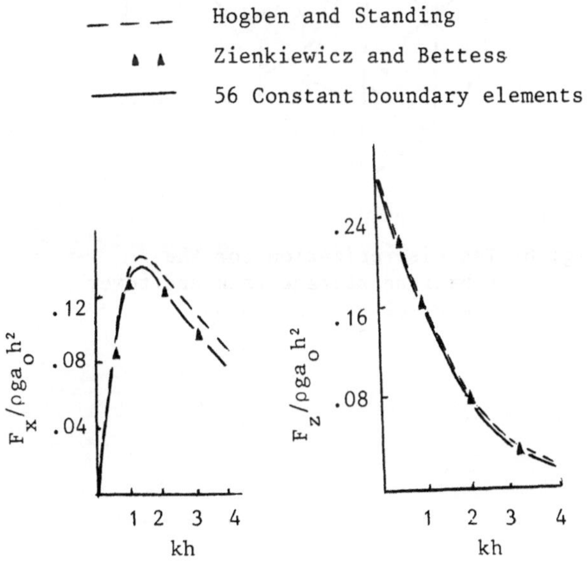

Fig. 5 The horizontal and the vertical forces on the submerged storage tank and tower

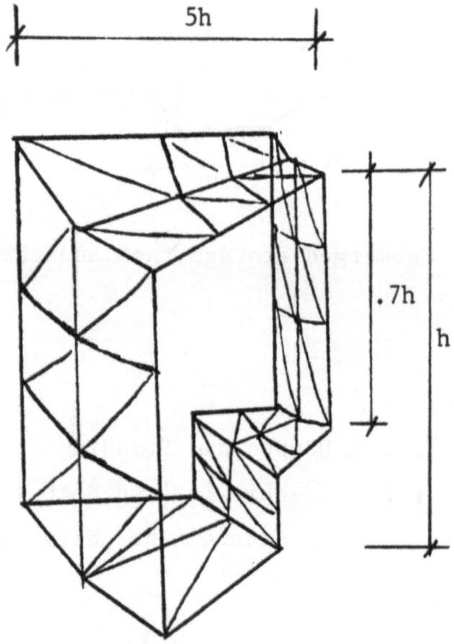

Fig. 6 The discretization for the submerged storage tank and tower Number of elements = 56

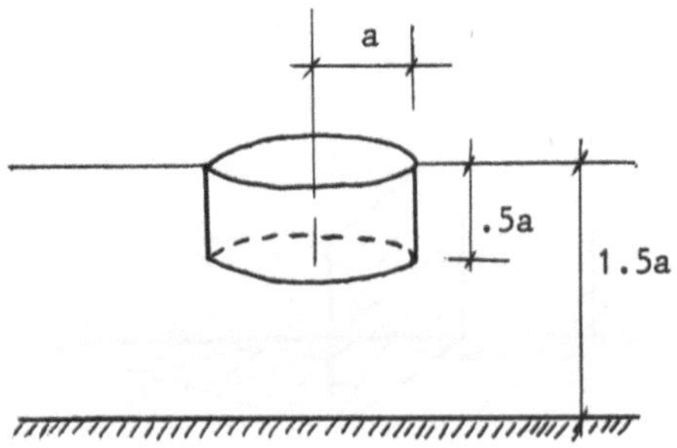

Fig. 7 Floating vertical cylinder

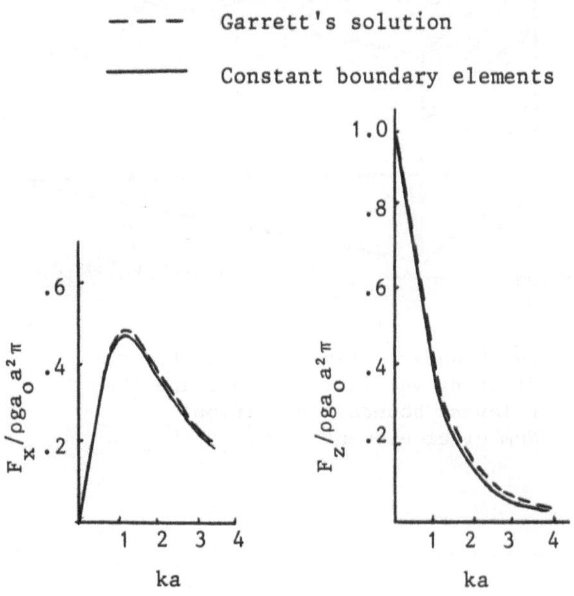

Fig. 8 The horizontal and the vertical forces on the floating vertical cylinder

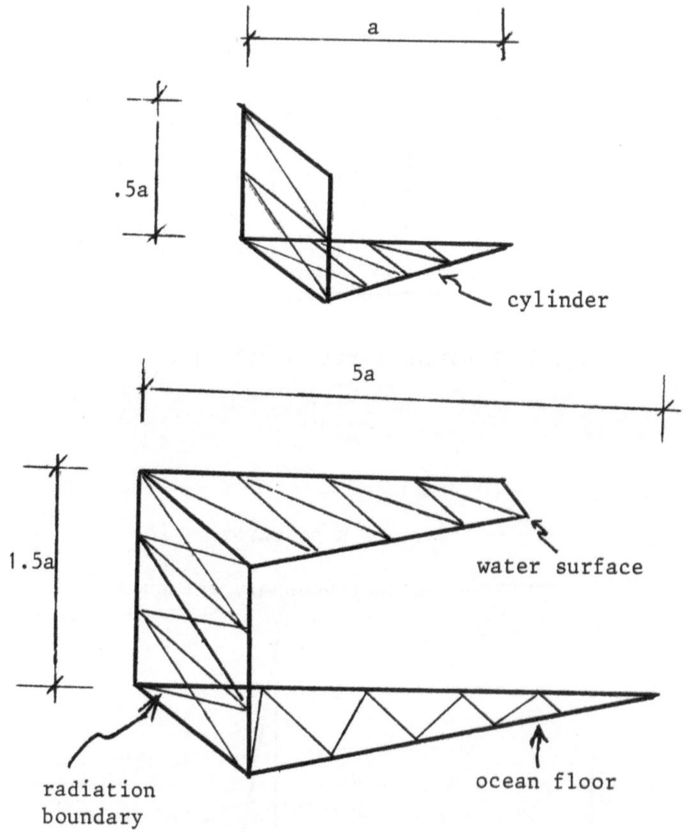

Fig. 9 The boundary element meshes for the floating vertical cylinder and the relevant boundary condition
Number of element = 40

126 elements were used to produce the result for these forces shown in figure (8) and the resulting forces are in agreement with those given by Garrett [14].

Example (iii) Submerged Hemisphere on the Ocean Floor

To illustrate the application of higher order elements quadratic six noded triangular elements were developed and used in this example. The geometry of the submerged hemisphere is given in figure (10).

For the constant and the linear element, the integrals over the element with the singularity can be obtained analytically but this is not possible for the quadratic element. The 1/R singularity was overcome by employing the numerical integration formulae developed by Pina, Fernandes and Brebbia [15].

For comparison purpose, the asymptotic solution given by Chakrabarti and Naftzger [16] was considered. Although this is an asymptotic solution, its accuracy is reasonable when compared against the available experimental data. Two planes of symmetry and 10 quadratic elements were used to compute the numerical solution. Both the horizontal and the vertical forces are in excellent agreement with those reported by Chakrabarti and Naftzger as shown in figure (11). For the purpose of illustrating the convergence of the results, the numerical values obtained by 8 and 10 elements were given in table 1 together with those due to Chakrabarti and Naftzger [16].

7. CONCLUSIONS

The boundary element method has been successfully applied to determine the wave forces on large offshore structures. The procedure adopted uses the simple fundamental solution for Laplace's equation (i.e. type 1/R) which means that the same governing equations can be used for different exciting frequencies.

The paper also presents a technique for applying symmetry which considerably reduces the computational time needed for the analysis of three dimensional structures.

REFERENCES

1. GARRISON, C.J. "Hydrodynamics of large objects in the sea, Part I - Hydrodynamic analysis", J. of Hydronautics, 8, 5-12.

2. HOGBEN, N. and STANDING, R.G."Wave loads on large bodies" Proc. Int. Sym. on Dynamics of Marine Vehicles and Structures in Waves. Published by I. Mech. E., London 1974.

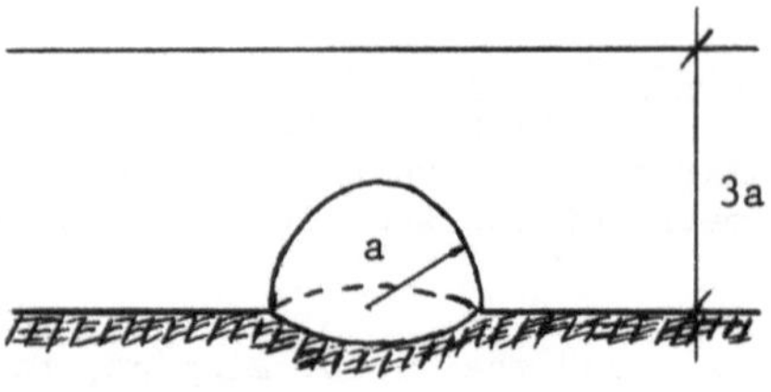

Fig. 10 Submerged hemisphere on the ocean floor

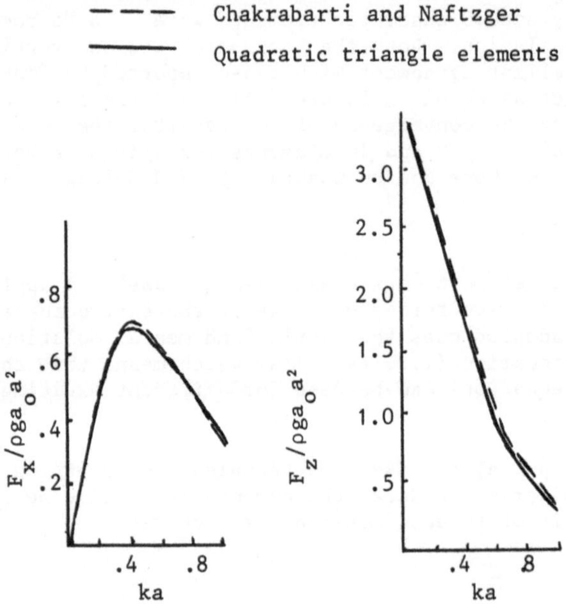

Fig. 11 The horizontal and the vertical forces on the submerged hemisphere

3. FALTINSEN, O.M. and MICHELSEN, F.C. "Motions of large structures in waves at zero Froude number" Proc. Int. Symposium Dynamics of Marine Vehicles and Structures in Waves. 1974, No.11, I. Mech. Eng.

4. TAYLOR, R.E. "Generalised hydrodynamic forces on vibrating offshore structures by wave diffraction techniques", in the "Offshore Structures Engineering" edited by F.L.L.B. Carneiro, A.J. Ferrante and C.A. Brebbia. 1977 Pentech Press, London.

5. JOHN, F. "On the motion of floating bodies", 1949, Comm. in Pure and Applied Math., Part I.

6. WEHAUSEN J.V. and E.V. LAITONE "Surface Waves", Encyclopedia of Physics, Vol.9, 1960, Springer-Verlag, Berlin, 446-778.

7. GARRISON, C.J. "Hydrodynamic Loading of Large Offshore Structures: Three-dimensional source distribution methods" Ch.3, Numerical Method in Offshore Engineering, edited by O.C. Zienkiewicz, R.W. Lewis and K.G. Stagg, Wiley-Interscience publication (1978)

8. BERKHOFF, J.C.W. "Linear wave propagation problems and the finite element method", 251-280, Finite elements in fluids, Vol. 1, edited by R.H. Gallagher et al., Wiley, London.

9. YUE, D.K.P., CHEN, H.S. and MEI, C.C. "Hybrid element method for diffraction of water waves by three dimensional bodies" In. J. Num. Meth., Eng., Vol.12, 245-266, (1978).

10. ZIEKNIEWICZ, O.C. and BETTESS, P. "Diffraction and refraction of surface waves using finite and infinite elements", In. J. Num. Meth., Eng., 11, 1271-1290, (1972).

11. BREBBIA, C.A. and WALKER, S. "Boundary Element Techniques in Engineering". Butterworths, London, 1980.

12. BREBBIA, C.A. and AU, M.C. "Diffraction of water waves for vertical cylinders by boundary elements", Applied Mathematical Modelling, Vol. No. , 1982.

13. ZIENKIEWICZ, O.C. and BETTESS, P. "Fluid-structure dynamic interaction and wave forces. An introduction to numerical treatment". In. J. Num. Meth. Eng., 1978, 13, 1-16.

14. GARRETT, C.J.R. "Wave forces on a circular dock" J. Fluid Mech., 1971, 46, Pt. 1, 129-139.

	$F_x/\rho g a_o a^2$			$F_z/\rho g a_o a^2$		
ka	8 elements	10 elements	Chakrabarti	8 elements	10 elements	Chakrabarti
0.2	0.517	0.522	0.530	2.639	2.658	2.671
0.4	0.675	0.682	0.694	1.740	1.763	1.792
0.6	0.589	0.597	0.607	1.024	1.047	1.088
0.8	0.443	0.452	0.452	0.578	0.599	0.640
1.0	0.310	0.318	0.312	0.320	0.338	0.376

Table 1 The horizontal and the vertical forces on the submerged hemisphere on the ocean floor

15. PINA, H.C.G., FERNANDES, J.L.M. and BREBBIA, C.A. "Some numerical integration formulae over triangles and squares with a 1/R singularity" Appl. Math. Modelling, June 1981, Vol.5, 209-211.

16. SUBRATA, K., CHAKRABARTI, and NAFTZGER, R.A. "Non-linear wave forces on half-cylinder and hemisphere" J. Waterways Harbors and Coastal Eng., Div. ASCE, Aug. 1974, 100, 189-204.

APPENDIX

The following three symmetrical cases were used in the examples presented in this paper. As the same pattern applies for [H*] and [G*] matrices only the form of [H*] is given.

(i) Two-axes symmetry

Relative locations

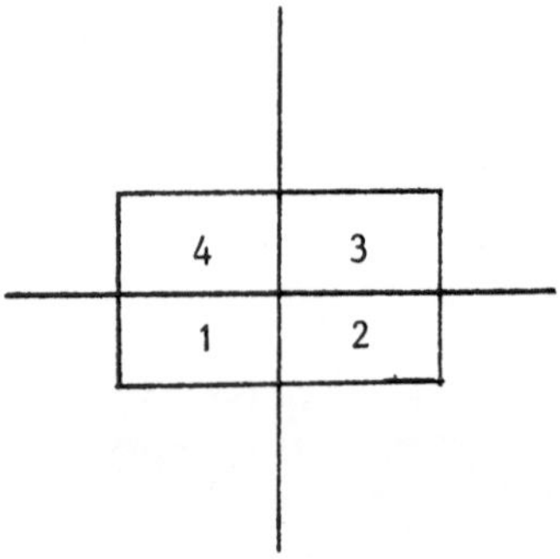

$\beta = 4$

$$R = \begin{bmatrix} 1 & 1 & 1 & 1 \\ 1 & -1 & 1 & -1 \\ 1 & 1 & -1 & -1 \\ 1 & -1 & -1 & 1 \end{bmatrix}$$

R*	h_1	h_2	h_3	h_4
H_1^*	1	1	1	1
H_2^*	1	-1	1	-1
H_3^*	1	1	-1	-1
H_4^*	1	-1	-1	1

(ii) Three-axes symmetry

Relative locations

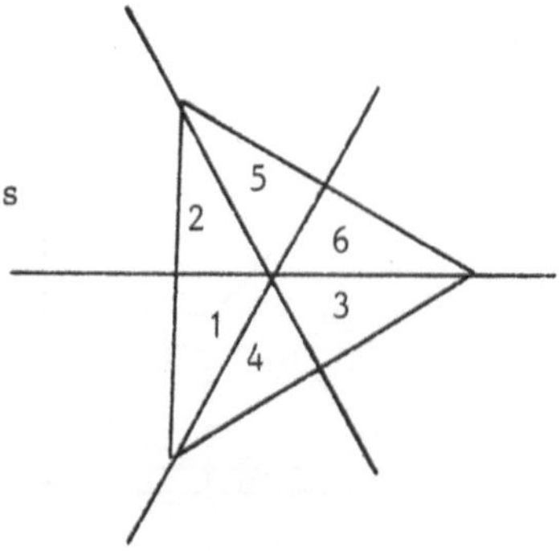

$\beta = 6$

$$R = \begin{bmatrix} 1 & 1 & -1 & -1 & -1 & 1 \\ 1 & -1 & -1 & -1 & 1 & -1 \\ 1 & 1 & 2 & -1 & -1 & -2 \\ 1 & -1 & -1 & 2 & -2 & -1 \\ 1 & 1 & -1 & 2 & 2 & 1 \\ 1 & -1 & 2 & -1 & 1 & 2 \end{bmatrix}$$

R^*	h_1	h_2	h_3	h_4	h_5	h_6	h_1	h_2	h_3	h_4	h_5	h_6
H_1^*	1	1	1	1	1	1						
H_2^*	1	-1	1	-1	1	-1						
H_3^*	2	-1	-1	2	-1	-1	-1	2	2	-1	-1	-1
	-1	2	-1	-1	2	-1	2	-1	-1	-1	-1	2
H_4^*	2	1	-1	1	-1	-2	1	2	1	-1	-2	-1
	1	2	-2	-1	1	-1	2	1	-1	-2	-1	1

(iii) Four-axes symmetry

Relative Locations

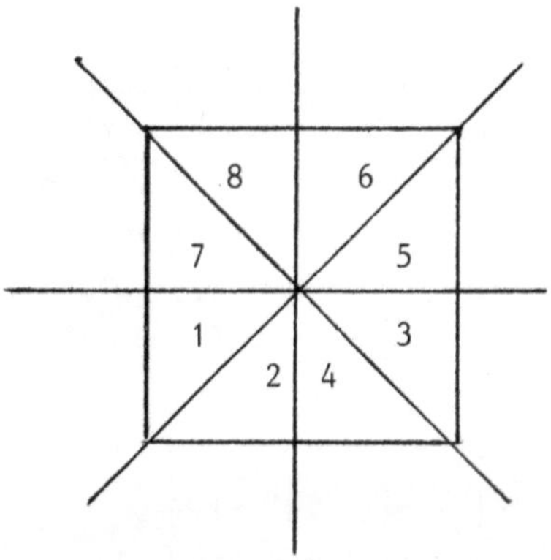

$\beta = 8$

$c = \sqrt{2}$

$$R = \begin{bmatrix} 1 & 1 & 1 & 1 & c & 0 & c & 0 \\ 1 & -1 & 1 & -1 & 0 & c & 0 & c \\ 1 & 1 & -1 & -1 & c & 0 & -c & 0 \\ 1 & -1 & -1 & 1 & 0 & c & 0 & -c \\ 1 & 1 & 1 & 1 & -c & 0 & -c & 0 \\ 1 & -1 & 1 & -1 & 0 & -c & 0 & -c \\ 1 & 1 & -1 & -1 & -c & 0 & c & 0 \\ 1 & -1 & -1 & 1 & 0 & -c & 0 & c \end{bmatrix}$$

	h_1	h_2	h_3	h_4	h_5	h_6	h_7	h_8	h_1	h_2	h_3	h_4	h_5	h_6	h_7	h_8
H_1^*	1	1	1	1	1	1	1	1								
H_2^*	1	−1	1	−1	1	−1	1	−1								
H_3^*	1	1	−1	−1	1	1	−1	−1								
H_4^*	1	−1	−1	1	1	−1	−1	1								
H_5^*	1	0	1	0	−1	0	−1	0	0	1	0	1	0	−1	0	−1
	0	1	0	−1	0	−1	0	1	1	0	−1	0	−1	0	1	0
H_6^*	1	0	−1	0	−1	0	1	0	0	1	0	−1	0	−1	0	1
	0	1	0	1	0	−1	0	−1	1	0	1	0	−1	0	−1	0

MODIFIED INTEGRAL EQUATION SOLUTION OF STEADY VISCOUS FLOW IN A BIFURCATING CHANNEL

D.B. Ingham and M.A. Kelmanson
Department of Applied Mathematical Studies
University of Leeds
Leeds LS2 9JT
England

ABSTRACT

Numerical solutions of the biharmonic equation are given for steady two dimensional viscous flow through a channel formed by two infinite parallel stationary planes. These planes are separated along the channel centreline by a third semi-infinite parallel stationary plane around one end of which the flow bifurcates. The flow proceeds from a parabolic velocity profile far upstream of the bifurcation to two identical parabolic profiles far downstream.

A biharmonic boundary integral equation method (BBIE) is used to solve for the flow in the region near the bifurcation. Because there is no slip velocity on the planes the point of bifurcation is a mathematical singularity of the solution domain and at this point the vorticity becomes unbounded. Solution of this problem using these unbounded values invariably causes inaccuracies. Hence a modified BBIE (MBBIE) is presented which takes into account the analytical nature of the singularity. The results indicate much improved convergence on using this MBBIE. An analytic expansion is found for the velocity and vorticity in the neighbourhood of the singularity and a graphical comparison of the streamlines near the singularity is given for each method.

INTRODUCTION

In recent years there has been a rapid increase in the application of integral techniques to problems in the fields of elastostatics, heat transfer and potential flow. Comprehensive summaries of such work have been given by Jaswon and Symm (1977), Brebbia (1978) and Banerjee and Butterfield (1981). Although several authors have solved viscous flow problems using finite element (FE) and finite differences (FD) techniques (Burggraf (1966) and Pan and Acrivos (1967) being notable examples), there seems to be very little previous work on the solution of such problems using BIE methods. Kelmanson (1982) formulates and applies a BBIE on a viscous flow problem and then formulates a MBBIE which incorporates the analytic nature of a singularity arising in the solution domain for the problem. The results from the MBBIE are in excellent agreement with those produced by the analytical methods of previous authors.

Several fundamental viscous flow problems occur in which, for a variety of reasons, the vorticity at certain points in the solution domain becomes unbounded. This is generally due to a discontinuity in boundary conditions: the point of discontinuity is referred to as a mathematical singularity of the flow field. Effects of singularities in viscous flow problems described by the biharmonic equation have been investigated analytically by, for example, Richardson (1970) and Watson (1981). Numerical investigations have been performed by Dennis and Smith (1980), Coleman (1981) etc. Dennis and Smith (1980) employ a FD technique which excludes the mesh point at which a singularity occurs in their problem - indeed, they remark that it is not really known what influence the singularity has on the solution. Coleman (1981) uses a discretized contour integral approach and tries to take account of the singularity in his problem by a 'packing of boundary nodes' into the troublesome region. Neither method accounts satisfactorily for the presence of the singularity and in the absence of an analytic solution we must treat such neglect with some caution.

In the present work we not only take account of the presence of the singularity but also determine its analytic nature in the form of a truncated series expansion which is valid in the neighbourhood of the singularity. We investigate the stream function ψ for viscous flow which satisfies the biharmonic equation

$$\nabla^4\psi = 0$$

in a channel between two infinite parallel plates which are separated along the channel centreline by a third semi-infinite parallel plate. The flow bifurcates around the upstream end of this semi-infinite plate: this point is a singularity of the solution domain. Far upstream of the bifurcation the velocity profile is parabolic; far downstream it has split into two symmetric flows about the centre plate, each with a parabolic velocity profile. This problem has several physical applications, for example, the slow flow of blood along bifurcating arteries.

We shall later see how application of the MBBIE to this problem appreciably improves the rate of convergence of the solution near the singular point at the bifurcation.

FORMULATION

To solve for the biharmonic stream potential ψ and vorticity ω

in the region Ω enclosed by boundary $\partial\Omega$ we first split the biharmonic equation into its coupled form

$$\nabla^2\psi = \omega \tag{1}$$

$$\nabla^2\omega = 0 \tag{2}$$

so that the problem reduces to solving two simultaneous elliptic equations. Defining the functions G_1 and G_2 by

$$G_1(p,q) = \log|p - q| \tag{3}$$

and

$$G_2(p,q) = \tfrac{1}{4}|p - q|^2 \{\log|p - q| - 1\} \tag{4}$$

where $p \in \Omega + \partial\Omega$ and $q \in \partial\Omega$, we have

$$\nabla^2 G_i(p,q) = \eta(p)\delta(p,q) \qquad i = 1, 2 \tag{5}$$

where

$$\delta(p,q) = \begin{cases} 1 \text{ if } |p - q| = 0 \\ 0 \text{ if } |p - q| \neq 0 \end{cases}$$

and

$$\eta(p) = \begin{cases} 0 \text{ if } p \notin \Omega + \partial\Omega \\ \text{internal angle included between the tangents to } \partial\Omega \text{ on either side of } p \text{ if } p \in \partial\Omega \\ 2\pi \text{ if } p \in \Omega . \end{cases}$$

Invoking Green's Theorem on Equations (1) and (2) and using relations (3), (4) and (5) gives the following expressions at the general field point p.

$$\eta(p)\psi(p) = \int_{\partial\Omega} \{\psi(q)G_1'(p,q) - \psi'(q)G_1(p,q) + \omega(q)G_2'(p,q) - \omega'(q)G_2(p,q)\}dq \tag{6}$$

$$\eta(p)\omega(p) = \int_{\partial\Omega} \{\omega(q)G_1'(p,q) - \omega'(q)G_1(p,q)\}\, dq \tag{7}$$

where

(i) $p \in \Omega + \partial\Omega$, $q \in \partial\Omega$
(ii) dq denotes the differential increments of $\partial\Omega$ at q.
(iii) the prime ' refers to differentiation with respect to the outward normal to $\partial\Omega$ at q.

From Equations (6) and (7) it is apparent that given any two of ψ, ψ', ω and ω' (or a linear combination of them) at each point $q\in\partial\Omega$ we can obtain an explicit integral for $\psi(p)$ and $\omega(p)$ at the general field point $p\in\Omega + \partial\Omega$. In practice Equations (6) and (7) can rarely be integrated analytically. Solution of these equations is effected by dividing $\partial\Omega$ into N smooth straight line segments $\partial\Omega_j$, j=1, . . . , N at the centre of each of which lies boundary nodes q_j, j=1, . . ., N. Over each interval $\partial\Omega_j$ we assume that ψ, ψ', ω and ω' have the piecewise constant values ψ_j, ψ_j', ω_j and ω_j', $j=1, \ldots, N$. A discretized form of Equations (6) and (7) are then applied at the midpoint $p \equiv q_i$, $i=1, \ldots, N$ of each interval.

It should be noted that the integrals in the discretized equations were evaluated analytically in the present study.

Previous BBIE formulations evaluate the integrals

$$\int_{q\varepsilon\partial\Omega_j} G_2(p,q)dq$$

and

$$\int_{q\varepsilon\partial\Omega_j} G_2'(p,q)dq$$

using numerical quadrature (cf. Maiti and Chakrabarty (1974)). The present formulation gives computer cpu time savings of up to 40% over numerical quadrature and is, of course more accurate.

Solution of the discretized equations generates the two unknown boundary conditions at each point. At this stage, ψ_j, ψ_j', ω_j and ω_j' are known for each $j=1, \ldots, N$. Hence applying a discretized form of Equations (6) and (7) at the general point $p \varepsilon \Omega + \partial\Omega$ provides $\psi(p)$ and $\omega(p)$ everywhere in the solution domain. In order to describe the process by which the BBIE is modified to incorporate analytically any singular behaviour in the solution domain we consider the following problem.

We solve for the stream function ψ describing viscous flow which satisfies

$$\nabla^4\psi = 0 \tag{8}$$

in the infinite channel $-1 \leq y \leq 1$ containing the semi-infinite strip $y = 0$, $x < 0$. Defining the x and y components of velocity by

$$u = \frac{\partial\psi}{\partial y} \tag{9a}$$

and

$$v = -\frac{\partial\psi}{\partial x} \tag{9b}$$

respectively, the boundary conditions for the bifurcating flow problem are

$$\psi = -1 \quad , \quad \psi_y = 0 \quad \text{on } y = +1 \tag{10a}$$

$$\psi = 0 \quad , \quad \psi_y = 0 \quad \text{on } y = 0,\ x < 0 \tag{10b}$$

$$\psi = 1 \quad , \quad \psi_y = 0 \quad \text{on } y = -1 \tag{10c}$$

The flow is from right to left and is generated by a parabolic velocity profile as $x \to +\infty$. As $x \to -\infty$, the velocity profile consists of two identical parabolic profiles, and the flow in the whole channel is symmetric about $y = 0$. Hence we need only solve for ψ in the lower half channel $-1 \le y \le 0$. The parabolic velocity profiles were applied a distance $x = \pm x_m$ both upstream and downstream of the bifurcation. Taking $x_m = 2$ was found to be sufficiently large. Hence the boundary conditions to impose on the region $-2 \le x \le 2$, $-1 \le y \le 0$ are

$$\psi = 0 \quad , \quad \psi_y = 0 \quad \text{on } y = 0,\ x < 0 \tag{11a}$$

$$\psi = 0 \quad , \quad \psi_{yy} = 0 \quad \text{on } y = 0,\ x > 0 \tag{11b}$$

$$\psi = 1 \quad , \quad \psi_y = 0 \quad \text{on } y = -1 \tag{11c}$$

$$\psi_y = \frac{3}{2}(y^2 - 1) \quad , \quad \psi_x = 0 \quad \text{on } x = +2 \tag{11d}$$

$$\psi_y = 6y(y + 1) \quad , \quad \psi_x = 0 \quad \text{on } x = -2. \tag{11e}$$

Condition (11b) arises from the absence of any shear stress on the channel centreline in $x > 0$. The parabolic profiles in conditions (11d) and (11e) are those which allow unit mass flow past $x = \pm 2$ in unit time. Figure 1 illustrates the geometry of the problem.

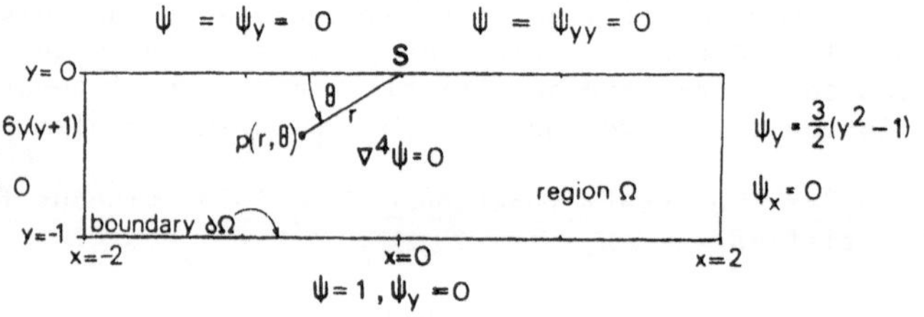

Figure 1. Geometry and boundary conditions for the problem

Solutions to the problem were obtained using BBIE's employing discretizations of 70, 140 and 280 boundary segments of equal length. The solution converged rapidly except in the vicinity of the singularity at the pont (0,0) which we shall hereafter refer to as S. This is because unbounded values of ω near S are affecting the stream function solution via Equations (6) and (7).

Modification of the BBIE is necessary if accurate solutions are to be obtained throughout the entire solution domain. This is achieved by incorporating into the BBIE the analytic behaviour of the solution in the vicinity of the singularity. Separated solutions of the biharmonic equation are obtained in polar coordinates (r,θ) centred on S, see Figure 1, and these are used to generate a truncated series expansion for the stream function which is valid for small r. If we denote these separated solutions by

$$\psi(r,\theta) = r^{\lambda+1} f_{\lambda}(\theta) \tag{12}$$

then it is found that (Moffatt (1964))

$$f_o(\theta) = A\cos\theta + B\sin\theta + C\theta\cos\theta + D\theta\sin\theta \tag{13a}$$

$$f_1(\theta) = A\cos 2\theta + B\sin 2\theta + C\theta + D \tag{13b}$$

$$f_{\lambda}(\theta) = A\cos(\lambda+1)\theta + B\sin(\lambda+1)\theta + C\cos(\lambda-1)\theta + D\sin(\lambda-1)\theta \tag{13c}$$

The constants A, B, C and D in Equations (13) are found from the four boundary conditions given in (11a) and (11b). Combining Equations (13) with these conditions gives

$$f_o(0) = f_o'(0) = f_o(\pi) = f_o''(\pi) = 0 \tag{14a}$$

$$f_1(0) = f_1'(0) = f_1(\pi) = f_1''(\pi) = 0 \tag{14b}$$

$$f_{\lambda}(0) = f_{\lambda}'(0) = f_{\lambda}(\pi) = f_{\lambda}''(\pi) = 0 \tag{14c}$$

where primes refer to derivatives with respect to θ. Relations (14) imply that

$$f_o(\theta) = f_1(\theta) = 0$$

$$f_{\lambda}(\theta) = \begin{cases} a_{\lambda}\{\cos(\lambda+1)\theta - \cos(\lambda-1)\theta\} \quad \text{for } \lambda = \frac{1}{2}, \frac{3}{2}, \frac{5}{2}, \ldots & (15a) \\ b_{\lambda}\{(\lambda-1)\sin(\lambda+1)\theta - (\lambda+1)\sin(\lambda-1)\theta\} \text{for } = 2,3,4,\ldots & (15b) \end{cases}$$

where a_{λ} and b_{λ} are arbitrary constants. Using Equation (12) and the linearity of the governing equations we have

$$\psi(r,\theta) = \sum_{k=1}^{\infty} \beta_k r^{\lambda_k+1} f_{\lambda_k}(\theta) \tag{16}$$

as the analytic solution near the singularity at (0,0). The a_{λ}'s and b_{λ}'s in expression (15a) and (15b) are now absorbed into the β_k in Equation (16); as yet these β_k are unknown. Defining M to be the integer such that $\lambda_M < 3$ but $\lambda_{M+1} \geq 3$ and the functions g and χ by

$$g(r,\theta) = \sum_{k=1}^{M} \beta_k r^{\lambda_k+1} f_{\lambda_k}(\theta) \tag{17}$$

$$\psi = \chi + g \tag{18}$$

means that χ is a biharmonic function whose derivatives remain bounded up to fourth order. In fact, Equations (17) and (18) imply that χ and its derivatives up to fourth order all tend to zero as $r \to 0$ i.e. as we approach S. Now defining the function h by

$$\nabla^2\chi = h \tag{19}$$

gives

$$\nabla^2 h = 0 \quad . \tag{20}$$

Whereas ω and ω' became unbounded as $r \to 0$ in the BBIE, the functions h and h' tend to zero as $r \to 0$ in the MBBIE, as the analytic nature of the singularity has been 'removed' by the introduction of Equations (17) and (18). Equations (15) and (17) imply that, for this problem, the function g is given by

$$g(r,\theta) = \beta_1 r^{\frac{3}{2}} \{\cos \frac{3\theta}{2} - \cos \frac{\theta}{2}\} + \beta_2 r^{\frac{5}{2}} \{\cos\frac{5\theta}{2} - \cos\frac{\theta}{2}\}$$

$$+ \beta_3 r^3\{\sin 3\theta - 3\sin\theta\} + \beta_4 r^{\frac{7}{2}} \{\cos\frac{7\theta}{2} - \cos\frac{3\theta}{2}\} \tag{21}$$

where the $\beta_1, \ldots, \beta_4$ are unknown. In solving Equations (19) and (20) the MBBIE generates 2N equations for the 2N unknown boundary values of χ, χ', h and h'. Since $\beta_1, \ldots, \beta_4$ are unknown, there are 2N + 4 unknowns in total. We now use the fact that $h' \to 0$ as $r \to 0$ in the MBBIE and so we prescribe

$$h_j' = 0$$

at the four boundary segments $\partial\Omega_j$ nearest to S, thus reducing the number of unknowns to 2N.

We now proceed to solve the coupled systems of Equations (19) and (20) in a manner analogous to the solution of Equations (6) and (7) as described previously. This gives χ_p and h_p at the general point $p \in \Omega + \partial\Omega$. Equation (21) and the values of $\beta_1, \ldots, \beta_4$ enable us to evaluate g_p at this same field point. Then Equation (18) provides the MBBIE stream potential at p.

RESULTS AND DISCUSSION

The BBIE and MBBIE were applied to the bifurcating flow problem for three discretizations employing 70, 140 and 280 boundary segments of equal length. Convergence of the results was seen to be good except in the vicinity of the singularity at S. In fact, results produced by the same discretization from each method differed only by an order of 1% in the region away from the singularity. However, near to S the results from the BBIE were very slow to converge. Table 1 shows the BBIE solution for the stream function for the three discretizations in the region $-0.1 \le x \le 0.1$, $-0.1 \le y \le 0$. Results are displayed at eighty-one equally spaced points in this region. Comparison of figures at each point shows the slow rate of convergence of the solution near S. Table 2 shows an equivalent distribution of MBBIE results in the same region. Note that by incorporating the analytic nature of the singularity into the method, the convergence of results has been rapidly accelerated. Further, the MBBIE requires only about 5% more execution time than the corresponding BBIE, and hence the accuracy of results has been noticeably improved without appreciably lengthening the method.

S

0.0081 0.0004 0.0000	0.0064 -0.0007 0.0000	0.0008 0.0031 0.0000	-0.0112 0.0002 0.0002	-0.0368 -0.0146 0.0012	-0.0072 -0.0028 -0.0057	-0.0068 0.0017 0.0007	-0.0025 -0.0017 -0.0003	0.0043 -0.0007 -0.0003
0.0060 0.0014 0.0007	0.0074 0.0022 0.0008	0.0086 0.0033 0.0013	0.0089 0.0045 0.0024	0.0071 0.0050 0.0039	0.0054 0.0056 0.0056	0.0059 0.0072 0.0072	0.0073 0.0085 0.0085	0.0089 0.0097 0.0095
0.0088 0.0037 0.0028	0.0105 0.0048 0.0033	0.0123 0.0065 0.0043	0.0135 0.0086 0.0061	0.0138 0.0107 0.0089	0.0142 0.0130 0.0120	0.0156 0.0155 0.0149	0.0177 0.0178 0.0172	0.0198 0.0198 0.0192
0.0128 0.0074 0.0063	0.0150 0.0090 0.0072	0.0173 0.0112 0.0088	0.0195 0.0140 0.0113	0.0213 0.0173 0.0150	0.0231 0.0207 0.0190	0.0255 0.0242 0.0229	0.0282 0.0273 0.0262	0.0309 0.0301 0.0290
0.0180 0.0124 0.0110	0.0207 0.0144 0.0124	0.0236 0.0172 0.0146	0.0265 0.0207 0.0178	0.0294 0.0247 0.0220	0.0324 0.0290 0.0268	0.0356 0.0332 0.0313	0.0389 0.0370 0.0354	0.0422 0.0405 0.0391
0.0243 0.0184 0.0169	0.0274 0.0209 0.0188	0.0308 0.0242 0.0215	0.0344 0.0282 0.0252	0.0382 0.0329 0.0299	0.0420 0.0378 0.0352	0.0459 0.0426 0.0403	0.0499 0.0471 0.0451	0.0536 0.0513 0.0494
0.0316 0.0255 0.0238	0.0351 0.0285 0.0261	0.0390 0.0322 0.0293	0.0432 0.0367 0.0335	0.0476 0.0418 0.0387	0.0520 0.0472 0.0443	0.0566 0.0525 0.0499	0.0611 0.0576 0.0552	0.0653 0.0623 0.0601
0.0399 0.0336 0.0317	0.0437 0.0369 0.0344	0.0480 0.0410 0.0381	0.0527 0.0459 0.0427	0.0576 0.0514 0.0481	0.0626 0.0572 0.0540	0.0676 0.0629 0.0600	0.0726 0.0685 0.0658	0.0773 0.0736 0.0712
0.0489 0.0425 0.0404	0.0531 0.0462 0.0436	0.0578 0.0507 0.0476	0.0628 0.0559 0.0525	0.0681 0.0616 0.0582	0.0736 0.0677 0.0644	0.0790 0.0738 0.0707	0.0844 0.0797 0.0768	0.0895 0.0853 0.0826

Nodes
70 140 280

Table 1. BBIE stream function ψ in the region $-0.1 \le x \le 0.1$, $-0.1 \le y \le 0$

S

0.0000 0.0000 0.0000	0.0000 0.0000 0.0000	0.0000 0.0000 0.0000	0.0000 0.0000 0.0000	-0.0001 0.0000 0.0000	-0.0001 0.0000 0.0000	-0.0001 0.0000 0.0000	0.0000 0.0000 0.0000	0.0001 0.0000 0.0000
0.0006 0.0006 0.0006	0.0006 0.0007 0.0007	0.0008 0.0008 0.0008	0.0011 0.0011 0.0011	0.0023 0.0023 0.0023	0.0047 0.0048 0.0048	0.0065 0.0065 0.0065	0.0079 0.0079 0.0079	0.0090 0.0090 0.0090
0.0025 0.0025 0.0025	0.0028 0.0028 0.0028	0.0033 0.0033 0.0034	0.0043 0.0043 0.0043	0.0066 0.0066 0.0066	0.0101 0.0102 0.0102	0.0133 0.0134 0.0134	0.0159 0.0159 0.0159	0.0181 0.0181 0.0181
0.0057 0.0058 0.0058	0.0063 0.0064 0.0064	0.0073 0.0074 0.0074	0.0091 0.0092 0.0092	0.0122 0.0123 0.0123	0.0165 0.0165 0.0166	0.0206 0.0207 0.0207	0.0242 0.0243 0.0243	0.0273 0.0274 0.0274
0.0101 0.0102 0.0102	0.0111 0.0112 0.0112	0.0127 0.0128 0.0128	0.0152 0.0153 0.0153	0.0189 0.0190 0.0190	0.0237 0.0238 0.0238	0.0286 0.0286 0.0287	0.0330 0.0331 0.0331	0.0369 0.0370 0.0370
0.0156 0.0158 0.0158	0.0171 0.0172 0.0172	0.0192 0.0193 0.0194	0.0223 0.0224 0.0224	0.0266 0.0267 0.0267	0.0317 0.0318 0.0319	0.0371 0.0372 0.0372	0.0422 0.0423 0.0423	0.0468 0.0469 0.0469
0.0222 0.0224 0.0224	0.0241 0.0243 0.0243	0.0268 0.0269 0.0269	0.0304 0.0305 0.0305	0.0350 0.0352 0.0352	0.0406 0.0407 0.0407	0.0464 0.0465 0.0465	0.0520 0.0521 0.0521	0.0571 0.0572 0.0573
0.0298 0.0300 0.0300	0.0321 0.0323 0.0323	0.0352 0.0354 0.0354	0.0392 0.0394 0.0394	0.0443 0.0444 0.0444	0.0501 0.0502 0.0502	0.0562 0.0563 0.0563	0.0622 0.0623 0.0623	0.0679 0.0680 0.0680
0.0382 0.0384 0.0385	0.0410 0.0412 0.0412	0.0445 0.0447 0.0447	0.0489 0.0491 0.0491	0.0542 0.0544 0.0544	0.0602 0.0604 0.0604	0.0666 0.0668 0.0668	0.0729 0.0731 0.0731	0.0790 0.0791 0.0792

Nodes

70
140
280

Table 2. MBBIE stream function ψ in the region $-0.1 \leq x \leq 0.1$, $-0.1 \leq y \leq 0$

In Table 3 we present the coefficients $\beta_1, \ldots, \beta_4$ in the analytic expansion for the function g. Results are again presented for the three different MBBIE discretizations and then extrapolated by Richardson's method (Smith (1978)). Notice that as is usual in these types of extrapolation the β_k's converge more rapidly the smaller the value of k.

N	β_1	β_2	β_3	β_4
70	-1.195830	-0.671268	0.416963	0.190976
140	-1.200898	-0.629447	0.393341	0.118114
280	-1.201589	-0.613674	0.380281	0.061327
Extrapolated values	-1.201659	-0.604126	0.364132	-0.139282

Table 3. A comparison of singularity expansion coefficients

Observe that these values of β_k provide us with seies expansions for the fluid velocity near the singularity. Under the polar system employed as in Figure 1, the components of fluid velocity near S are given by

$$u(r,\theta) \sim \frac{\partial g}{\partial y} = \sum_{k=1}^{4} \beta_k r^{\lambda_k} \{-(\lambda_k+1)\sin\theta f_{\lambda_k}(\theta) - \cos\theta f_{\lambda_k}'(\theta)\} \quad (22)$$

$$v(r,\theta) \sim \frac{-\partial g}{\partial x} = \sum_{k=1}^{4} \beta_k r^{\lambda_k} \{(\lambda_k+1)\cos\theta f_{\lambda_k}(\theta) - \sin\theta f_{\lambda_k}'(\theta)\} . \quad (23)$$

Since the β_k, λ_k and f_{λ_k} are all known for $k=1, \ldots, 4$ then expressions (22) and (23) provide the fluid velocity near the singularity. For the special case of fluid velocity on the centreline $y = 0$ of the channel, in $x > 0$ we have

$$u(x,0) \sim \left.\frac{\partial g}{\partial y}\right|_{y=o} = \left.\frac{1}{r}\frac{\partial g}{\partial \theta}\right|_{\substack{r=x \\ \theta=\pi}} = 2\beta_1 x^{1/2} - 2\beta_2 x^{3/2} + 2\beta\, x^{5/2} \quad (24)$$

as the velocity of particles approaching the point of bifurcation along the streamline $\psi = 0$. Note also that the vorticity in the neighbourhood of S is given by

$$\omega(r,\theta) \sim \nabla^2 g = \sum_{k=1}^{4} \beta_k\, r^{\lambda_k - 1} \{(\lambda_k+1)^2 f_{\lambda_k}(\theta) + f_{\lambda_k}''(\theta)\} \quad (25)$$

and so even though the vorticity is unbounded as $r \to \infty$ we can determine accurately the nature of the vorticity behaviour in this region.

In Figure 2 we display a contour plot comparing streamlines ψ = constant generated by the BBIE and MBBIE in the vicinity of the singularity S. It is readily apparent that as we approach S that the accuracy of the BBIE deteriorates whereas the streamlines retain their continuous form when using MBBIE. The plots were generated using the boundary information from the discretizations employing 280 boundary nodes.

CONCLUSIONS

By incorporating the analytic form of the singularity into the MBBIE accurate solutions to a singular viscous flow problem has been obtained using a direct boundary integral approach. The employment of analytic integration of the kernel functions in the discretized equations has enabled more accurate results to be obtained via more efficient methods than previously presented BBIE's.

A MBBIE provides solutions of greater accuracy to those produced by a BBIE employing several times as many boundary nodes.

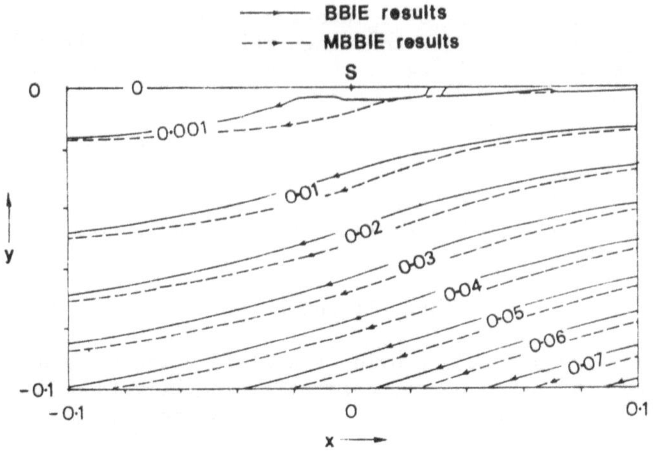

Figure 2. A comparison of streamlines near the singularity S.

Although in the present work a simple flow geometry and singularity form have been considered it is hoped to extend the MBBIE to solve problems involving viscous flows near sharp corners in re-entrant solution domains. In such geometries some of the eigenvalues in the singularity expansion are complex and a further modification of the method will be required.

The methods presented in this work can also be applied to general industrial problem involving viscous flows in channels containing discontinuities in cross-section as, for example, occurs in the junction between two pipes of different size. The authors are presently engaged on the problem of flows in constricting channels.

ACKNOWLEDGEMENTS

The financial support received by M.A. Kelmanson from the Science and Engineering Research Council is gratefully acknowledged.

REFERENCES

Banerjee and Butterfield (1981) Boundary Element Methods in Engineering Science . McGraw-Hill, U.K.
Brebbia, C.A. (1978) Recent Advances in Boundary Element Methods . Pentech Press, London.
Burggraf, O.R. (1966) Analytical and numerical studies on the structure of steady separated flows . J. Fluid Mech. 24, 1:113-151.

Coleman, C.J. (1981) A contour integral formulation of plane creeping Newtonian flow. Q.J. Mech. Appl. Math. 34, 4:453-464.
Dennis, S.C.R. and Smith, F.T. (1980) Steady flow through a channel with a symmetrical constriction in the form of a step. Proc. R. Soc. Lond. A372:393-414.
Jaswon, M.A. and Symm, G.T. (1977) Integral Equation Methods in Potential Theory and Elastostatics. Academic Press, London.
Kelmanson, M.A. (1982) An integral equation method for the solution of singular flow problems. Submitted for publication.
Maiti, M. and Chakrabarty, S.K. (1974) Integral equation solutions for simply supported polygonal plates. Int. J. Enging Sci. 12:793-806.
Moffatt, H.K. (1964) Viscous and resistive eddies near a sharp corner. J. Fluid Mech. 18, 1:1-18.
Pan, F. and Acrivos, A. (1967) Steady flows in rectangular cavities.J. Fluid Mech. 28, 4:643-655.
Richardson, S. (1970) A 'stick-slip' problem related to the motion of a free jet at low Reynolds numbers. Proc. Camb. Phil. Soc. 67:477-489.
Smith, G.D. (1978) Numerical Solution of P.D.E.'s: Finite Difference Methods. Clarendon Press, Oxford: 217.
Watson, E. (1981) Private communication.

Coleman, C.J. (1981) A contour integral formulation of plane creeping Newtonian flow. Q.J. Mech. Appl. Math. 34, 453-464.
Dennis, S.C.R. and Smith, F.T. (1980) Steady flow through a channel with a symmetrical constriction in the form of a step. Proc. R. Soc. Lond. A372, 393-414.
Jaswon, M.A. and Symm, G.T. (1977) Integral Equation Methods in Potential Theory and Elastostatics. Academic Press, London.
Kelmanson, M.A. (1983) An integral equation method for the solution of singular slow flow problems. [illegible]
[illegible] (1974) [illegible] equation [illegible] supported polygonal [illegible] [illegible] 105-[illegible]
Moffatt, H.K. (1964) Viscous and resistive eddies near a sharp corner. J. Fluid Mech. 18, 1-18.
Pan, F. and Acrivos, A. (1967) Steady flows in rectangular cavities. J. Fluid Mech. 28, 643-655.
Richardson, S. (1970) A 'stick-slip' problem related to the motion of a free jet at low Reynolds numbers. Proc. Camb. Phil. Soc. 67, 477-489.
Smith, G.D. (1978) Numerical Solution of Partial Differential Equations: Finite Difference Methods. Clarendon Press, Oxford.
[illegible] (1981) Private communication.

Session III
Solid Mechanics

AN IMPLEMENTATION OF STRESS DISCONTINUITY IN THE BOUNDARY ELEMENT METHOD

A. Gakwaya, G. Dhatt, A. Cardou

Université Laval, Québec, Canada G1K 7P4

1. INTRODUCTION

Since the last two decades, continuous developments in digital computers have been influencing every aspect of engineering analysis. We thus find that the numerical methods are presently playing a dominant role for solving practical engineering problems. The finite element method, the development of which is closely related to progress in computer technology, is considered today as a powerful analytical tool for an engineer in industry.

In the present study, we are interested in employing the boundary element method - BEM, (also known as the boundary integral equation method) for solving 3-dimensional elasticity problems. In this method, the partial differential equations over a region are first transformed to integral equations over the boundary of the region through the use of their basic fundamental solutions (or singular functions). The integral expression is then discretized by using finite element approximations to obtain a set of algebraic equations. Since the dimensions of a problem are reduced by one by the boundary integral equation formulation, this method may in certain situations lead to computational economy as compared with the FEM. Moreover, the rapid variations of stresses are better approximated by the boundary integral equation method due to the special nature of weighting functions employed.

This method has been successfully employed in a wide variety of engineering problems in fields of elastostatics [3,4, 5, 10, 11], elastodynamics [16, 17] and thermoelasticity [9]. It has been applied as well to certain problems of plasticity, viscoplasticity [12,13,14,15], and fracture mechanics [15,17], etc. The efficiency of the BEM as compared with the FEM depends entirely on the type of problems studied and the nature of variations of gradients in the zone of the interest.

Certain authors e.g.[11,12,20] claim that the BEM requires relatively less computational efforts than the FEM. However, we believe that the validity of such a claim is highly dependent on the type of problems studied especially due to unfavorable facts: significant computational efforts for integrating the singularity terms, unsymmetrical and full matrices to be solved (one may employ subregion approach [10,11] to obtain some sort of block banded matrices), enormous programming efforts and some difficulty to generalize this method to general non-linear problems.

In this paper, we apply the BEM and FEM to several test problems of 3-dimensional linear elasticity in order to assess the relative efficiency of the two methods and to study the effects of stress discontinuity along edges in the BEM on the accuracy of results. Though the problem of stress discontinuity has been identified by Cruze [3,4], its effects on deterioration of accuracy and consequent remedies have not been adequately studied in the literature. When one employs constant elements (unknowns are defined on the interior of an element), the problem of edge discontinuity does not arise. However, use of isoparametric elements (unknowns are defined on the corner and mid-point nodes which are shared by various elements) over regions with discontinuous boundary surfaces, forming edges and corners, requires that the discontinuity of surface tractions must be properly taken into account in order to obtain reasonable precision. We thus present a new scheme (or algorithm) to introduce properly the discontinuous traction forces without taking recourse to uneconomical technique of node separation [21] along edges and corners. A number of test examples dealing with cube and thick cylinder are presented to demonstrate the effectiveness of the present algorithm for introducing discontinuous boundary tractions. The results presented demonstrate that a high accuracy may be achieved using relatively small number of elements if discontinuity of tractions and continuity of displacements are correctly implemented.

2. ELASTICITY EQUATIONS

Consider a body occupying a finite domain V with boundary S (assumed to be piecewise smooth) in a three dimensional Euclidean space R^3. For a homogeneous and isotropic domain with density ρ, and Lamé's constant λ, μ such that $\rho > 0$, $\mu > 0$, $3\lambda + 2\mu > 0$ (ellipticity), the elastostatic state under body forces $\{F\}$ is defined by the pair ($\{u\}$, $[\sigma]$) such that [6]:

a) $[\sigma] \in C^1(V)$, $\{u\} \in C^2(V) \cdot C^1(\overline{V})$ (1)

where C^1, C^2 denote the set of continuous differentiable functions up to order 1 and 2 respectively, with • representing the intersection. $\overline{V}$ is the domain with its frontiers, $[\sigma]$ is the stress tensor and $\{u\}$ is the displacement field.

b) Equilibrium equations in V are:

$$\sum_{j=1}^{3} \frac{\partial\sigma_{ij}}{\partial x_j} + \rho F_i = 0 \qquad i = 1, 2, 3 \tag{2}$$

c) Constitutive equations for Hookean materials are

$$[\sigma] = \lambda \mathrm{tr}(\nabla\{u\})[I] + 2\mu[E] \tag{3}$$

where $\{u\}^T = \langle u\rangle = \langle u_1, u_2, u_3\rangle$

$\mathrm{tr}(\nabla\{u\})$: trace of the vector gradient

E = linear strain tensor with

$$E_{ij} = \frac{1}{2}\left(\frac{\partial u_i}{\partial x_j} + \frac{\partial u_j}{\partial x_i}\right) \tag{4}$$

I = Identity matrix

x_i = cartesian coordinates x_1, x_2, x_3

Using eqs.(2) and (3), we obtain

$$\mu\, \Delta u_i + (\lambda+\mu)\frac{\partial}{\partial x_i}(\mathrm{tr}\, \nabla\{u\}) + \rho F_i = 0 \qquad i = 1,2,3 \tag{5}$$

where

$$\Delta u_i = \frac{\partial^2 u_i}{\partial x_1^2} + \frac{\partial^2 u_i}{\partial x_2^2} + \frac{\partial^2 u_i}{\partial x_3^2} \tag{6}$$

$$\mathrm{tr}(\nabla\{u\}) = \frac{\partial u_1}{\partial x_1} + \frac{\partial u_2}{\partial x_2} + \frac{\partial u_3}{\partial x_3} = \mathrm{Div.}\ \{u\}$$

Equation (5) may be written in a matrix form as:

$$[L]\{u\} + \rho\{F\} = 0 \tag{7}$$

where

$$[L]_{3\times 3} = \mu\Delta[I] + (\mu+\lambda)\left[\frac{\partial^2}{\partial x_i \partial x_j},\ i,j = 1,2,3\right] \tag{8}$$

The surface traction on a point y on the boundary S is given by

$$\{t(y)\} = [\sigma]\{n\} \tag{9}$$

where $\{n\}$ represents the direction cosines of the outward normal to the boundary at y. By using constitutive relations (3), we may write eq. (9) in terms of displacements:

$$\{t\} = [T]\{u\} \qquad \text{on } S \tag{10}$$

where

$$[T]_{3\times 3} = (\lambda+\mu)[A] + \mu[A]^T \tag{11}$$

$$[A] = \begin{Bmatrix} n_1 \\ n_2 \\ n_3 \end{Bmatrix} \left\langle \frac{\partial}{\partial x_1} \quad \frac{\partial}{\partial x_2} \quad \frac{\partial}{\partial x_3} \right\rangle$$

or

$$[T] = (\lambda+2\mu)[A] + \mu[A^T - A] \tag{12}$$

The boundary conditions are divided into two parts:

Geometrical conditions:

$$u_i(y) = \bar{u}_i(y) \quad \text{for a point } y \text{ on } S_1 \tag{13}$$

Force conditions:

$$t_i(y) = T_{ij}u_j = \bar{t}_i(y) \quad \text{for a point } y \text{ on } S_2 \tag{14}$$

where $S_1 + S_2 = S$ and S_1 and S_2 represent parts of the boundary S having geometrical and force conditions respectively. In certain situations either S_1 or S_2 may be void. $\bar{u}_i$ and $\bar{t}_i$ are specified values of displacements and traction forces respectively.

The problem of elastostatics consists thus in finding the pair (u, σ) which satisfies eqs. (2), (3) over V and eqs. (13), (14) over S_1 and S_2.

3. INTEGRAL EQUATION FORMULATION

We may formulate the problem of elastostatics in the form of integral equations if it is possible to define a kernel function $[U^*(x,Q)]$ such that, for eq. (7):

$$[L][U^*(x,Q)] = \delta(Q)[I] \quad \text{over } V \tag{15}$$

where Q is any point within V and δ is the Dirac function at point Q. The kernel function $[U^*]$ has components [1].

$$U^*_{ij} = \frac{1}{4\pi\mu(\lambda+2\mu)}\frac{1}{r}\left[(\lambda+3\mu)\delta ij + (\lambda+\mu)\frac{\partial r}{\partial x_i}\frac{\partial r}{\partial x_j}\right] \tag{16}$$

where

$$r = |x - y(Q)|$$

$$\frac{\partial r}{\partial x_i} = \frac{x_i - y_i(Q)}{r} = -\frac{\partial r}{\partial y_i(Q)} \tag{17}$$

If $\{u\}$ is any solution of eq. (7), one obtains:

$$\int_V [U^*(x,Q)]^T([L]\{u\} + \rho\{F\})dV \equiv 0 \tag{18}$$

By applying two successive integrations by parts, we obtain:

$$\int_V ([L][U^*]^T\{u\}dV + \int_V [U^*]^T\{\rho F\}dV$$

$$= \int_S [U^*]^T([T]\{u\})dS - \int_S ([T][U^*])^T\{u\}dS \tag{19}$$

Using eqs. (10) and (15) we arrive at:

$$u_i(Q) = \int_V \langle a\rangle\{\rho F\}dV + \int_S (\langle a\rangle\{t\} - \langle t^*_a\rangle\{u\})dS \qquad i = 1,2,3 \tag{20}$$

where

$$\langle a\rangle = \langle U^*_{1i}, U^*_{2i}, U^*_{3i}\rangle$$

$$\{t^*_a\} = [T]\{a\}$$

If the body forces $\{F\}$ are zero, then

$$u_i(Q) = \int_S (\langle a\rangle\{t\} - \langle t^*_a\rangle\{u\})dS \tag{21}$$

for **x** interior to domain V.

The problem of elastostatics in the form of integral equations is defined by eq. (20) or (21). Equation (21) is the boundary integral equation we shall exploit in the sequel. We seek the displacement field {u} such that eq. (19) is satisfied for all boundary values on S.

If we choose the point Q=P on the boundary S, one obtains through a limiting process:

$$C_{ij}u_j(P) = \int_S (\langle a\rangle\{t\} - \langle t^*_a\rangle\{u\})dS \qquad i = 1,2,3 \tag{22}$$

where C_{ij} are the elements of the "characteristic function" of the region V with respect to the operator [L] at the boundary point P[7]. If the surface is smooth (continuous tangent plane) at P then $C_{ij} = \frac{1}{2}\delta_{ij}$ as discussed for example by Brebbia [1,2].

The solution of eq. (22) leads to the evaluation of {u} and {t} on all boundary points. This solution is then employed to calculate the displacement field within the domain V by using eq. (20). Internal stresses are evaluated by applying the operator [T] to eq. (20) [see 1, 2 for details].

4. DISCRETIZATION BY FINITE ELEMENTS

An approximate solution of eq. (22) is obtained by choosing a finite element approximation for the unknowns {u} and {t} on the boundary and employing a point collocation method for the integral equation. The collocation points P are the nodal points of boundary finite elements leading thus to a square matrix. We choose an 8 nodes isoparametric element with approximations:

$$\begin{aligned} u_i(\xi) &= \langle N(\xi,\eta)\rangle\{u_n\} \\ t_i(\xi) &= \langle N\rangle\{t_n\} \qquad \text{with } y = y\,(\xi,n) \end{aligned} \tag{23}$$

where N and (ξ,η) are respectively the interpolation functions and intrinsic coordinates on the reference elements given in standard books on finite elements [8, also 1], and $\{u_n\}$ and $\{t_n\}$ represents nodal variables for u_i or t_i.

The discretized equation for a collocation point P_k corresponding to noke k are:

$$C_i u_j(P_k) = \sum_{\text{Elements}} \int_{-1}^{1}\int_{-1}^{1} (\langle a(P_k,y)\rangle[N]\{t_n\} - \langle t^*_a(P_k,y)\rangle[N]\{u_n\})\det J d\xi d\eta \tag{24}$$

$$\text{with} \quad i = 1,2,3$$
$$k = 1,2 \ldots n_t$$

where n_t is the total number of nodes, [N] is obtained from eq. (23) by reorganizing for all element nodal variables. Σ represents assemblage over all elements, y is a point on surface elements and J is the jacobian matrix of the transformation [see Ref.1,8]. Equation (24) can also be written as:

$$[C]\{u(P_k)\} = \sum^{e} \int_{-1}^{1} \int_{-1}^{1} ([U^*(P_k,y)][N]\{t_n\} - [T^*(P_k,y)][N]\{u_n\}) \det J \, d\xi d\eta \tag{25}$$

Application of equation (25) at all collocation nodes leads to a set of full matrix equations

$$[H]\{U_n\} = [G]\{T_n\} \tag{26}$$

where [H] is obtained from the left hand side and the second term of the right hand side of eq. (25), and [G] from the first term of the right hand side. $\{U_n\}$ and $\{T_n\}$ include all surface nodal displacements ($n = 3 \times n_t$) and nodal tractions.

By introducing boundary conditions on u_i and t_i given in eqs. (13) and (14), we obtain after rearrangement of matrix coefficients, the following system of equations with n unknowns:

$$[K]\{X\} = \{F_n\} \tag{27}$$

where $\{X\}$ groups all unknowns in $\{U_n\}$ and $\{T_n\}$, and $\{F_n\}$ is equivalent to a load vector.

5. NUMERICAL TREATMENT OF INTEGRAL EQUATIONS

In practical situations, integrals appearing in (24), cannot be evaluated analytically. One is required to employ numerical integration to evaluate integrals of the form:

$$I = \int_{-1}^{1} \int_{-1}^{1} K \cdot d\xi \, d\eta \tag{28}$$

where $K = (a_1(P_k,y) \text{ or } a_2(P_k,y) \text{ or } a_3) \, N_{ij} \cdot (\det J)$

or

$$K = (t^*_{a_1} \text{ or } t^*_{a_2}, \text{ or } t^*_{a_3}) \, N_{ij} \quad (\det J) \tag{29}$$

If the expressions in (29) are not singular, the integration is straightforward from the Gauss integration scheme:

$$I = \sum_{j=1}^{n_1} \sum_{i=1}^{n_2} w_i \, w_j \, K(\xi_i, \eta_j) \tag{30}$$

where $(n_1 \times n_2)$ is the number of integration points, w_i are weight coefficients and ξ_i, η_i are the coordinates of integration points. Unfortunately, in contrast to the integrals arising in the finite element method for elastostatic problem, some of the integrals in the BEM are singular. Such integrals arise when the integrand involves $1/r^2(P_k,y)$ and the point y belongs to the element $S^{(e)}$ over which is located. As has been pointed out in the literature [7,10,12,15,18], the success of a BEM program will greatly depend on the efficiency with which the calculation of the matrix elements in eq. (24) is performed and thus it is a crucial item in the whole process!

There arise two different cases:

(i) $P_k \notin S^{(e)}$ and (ii) $P_k \in S^{(e)}$

Since in the case (ii), some of the kernels are of $0(\frac{1}{r^2})$, and others at least of $0(\frac{1}{r})$, the integrands while integrable require special treatment.

Case (i), $P_k \notin S^{(e)}$ [see 7,10]: This is a more straightforward case, and we can use standard Gaussian quadrature formula such as eq. (30). The choice of order of integration n_1, n_2 is a more or less sophisticated process [10,11]. For simplicity of programming, we decided to choose n_1, n_2 based on numerical experiment and do not subdivide the element $S^{(e)}$ as in [11].

Case (ii), $P_k \in S^{(e)}$: In the present situation, the integral I requires special attention due to singularities of the integrands. Moreover since these elements represent contributions generally of greatest numerical value to the coefficient matrices, they should be computed with extra care to obtain any accuracy [7]. In the method we decided to utilize, the element is subdivided into triangular subelements, according to the location of the point P_k (fig. 1).

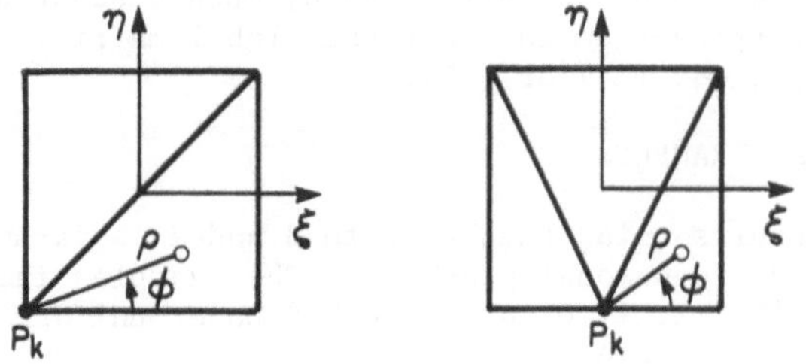

Fig. 1. Subdivision of Element into subelements

We introduce a local polar coordinate system (ρ,ϕ) with origin at P_k such that the element area $d\xi d\eta$ becomes $\rho d\rho d\phi$. This transformation is such that for sufficiently small value of $r = \sqrt{\sum_i (x_i(P_k)-y_i)^2}$, the strong singularity of kernel t_a^*

is reduced and the corresponding integral can be evaluated using the same type of Gaussian integration as for case (i). It was observed that a relatively low number (3 or 4 Gauss points) of integration points was required to achieve a reasonable accuracy. Now the integral over the element is made up of integrals over each subelement, i.e. for kernel t^*_a for example:

$$\int_{S^{(e)}} [K] d\xi d\eta = \sum_{i=1}^{\Delta} \iint_{\Delta i} [K] \rho d\rho d\phi \tag{31}$$

where now ρ, ϕ are the explicit Gauss variables of integration and Δi represents a triangle subelement as represented in Fig.1.

6. SCHEME FOR STRESS DISCONTINUITY

The discontinuity of traction at points located at edges (normal discontinuity) with imposed displacements has not been treated in a clear manner in the literature dealing with 3-dimensional problems. It may be mentioned that the present authors are aware of works in references [3,4,11,12,13,15,18, 22] where no satisfactory answer to this problem has been given. One aim of this study is to investigate the effect of such discontinuities on accuracy and choice of mesh sizes.

We assume that the displacement field is continuous throughout the domain. Thus the boundary displacements are treated in a global form. However for tractions, in addition to the global physical tractions defining the problem by eq. (26), we introduce local boundary tractions pertinent to each element called in the sequel "element stress discontinuities". If an element does not present such discontinuity, its tractions boundary conditions are given by the global boundary conditions.

Thus in the assembling process, the coefficients corresponding to such nodes are multiplied by such discontinuity data, before transferring them to the final global matrix. That is at the level of eq. (24) or (25).

7. NUMERICAL EXAMPLES

We present results using the BEM with 8 nodes isoparametric elements for 3-dimensional problems. The examples include:

- A cube (±1, ±1, ℓ) with a length ℓ under uniform tension and bending,
- A thick cylindrical shell under internal pressure,
- A practical problem of a spatial gear tooth under static loading.

The results are presented to show comparative influence of traction boundary conditions on points located on edges and corner assuming that:

- tractions are continuous
- tractions are discontinuous (introduced as discussed in section 6).

The relative efficiency of BEM and FEM is compared as well for the cases studied.

7.1 Cube with traction continuity

The convergence of traction values are studied using various mesh sizes for a cubic beam (±1, ±1, ±1) as shown in Fig. 2. The beam is fixed at one end (x_1=-1) and is subject to uniform tension at x_1=1 by imposing a displacement u_1=1.

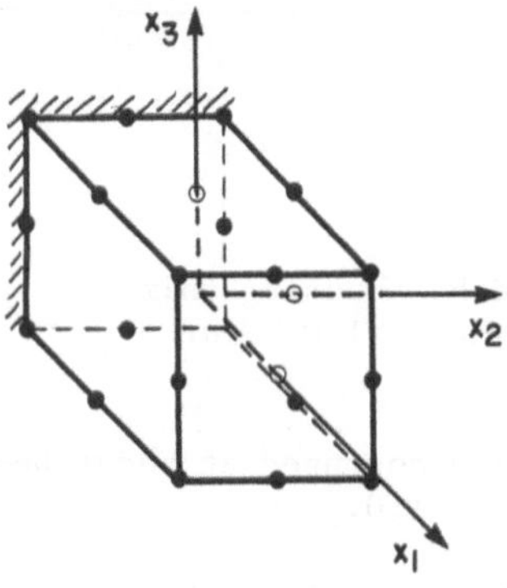

$E = 1.0\ N/mm^2,\ \nu = 0$

(u_i, x_i[mm], t_i[N/mm²])

Boundary conditions:

$u_1 = 1.0$ at $x_1 = 1.0$

$u_1 = u_2 = u_3 = 0$ at $x_1 = -1.0$

$u_2 = u_3 = 0$ elsewhere

Figure 2.a Cubic beam under uniform tension (6 elements)

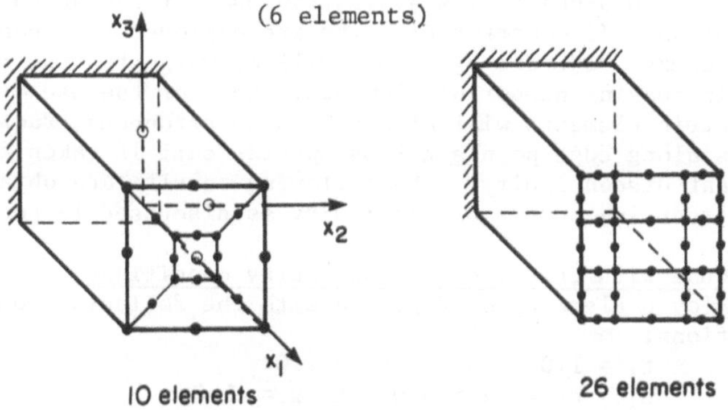

Figure 2.b Cubic beam under uniform tension

Face x_1=1 is subdivided respectively into 1 element (for a total of 6 elements for the whole boundary), 5 elements (for a total of 10 surface elements) and 9 elements (for a total of 26 elements) according to Figs.2.a and 2.b.

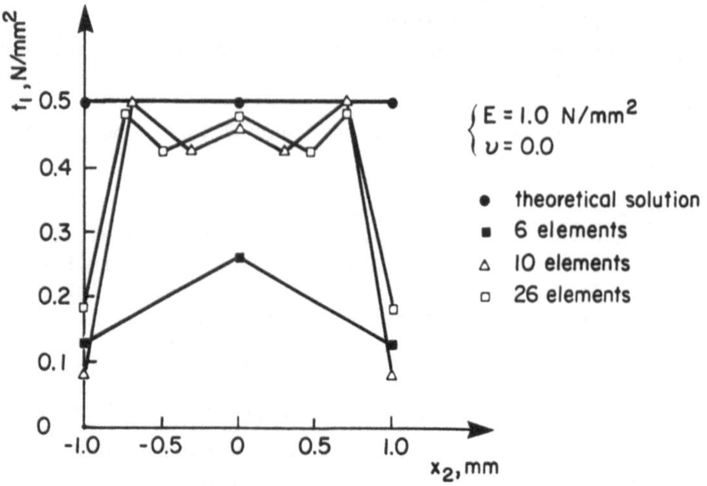

Figure 2.c Tractions obtained by imposing unit displacement at face x_1=1.0 for various meshes

Results plotted in Fig. 2.c were computed at nodes belonging to elements on the face x_1 = 1.0.

The results of Figure 2.c demonstrate that the traction t_1 along the edge points will never converge to the true value $t_1 = 0.5$ due to its evaluation under the continuity assumption. If constant elements are employed for such problems [Cruze, 22], the t_1 values will converge to true value with increasing number of elements. However the use of isoparametric elements will always lead to erroneous traction values along edge points unless special care is taken for the inherent discontinuity. The following results are obtained by introducing stress discontinuity as discussed in section 6.

7.2 Cube with tractions discontinuity conditions

We employ 6 elements of Fig. 2.a with the following boundary conditions:

a) $t_1 = 1.0$ at $x_1 = 1.0$

$u_1 = u_2 = u_3 = 0.0$ at $x = -1.0$

$t_1 = t_2 = t_3 = 0.0$ at other nodes

The following simple example shows clearly the influence of proper consideration of stress conditions on precision of results. (Table 1). Reference [19] reports an error of 20.2% in axial stress.

Discontinuous stress case considers the stress values element wise.

Results at $x_1 = 1.0$ and $x_1 = -1.0$:

	Displacement u_1		
Nodes	Discontinuous stress conditions	Continuous stress conditions	Theoretical
$x_1 = 1.0$ corner nodes mid-nodes	 2.0012 1.9985	 4.4026 3.4186	 2.0 2.0
	Reactions t_1 at $x_1 = -1.0$		
$x_1 = -1.0$ corner nodes mid-nodes	 -0.99489 -0.99859	 -0.32947 -0.71414	 -1.0 -1.0

TABLE 1

The following example studies the same problem but with displacement boundary conditions. The unknown stresses are retained in such a way that at an edge node, only tractions belonging to one face are kept as unknown with tractions at other face being introduced as boundary conditions.

b) <u>Displacement boundary conditions</u>: $E = 1.0 (N/mm^2)$ $\nu = 0.0$

At $x_1 = 1.0$, $u_1 = 2.0$
$x_1 = -1.0$ $u_1 = u_2 = u_3 = 0.0$
$u_2 = u_3 = 0.0$ otherwise

The tractions results are given in the following Table 2:

Nodes	Discontinuous stress conditions	Continuous stress conditions	Theoretical
$x = 1.0$			
•corner			
nodes t_1 =	0.99644	0.25769	1.0
t_2 =	0.000259	0.036375	0.0
t_3 =	0.000292	0.036375	0.0
•mid-nodes			
t_1 =	0.99981	0.52257	1.0
t_2 =	0.0	0.0	0.0
t_3 =	0.00117	0.013966	0.0
$x_1 = 0.0$			
t_1 =	0.0	0.001284	0.0
t_2 =	0.000288	0.003236	0.0
t_3 =	0.000376	0.003236	0.0

TABLE 2

These results again demonstrate that high precision is attained by properly considering stress discontinuities.

7.3 Cubic beam under shear bending.

We consider the same cube as before but under a uniform shear with following boundary conditions:

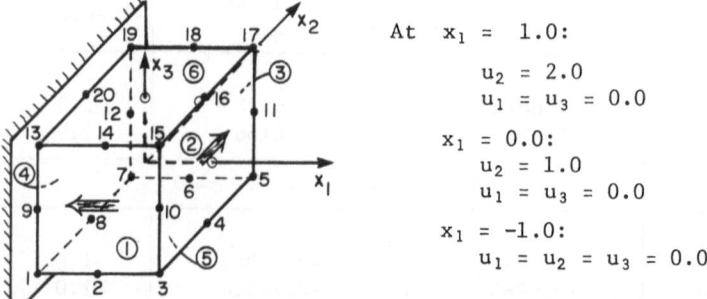

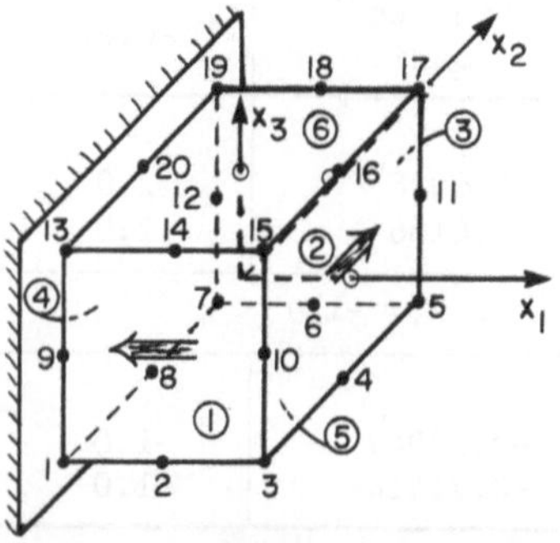

At $x_1 = 1.0$:

$u_2 = 2.0$
$u_1 = u_3 = 0.0$

$x_1 = 0.0$:
$u_2 = 1.0$
$u_1 = u_3 = 0.0$

$x_1 = -1.0$:
$u_1 = u_2 = u_3 = 0.0$

Figure 3 Unit cube under shear bending

The effect of various tractions discontinuity conditions on the accuracy is studied as follows:

Case 1: Discontinuity is assumed at nodes shared by elements ② and ④ and the other elements. However for nodes shared by element ①, ③, ⑤ and ⑥, we consider several combinations of tractions discontinuities at the middle nodes 2, 6, 14 and 18 of the lateral edges. That is elements ①, ③, ⑤ and ⑥ are discontinuous elements. In the first case, we assume continuity of tractions at these nodes. The results are given in the first column of Table 3. Clearly, the low accuracy and large scattering in results are shown up.

Case 2: In this case, we consider also the same 4 discontinuous elements as in the preceding situation, but now with discontinuity at nodes 2, 6, 14, 18 (i.e. 3 tractions are unknown at each of these nodes when considered to belong to elements ① and ③ and they are zero at the same nodes when considered to belong to elements ⑤ and ⑥.

Case 3: Here we take account for the traction discontinuities on all the 6 elements.

From the results given in Table 3, we see how improvement in accuracy can be achieved by a refinement in traction discontinuity model representation. The error with respect to theoretical values passes from about 38% in representation to 0.26% for the third modelling which is the best we can achieve with such a coarse discretization.

Nodes	Theoretical values	I	II	III
		4 Discontinuous elements		6 Discontinuous Elements generalized Discontinuities
		Continuity at nodes 2,6,14,18	Discontinuity at nodes 2,6,14,18	
3,5 15,17	$t_1 = \pm 0.5$ $t_2 = 0.5$ $t_3 = 0.0$	{± 0.40559 0.41747 0.13013	{± 0.4405 0.48564 0.06708	{± 0.49909 0.49869 0.0001737
4,10* 11*,16	$t_1 = \{ \begin{matrix} 0.0 \\ \pm 0.5 \end{matrix}$ $t_2 = 0.5$ $t_3 = 0.0$	{ 0.0 ±0.69150* { 0.48557 0.46859* 0.0	{ 0.0 ±0.44342* { 0.49169 0.49961* 0.0	{ 0.0 ±0.50035* { 0.49923 0.50062* 0.0
2, 6 14, 18	$t_1 = \pm 0.5$ $t_2 = 0.0$ $t_3 = 0.0$	±0.31140 0.0 0.0	±0.53639 0.0 0.0	±0.49957 0.0 0.0
Maximum error w.r.t. theor. value		38%	12%	0.26%

TABLE 3: Effect of traction discontinuity models on result accuracy (shear bending problem).

7.4 Bending of a cantilever beam under end load

We consider a cantilever beam (±2, ±1, ±1) under end load (Fig. 4). $E = 1.0\ N/mm^2$, $\nu = 0.0$ and with boundary conditions:

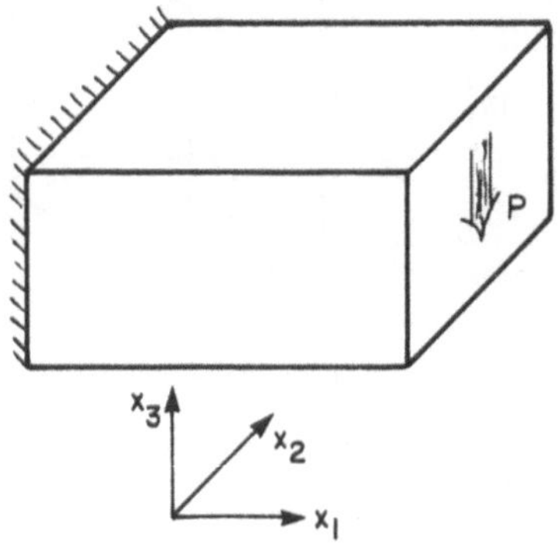

at $x_1 = 2.0$

$t_3 = -1.0 \Rightarrow P_{tot} = -4.0\ N$

$t_1 = t_2 = 0.0$

at $x = -2.0$

$u_1 = u_2 = u_3 = 0.0;$

elsewhere we impose:

$t_1 = t_2 = t_3 = 0.0$

and compute the displacement field.

Figure 4: Cantilever beam

The value of tip deflection is given by [Ref. 4]:

$$u_3 = \frac{P\ell^3}{3EI}\left[1 + .71\left(\frac{h}{\ell}\right)^2 + .10\left(\frac{h}{\ell}\right)^3\right]$$

with $\ell = 4.0$ mm, $h = 2.0$ mm, $I = \frac{4}{3}$ mm^4

$$P = \int t_3 dx_2 dx_3 = -4.0 \text{ N}$$

so that $u_{3_{theor}} = -76.16$ mm (from the above formula)

The calculated deflections using BEM and FEM programs are as follows:

Method	Tip Deflection	Number of Elements	CPU time	System
BEM	-71.787	10 (32 nodes)	74 sec	IBM370-158
FEM	-72.000	2 (using a 20-nodes 3-D isoparametric element)	5 sec	IBM370-158

TABLE 4

7.5 Thick Cylinder

We consider a thick circular cylinder under uniform internal pressure $P_i = 20$ N/mm^2 with internal radius $R_i = 10$ mm, external radius $R_e = 20$ mm, $E = 210\,000$ N/mm^2, $\nu = 0.3$ (see Figure 5).

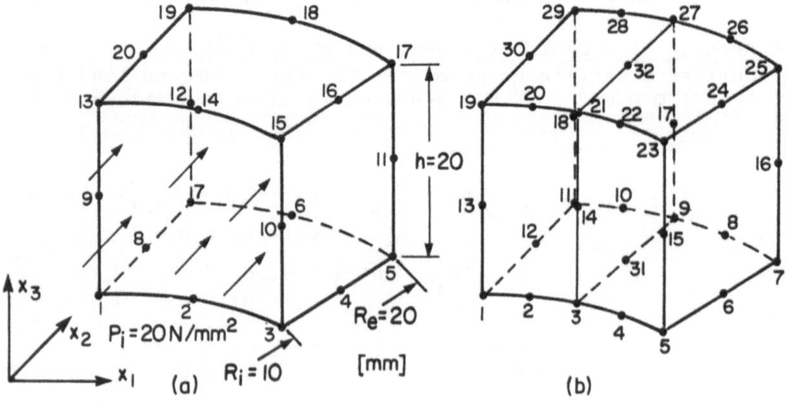

Figure 5 Thick cylinder meshes:
a) 6 elements, b) 10 elements

Along x_3, we have taken a length of 20 mm to obtain balanced element sizes. Due to double symmetry, only one quarter of the cylinder is studied as in Fig. 5. We obtain results using 6 elements (see Table 5) and 10 elements (see Table 6).

The results with traction or displacement boundary conditions are shown in Tables 5 and 6. Values obtained by discontinuous and continuous stress conditions are given in the same tables as well. One may observe the influence of discontinuous stress conditions and excellent precision of results obtained using 6 or 10 elements. In Table 5, we also give results obtained by the FEM using 4 brick elements (20 nodes quadratic isoparametric elements). Comparison between the displacements given by the two methods is shown below in Table 7. All examples were run on the computer IBM-370-158.

	BEM		FEM
	6 elements discontinuous stress	10 elements discontinuous stress	4 elements
CPU time	33.7 sec	73.6 sec	12.5 sec
Max. error	10.8%	8.6%	6.7%

TABLE 7

One should note that the purpose of this example is not to obtain exact results by these numerical methods, but to show the effect of discontinuous tractions on precision and for simple cases.
The relatively high CPU time for the BEM is due mainly to the fact that we did not use the optimal integration scheme as described in [10,11,23], but a fixed number of integration points.

7.6 Spatial Gear Tooth

Our final example is a real life one. We consider a spur gear having the following characteristics:

Pressure angle	$\phi = 20$ deg
Number of teeth	$N = 20$
Diametral pitch	$P = 1$
Addendum	$a = 1.0/P$
Dedendum	$b = 1.25/P$
Fillet radius	$r_f = .380/P$
Face width	$1/P$
Material	Steel ($E = 30\times10^6$psi, $\nu = 0.3$)

Figure 6 shows the BEM discretization used and the applied load.

Nodes (BEM)	BEM Tractions Imposed, Calculated Displacements						FEM 4 Elements (20 Nodes isoparametric brick element)		BEM Theoretical Displacements Imposed, Calculated Tractions			
	Continuous		Theoretical (mm)		Discontinuous				Discontinuous		Theoretical	
	u_1 $\times 10^{-3}$	u_2 $\times 10^{-3}$	u_1 $\times 10^{-3}$	u_2 $\times 10^{-3}$	u_1 $\times 10^{-3}$	u_2 $\times 10^{-3}$	u_1 $\times 10^{-3}$	u_2 $\times 10^{-3}$	t_1 (N/mm^2)	t_2	t_1	t_2
1		3.05		1.87		1.78		1.77	-33.61	19.98	-33.33	20.0
2	1.67	1.67	1.32	1.32	1.26	1.26	1.25	1.25	12.70	12.70	14.14	14.14
3	3.05		1.87		1.78		1.77		19.98	-33.61	20.0	-33.33
4	2.01		1.43		1.31		1.33		0.62	-19.70		-18.52
5	1.76		1.27		1.13		1.18		1.13	-14.13		-13.33
6	0.91	0.91	0.90	0.90	0.80	0.80	0.84	0.84	0.36	0.36		
7		1.76		1.27		1.13		1.18	-14.13	- 1.13	-13.33	
8		2.01		1.43		1.31		1.33	-19.70	0.62	-18.52	
9		2.60		1.87		1.78		1.77	-35.10	18.43	-33.33	20.0
10	2.60		1.87		1.78		1.77		18.43	-35.10	20.0	-33.33
11	1.61		1.27		1.13		1.18		2.74	-15.28		-13.33
12		1.61		1.27		1.13		1.18	-15.28	2.74	-13.33	
Hoop Stress (N/mm^2)												
σ_θ (r = 10 mm)		-6.02	-33.33		-33.94				-33.61			
σ_θ (r = 15 mm)		-9.94	-18.52		-17.70				-19.70			
σ_θ (r = 20 mm)		-4.57	-13.33		-13.14				-14.13			

Table 5. Quarter thick cylinder under internal pressure (6 elements - BEM)

Nodes (BEM)	Tractions Imposed, Calculated Displacements						Displacement Imposed	
	Continuous		Theoretical (mm)		Discontinuous		Discontinuous	
	u_1 $\times 10^{-3}$	u_2 $\times 10^{-3}$	u_1 $\times 10^{-3}$	u_2 $\times 10^{-3}$	u_1 $\times 10^{-3}$	u_2 $\times 10^{-3}$	t_1	t_2
1		3.29		1.87		1.81	-35.63	18.03
2	0.93	2.68	0.72	1.73	0.68	1.68	7.70	16.10
3	1.94	1.94	1.32	1.32	1.28	1.28	12.34	12.34
4	2.68	0.93	1.73	0.72	1.68	0.68	16.10	7.70
5	3.29		1.87		1.81		18.03	-35.63
6	2.22		1.43		1.36		0.09	-21.23
7	1.89		1.27		1.18		2.31	-15.43
8	1.55	0.62	1.17	0.49	1.07	0.45	2.12	0.86
9	1.11	1.11	0.90	0.90	0.82	0.82	1.68	1.68
10	0.62	1.55	0.49	1.17	0.45	1.07	0.86	2.12
11		1.89		1.27		1.18	-15.43	2.31
12		2.22		1.43		1.36	-21.23	0.09
13		2.89		1.87		1.82	-34.63	18.08
14	1.60	1.60	1.32	1.32	1.28	1.28	11.95	11.95
15	2.89		1.87		1.82		18.08	-34.63
16	1.75		1.27		1.16		2.43	-15.79
17	1.09	1.09	0.90	0.90	0.81	0.81	1.63	1.63
18		1.75		1.27		1.28	-15.79	2.43
Hoop Stress (N/mm^2)								
σ_θ (r = 10 mm)	-10.64		-33.33		-31.66		-35.63	
σ_θ (r = 15 mm)	-14.11		-18.52		-18.63		-21.23	
σ_θ (r = 20 mm)	- 6.07		-13.33		-13.09		-15.43	

Table 6. Quarter thick cylinder under internal pressure (10 elements)

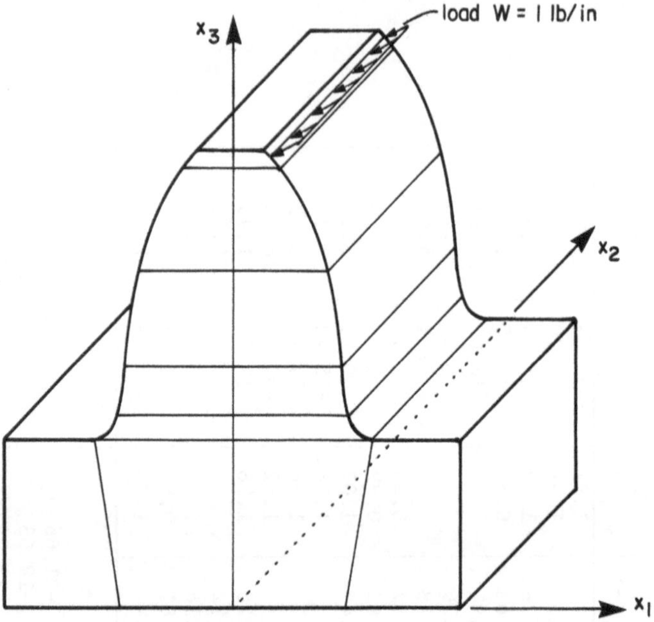

Figure 6 Spatial Gear Tooth,
Boundary element discretization

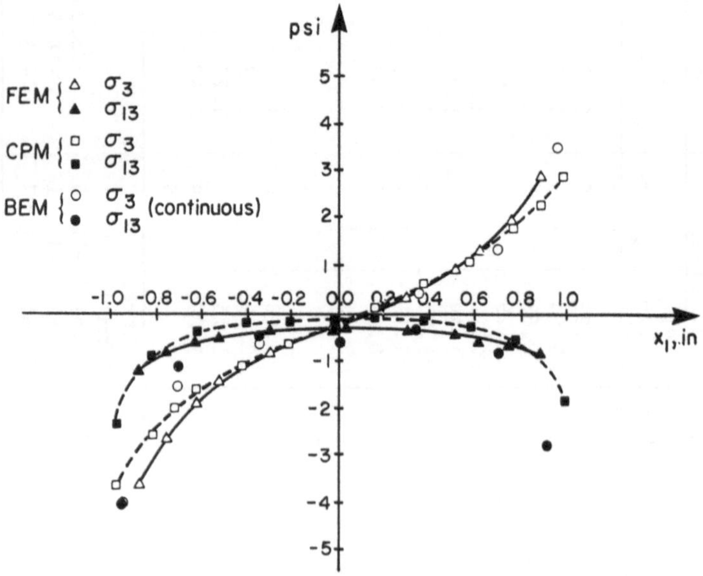

Figure 7 Critical section, stresses σ_1 and σ_{13}

This example has been treated in Reference [24] as a two-dimensional problem by the complex potential method (C.P.M.). Figure 7 shows the stress distribution obtained by the CPM, the FEM and the BEM (using the continuity of tractions hypothesis). In general the BEM gives the same distribution of stresses as the other two methods but the magnitudes are different. Further studies are still going on.

8. CONCLUSION

In this study, we employed an 8-nodes isoparametric element to study 3-dimensional problems of elasticity. The problem of stress discontinuity for edge nodes employing such element is discussed in details. A scheme has been presented by which one may properly introduce stress discontinuity conditions without using the uneconomical node separation technique. Results presented show clearly that it is essential that discontinuity of stress be rightfully introduced for obtaining accurate results with reasonable number of elements. A number of examples including a spatial gear tooth has been studied by the BEM to show the significant influence of traction discontinuity conditions on precision.

REFERENCES

1. BREBBIA, C.A. and WALKER, S. "Boundary Element Techniques in Engineering", Newness-Butterworth, London, 1980.
2. BREBBIA, C.A. "The Boundary Element Method for Engineers", John Wiley, New York, 1978.
3. CRUZE, T.A. "Application of the Boundary Integral Equation Method to 3-dimensional Stress Analysis", Computer & Structures, Vol.3, pp.509-527, 1973.
4. CRUZE, T.A. "An Improved Boundary Integral Equation Method for Three-dimensional Elastic Stress Analysis", Computer & Structures, Vol.4, pp.741-754, 1974.
5. CRUZE, T.A. "Mathematical Foundations of the Boundary Integral Equation Method in Solid Mechanics", Airforce Office of Scientific Research, Report No AFOSR-TR-1002, 1977.
6. KUPRADZE, V.D. "Three-dimensional Problems of the Mathematical Theory of Elasticity", North-Holland, Amsterdam, 1979.
7. HARTMANN, F. "The Somigliana Identity on Piece-wise Smooth Surface", J. of Elasticity, Vol.11, No.4, 1981.
8. DHATT, G. and TOUZOT, G. "Une présentation de la méthode des éléments finis", Maloine- Les Presses de l'Université Laval, Paris, 1981.
9. RIZZO, F.J. and SHIPPY, D.J. "An Advanced Boundary Integral Equation Method for Three-dimensional Thermoelasticity", Int. J. for Num. Methods in Engng, Vol.11, pp.1753-1768, 1977.

10. LACHAT, J.C. and WATSON, J.O. "Effective Numerical Treatment of Boundary Integral Equation: A Formulation for Three-dimensional Elastostatics", Int.J., Num. Methods in Engng, Vol.10, pp. 991-1005, 1976.

11. LACHAT, J.C. "A Further Development of the Boundary Integral Technique for Elastostatics", Ph.D. Dissertation, University of Southampton, 1975.

12. BREBBIA, C.A., Editor "Recent Advances in Boundary Element Methods", Pentech Press, 1978.

13. CHAUDONNERET, M. "On the Discontinuity of the Stress Vector in the Boundary Integral Equation Method for Elastic Analysis", in [12], pp. 185-194, 1978.

14. CHAUDONNERET, M. "Méthode des équations intégrales appliquée à la résolution de problèmes de viscoplasticité", J. de Mécanique appliquée, Vol.1, pp.113-132, 1977.

15. BANERJEE, P.K. and BUTTERFIELD, P., Editors "Development in Boundary Element Methods - 1", Applied Sciences Publishers, London, 1979.

16. MANOLIS, G.D. "Dynamic Response of Underground Structures" Ph.D. Thesis, University of Minnesota, 1980.

17. CRUZE, T.A. and RIZZO, F.J., Editors "Boundary Integral Equation Method: Computational Applications in Applied Mechanics", ASME, AMD, Vol.11, 1975.

18. BREBBIA, C.A., Editor "Boundary Element Methods", Proc. of Third International Seminar, Irvine, California, July 1981, Springer Verlag, 1981.

19. SEABRA PEREIRA, M.F., MOTA SOARES, G.A., and OLIVEIRA FARIA, L.M. "A Comparative Study of Several Boundary Elements in Elasticity", Proc. of Third International Seminar, Irvine, Calif., July 1981, p.123, Springer Verlag, 1981.

20. MORJARIA, M., SARIHAN, V. and MUKHERJEE, S. "Comparison of Boundary Element and Finite Element Methods in Two-dimensional Inelastic Analysis", Int. J. of Structural Mechanics and Material Sciences, Vol.1, No.1, pp.3-20, 1980.

21. ALARCON, E., MARTIN, A., and PARIS, F. "Boundary Element in Potential and Elasticity Theory", Computer and Structures, Vol. 10, pp. 351-362, 1979.

22. CRUZE, T.A. "Numerical Solutions in Three-Dimensional Elastostatics", Int. J. of Solids, Structures, Vol.5, pp. 1259-1274, 1969.

23. BOLTEUS, L. and TULLBERG, O. "BEMSTAT - A New Type of Boundary Element Program for 2-dimensional Elasticity Problems, Proc. of Third Int. Seminar, Irvine, Calif., July 1981, pp. 518-537, Springer Verlag, 1981.

24. CARDOU, A. and TORDION, G. "Calcul des contraintes dans les engrenages droits par la méthode des potentiels complexes" Laboratoire d'Eléments de Machines, Rapp. technique No. EM-25, Département de Génie mécanique, Université Laval, 1980.

APPLICATION OF INDIRECT BOUNDARY INTEGRAL METHOD TO THREE-DIMENSIONAL PROBLEMS IN THE LINEAR, COUPLE-STRESS THEORY

M. Kishida and K. Sasaki

Department of Mechanical Engineering, Hokkaido University, North 13, West 8, Kita-ku, Sapporo 060, Japan

ABSTRACT

This paper describes a numerical approach by the indirect boundary integral method to three-dimensional boundary value problems in the couple-stress theory of elasticity. As the concrete examples, the stress concentration problems are treated for a circular cylinder with arc-shaped annular groove under uniform tension and simple torsion, respectively. By the results, the influence of parameters, such as Poisson's ratio ν, characteristic length ℓ, and the ratio η_r of bending-twisting moduli, on stress concentration factor is made clear.

INTRODUCTION

One of authors and his collaborators have been analysing many boundary value problems in the linear classical elastostatics by the use of the indirect boundary integral method [e.g. Jaswon and Symm (1977)] together with the fictitious-boundary method introduced by Oliveira (1968) and Kishida(1980a), what is called, the indirect fictitious-boundary integral method (IFBIM). Namely, Saint-Venant's torsion problems (thin I, L, and cruciform beams [(1980b)]), two-dimensional problems (tension of a notched plate [(1979)] and compression of a cylinder between two rigid plates [(1982c)]), and three-dimensional problems (cylindrical pressure vessels with torispherical drumheads [(1981)] and axisymmetric bending of thick plate [(1982d)]). As the solution for the above three-dimensional problems, the vector potential proposed by the authors [Kishida, et al. (1980a and 1982a)] are taken.

Recently, the authors [Kishida, et al. (1982b)] proposed one numerical approach on the basis of foregoing IFBIM to the boundary value problems in the linear, couple-stress theory, which was presented by Mindlin and Tiersten (1962) and Koiter (1964) to explain some discrapancies between theoretical predictions and experimental results. Many researchers have applied this theory to the fundamental boundary value problems, restricted to two and three-dimensional infinite media. But, we must treat the three-dimensional problems not for the infinite but for the finite media, where the analytical results can be easily compared with experimental ones, because the usefulness of this theory depends on whether it is possible to determine the new material constants such as the characteristic length ℓ and the ratio η_r of bending-twisting moduli.

In the present paper, we treat the uniform tension and the simple torsion problems for a circular cylinder with arc-shaped annular groove by the use of the IFBIM to investigate the influence of the various parameters on a stress concentration factor and the relationship between couple-stresses and stress gradients.

FUNDAMENTAL EQUATIONS

The fundamental equations for a homogeneous and isotropic medium without body force and body couple in the linear, couple-stress theory are given by Mindlin and Tiersten (1962).

The stress equation of equilibrium is

$$\overset{S}{\tau}_{ji,j} + \frac{1}{2}\,\varepsilon_{ijk}\,\overset{D}{\mu}_{lk,lj} = 0 \tag{1}$$

where $\overset{S}{\tau}_{ij}$ is the symmetric part of a force-stress tensor τ_{ij}, $\overset{D}{\mu}_{ij}$ is the deviatric part of a couple-stress tensor μ_{ij}, ε_{ijk} is the Eddington's epsilon, and (,i) denotes differentiation with respect to x_i. Hereafter, we assume the summation convention over repeated indices.

The stresses $\overset{S}{\tau}_{ij}$ and $\overset{D}{\mu}_{ij}$ are represented by the displacement u_i as

$$\overset{S}{\tau}_{ij} = \lambda\,u_{k,k}\,\delta_{ij} + \mu\,(u_{i,j} + u_{j,i}) \tag{2}$$

$$\overset{D}{\mu}_{ij} = 2\mu\ell^2(\varepsilon_{jlm}\,u_{m,li} + \eta_r\,\varepsilon_{ilm}\,u_{m,lj}) \tag{3}$$

where λ and μ are Lamé's constants, ℓ and η_r are the material constants adopted in the couple-stress theory, and δ_{ij} is the Kronecker's delta. Also, ℓ has the dimension of length and η_r is dimensionless.

Substituting equations (2) and (3) into equation (1), the displacement equation of equilibrium is

$$(\lambda + 2\mu)\,u_{j,ji} - \mu\,(1 - \ell^2\nabla^2)(u_{j,ji} - u_{i,jj}) = 0 \tag{4}$$

Any solution u_i is representable as [Mindlin and Tiersten (1962)]

$$u_i = B_i - \ell^2 B_{j,ji} - \frac{(\lambda + \mu)}{2(\lambda + 2\mu)}\,[x_j(1 - \ell^2\nabla^2)B_j + B_0]_{,i} \tag{5}$$

where B_i and B_0 are the vector and the scalar functions, respectively, characterized by

$$(1 - \ell^2\nabla^2)\nabla^2\,B_i = 0 \quad , \quad \nabla^2\,B_0 = 0 \tag{6}$$

Now, let τ^A_{ij} be the anti-symmetric part of τ_{ij}, and $\mu_{(S)}$ ($= \mu_{kk}/3$) be the scalar of μ_{ij}. Neither of them appears in the equation (1), and they are only related by

$$\tau^A_{ij} = \mu\ell^2\nabla^2(u_{i,j} - u_{j,i}) - \frac{1}{2}\varepsilon_{ijk}\,\mu_{(S),k} \tag{7}$$

Namely, τ^A_{ij} and $\mu_{(S)}$ cannot be determined independently, and we can take $\mu_{(S)}$ to be identically zero without loss in physical generality [Koiter (1964)] because of the irrelativeness to the statical equilibrium.

ANALYSIS

Consider a homogeneous and isotropic medium that occupies a "real-domain" R bounded by a closed smooth "real-boundary surface" S, and is subjected to a force traction T_i and a couple traction M_i on S. We now suppose a closed Liapunov "fictitious-boundary surface" S* outside S not intersecting each other, and extend a domain from R to R* bounded by S*. Hereafter, we asterisk the quantities defined on S*. A displacement u_i satisfying equation (4) in R* is given by equation (5). When $\ell = 0$, equations (4) and (5) agree with a well-known equation of equilibrium and its Boussinesq-Papkovich solution in classical elasticity, respectively.

A continuous distribution of surface density vectors $\vec{\xi}^*$ and $\vec{\zeta}^*$, corresponding to concentrated force and couple in an infinite medium, over a surface S* generates the potential $\vec{B}$ for a present solution [Kishida, et al. (1982b)];

$$\vec{B}(\vec{P}) = \int_{S^*}\{\vec{\xi}^*(\vec{q}^*)[1 - \exp(-r_0/\ell)]/r_0\}dS^*(\vec{q}^*)$$
$$+ \int_{S^*}\vec{\zeta}^*(\vec{q}^*)\times\nabla^*\{[1 - \exp(-r_0/\ell)]/r_0\}dS^*(\vec{q}^*) \tag{8}$$
$$B_0(\vec{P}) = 0$$

Here, $\vec{\xi}^*(\vec{q}^*)$ and $\vec{\zeta}^*(\vec{q}^*)$ are Hölder continuous density vectors defined on S*, $\vec{P}$ and $\vec{q}^*$ are the vector variables specifying poits in R and on S*, respectively, r_0 is the distance between $\vec{P}$ and $\vec{q}^*$, and ∇^* is the spacial gradient with respect to $\vec{q}^*$.

Thus substituting equations (8) into equation (5) and , further, equation (5) into equations (2), (3) and (7), we can obtain a force-stress and a couple-stress in the forms of integral equations for $\vec{\xi}^*$ and $\vec{\zeta}^*$;

$$\tau_{ij} = \int_{S^*}[\xi^*_k(\vec{q}^*)\,E_{kij} + \zeta^*_k(\vec{q}^*)\,F_{kij}]\,dS^*(\vec{q}^*) \tag{9}$$

$$\mu_{ij} = \int_{S^*}[\xi^*_k(\vec{q}^*)\,G_{kij} + \zeta^*_k(\vec{q}^*)\,H_{kij}]\,dS^*(\vec{q}^*) \tag{10}$$

where ξ^*_k and ζ^*_k are the components of $\vec{\xi}^*$ and $\vec{\zeta}^*$, respectively, and E_{kij}, F_{kij}, G_{kij} and H_{kij} are given in the Appendix.

In this way, the present problem is reduced to solving the simultaneous Fredholm type integral equations of the first kind for $\vec{\xi}^*$ and $\vec{\zeta}^*$ under the conditions

$$n_i \tau_{ij} = T_j \quad , \quad n_i \mu_{ij} = M_j \qquad \text{on } S \qquad (11)$$

where n_i is the component of outward unit normal vector on S. Since $\vec{P}$ and $\vec{q}^*$ never coincide with each other, the relevant integral equations have no singularity. To solve these equations analytically is, generally speaking, very difficult, so it is rather convenient to solve them numerically.

NUMERICAL ANALYSIS

Three-dimensional axisymmetric problem

In order to treat three-dimensional axisymmetric problems, we take a cylindrical coordinate system as shown in Figure 1, and

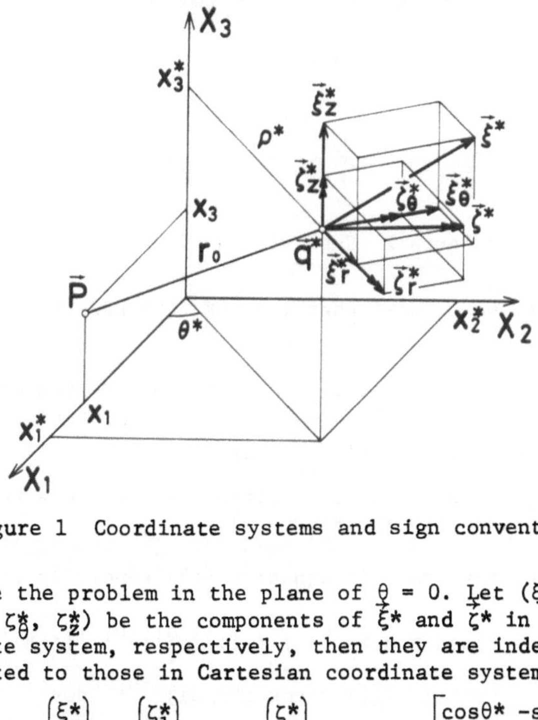

Figure 1 Coordinate systems and sign convention

formulate the problem in the plane of $\theta = 0$. Let $(\xi_r^*, \xi_\theta^*, \xi_z^*)$ and $(\zeta_r^*, \zeta_\theta^*, \zeta_z^*)$ be the components of $\vec{\xi}^*$ and $\vec{\zeta}^*$ in cylindrical coordinate system, respectively, then they are independent on θ^* and related to those in Cartesian coordinate system by

$$\begin{Bmatrix} \xi_1^* \\ \xi_2^* \\ \xi_3^* \end{Bmatrix} = [N] \begin{Bmatrix} \xi_r^* \\ \xi_\theta^* \\ \xi_z^* \end{Bmatrix} , \quad \begin{Bmatrix} \zeta_1^* \\ \zeta_2^* \\ \zeta_3^* \end{Bmatrix} = [N] \begin{Bmatrix} \zeta_r^* \\ \zeta_\theta^* \\ \zeta_z^* \end{Bmatrix} , \quad [N] = \begin{bmatrix} \cos\theta^* & -\sin\theta^* & 0 \\ \sin\theta^* & \cos\theta^* & 0 \\ 0 & 0 & 1 \end{bmatrix} \qquad (12)$$

We define Γ and Γ^* to be the contours of S and S* in the plane of $\theta = 0$, respectively. For convenience's sake, we divide Γ into m line-segments with equal length h, and then set Γ^* at distance $t_i = \alpha_i \times h$ from Γ [Kishida (1980a)], where α_i is called the coefficient of a fictitious-boundary distance and is the most important parameter affecting the accuracy of solution. For the numerical calculation, we divide Γ^* into m line-segments, the i-th segment being the length of h_i^* and centred about a nodal point $\vec{q}_i^*$ midway between the interval points $\vec{q}_{i\pm 1/2}^*$, and obtain the i-th circular surface-segment ΔS_i^* by turnig its i-th line-segment on Γ^* around axis fully once (Figure 2). On each

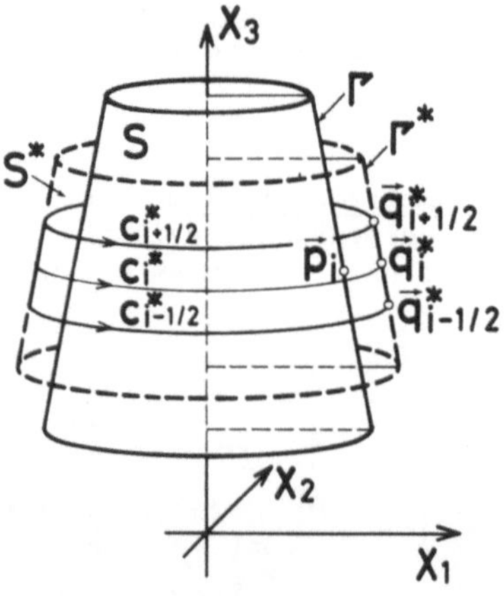

Figure 2 Subdivision of S*

surface-segment, we suppose that $(\xi_r^*, \xi_\theta^*, \xi_z^*)$ and $(\zeta_r^*, \zeta_\theta^*, \zeta_z^*)$ are constant.

The surface integral around the i-th surface-segment can be calculated numerically by Simpson's quadrature formula

$$\int_{\Delta S_i^*} f^*(\vec{q}^*)\, dS^* = (h_i^*/6)(f_{i-1/2}^* + 4f_i^* + f_{i+1/2}^*) \tag{13}$$

where f_i^* and $f_{i\pm 1/2}^*$ are contour integral values along the circles c_i^* and $c_{i\pm 1/2}^*$, respectively (Figure 2).

The contour integrals in equation (13) generally take the forms

$$\int_{-\pi}^{\pi}[\cos^{n_1}\theta^*/r_0^{\,n_2}]\rho^* d\theta^* = 2B^{-n_1}\sum_{i=0}^{n_1}(-1)^i\binom{n_1}{i}A^{n_1-i}(A+B)^{i-n_2/2}\rho^* \times\int_0^{\pi}[1-k^2\cos^2(\theta^*/2)]^{i-n_2/2}d\theta^* \tag{14}$$

$$\int_{-\pi}^{\pi}[\cos^{n_1}\theta^*/r_0^{\,n_3}]\exp(-r_0/\ell)\rho^* d\theta^* = 2B^{-n_1}\sum_{i=0}^{n_1}(-1)^i\binom{n_1}{i}A^{n_1-i} \times(A+B)^{i-n_3/2}\rho^*\int_0^{\pi}[1-k^2\cos^2(\theta^*/2)]^{i-n_3/2}\exp(-r_0/\ell)d\theta^* \tag{15}$$

where $n_1 = 0,1,2,3,4$, $n_2 = 3,5,7$, $n_3 = 1,2,3,4,5,6,7$, $r_0 = \{(A+B)[1-k^2\cos^2(\theta^*/2)]\}^{1/2}$, $A = x_1^{\,2} + \rho^{*2} + (x_3-x_3^*)^2$, $B = 2x_1\rho^*$ and $k^2 = 2B/(A+B)$.

The integral in the right hand side of equation (14)

$$I_n = \int_0^{\pi} [1 - k^2\cos^2(\phi/2)]^{-n/2} d\phi \quad (n : \text{odd number}) \tag{16}$$

can be reduced to the sum of complete elliptic integrals of the first kind I_1 and the second kind I_{-1} by successive use of the recurrence formula

$$I_n = [(n-2)(k^2-1)]^{-1}[(n-3)(k^2-2)I_{n-2} + (n-4)I_{n-4}] \tag{17}$$

Similarly, the integral in the right hand side of equation (15)

$$H_{n'} = \int_0^{\pi} [1 - k^2\cos^2(\phi/2)]^{-n'/2}\exp(-r_0/\ell)d\phi \tag{18}$$

can be reduced to the sum of H_{-2}, H_{-1}, H_0, H_1 and H_2 by

$$H_{n'} = [(n'-2)(k^2-1)]^{-1}\{-k'(k^2-1)H_{n'-1} + (k^2-2)[(n'-3)H_{n'-2} + k'H_{n'-3}] + (n'-4)H_{n'-4} + k'H_{n'-5}\} \tag{19}$$

where $k' = (A+B)^{1/2}/\ell$. We obtain these five integrals, $H_{-2} \sim H_2$, numerically.

Surface density vector

In the couple-stress theory, the boundary conditions are reduced from six [equations (11)] to five [Koiter (1964)] ;

$$\overline{T}_j = T_j + \frac{1}{2}\varepsilon_{jkl} M_{(n),k} n_l \quad , \quad \overline{M}_j = M_j - M_{(n)} n_j \quad \text{on } S \tag{20}$$

where $M_{(n)}$ is the normal component of couple traction. In the case of axisymmetric problem, r, θ and z components of force traction, and θ component and tangential component to the contour Γ of couple traction are given. Therefore, r, θ and z components of the density vector $\vec{\xi}^*$, and θ component and tangential component to the contour Γ^* of $\vec{\zeta}^*$ can be considered as the unknowns.

Thus, the numerical computation for a present problem is reduced to the collocation method so as to satisfy the conditions (20) at m points on Γ. Here we determine the collocation point $\vec{p}_i$ such as a point of intersection of normal line at the nodal point $\vec{q}_i^*$ and Γ.

Accuracy for analysis

In the BIM with a fictitious-boundary method, the lack of strong diagonal dominance may give rise to a little difficulty in solving simultaneous algebraic equations; therefore, we use double precision, which gives satisfactory results.

The estimates of accuracy in a numerical computation are done by

$$|T_j^{\,n} - T_j^{\,e}|/\,\overline{\tau} \leqq 1.0\times10^{-2} \quad , \quad |M_j^{\,n} - M_j^{\,e}|/[\overline{\tau}\times(D/2)] \leqq 1.0\times10^{-2} \tag{21}$$

midway between neighbouring collocation points on Γ. Here $\overline{\tau}$ is the standard stress, $T_j^{\,n}$ and $M_j^{\,n}$ are the components of force and couple tractions obtained numerically, $T_j^{\,e}$ and $M_j^{\,e}$ are those given on Γ, and D is a diameter of a cylinder.

NUMERICAL RESULTS

Uniform tension case

Consider a circular cylinder (Figure 3) subjected to uniform tension τ_{load} parallel to X_3 axis at the ends. Here,

$L/D = 3.0$, $a'/(D/2) = 0.1$

(i) semicircular groove :
$[a/(D/2) = b/(D/2)]$
$b/(D/2) = 0.5, 0.4, 0.3, 0.2$

(ii) arc-shaped groove :
$[a/(D/2) = 0.5$ (const.)$]$
$b/(D/2) = 0.4, 0.3, 0.2$

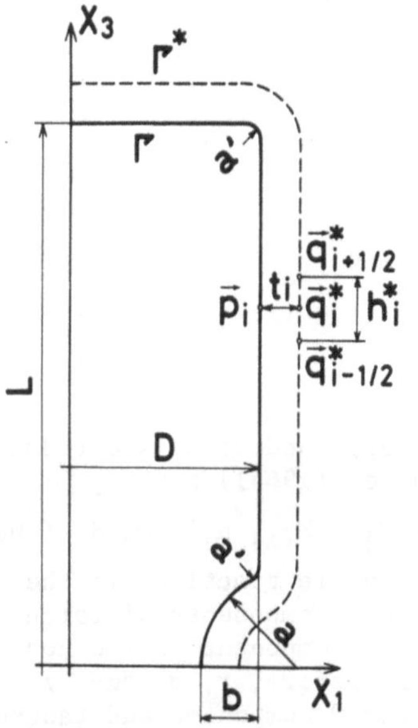

Figure 3 Geometry of example problem

In the axisymmetric tension problem, we can take the density functions related to the torsion problem, ξ_θ^*, ζ_r^* and ζ_z^*, to be identically zero. As the standard stress $\overline{\tau}$, we take a nominal stress on a net section of a cylinder.

First of all, to determine the suitable value of α_i, we analyse a simple problem, e.g. a uniform bar under uniform tension. In Table 1, the results in the case of $m/2 = 63$ and $\alpha_i \fallingdotseq 3.5$ are shown. In this case, as there exist no stress gradient and no couple-stress, the results agree well with those in classical elasticity.

Next, to investigate the influence of each parameter on stress concentration, a cylinder with $a/(D/2) = b/(D/2) = 0.5$ is analysed varying the parameters ν, ℓ and η_r. In Table 2, the stress concentration factors $\alpha = \tau_{33max.}/\overline{\tau}$ are shown. In Figure 4, α against $(D/2)/\ell$ for various values of η_r with $\nu = 0.3$, and in Figure 5, α against ν with $\eta_r = 0.0$ are shown.

In Table 3, the stress concentration factors α for various shapes are shown with $\nu = 0.3$ and $\eta_r = 0.0$. Comparing the results for two shapes with the same depth of groove, it is found that the reduction of α to the value in classical elasticity is greater for semicircular groove than for arc-shaped groove.

In Figure 6, to clarify the asymmetry of shear stresses by the existence of couple-stresses, τ_{31}/τ_{load} is shown along a surface of a cylinder with $a/(D/2) = b/(D/2) = 0.3$, $\nu = 0.3$, $\eta_r = 0.0$ and $(D/2)/\ell = 10$. On the surface, τ_{13} is identically zero, however, τ_{31} is not zero near the shoulder of groove and becomes asymptotically zero near the end, where the stress gradient is considered not to exist. This makes it clear that the couple-stress has a close relation with stress gradient.

Table 1 Force and couple-stresses for uniform bar under uniform tension [$\nu = 0.3$, $\eta_r = 0.0$, $(D/2)/\ell = 10$]

$x_1/(D/2)$	$\tau_{11}/\bar{\tau}$	$\tau_{22}/\bar{\tau}$	$\tau_{33}/\bar{\tau}$	$\mu_{32}/[\tau\times(D/2)]$
	$\times10^{-5}$	$\times10^{-5}$	$\times10^{0}$	$\times10^{-7}$
0.000	-1.264	-1.264	1.000	0
0.125	-1.257	-1.263	1.000	0.114
0.250	-1.234	-1.258	1.000	0.230
0.375	-1.197	-1.251	1.000	0.353
0.500	-1.144	-1.241	1.000	0.492
0.625	-1.075	-1.228	1.000	0.676
0.750	-0.980	-1.211	1.000	0.990
0.875	-0.779	-1.223	1.000	-1.223
1.000	-0.201	-1.261	1.000	-6.951

Table 2 Stress concentration factors for various values of parameters [$a/(D/2) = b/(D/2) = 0.5$]

	$(D/2)/\ell$	$\eta_r = -1.0$	$\eta_r = 0.0$	$\eta_r = 1.0$
$\nu = 0.1$	∞	1.451	—	—
	10	1.328	1.350	1.380
	8	1.295	1.318	1.352
	5	1.224	1.246	1.284
	2	1.137	1.155	1.178
$\nu = 0.3$	∞	1.411	—	—
	10	1.336	1.351	1.372
	8	1.313	1.329	1.353
	5	1.260	1.276	1.304
	2	1.193	1.201	1.222
$\nu = 0.45$	∞	1.390	—	—
	10	1.339	1.350	1.365
	8	1.321	1.333	1.351
	5	1.282	1.294	1.315
	2	1.241	1.246	1.259

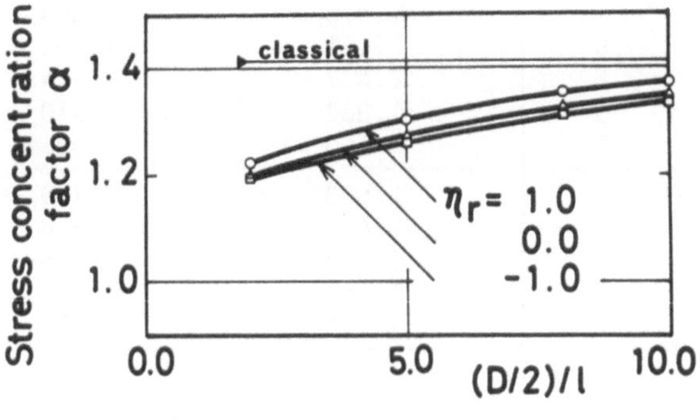

Figure 4 Stress concentration factor versus $(D/2)/\ell$ for various values of η_r [$a/(D/2) = b/(D/2) = 0.5$, $\nu = 0.3$]

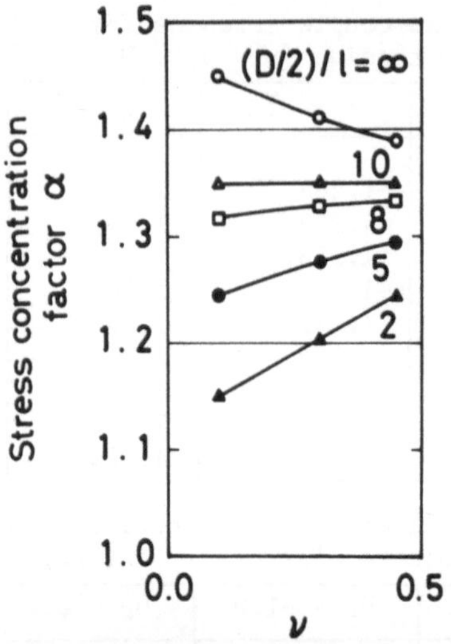

Figure 5 Stress concentration factor versus ν for various values of (D/2)/ℓ [a/(D/2) = b/(D/2) = 0.5, η_r = 0.0]

Table 3 Stress concentration factors for various shapes [ν = 0.3, η_r = 0.0]

b/(D/2)	(D/2)/ℓ	a/(D/2) = 0.5	a/(D/2) = b/(D/2)
0.4	∞	1.501	1.609
	10	1.421	1.496
	8	1.393	1.462
	5	1.328	1.385
	2	1.234	1.280
0.3	∞	1.578	1.868
	10	1.471	1.666
	8	1.437	1.614
	5	1.362	1.510
	2	1.260	1.381
0.2	∞	1.623	2.189
	10	1.478	1.828
	8	1.439	1.756
	5	1.361	1.631
	2	1.266	1.507

Finally, in Figure 7, the stress distribution on a net section of a cylinder is shown with $a/(D/2) = b/(D/2) = 0.3$, $\nu = 0.3$ and $\eta_r = 0.0$.

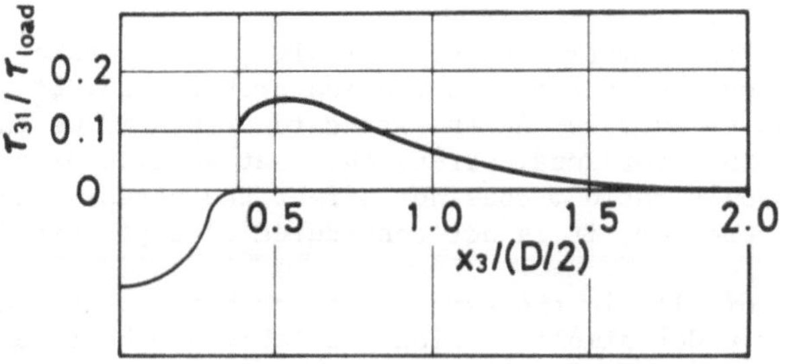

Figure 6 Stress τ_{31}/τ_{load} along a surface of cylinder $[a/(D/2) = b/(D/2) = 0.3, \nu = 0.3, \eta_r = 0.0, (D/2)/\ell = 10]$

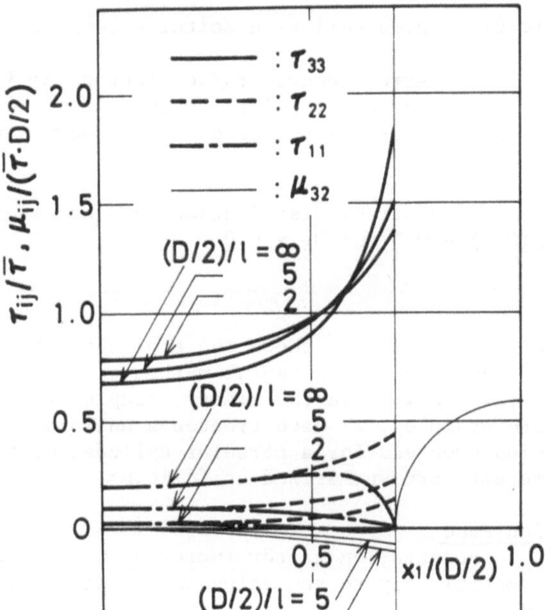

Figure 7 Stress distribution on a net section of cylinder $[a/(D/2) = b/(D/2) = 0.3, \nu = 0.3, \eta_r = 0.0]$

Simple torsion case

Consider a circular cylinder (Figure 3) subjected to simple torsion M_{load} at the ends. Here,

$L/D = 3.0$, $a'/(D/2) = 0.1$
$a/(D/2) = 0.5$, $b/(D/2) = 0.3, 0.2$

In the axisymmetric torsion problem, we can take the density functions related to the tension problem, ξ_r^*, ξ_z^* and ζ_θ^*, to be identically zero. As the standard stress $\overline{\tau}$, we take a maximum nominal torsional stress on a net section of a cylinder. Since Poisson's ratio ν does not affect the stress concentration in torsion problem, it is not considered as a parameter.

First of all, in the same manner as before, we analyse a uniform bar under simple torsion. In Table 4, the results in the case of $m/2 = 71$ and $\alpha_i \doteqdot 4.1$ are shown. In connection with this problem, Koiter (1964) has analysed Saint-Venant's torsion problem of a bar; the force and couple-stresses on X_1 axis are given by

$$\tau_{32}/\overline{\tau} = x_1/(D/2), \quad \mu_{11}/[\overline{\tau}\times(D/2)] = -2\mu[\ell/(D/2)]^2(1+\eta_r)$$
$$\mu_{33}/[\overline{\tau}\times(D/2)] = 4\mu[\ell/(D/2)]^2(1+\eta_r)$$

The present results agree well with Koiter's results.

In Table 5, the stress concentration factors $\alpha = \tau_{32max.}/\overline{\tau}$ for a cylinder with annular groove are shown. Also, in Figure 8, α against $(D/2)/\ell$ for various values of η_r is shown with $b/(D/2) = 0.3$.

In Figure 9, the stress distribution on a net section is shown with $b/(D/2) = 0.3$ and $\eta_r = 0.0$.

CONCLUSIONS

In the present paper, we have shown the applicability of the indirect fictitious-boundary integral method to three-dimensional boundary value problems in the linear, couple-stress theory. As the concrete examples, we have treated a uniform tension and a simple torsion problems for a circular cylinder with annular groove. The results are summarized as follows:

Uniform tension case

(i) The stress concentration factor increases with $(D/2)/\ell$, and approaches asymptotically to the value in classical elasticity, when $(D/2)/\ell \to \infty$.

(ii) The influence of couple-stress on stress concentration increases with decreasing ν and η_r.

(iii) The change of α with η_r is substantially small, in fact, ±2 ~ 3% to the value of α with $\eta_r = 0.0$.

Table 4 Force and couple-stresses for uniform bar under simple torsion [η_r = 0.0, (D/2)/ℓ = 15]

$x_1/(D/2)$	$\tau_{32}/\bar{\tau}$	$\mu_{11}/[\bar{\tau}\times(D/2)]$	$\mu_{33}/[\bar{\tau}\times(D/2)]$
	$\times 10^0$	$\times 10^{-2}$	$\times 10^{-1}$
0.0	0	-0.887	0.177
0.1	0.100	-0.887	0.177
0.2	0.200	-0.887	0.177
0.3	0.299	-0.887	0.177
0.4	0.399	-0.887	0.177
0.5	0.499	-0.887	0.177
0.6	0.599	-0.887	0.177
0.7	0.698	-0.887	0.177
0.8	0.798	-0.887	0.177
0.9	0.898	-0.886	0.177
1.0	1.001	-0.873	0.176

Table 5 Stress concentration factors for various values of η_r and b/(D/2) [a/(D/2) = 0.5]

b/(D/2)	(D/2)/ℓ	η_r = -0.3	η_r = 0.0	η_r = 0.3
0.3	∞	1.195	—	—
	20.0	1.154	1.142	1.124
	15.0	1.153	1.130	1.079
	10.0	1.124	1.070	0.995
0.2	∞	1.214	—	—
	20.0	1.251	1.240	1.229
	17.5	1.232	1.215	1.199
	15.0	1.207	1.181	1.159
	10.0	1.119	1.055	1.010

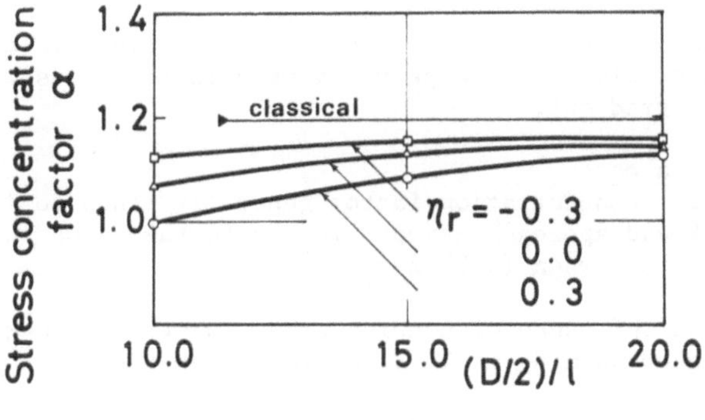

Figure 8 Stress concentration factor versus (D/2)/ℓ for various values of η_r [a/(D/2) = 0.5, b/(D/2) = 0.3]

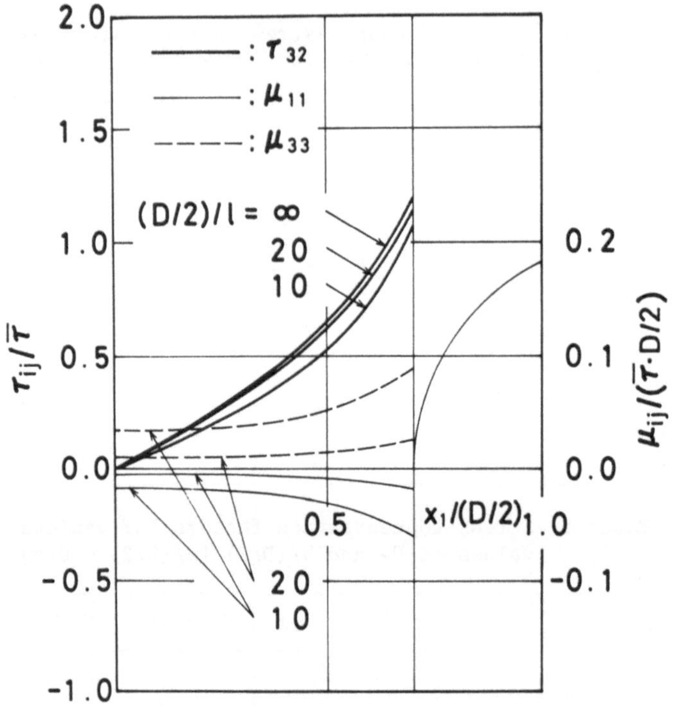

Figure 9 Stress distribution on a net section of cylinder [$a/(D/2) = 0.5$, $b/(D/2) = 0.3$, $\eta_r = 0.0$]

(iv) In classical elasticity, α decreases with increasing ν ; on the other hand, in the couple-stress theory, α increases with ν under a certain value of $(D/2)/\ell$.

(v) The influence of couple-stress has a close relation with stress gradient, and appears largely in the cases with large stress gradient.

<u>Simple torsion case</u>

(i) The stress concentration factor generally tends to increase with $(D/2)/\ell$ and approach asymptotically to the value in classical elasticity, when $(D/2)/\ell \to \infty$.

(ii) The influence of couple-stress on stress concentration does not appear when $\eta_r = -1.0$, and increases with η_r.

(iii) Poisson's ratio ν does not affect the stress concentration, and η_r affects it significantly.

REFERENCES

1. Jaswon, M.A. and Symm, G.T. (1977) Integral Equation Methods in Potential Theory and Elastostatics. London, Academic Press.

2. Kishida, M. et al. (1979) On the Interference of Circular Notches and Loaded Ends (1st Report, Plane Stress Problems). Trans. JSME, 45 - 391A: 245 - 251 (In Japanese).

3. Kishida, M. (1980a) On Fictitious-Boundary Method in Boundary-Integral Methods. Theoretical & Appl. Mech., 28: 139 - 151.

4. Kishida, M. et al. (1980b) Application of the Fictitious-Boundary Method — Integral Equation Method to the Torsion Problems (Thin I, L and Cruciform Beams). Trans. JSME, 46 - 408A: 901 - 906 (In Japanese).

5. Kishida, M. and Ozawa, H. (1981) Three-Dimensional Elastic Stresses in Pressure Vessels with Torispherical Drumheads (Comparison of Elasticity, Photoelasticity and Shell Theory Solutions). Trans. JSME, 47 - 418A: 626 - 634 (In Japanese).

6. Kishida, M. and Fujimura, K. (1982a) One Solution of Three-Dimensional Boundary Value Problems in Classical Elasticity (Indirect Fictitious-Boundary Integral Method). Trans. JSME, 48 - 427A: 358 - 366 (In Japanese).

7. Kishida, M. et al. (1982b) One Solution of Three-Dimensional Boundary Value Problems in the Couple-Stress Theory of Elasticity. Trans. ASME, J. Appl. Mech. (In print).

8. Kishida, M. and Honma, K. (1982c) Compression of an Elastic Cylinder Between Two Rigid Plates (Application of the indirect Fictitious-Boundary Integral Method to the Contact Problems). Trans. JSME, 48 - 433A (In Japanese).

9. Kishida, M. and Matsumoto, N. (1982d) Three-Dimensional Elastic Stresses in Thick Plate (Comparison with Classical Plate Theory Solutions). Trans. JSME (In contribution).

10. Koiter, W.T. (1964) Couple-Stresses in the Theory of Elasticity. Proc. Koninkl. Nederl. Akad. Wetenschappen, B - 67: 17 - 44.

11. Mindlin, R.D. and Tiersten, H.F. (1962) Effects of Couple-Stresses in Linear Elasticity. Arch. Rational Mech. Anal., 11: 415 - 448.

APPENDIX

$$E_{kij} = \mu \Big\{ \delta_{ij} x'_k [-1+2\alpha-6\beta^{-2}+2(3\beta^{-2}+3\beta^{-1}+ 1)\exp(-\beta)]/r_0{}^3$$
$$+ (\delta_{ki} x'_j + \delta_{kj} x'_i)[-1+2\alpha'-6\beta^{-2}+(6\beta^{-2}+6\beta^{-1}+ 3+ \beta)\exp(-\beta)]/r_0{}^3$$
$$+ 2x'_k x'_i x'_j [15\beta^{-2}-(15\beta^{-2}+15\beta^{-1}+ 6+ \beta)\exp(-\beta)]/r_0{}^5$$
$$+ 2\alpha' x_k (\delta_{ij} - 3x'_i x'_j / r_0{}^2)/r_0{}^3$$
$$+ (\delta_{ki} x'_j - \delta_{kj} x'_i)(1 + \beta)\exp(-\beta)/r_0{}^3 \Big\}$$

$$F_{kij} = \mu\, \varepsilon_{kmn} \Big\{ (\delta_{im} x'_j + \delta_{jm} x'_i) x'_n [3(1-2\alpha') - (3 +3\beta+\beta^2)\exp(-\beta)]/r_0{}^5$$
$$+ 6\alpha' x_m [-(\delta_{ij} x'_n + \delta_{jn} x'_i + \delta_{ni} x'_j) + 5x'_i x'_j x'_n / r_0{}^2]/r_0{}^5 \Big\}$$
$$+ \mu\, \varepsilon_{jim} \Big\{ \delta_{km} (1 + \beta+\beta^2)\exp(-\beta)/r_0{}^3$$
$$- x'_k x'_m (3 +3\beta+\beta^2)\exp(-\beta)/r_0{}^5 \Big\}$$

$$G_{kij} = 2 \Big\{ \varepsilon_{kij} (\eta - \eta')[1 -(1 + \beta)\exp(-\beta)]/r_0{}^3$$
$$+ (\eta\, \varepsilon_{kjm} x'_i + \eta' \varepsilon_{kim} x'_j) x'_m [3 -(3 +3\beta+\beta^2)\exp(-\beta)]/r_0{}^5 \Big\}$$

$$H_{kij} = -2(\eta\, \delta_{kj} x'_i + \eta' \delta_{ki} x'_j)(1 + \beta)\beta^2 \exp(-\beta)/r_0{}^5$$
$$+ 2(\eta + \eta') \Big\{ (\delta_{ki} x'_j + \delta_{ij} x'_k + \delta_{jk} x'_i)[3 -(3 +3\beta+\beta^2)\exp(-\beta)]/r_0{}^5$$
$$+ x'_k x'_i x'_j [-15+(15+15\beta+6\beta^2+\beta^3)\exp(-\beta)]/r_0{}^7 \Big\}$$

where δ_{ij} is the Kronecker's delta, ε_{ijk} is the Eddington's epsilon, $\alpha = \mu/(\lambda + 2\mu)$, $\alpha' = (\lambda + \mu)/[2(\lambda + 2\mu)]$, $\beta = r_0/\ell$, $\eta = \mu\ell^2$, $\eta' = \mu\ell^2\eta_r$, $x'_i = x_i - x^*_i$ (i=1,2,3), x_i and x^*_i are the components of the vector variables $\vec{P}$ and $\vec{q}^*$, respectively.

APPLICATIONS OF THE BOUNDARY ELEMENT METHOD TO TWO AND THREE DIMENSIONAL STRESS ANALYSIS AND PLATE BENDING PROBLEMS IN ELASTICITY

Q.H. DU and Z.H. YAO

Tsing Hua University, Beijing, CHINA

SUMMARY

Some fundamental aspects of the boundary integral equation -boundary element method in elasticity are presented by using weighted residual processes while the equivalent variational functionals are also given. By proper treatment of the specific fundamental solutions and using bilinear and second order boundary elements a number of concrete problems have been performed. The results of investigation include the following cases: the two dimensional stress concentration calculations for the torsion of grooved shafts, the square shouldered shafts with fillet, the shaft with complex profile (a turbo-jet engine shaft); the axially loaded grooved shafts and square shouldered shafts with fillet; the torsion and axial loading of shafts(solid or hollow) with transverse hole;and the bending problems for cases of simply-supported square plate and cantilevered square plate. The investigations showed that the numerical schemes seem to be satisfactory for the above mentioned problems.

INTRODUCTION

The basic principle of boundary element method was for the first time introduced in[1]. And since [2] and [3] ,it can be considered as well adapted to some stress analysis problems. [4] , [5] and [6] were known to the authors as it gains increasing attentions recently. And [11] has been introduced as an engineering text for B.E.M. Authors of this paper and their group started their works on B.E.M.in Tsing Hua University since 1979. Some previous works have been reported in [7] and [8] . In this paper, the boundary element method is reviewed by weighted residual processes with its variational equivalent formulations. The method is then applied to two and three dimensional problems.

They are the axialor twist loaded shafts with variable diameter, and also axial or twist loaded shafts (solid or hollow) with transverse hole. And for the Kirchhoff plate bending problems, the simply-supported square plate and the cantilevered square plate have been worked out as examples.

In what follows some specific symbols are used, e.g., the $\langle \; , \; \rangle$ stands for the integration inner product, $(\_)$ for vector, $(\bar{\;})$ for given value. For brevity, other conventionally used symbols are introduced without further explanations.

BOUNDARY INTEGRAL EQUATION OF LINEAR ELASTICITY

Governing Equations of Linear Elasticity and Boundary Conditions

The governing equations of linear elasticity are:

$$\sigma_{ji,j} + f_i = 0 , \qquad \sigma_{ij} = \sigma_{ji} \qquad \forall(\underline{x}) \in V \tag{1}$$

$$\varepsilon_{ij} = \varepsilon_{ji} = \tfrac{1}{2}(u_{i,j} + u_{j,i}) \qquad \forall(\underline{x}) \in V \tag{2}$$

$$\sigma_{ij} = E_{ijkl}\varepsilon_{kl} , \quad E_{ijkl} = E_{jikl} = E_{ijlk} = E_{klij} \qquad \forall(\underline{x}) \in V \tag{3}$$

The boundary conditions are:

$$\left.\begin{aligned} & u_i - \bar{u}_i = 0 && \forall(\underline{x}) \in S^u \\ & t_i - \bar{t}_i \equiv \sigma_{ji} n_j - \bar{t}_i = 0 && \forall(\underline{x}) \in S^t \end{aligned}\right\} \tag{4}$$

$$S^u \cup S^t = S , \qquad S^u \cap S^t = \emptyset \tag{4a}$$

For displacement solution,

$$E_{ijkl} u_{k,lj} + f_i = 0 \qquad \forall(\underline{x}) \in V \tag{5}$$

$$u_i - \bar{u}_i = 0 \qquad \forall(\underline{x}) \in S^u , \quad n_j E_{ijkl} u_{k,l} - \bar{t}_i = 0 \quad \forall(\underline{x}) \in S^t \tag{6}$$

In case of isotropic solids,

$$\sigma_{ij} = \frac{2G\nu}{1-2\nu}\delta_{ij} u_{k,k} + G(u_{i,j} + u_{j,i}) \tag{7}$$

Then the basic equations and boundary conditions are:

$$\sigma_{ji,j}(u) + f_i \equiv G(u_{i,jj} + \tfrac{1}{1-2\nu} u_{j,ji}) + f_i = 0 \qquad \forall(\underline{x}) \in V \tag{8}$$

$$\left.\begin{aligned} & u_i - \bar{u}_i = 0 \qquad \forall(\underline{x}) \in S^u \\ & t_i(u) - \bar{t}_i \equiv G[(u_{i,j} + u_{j,i}) + \tfrac{2\nu}{1-2\nu}\delta_{ij} u_{k,k}] n_j - \bar{t}_i = 0 \quad \forall(\underline{x}) \in S^t \end{aligned}\right\} \tag{9}$$

Governing Equations of Plate Bending (Kirchhoff) and Boundary Conditions

The governing equation of plate deflection w is,

$$D\nabla^2\nabla^2 w - \bar{p}(x,y) = 0 \qquad \forall(x,y) \in \Omega \tag{10}$$

where $\bar{p}(x,y)$ stands for pressure loading and D for the plate flexurel rigidity.

The boundary conditions for plate with smooth (Liapunov) boundary curve Γ are:

$$\left.\begin{aligned} & w(s) - \bar{w}(s) = 0 && \forall(x,y) \in \Gamma^w \\ & \theta(s) - \bar{\theta}(s) = 0 && \forall(x,y) \in \Gamma^\theta \\ & M(s) - \bar{M}(s) = 0 && \forall(x,y) \in \Gamma^M \\ & V(s) - \bar{V}(s) = 0 && \forall(x,y) \in \Gamma^V \end{aligned}\right\} \tag{11}$$

where w , θ , M , V stand for deflection, angle of rotation in normal direction, bending moment and equivalent transverse shear force of the plate element. Then,

$$\left.\begin{aligned} &\theta(s)=\frac{\partial w}{\partial n} \\ &M(s)=-D\left(\frac{\partial^2}{\partial n^2}+\nu\frac{\partial^2}{\partial t^2}\right)w=-D\left[\frac{\partial^2}{\partial n^2}+\nu\left(\frac{\partial^2}{\partial s^2}+\frac{1}{\rho}\frac{\partial}{\partial n}\right)\right]w \\ &V(s)=Q_n+\frac{\partial T}{\partial s}=-D\left[\frac{\partial^3}{\partial n^3}+\frac{\partial^3}{\partial n\partial s^2}+\frac{1}{\rho}\frac{\partial^2}{\partial n^2}+(1-\nu)\frac{\partial}{\partial s}\left(\frac{\partial^2}{\partial s\partial n}-\frac{1}{\rho}\frac{\partial}{\partial s}\right)\right]w \end{aligned}\right\} \quad (12)$$

where,

$$T(s)=-D(1-\nu)\frac{\partial^2 w}{\partial n\partial t}=-D(1-\nu)\left[\frac{\partial^2 w}{\partial s\partial n}-\frac{1}{\rho}\frac{\partial w}{\partial s}\right]$$

$$Q_n=-D\frac{\partial}{\partial n}\nabla^2 w$$

It should be noted that,

$$\Gamma^w\cup\Gamma^V=\Gamma^\theta\cup\Gamma^M=\Gamma \;,\; \Gamma^w\cap\Gamma^V=\Gamma^\theta\cap\Gamma^M=\emptyset \qquad (13)$$

While the conventional clamped, simply-supported and free edge can be listed as:

$$\left.\begin{aligned} &w(s)=0,\ \theta(s)=0 && \forall(x,y)\in\Gamma^c \text{ (clamped)} \\ &w(s)=0,\ M(s)=0 && \forall(x,y)\in\Gamma^s \text{ (simply-supported)} \\ &M(s)=0,\ V(s)=0 && \forall(x,y)\in\Gamma^f \text{ (free)} \end{aligned}\right\} \quad (14)$$

These are the cases of the nature, $\overline{w}=\overline{\theta}=\overline{M}=\overline{V}=0$, and for such cases:

$$\Gamma^c=\Gamma^w\cap\Gamma^\theta,\ \Gamma^s=\Gamma^w\cap\Gamma^M,\ \Gamma^f=\Gamma^M\cap\Gamma^V \qquad (15)$$

Weighted Residual Processes and Basic Boundary Integral Equation

Elastostatic problems The weighted residual process asks for,

$$\langle\sigma_{ji,j}(u)+f_i,W_i\rangle_V+\langle u_i-\overline{u}_i,W_i\rangle_{S^u}+\langle t_i(u)-\overline{t}_i,W_i\rangle_{S^t}=0 \qquad (16)$$

For the sake of consistency with the results of volume-surface integration transformation, or V-S integration formula, the weighting function W should be,

$$W_i=\begin{cases} v_i & \forall(\underline{x})\in V \\ t_i(v) & \forall(\underline{x})\in S^u \\ -v_i & \forall(\underline{x})\in S^t \end{cases} \qquad (17)$$

Then the linearity in stress strain relations provides $\sigma_{ij}(u)v_{i,j}=\sigma_{ij}(v)u_{i,j}$ so that it leads to,

$$\langle\sigma_{ij,j}(v),u_i\rangle_V+\langle f_i,v_i\rangle_V+\langle t_i(u),v_i\rangle_{S^u}+\langle\overline{t}_i,v_i\rangle_{S^t}-\langle t_i(v),u_i\rangle_{S^t}-\langle t_i(v),\overline{u}_i\rangle_{S^u}=0 \quad (18a)$$

Thus eqn(16) results in Betti identity [eqn(18a)].
Eqn (18a) can be obtained in form of,

$$\langle\sigma_{ij,j}(v),u_i\rangle_V+\langle f_i,v_i\rangle_V+\langle t_i(u),v_i\rangle_S-\langle t_i(v),u_i\rangle_S=0 \qquad (18b)$$

which can be obtained directly from V-S integration formula. In case of v being chosen from

$$\sigma_{ji,j}(v)+f_i^v=0 \qquad (19)$$

and for the case of $f_i^v=\Delta_k(P)\delta_{ik}$, i.e.,

$$\sigma_{ji,j}(v)+\Delta_k(P)\delta_{ik}=0 \qquad (20)$$

where $\Delta_k(P)$ is a Dirac function which is singular in point P, and k indicates the direction of unit concentric force. But this is the Kelvin's solution, or the fundamental solution for the operator; so for three dimensional elastostatic problem,

$$u^s_{ki}(Q,P) = \frac{1}{16\pi G(1-\nu)r}\left[(3-4\nu)\delta_{ki} + r_{,k}r_{,i}\right] \tag{21}$$

and the boundary traction it can be derived from eqn(21)

$$t_i(u^s_{ki}) \equiv t^s_{ki}(q,P) = -\frac{1}{8\pi(1-\nu)r^2}\left\{n_j r_{,j}\left[(1-2\nu)\delta_{ki} + 3r_{,k}r_{,i}\right] - (1-2\nu)\left[n_i r_{,k} - n_k r_{,i}\right]\right\} \tag{22}$$

where P,Q always stand for points inside V and p,q stand for points on S.
So for P inside V,

$$u_k(P) = \langle f_i, u^s_{ki}\rangle_V + \langle t_i(u), u^s_{ki}\rangle_S - \langle t^s_{ki}, u_i\rangle_S \tag{23}$$

But as P→p, namely, for the boundary points, in case of smooth surface(Liapunov),

$$\frac{1}{2}u_k(p) = \langle f_i, u^s_{ki}\rangle_V + \langle t_i(u), u^s_{ki}\rangle_{S_u} + \langle \bar{t}_i, u^s_{ki}\rangle_{S_t} - \langle t^s_{ki}, u_i\rangle_{S_t} - \langle t^s_{ki}, \bar{u}_i\rangle_{S_u} \tag{24}$$

Plate bending problems(Kirchhoff) The weighted residual process asks for,

$$\langle (D\nabla^2\nabla^2 w - \bar{p}), W\rangle_\Omega + \langle (w-\bar{w}), W\rangle_{\Gamma^w} + \langle (\theta(w)-\bar{\theta}), W\rangle_{\Gamma^\theta} + \langle (M(w)-\bar{M}), W\rangle_{\Gamma^M} + \langle (V(w)-\bar{V}), W\rangle_{\Gamma^V} = 0 \tag{25}$$

And for the sake of consistency with the results of Ω-Γ integration formula for the bilinear function,

$$B(w,v) \equiv \langle \nabla^2 w, \nabla^2 v\rangle_\Omega - (1-\nu)\int_\Omega \lozenge^4(w,v)\,d\Omega \tag{26}$$

where

$$\lozenge^4(w,v) \equiv \frac{1}{2}\left[\nabla^2 w\nabla^2 v + \nabla^2(w\nabla^2 v + v\nabla^2 w) - \frac{1}{2}(\nabla^4(wv) + w\nabla^4 v + v\nabla^4 w)\right] \equiv w_{,xx}v_{,yy} - 2w_{,xy}v_{,xy} + w_{,yy}v_{,xx}$$

and $\lozenge^4(\ )$ is an invariant operator. (this should be refered to [9] and [10])
That is, for smooth boundary Γ,

$$\langle D\nabla^2\nabla^2 w, v\rangle_\Omega - \langle w, D\nabla^2\nabla^2 v\rangle_\Omega = \langle w, V(v)\rangle_\Gamma - \langle \theta(w), M(v)\rangle_\Gamma + \langle M(w), \theta(v)\rangle_\Gamma - \langle V(w), v\rangle_\Gamma \tag{27}$$

So the weighting function should be of the form,

$$W = \begin{cases} v & \forall (x,y)\in\Omega \\ -V(v) & \forall (x,y)\in\Gamma^w \\ v & \forall (x,y)\in\Gamma^V \\ M(v) & \forall (x,y)\in\Gamma^\theta \\ -\theta(v) & \forall (x,y)\in\Gamma^M \end{cases} \tag{28}$$

Put W of eqn(28) into eqn(27), then it leads to,

$$\langle \bar{p}, v\rangle_\Omega - \langle \bar{w}, V(v)\rangle_{\Gamma^w} + \langle \bar{\theta}, M(v)\rangle_{\Gamma^\theta} - \langle \bar{M}, \theta(v)\rangle_{\Gamma^M} + \langle \bar{V}, v\rangle_{\Gamma^V} = \langle D\nabla^2\nabla^2 w, v\rangle_\Omega - \langle w, V(v)\rangle_{\Gamma^w} + \langle \theta(w), M(v)\rangle_{\Gamma^\theta} - \langle M(w), \theta(v)\rangle_{\Gamma^M} + \langle V(w), v\rangle_{\Gamma^V} \tag{29a}$$

or,

$$\langle \bar{p}, v\rangle_\Omega - \langle w, D\nabla^2\nabla^2 v\rangle_\Omega - \langle \bar{w}, V(v)\rangle_{\Gamma^w} - \langle w, V(v)\rangle_{\Gamma^V} + \langle \bar{\theta}, M(v)\rangle_{\Gamma^\theta} + \langle \theta(w), M(v)\rangle_{\Gamma^M} - \langle \bar{M}, \theta(v)\rangle_{\Gamma^M} - \langle M(w), \theta(v)\rangle_{\Gamma^\theta} + \langle \bar{V}, v\rangle_{\Gamma^V} + \langle V(w), v\rangle_{\Gamma^w} = 0 \tag{29b}$$

or,

$$\langle \bar{p}, v\rangle_\Omega + \langle V(w), v\rangle_\Gamma - \langle M(w), \theta(v)\rangle_\Gamma = \langle D\nabla^2\nabla^2 v, w\rangle_\Omega + \langle V(v), w\rangle_\Gamma - \langle M(v), \theta(w)\rangle_\Gamma \tag{29c}$$

Now $w^s(Q,P)$, the fundamental solution is taken as v, such that,

$$\nabla^2\nabla^2 w^s = \Delta(P) \tag{30}$$

and for the biharmonic operator,

$$w^s(Q,P) = \frac{1}{8\pi}r^2\ln r \tag{31}$$

So for P inside Ω ,(or from outside),

$$D w(P)=\langle \bar{p}, w^s\rangle_\Omega+\langle V(w), w^s\rangle_\Gamma-\langle M(w), \theta(w^s)\rangle_\Gamma+\langle \theta(w), M(w^s)\rangle_\Gamma-\langle w, V(w^s)\rangle_\Gamma \quad (32)$$

And for P $\rightarrow$ p, a point on the smooth boundary,

$$\frac{1}{2} D w(p)=\langle \bar{p}, w^s\rangle_\Omega+\text{P.B.B.T.} \quad (33)$$

where
$$\text{P.B.B.T.}=\langle V(w), w^s\rangle_{\Gamma w}+\langle \bar{V}, w^s\rangle_{\Gamma V}-\langle M(w), \theta(w^s)\rangle_{\Gamma\theta}-\langle \bar{M}, \theta(w^s)\rangle_{\Gamma M}$$
$$+\langle \theta(w), M(w^s)\rangle_{\Gamma M}+\langle \bar{\theta}, M(w^s)\rangle_{\Gamma\theta}-\langle w, V(w^s)\rangle_{\Gamma V}-\langle \bar{w}, V(w^s)\rangle_{\Gamma w} \quad (33a)$$

it stands for the eight plate bending boundary terms.This is the basic boundary integral equation. However,for solving the plate bending problems the boundary conditions will require the differential forms of eqn(33). This can be done ,so long as the boundary conditions and so also the interpolation requirements have been considered. It will be shown in the SPECIFIC SOLUTIONS.

The Variational Formulations and Corner Point Requirements

Three dimensional elastostatic problems The above mentioned results can also be derived from the minimization of functional $J(u)$, thus by introducing

$$J(u)=\langle E_{ijkl} u_{i,j}, u_{k,l}\rangle_V-2\langle f_i, u_i\rangle_V-2\langle \bar{t}_i, u_i\rangle_{St}-2\langle (u_i-\bar{u}_i), n_j E_{ijkl} u_{k,l}\rangle_{Su} \quad (34)$$

Then from $\delta J=0$,

$$\langle (E_{ijkl} u_{k,lj}+f_i), \delta u_i\rangle_V+\langle (u_i-\bar{u}_i), n_j E_{ijkl}\delta u_{k,l}\rangle_{Su}-\langle (n_j E_{ijkl} u_{k,l}-\bar{t}_i), \delta u_i\rangle_{St}=0 \quad (35)$$

Thus it shows that if δu_i changes to v_i ,eqn(35) will be exactly the same as eqn(18). So, through using of fundamental solutions as the permissible displacement function the same basic boundary integral equation can be obtained. i.e.,

$$u_k(P)=\langle f_i, u^s_{ki}\rangle_V+\langle t_i(u), u^s_{ki}\rangle_S-\langle t^s_{ki}, u_i\rangle_S \quad (36)$$

By using eqn(36) and let P $\rightarrow$ p, now relaxing the boundary surface smooth requirements, then

$$C_{ki} u_i(p)=\langle f_i, u^s_{ki}\rangle_V+\langle t_i(u), u^s_{ki}\rangle_{Su}+\langle \bar{t}_i, u^s_{ki}\rangle_{St}-\langle t^s_{ki}, u_i\rangle_{St}-\langle t^s_{ki}, \bar{u}_i\rangle_{Su} \quad (37)$$

where

$$C_{ki}=\lim_{\varepsilon\to 0}\int_{S^\varepsilon} t^s_{ki}\, ds \quad (38)$$

and for smooth surface, $C_{ki}=\frac{1}{2}\delta_{ki}$.
As this is the results, it had been suggested by [3] .

Plate bending problems (Kirchhoff) The corresponding functional can be established as

$$J(w)=\langle D_{ijkl} w_{,ij} w_{,kl}\rangle_\Omega-2\langle \bar{p}, w\rangle_\Omega+2\langle \bar{M}, \theta(w)\rangle_{\Gamma M}-2\langle \bar{V}, w\rangle_{\Gamma V}$$
$$+2\langle (\theta(w)-\bar{\theta}), M(w)\rangle_{\Gamma\theta}-2\langle (w-\bar{w}), V(w)\rangle_{\Gamma w} \quad (39)$$

where
$$D_{ijkl}=D[\nu\delta_{ij}\delta_{kl}+(1-\nu)\delta_{ik}\delta_{jl}] \quad (39a)$$

Then from $\delta J=0$,

$$\langle (D\nabla^2\nabla^2 w-\bar{p}), \delta w\rangle_\Omega-\langle (w-\bar{w}), V(\delta w)\rangle_{\Gamma w}+\langle (\theta(w)-\bar{\theta}), M(\delta w)\rangle_{\Gamma\theta}-\langle (M(w)-\bar{M}), \theta(\delta w)\rangle_{\Gamma M}+\langle (V(w)-\bar{V}), \delta w\rangle_{\Gamma V}=0 \quad (40)$$

This shows that $\delta w=v$ will make eqn(40) just the same as eqn(27).
By properly introducing the boundary requirements of corner points or for the jump of loading (the concentric

force acting on the boundary) to the functional J and forming the new functional,

$$J_c(w) = J(w) - 2\sum_{i=1}^{m_1}(w\bar{F})_{\xi_i} - 2\sum_{i=m_1+1}^{m}\left((w-\bar{w})[\![T(w)]\!]\right)_{\xi_i} \qquad (41)$$

where $J(w)$ is defined in eqn(39) and $\bar{F}$ stand for concentric forces at ξ_i ($i=1,\cdots,m_1$) and

$$[\![T]\!]_{\xi_i} \equiv T(\xi_i^+) - T(\xi_i^-)$$

it stands for jump of T at ξ_i ($i=m_1+1,\cdots,m$), all $\bar{F}$ and $[\![T]\!]$ are situated on Γ^V.

Then from $\delta J_c = 0$,

$$\langle\langle D\nabla^2\nabla^2 w-\bar{p}), \delta w\rangle_\Omega - \langle\langle (w-\bar{w}), V(\delta w)\rangle_{\Gamma w} + \langle\langle \theta(w)-\bar{\theta}), M(\delta w)\rangle_{\Gamma\theta} - \langle\langle M(w)-\bar{M}), \theta(\delta w)\rangle_{\Gamma M}$$
$$+ \langle\langle V(w)-\bar{V}), \delta w\rangle_{\Gamma V} + \sum_{i=1}^{m_1}\left([\![T(w)]\!]_{\xi_i} - \bar{F}(\xi_i)\right)\delta w(\xi_i) - \sum_{i=m_1+1}^{m}\left((w-\bar{w})[\![T(\delta w)]\!]\right)_{\xi_i} = 0 \quad (42)$$

This results of variational formulation shows that in addition to eqn(40) which is for smooth boundary, there are

$$\left.\begin{aligned} &[\![T(w)]\!] - \bar{F} = 0 && \forall \xi = \xi_i \ (i=1,2,\cdots,m_1) \\ &w - \bar{w} = 0 && \forall \xi = \xi_i \ (i=m_1+1,\cdots,m) \end{aligned}\right\} \qquad (43)$$

Change the δw in eqn(42) to $v=w^s$ then it leads to,

$$DC(p)w(p) = \langle \bar{p}, w^s\rangle_\Omega + \text{P.B.B.T.}$$
$$+ \sum_{i=1}^{m_1}\left(\bar{F}w^s - w[\![T(w^s)]\!]\right)_{\xi_i} - \sum_{i=m_1+1}^{m}\left(\bar{w}[\![T(w^s)]\!] - w^s[\![T(w)]\!]\right)_{\xi_i} \qquad (44)$$

This is the general reciprocal relation for basic boundary integral equation of plate bending with corner points and concentric boundary forces, where $C(p)$ stands for characteristic coefficient of geometry and P.B.B.T. as defined in eqn(33a).

SOME SPECIFIC SOLUTIONS

As some specific solutions of different kind of concrete problems might be obtained by proper integration or differentiation of the basic fundamental solutions.

1. Torsion of Axi-symmetric Bodies

The cylindrical coordinates (R,θ,Z) have been introduced. The basic equations and boundary conditions are expressed in two dimensional forms (R,Z) i.e.,

$$\left(\frac{\partial^2}{\partial R^2} + \frac{1}{R}\frac{\partial}{\partial R} - \frac{1}{R^2} + \frac{\partial^2}{\partial Z^2}\right)v = 0 \qquad \forall (R,Z)\in\Omega \qquad (45)$$

$$\left.\begin{aligned} &v - \bar{v} = 0 && \forall (R,Z)\in\Gamma^v \\ &t(v) - \bar{t} = 0 && \forall (R,Z)\in\Gamma^t \end{aligned}\right\} \qquad (46)$$

where v stands for θ directional displacement and $t = \sigma_{n\theta}$ stands for shear stress.

By integration of the tangentially oriented Kelvin's solutions around a concentric circular rings the following fundamental solutions can be obtained,

$$v^s(\xi,p) = \frac{1}{2\pi G R(\xi)}\left[\frac{H}{a}K\left(\frac{b}{a}\right) - aE\left(\frac{b}{a}\right)\right] \qquad (47)$$

$$t^s(\xi,p) = G\left[n_1\left(\frac{\partial}{\partial R} - \frac{1}{R}\right) + n_2\frac{\partial}{\partial z}\right]v^s(\xi,p)$$
$$= \frac{R(p)}{\pi}\left\{n_1\left[\frac{R^2(\xi)-2H}{2aBR(\xi)}K\left(\frac{b}{a}\right) + \left(\frac{R(p)+R(\xi)}{ar^2} - \frac{aR(\xi)}{2Br^2}\right.\right.\right.$$
$$\left.\left.+\frac{a}{BR(\xi)}\right)E\left(\frac{b}{a}\right)\right] + n_2\left(Z(\xi)-Z(p)\right)\left[\frac{1}{2Ba}K\left(\frac{b}{a}\right) - \left(\frac{1}{ar^2} + \frac{1}{2Ba}\right)E\left(\frac{b}{a}\right)\right]\Bigg\}$$
$$(48)$$

Thus the boundary integration formula is,

$$C(p)v(p)=\int_{\Gamma}\frac{R(q)}{R(p)}\left[t(q)v^{s}(q,p)-v(q)t^{s}(q,p)\right]ds \qquad (49)$$

where

$$\left.\begin{array}{l} K(\frac{b}{a})=\int_{0}^{\pi/2}\frac{d\varphi}{[1-(\frac{b}{a})^{2}\sin^{2}\varphi]^{1/2}}, \qquad E(\frac{b}{a})=\int_{0}^{\pi/2}[1-(\frac{b}{a})^{2}\sin^{2}\varphi]^{1/2}d\varphi \\ a^{2}=(R(q)+R(p))^{2}+(Z(q)-Z(p))^{2}, \quad b^{2}=4R(p)R(q), \quad r^{2}=a^{2}-b^{2} \\ H=a^{2}-b^{2}/2, \quad B=b^{2}/4, \quad n_{1}=\cos(n,R), \quad n_{2}=\cos(n,Z) \end{array}\right\} \qquad (49a)$$

Since in case of 2. and 3. , the idea of obtaining solutions is similar to 1. For the limited length of the paper only brief discussions are given below.

2. Axially Loaded Axi-symmetric Bodies

By using (R,Z) denotes 1,2 directions, the integration of the circular symmetric R and Z directed Kelvin's terms it leads to the following displacement formulas,

$$\left.\begin{array}{l} u_{11}^{s}(q,p)=\frac{2A}{aR(q)}\left\{\left[4H(1-\nu)-(R^{2}(q)+R^{2}(p))\right]K(\frac{b}{a})-\left[a^{2}(3-4\nu)+H\frac{(Z(q)-Z(p))^{2}}{r^{2}}\right]E(\frac{b}{a})\right\} \\ u_{12}^{s}(q,p)=\frac{2A}{a}(Z(q)-Z(p))\left\{-K(\frac{b}{a})+\frac{H-2R^{2}(p)}{r^{2}}E(\frac{b}{a})\right\} \\ u_{21}^{s}(q,p)=\frac{2AR(p)}{aR(q)}(Z(q)-Z(p))\left\{K(\frac{b}{a})-\frac{H-R^{2}(q)}{r^{2}}E(\frac{b}{a})\right\} \\ u_{22}^{s}(q,p)=\frac{4AR(p)}{a}\left\{(3-4\nu)K(\frac{b}{a})+\frac{1}{r^{2}}(Z(q)-Z(p))^{2}E(\frac{b}{a})\right\} \end{array}\right\} \qquad (50)$$

where $A=[16\pi G(1-\nu)]^{-1}$. K.E. etc are same as eqn(49a).
And t_{ij}^{s} can be obtained by proper manipulation on eqn(50).
Then the boundary integral equation can be written as:

$$R(p)C_{ij}(p)u_{j}(p)=\int_{\Gamma}\left[t_{j}(q)u_{ij}^{s}(q,p)-u_{j}(q)t_{ij}^{s}(q,p)\right]R(q)ds(q) \qquad (51)$$
$$(i,j=1,2)$$

3. Bending of Axi-symmetric Bodies

Similar formulations for this case had been obtained in [7] . However, the computations have not yet completed. So no working example can be afforded.

4. Three Dimensional Problems

The Kelvin's solution can be used directly. So the main tasks will be the numerical treatments on discretization, interpolation and proper integrations. As it had been given in [8] , the details of treatment have been omitted and only final stress calculations are given in RESULTS OF COMPUTATIONS.

5. Plate Bending Problems (Kirchhoff)

Because there are two boundary conditions for the problem. And some of them are high order derivatives of w , so for the purpose of proper discretization and interpolation some special schemes have been suggested. In this paper the first order Hermitian interpolation is used for w and zero order Hermitian interpolation is used for θ , M and V . Thus for each node on clamped and simply-supported edge two equations should be taken. While for each node on free edge three equations , namely, w , $\frac{\partial w}{\partial t}$ and $\theta=\frac{\partial w}{\partial n}$ should be taken. Then for each node there are three independent boundary integral equations which can be derived from the fundamental solutions of the displacement field for infinite plate due to the generalized forces corresponding to the generalized displacements w , $\frac{\partial w}{\partial t}$ and $\frac{\partial w}{\partial n}$. They are:

$$\left.\begin{aligned} w_1^s &= v_{11}^s(q,p) = \frac{1}{8\pi} r^2 \ln r \\ w_2^s &= v_{21}^s(q,p) = \left[\frac{\partial}{\partial n}\right]_{(p)} v_{11}^s(q,p) = \frac{1}{8\pi} r(2\ln r+1)\frac{\partial r}{\partial n_{(p)}} \\ w_3^s &= v_{31}^s(q,p) = \left[\frac{\partial}{\partial t}\right]_{(p)} v_{11}^s(q,p) = \frac{1}{8\pi} r(2\ln r+1)\frac{\partial r}{\partial t_{(p)}} \end{aligned}\right\} \quad (52)$$

However, for the purpose of neatness, w, θ and V , M are replaced by number ordered boundary variables v_1 , v_2 and τ_1 ,$-\tau_2$, so that w_k^s , θ_k^s , V_k^s , M_k^s are denoted by v_{k1}^s , v_{k2}^s , τ_{k1}^s and $-\tau_{k2}^s$; and $[\ \ ]_{(p)}$ denotes that the differential operator is acted on point p . Thus the corresponding integral equations are:

$$\int_\Gamma [\tau_i(q) v_{ki}^s(q,p) - v_i(q)\tau_{ki}^s(q,p)]ds(q) + \sum_{j=1}^{m}[F(q_j)v_{k1}^s(q_j,p) - v_1(q_j)F_k^s(q_j,p)] = 0 \quad (i=1,2;\ k=1,2,3) \quad (53)$$

where $\int_\Gamma ds(q)$ should be taken as the limit case of point P moves to p from outside. For eqn(53) is a format, so it will contain those singular terms with no Cauchy's principal value; and q_j will contain all corner points and boundary points with concentric forces, and

$$\left.\begin{aligned} &v_{k2}^s(q,p) = \left[\frac{\partial}{\partial n}\right]_{(q)} v_{k1}^s(q,p), \quad && \tau_{k1}^s(q,p) = -D\left[\frac{\partial}{\partial n}\nabla^2 + (1-\nu)\frac{\partial}{\partial s}\frac{\partial^2}{\partial n\partial t}\right]_{(q)} v_{k1}^s(q,p) \\ &\tau_{k2}^s(q,p) = D\left[\nabla^2 - (1-\nu)\frac{\partial^2}{\partial t^2}\right]_{(q)} v_{k1}^s(q,p), \quad && F_k^s(q_j,p) = -(1-\nu)D\left[\!\left[\left[\frac{\partial^2}{\partial n\partial t}\right]_{(q)} v_{k1}^s(q,p)\right]\!\right]_{q_j} \end{aligned}\right\} (54)$$

where $[\ \ ]_{(q)}$ denotes that the differential operator is acted on point q . For each node on free edge or on the symmetric line, three equations are required ($k=1,2,3$), while for each node on clamped edge or on simply-supported edge only two equations are required ($k=1,2$). For the boundary corner node by using the relations existed between the first and second order partial derivatives of the two sides of the corner then the number of independent boundary equations are just the same as node of smooth boundary. Some typical corner node, its boundary variables involved and its number are listed in the following table.

Type of corner	boundary variables	k
C – C	$\tau_1(q^+)$, $\tau_1(q^-)$	1, 2
C – S	$\tau_1(q^+)$, $\tau_1(q^-)$	1, 2
C – f	$\tau_1(C)$	1
S – S	$\tau_1(q^+)$, $\tau_1(q^-)$, $(F)^*$	1, 2, $(3)^*$
S – f	$\tau_1(S)$, $v_2(S)$, $(F)^*$	1, 2, $(3)^*$
f – f	v_1 , $v_2(q^+)$, $\frac{\partial v_1}{\partial t}(q^+)$	1, 2, 3

C: Clamped, S: Simply-supported, f : Free

( )* : Corner point with two outward normals at 90^0

The above mentioned three-equation scheme is a conforming interpolation. This is used for Fig. 8 and Fig. 9. In the five-equation scheme with higher conforming properties have been used previously, but the results are about the same. This justifies that three- equation scheme is appropriate for the chosen discretization. And this procedure seems to be comparable with [12] which has been noticed by the authors soon after the computations had done in1980 (using the five-equation scheme).

RESULTS OF COMPUTATIONS

Based upon the above mentioned boundary integral equations nine examples have been worked out. The principal results of computations have been shown in Fig.1 to Fig.9. For the descretization, interpolation and integration processes of boundary elements the bilinear and the second order elements and in some cases the linear elements and together with Gauss-Legendre integration formula have been used. As for the singularity included in the process for three dimensional problems the treatment is similar to the work of Lachat [3] and Watson [5] . In the examples of shafts transverse hole (Fig.6 and Fig.7) some special cylindrical boundary elements have been used. For Fig.3, the turbo-jet engine shaft, the subregion-technique has been used. The programs for Fig.4 and Fig.5 are in Algol. The other programs in Fortran IV include: B.I.E.-T(2).,for Fig.1, Fig.2 and Fig.3; B.I.E.-B.E.M.(3D).,for Fig.6 and Fig.7; and B.E.M.-P.B. for Fig.8 and Fig.9. Some other details can be found in [13], [14] , [15] and [16].

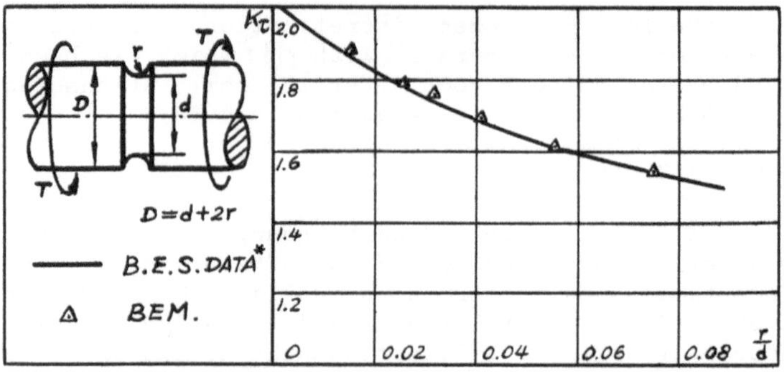

* British Engineering Sciences Data. Item No. 69021 (1970)

Fig. 1 Torsion of Grooved Shafts

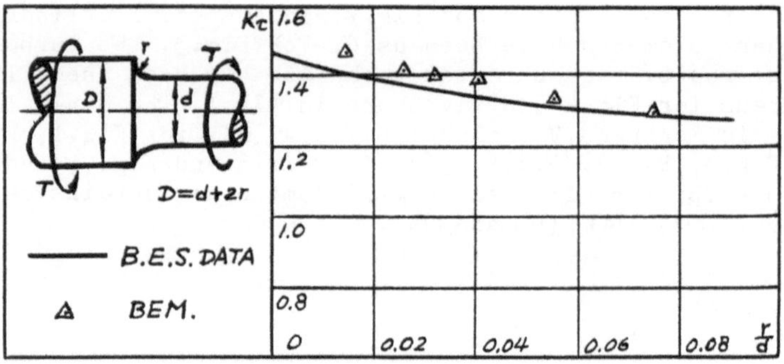

Fig. 2 Torsion of Square Shouldered Shafts with Fillet

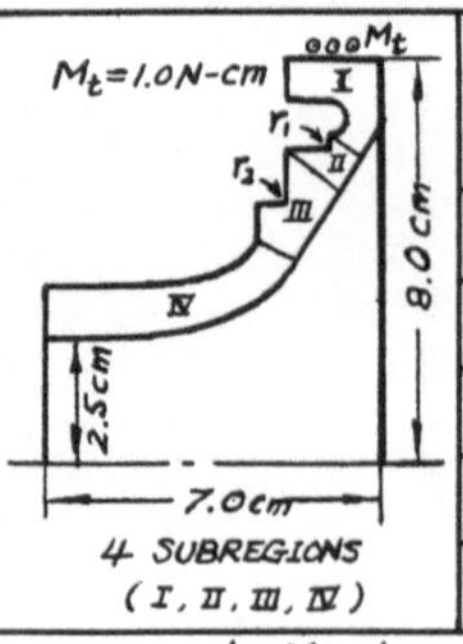

POSITION OF STRESS CONCENTRATION		MAXIMUM SHEAR STRESS N/mm^2	
		BIE-BE	FINITE DIFFERENCES *
r_1	0.15 mm	0.03212	0.0314
"	0.4 mm	0.01946	0.0191
"	1.0 mm	0.01215	0.01215
r_2	0.4 mm	0.0218	——
"	0.5 mm	——	0.0211

* Li Min-hua, Acta Mechanica Solida Sinica, 1980 No. 2

Fig. 3 Torsion of Turbo-jet Engine Shaft

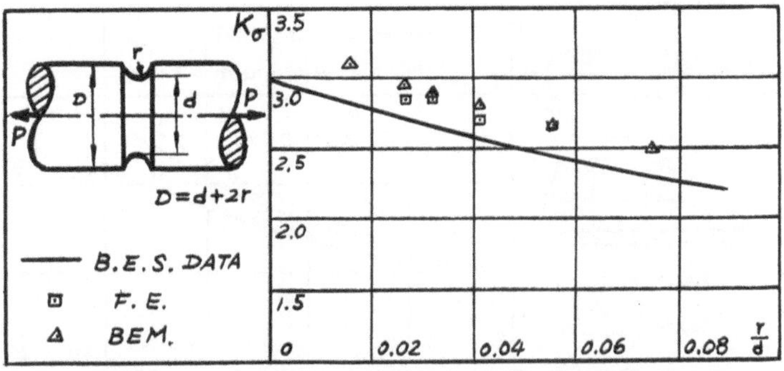

Fig. 4 Axially Loaded Grooved Shafts

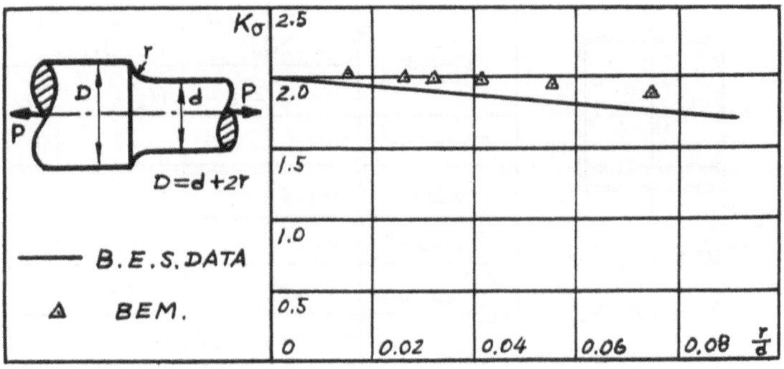

Fig. 5 Axially Loaded Square Shouldered Shafts with Fillet

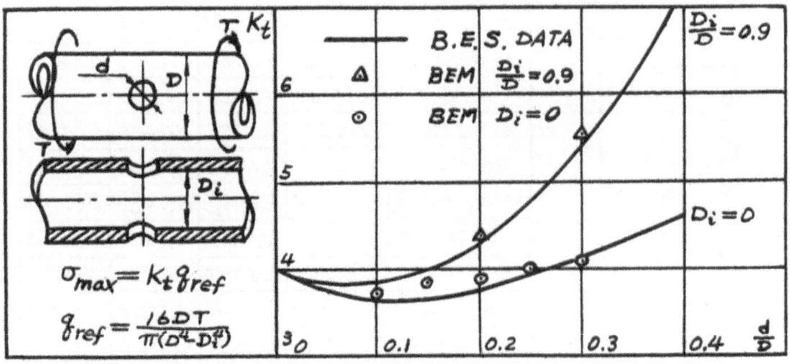

Fig. 6 Torsion of Shafts with Transverse Hole

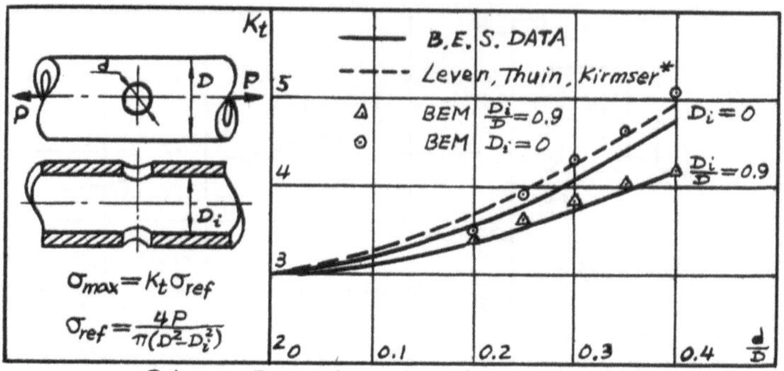

* Peterson, R.E., Stress Concentration Design Factors (1974)

Fig. 7 Axially Loaded Shafts with Transverse Hole

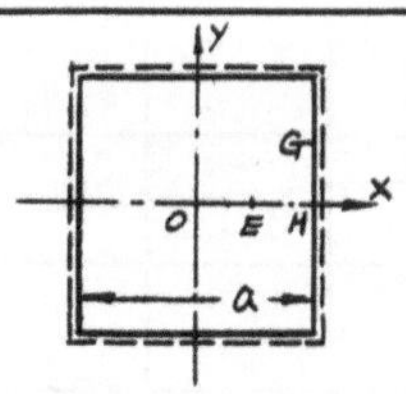

$$w=\frac{\bar{p}a^4}{Eh^3}\cdot K_w$$

$$V=\bar{p}a\cdot K_v$$

$\nu=0.3$	DEFLECTION K_w	SHEAR FORCE K_v		
	$O(0,0)$	$E(\frac{a}{4},0)$	$H(\frac{a}{2},0)$	$G(\frac{a}{2},\frac{a}{4})$
Вайнберг	0.0433	0.0320	0.420	0.354
Timoshenko	0.04434	—	0.420	—
Jaswan *	0.04586	—	—	—
BEM. 3 equations	0.04463	0.0323	0.424	0.352
BEM. 5 equations	0.04452	0.0322	0.423	0.358

* Jaswon, M.A. and Maiti, M., J. of Engng Math. Vol.2 (1968)

Fig. 8 Uniformly Distributive-loaded Square Plates with Simply-supported Edges

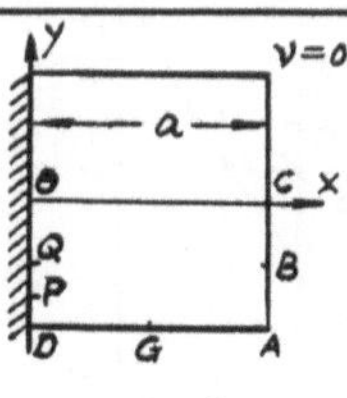

$\nu=0.3$

$$w=\frac{\bar{p}a^4}{D}\cdot K_w$$

$$M=\bar{p}a^2\cdot K_M$$

DEFLECTION K_w	$C(a,0)$	$B(a,\pm\frac{a}{4})$	$A(a,\pm\frac{a}{2})$	$G(\frac{a}{2},\pm\frac{a}{2})$
SERIES METH.	0.13102	0.13056	0.12933	0.04433
F. E.	0.12905	0.12851	0.12708	0.04322
BEM. 3 equations	0.12800	0.12752	0.12608	0.04279
BEM. 5 equations	0.12731	0.12666	0.12491	0.04242
BEND. MOMENT K_M	$D(0,\pm\frac{a}{2})$	$P(0,\pm\frac{3a}{8})$	$Q(0,\pm\frac{a}{4})$	$O(0,0)$
SERIES METH.	0	-0.51270	-0.53353	-0.53560
F. E.	-0.34571	-0.50399	-0.52760	-0.53092
BEM. 3 equations	-0.27508	-0.54452	-0.52102	-0.52829
BEM. 5 equations	0	-0.49690	-0.52568	-0.52847

Fig. 9 Uniformly Distributive-loaded Cantilevered Square Plate

REFERENCES

1. Rizzo,F.J.(1967) "An Integral Equation Approach to Boundary Value Problems of Classical Elastostatics" Q.Appl.Math.Vol.XXV
2. Cruse,T.A.and Rizzo,F.J.ed.(1975) "Boundary Integral Equation Method:Computation Applications in Applied Mechanics" AMD Vol.11 A.S.M.E.
3. Lachat,J.C.and Watson,J.O.(1976) "Effective Numerical Treatment of Boundary Integral Equations" Int. J.Num.Meth.Eng.Vol.10
4. Brebbia,C.A.ed.(1978) "Recent Advances in Boundary Element Methods" Pentech.
5. Banerjee,P.K.and Butterfield,R.ed.(1979) "Developments in Boundary Element Methods-I"App.Sci.Publisher.
6. Brebbia,C.A.ed.(1980) "Recent Advances in Boundary Element Methods" March,Southampton
7. Du,Q.H.and Yao,Z.H.(1982) "Some Basic Problems and Engineering Applications of B.I.E.-B.E.M. in Elasticity" Acta Mechanica Solida Sinica Feb. (in Chinese with English Abstract)
8. Du,Q.H.and Yao,Z.H.(1982) "Some Stress Concentration Problems by Boundary Integral Equation-Boundary Element Method" An International Conference on Finite Element Methods, Shanghai,China.(to be published in English)
9. Bergman,S.and Schiffer,M.(1953) "Kernel Functions and Elliptical Differential Equations in Mathematical Physics" Academic Press
10. Mansfield,E.H.(1962)"On the Analysis of Elastic Plate of Variable Thickness" Q.J. of Mechanics and Math. Vol.XV Part 2.
11. Brebbia,C.A.and Walker,S.(1980)"Boundary Element Techniques in Engineering" Newnes-Butterworths
12. Bezine,G.P.and Gamby,D.A.(1978) "A New Integral Equation Formulation for Plate Bending Problems" in Ref.4

The following references are four M.S. theses of Department of Engineering Mechanics, Tsing Hua University. They are in Chinese.

13. Zang Kun(1981) "Stress Concentration Problems for Bodies of Revolution byB.I.E.-B.E.M."
14. Lu Xi-lin(1982) "Three Dimensional Stress Analysis for Shafts with Transverse Hole by B.I.E.-B.E.M."
15. Shan Wen-wen(1982) "Stress Concentration Problem for Torsion of Shafts with Complex Profile by B.I.E.-B.E.M."
16. Song Guo-shu (1981) " Bending of Plates by B.I.E.-B.E.M."

ON THE INDETERMINANCY OF BIE SOLUTIONS FOR THE EXTERIOR PROBLEMS OF TIME-HARMONIC ELASTODYNAMICS AND INCOMPRESSIBLE ELASTOSTATICS

S. Kobayashi & N. Nishimura

Department of Civil Engineering, Kyoto University, Kyoto, Japan

ABSTRACT

The non-uniqueness of the BIE solution for exterior problems of elastodynamics is investigated and some numerical techniques to avoid this difficulty are discussed. Also, an analogous phenomenon and its remedies in incompressible elastostatics are described.

1. INTRODUCTION

BIEM is one of the most effective numerical methods of analysis for exterior or half-space problems because it is not of domain type. Especially, BIEM is suitable for exterior elastodynamics since it can handle infinite domains directly. This is in contrast with most of the conventional methods such as FEM. In addition, BIEM for exterior problems reduces the number of unknowns drastically compared with domain type methods. However, BIEM has one drawback called the problem of 'fictitious eigenfrequencies.'

Like other techniques, BIEM for elastodynamics can be classified into the time domain analysis and the frequency domain analysis. The latter, presumablly more practical than the former, converts transient problems into time-harmonic problems. Conventional BIEM successfully solves time-harmonic problems in most cases. For a certain set of frequencies called the 'fictitious eigenfrequencies', however, it loses uniqueness of the solutions, and yields very poor numerical results. This difficulty has long been known to mathematicians [Kupradze (1965)] and some techniques to avoid this difficulty are available in acoustics [Schenck (1968), Burton & Miller (1971), Ursell (1973), Jones (1974), Kleinman & Roach (1974), Meyer et al. (1978), Terai (1980)].

In this paper we discuss the problem of the 'fictitious eigenfrequencies' for exterior elastodynamics in detail. We first determine when the conventional BIE for mixed exterior boundary value problem loses its uniqueness of solutions. After investigating some particular cases, we discuss several methods to overcome this difficulty. Specifically, we modify some known techniques in acoustics in order to apply them to elastodynamics, and then examine their efficiency and accuracy via numerical examples. Finally, we remark that essentially the same phenomenon occurs to BIE for exterior incompressible elastostatics.

2. FICTITIOUS EIGENFREQUENCIES

2.1 Exterior problems

Let D^+ be a domain exterior to a boundary ∂D. Also let the interior of ∂D, denoted by D^-, be simply-connected. We intend to find a solution $\mathbf{u}$ of the equation

$$P(\nabla)\mathbf{u} = (\mu\Delta + (\lambda + \mu)\nabla\nabla\cdot)\mathbf{u} + \rho\omega^2\mathbf{u} = 0 \tag{1}$$

in D^+ subject to the boundary conditions

$$\mathbf{u} = \mathbf{u}^\circ \quad \text{on } \partial D_1, \qquad (\partial D_1 + \partial D_2 = \partial D)$$

$$T\mathbf{u} = \lambda\mathbf{n}\,\mathrm{tr}(\nabla\mathbf{u}) + \mu(\nabla\mathbf{u} + (\nabla\mathbf{u})^T)\mathbf{n} = \mathbf{t}^\circ \quad \text{on } \partial D_2, \tag{2}$$

and the radiation condition, where we used the notations $\mathbf{u}$ and $\mathbf{t}$ for displacement and traction, $\mathbf{u}^\circ$ and $\mathbf{t}^\circ$ for their prescribed boundary values, λ and μ for Lame's constants, ρ for density, ω for frequency, and $\mathbf{n}$ for the unit outward normal vector on ∂D, respectively. For definiteness, the time factor is assumed to be $e^{-i\omega t}$.

The usual argument gives an expression of $\mathbf{u}$ as a sum of two potentials (Uppercase Greek letters indicate vectors or tensors);

$$\tilde{\mathbf{u}}(\mathbf{x}) = \int_{\partial D}\Gamma_I(\mathbf{x}, \mathbf{y})\mathbf{u}(\mathbf{y})dS_y - \int_{\partial D}\Gamma(\mathbf{x}, \mathbf{y})\mathbf{t}(\mathbf{y})dS_y, \tag{3}$$

where $\tilde{\mathbf{u}}$ stands for a field on $R^n\backslash\partial D$ (n = 2 or 3) defined by

$$\tilde{\mathbf{u}}(\mathbf{x}) = \begin{cases} \mathbf{u}(\mathbf{x}) & \mathbf{x} \in D^+ \\ 0 & \mathbf{x} \in D^-, \end{cases} \tag{4}$$

and Γ, Γ_I denote the kernels of the simple layer (i.e., the fundamental solution) and the double layer potentials [Kupradze (1965)]. Γ_I can be written as

$$\Gamma_I(\mathbf{x}, \mathbf{y}) = \Gamma(\mathbf{x}, \mathbf{y})T_y,$$

using the operator defined by equation (2), where the subscript y means that the differentiation is with respect to $\mathbf{y}$. The conventional method of obtaining a boundary integral equation (BIE) consists in a limiting operation of letting the point $\mathbf{x} \in D^-$ approach ∂D, which yields for $\mathbf{x} \in \partial D$

$$0 = c^-(\mathbf{x})\mathbf{u}(\mathbf{x}) + ⨍_{\partial D}\Gamma_I(\mathbf{x}, \mathbf{y})\mathbf{u}(\mathbf{y})dS_y - \int_{\partial D}\Gamma(\mathbf{x}, \mathbf{y})\mathbf{t}(\mathbf{y})dS_y, \tag{5}$$

where $c^-(\mathbf{x})\mathbf{u}(\mathbf{x})$ (= $-\mathbf{u}(\mathbf{x})/2$ when ∂D is smooth near $\mathbf{x}$) is the non-integral term of the interior limit of the double layer potential. We then introduce the boudary conditions (equation (2)) and solve equation (5) numerically by using boundary elements.

We now investigate the uniqueness of the solutions of equation (5) taking for granted the existence. To this end, we introduce the following notations;

$U(D, \lambda, \mu, \rho, \omega)$: The set of solutions of equation (1) in D which satisfies the radiation condition if D is an exterior domain.

$U_o(D, \lambda, \mu, \rho, \omega)$: The set of the element of $U(D, \lambda, \mu, \rho, \omega)$ which satisfies $u(x) = 0$ on ∂D.

$\Omega(D, \lambda, \mu, \rho)$: The set of eigenfrequencies ω of the displacement boundary value problem for D.

It is known that the expression

$$v(x) = \int_{\partial D}\Gamma_I(x, y)u(y)dS_y - \int_{\partial D}\Gamma(x, y)t(y)dS_y$$

satisfies

$$v^+(x) = u(x) \quad \text{and} \quad Tu^+(x) = t(x) \qquad x \in \partial D \tag{6}$$

if and only if $v(x) = 0$ in D^- [Kupradze (1965)] where the superimposed + (-) indicates the limiting value from D^+ (D^-). Since equation (5) is equivalent to the statement $v(x) \in U_o(D^-, \lambda, \mu, \rho, \omega)$, we have equation (6) when $\omega \notin \Omega(D^-, \lambda, \mu, \rho)$. In this case, homogeneous boundary conditions and equation (6) give $v(x) = 0$ for $x \in D^+$ which, together with equation (6), yields $t(x) = 0$ on ∂D_1 and $u(x) = 0$ on ∂D_2. On the other hand, if $\omega \in \Omega(D^-, \lambda, \mu, \rho)$, we have

$$v(x) = u^*(x) \qquad x \in D^-$$

where $u^*(x)$ is a linear combination of nontrivial elements of $U_o(D^-, \lambda, \mu, \rho, \omega)$. (We will refer to this field simply as eigenmode in the sequel.) Accordingly, we have

$$v^+(x) = u(x) \quad \text{and} \quad Tv^+(x) - Tu^*(x) = t(x) \quad \text{on } \partial D, \tag{7}$$

and the homogeneous boundary condition gives

$$v^+(x) = 0 \text{ on } \partial D_1 \quad \text{and} \quad Tv^+(x) = Tu^*(x) \text{ on } \partial D_2. \tag{8}$$

Therefore, we have

$$t(x) = Tu^e(x) - Tu^*(x) \text{ on } \partial D_1 \quad \text{and} \quad u(x) = u^e(x) \text{ on } \partial D_2, \tag{9}$$

where $u^e \in U(D^+, \lambda, \mu, \rho, \omega)$ is a function which satisfies the boundary condtion of the form given in equation (8). Conversely, we can readily check that equation (9) indeed gives a non-trivial solution of equation (5). Thus we conclude that the 'fictitious eigenfrequencies' of the problem are $\Omega(D^-, \lambda, \mu, \rho)$ whatever the type of the boundary conditions may be, as long as we use equation (5).

This type of results appear repeatedly in literatures mostly in the context of Dirichlet or Neumann problems for the Helmholtz equation [e.g. Kleinman & Roach (1974)].

2.2 Miscellaneous examples

Consider the problem of determining the elastodynamic fields in $D^i \cup D^e$ (Fig. 1(a)) given the incident field, boundary conditions on S, and the connectivity conditions on S' (e.g. $u^+ = u^-$ and $t^+ = t^-$). If one uses the Green's formula (equation (3)) for the domains D^e (exterior to S') and D^i separately, the obtained BIE loses the uniqueness of the solutions when $\omega \in \Omega(D^o \cup S' \cup D^i, \lambda^e, \mu^e, \rho^e) \cup \Omega(D^o, \lambda^i, \mu^i, \rho^i)$ [Kobayashi & Nishimura (1982 a)].

The same phenomenon can occur even to a bounded domain when it has interior boundaries. The set of the fictitious eigenfrequencies for the annular domain D shown in Fig. 1(b) is $\Omega(D^o, \lambda, \mu, \rho)$. Of course there are infinite set of 'physical' eigenfrequencies in this case.

For the indented half-space shown in Fig. 1(c), we can reduce the boundary value problem to a BIE defined on S_1 by using a Green's function which satisfies, say, the traction free condition on $S_o + S_2$. We can economize the analysis at the price of the loss of uniqueness of the solutions when ω is equal to one of the eigenfrequencies of $D^o(\lambda, \mu, \rho)$ with the boundary conditions $u = 0$ on S_1 and $t = 0$ on S_o.

3. METHODS TO AVOID 'FICTITIOUS EIGENFREQUENCIES'

There are some integral equations for the exterior problems of

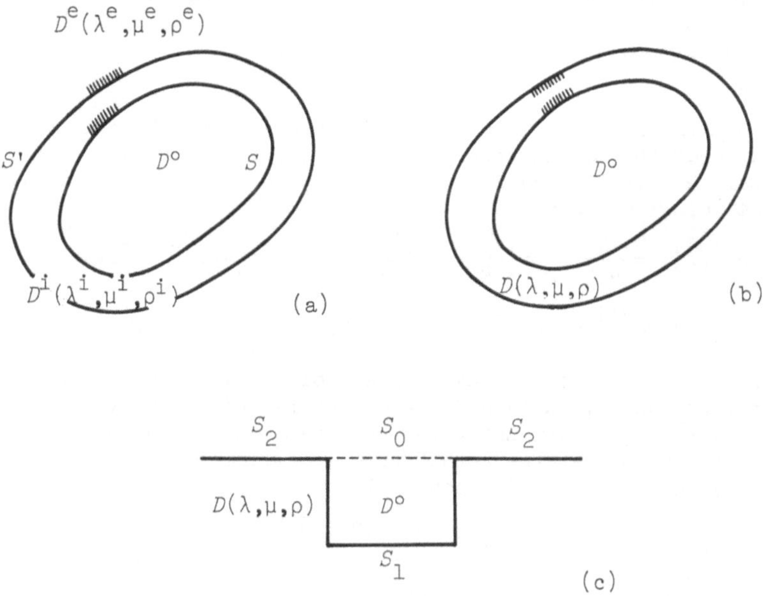

Figure 1. Domains for which the conventional BIE loses uniqueness of the solutions. (a) Lined hole (b) Annular domain (c) Indented half-space.

the Helmholtz equation which are known to have a unique solution. However, some of them have a strongly singular kernel [e.g. Burton & Miller (1971)] and its regularization does not seem to be practical. Especially when one wishes to calculate stresses around the boundary, it is not preferable that the integrand be strongly singular. It is because one has to use higher order boundary elements to this end which implies the use of numerical quadrature.

Jones [Jones (1974)] proposed two remedies for exterior Neumann problem of the Helmholtz equation in which the kernels do not become more singular than those of the conventional BIE. We here try to extend his idea to the mixed problems of elastodynamics. We restrict our attention to two dimensional problems for simplicity and describe the methods for the domains used in 2.1.

3.1 Some preliminaries

We prepare some formulae for the later use. We first introduce a polar coordinate (R, θ) which gives the cartesian coordinate of two points $\mathbf{x}$ and $\mathbf{y}$ as

$$\begin{pmatrix} x_1 \\ x_2 \end{pmatrix} = R\begin{pmatrix} \cos\theta \\ \sin\theta \end{pmatrix}, \quad \begin{pmatrix} y_1 \\ y_2 \end{pmatrix} = R_0\begin{pmatrix} \cos\theta_0 \\ \sin\theta_0 \end{pmatrix}.$$

The fundamental solution $\Gamma(\mathbf{x}, \mathbf{y})$ can easily be expanded in Fourier series on the circle R = const. by using the Fourier transform. It is well known that the Fourier transform on $\mathbf{x}$ of the fundamental solution $\Gamma(\mathbf{x}, \mathbf{y})$ of equation (1) is written as

$$-\mathrm{P}^{-1}(i\Xi)e^{-i\Xi\cdot\mathbf{y}},$$

where Ξ is the variable of the transformation and $\mathrm{P}^{-1}(i\Xi)$ is the inverse of the matrix obtained by replacing ∇ in $\mathrm{P}(\nabla)$ by $i\Xi$ [Dubois & Lachat(1973) for example]. Using the polar coordinate (r, ϕ) of Ξ, we have

$$-\mathrm{P}^{-1}(i\Xi) = \frac{1}{4\mu}\{(\frac{1}{r^2 - k_T^2} + (\frac{k_L}{k_T})^2 \frac{1}{r^2 - k_L^2})(\mathbf{v}\otimes\bar{\mathbf{v}} + \bar{\mathbf{v}}\otimes\mathbf{v})$$
$$- (\mathbf{v}\otimes\mathbf{v}e^{2i\phi} + \bar{\mathbf{v}}\otimes\bar{\mathbf{v}}e^{-2i\phi})(\frac{1}{r^2 - k_T^2} - (\frac{k_L}{k_T})^2 \frac{1}{r^2 - k_L^2})\},$$

where k_L and k_T are the wave numbers of the longitudinal and transverse waves, and $\mathbf{v}$ and $\bar{\mathbf{v}}$ are vectors defined by $\mathbf{v} = \mathbf{i}_1 - i\mathbf{i}_2$ and $\bar{\mathbf{v}} = \mathbf{i}_1 + i\mathbf{i}_2$, using the base vectors $\mathbf{i}_1$ and $\mathbf{i}_2$. Then it is easily seen that the n-th Fourier coefficient (of $e^{in\theta}$) of $\Gamma(\mathbf{x}, \mathbf{y})$ on the circle R = const. is

$$-\frac{1}{(2\pi)^2}\oint d\theta e^{-in\theta}\int_0^\infty dr\oint d\phi\, r e^{irR\cos(\phi - \theta)}\mathrm{P}^{-1}(i\Xi)e^{-irR_0\cos(\phi - \theta_0)}. \tag{10}$$

For $e^{-i\omega t}$ time-harmonic problem, we can calculate the integral in equation (10) assuming $\mathrm{Im}\, k_{L,T} > 0$. Use of some known formulae of the Bessel functions gives

$$\Gamma(\mathbf{x}, \mathbf{y}) = -\frac{i}{16\mu}\sum_{n=-\infty}^{\infty}\sum_{i=1}^{2}(-1)^n \Psi_H^{ni}(\mathbf{x})\otimes\Psi_J^{-ni}(\mathbf{y}) \tag{11}$$

when $R_0 < R$. When $R < R_0$ we must interchange the subscripts J and H in the above expression. The vectors $\Psi_H^{ni}(x)$ are defined by

$$\Psi_H^{n1}(x) = e^{in\theta}(-H_{n+1}^{(1)}(k_L R)V + H_{n-1}^{(1)}(k_L R)\overline{V})\frac{k_L}{k_T}i,$$

$$\Psi_H^{n2}(x) = e^{in\theta}(H_{n+1}^{(1)}(k_T R)V + H_{n-1}^{(1)}(k_T R)\overline{V}),$$

where $H_{\cdot}^{(1)}$ is the Hankel function of the first kind and $V = i_r - i i_\theta$ and $\overline{V} = i_r + i i_\theta$, i_r and i_θ being the unit vectors in the direction of r and θ coordinate at x, respectively. $\Psi_J^{ni}(x)$ have the same expressions as $\Psi_H^{ni}(x)$ except that $H_{\cdot}^{(1)}$ is replaced by J, i.e. the Bessel function. We next prove the identity

$$\int_{\partial D} T\Psi_J^{-mi}(x)\cdot\Psi_H^{nj}(x)dS - \int_{\partial D}\Psi_J^{-mi}(x)\cdot T\Psi_H^{nj}(x)dS = 16\mu i(-1)^m\delta_{mn}\delta_{ij},$$

$$-\infty < m,\ n < \infty,\ 1 \le i,\ j \le 2, \qquad (12)$$

which holds for any boundary ∂D to which the origin is interior. To see this we assume that x (y) is exterior (interior) to ∂D. Then equation (3) gives

$$\Gamma(x, y) = \int_{\partial D}\Gamma_I(x, z)\Gamma(z, y)dS_z - \int_{\partial D}\Gamma(x, z)\overrightarrow{T}_z\Gamma(z, y)dS_z.$$

By introducing equation (11) and comparing the coefficients of the terms $\Psi_H^{mi}(x)\otimes\Psi_J^{nj}(y)$ in the above equation we obtain equation (12).

These results might have been obtained by using the addition theorems for the Bessel functions [e.g. Shaw (1979) for acoustics]. However, the present method is simpler.

3.2 Elastodynamic counterparts of Jones' methods

In this section we discuss the elastodynamic version of Jones' methods [Jones (1974)]. As will be seen, only a slight change is required to his original proof of uniqueness etc. for the Neumann problems of the Helmholtz equation in order to extend it to mixed problems of elastodynamics.

Let the origin be in D^-. The elastodynamic counterpart of Jones' method may be stated as follows; To solve equation (5) subject to the conditions

$$I^{ni} = \int_{\partial D}\Psi_H^{ni}(x)\cdot t(x)dS - \int_{\partial D}T\Psi_H^{ni}(x)\cdot u(x)dS = 0, \qquad (13)$$

for $-N \le n \le N$, $1 \le i \le 2$, where N is a properly chosen integer. This method will be called the first method in the sequel. It is clear, when $\omega \notin \Omega(D^-, \lambda, \mu, \rho)$, that the unique solution of equation (5) satisfies equation (13).

We next interpret equation (13) when $\omega \in \Omega(D^-, \lambda, \mu, \rho)$. Since for the present case u and t are given by equation (7), equation (13) can be written as

$$\int_{\partial D}\Psi_H^{ni}(x)\cdot Tu^*(x)dS - \int_{\partial D}T\Psi_H^{ni}(x)\cdot u^*(x)dS = 0,$$

from which it follows

$$\int_C \Psi_H^{ni}(x)\cdot Tu^*(x)dS - \int_C T\Psi_H^{ni}(x)\cdot u^*(x)dS = 0 \tag{14}$$

for any circle C inside D^- and centered at the origin. Since u^* can be expanded in the interior of C as

$$u^*(x) = \sum_{n=-\infty}^{\infty} \sum_{i=1}^{2} C^{ni}\Psi_J^{ni}(x), \quad (C^{ni} : \text{const.}) \tag{15}$$

equation (14), together with equation (12) implies that

$$C^{ni} = 0 \quad \text{for } -N \le n \le N,\ 1 \le i \le 2. \tag{16}$$

Therefore, we conclude that $u^*(x) = 0$ in D^- unless we can find a nontrivial eigenmode u^* which satisfies equation (16). Since the lower eigenmodes generally contain terms of small $|n|$ in equation (15), this method determines u and t for given ω uniquely if N is sufficiently large.

Jones also showed that the equivalent result as above can be achieved by using a modified kernel in a simple layer potential formulation for the Neumann problem of the Helmholtz equation. The elastodynamic counterpart of this second method may be stated as follows, using Grees's formula instead; Replace the kernels in equations (3) and (5) by the following expressions

$$\Gamma(x, y) - \varepsilon \sum_{n=-N}^{N} \sum_{i=1}^{2} (-1)^n \Psi_H^{ni}(x) \otimes \Psi_H^{-ni}(y),$$

$$\Gamma_I(x, y) - \varepsilon \sum_{n=-N}^{N} \sum_{i=1}^{2} (-1)^n \Psi_H^{ni}(x) \otimes T\Psi_H^{-ni}(y),$$

respectively, and solve the modified version of equation (5), where $\varepsilon \neq 0$ is a constant such that $\text{Im}\,\varepsilon \le 0$.

The new integral equation is written for $x \in \partial D$ as

$$0 = [\int_{\partial D}\Gamma_I(x, y)u(y)dS - \int_{\partial D}\Gamma(x, y)t(y)dS]^- - \varepsilon \sum_{n=-N}^{N} \sum_{i=1}^{2} (-1)^n \times$$

$$\times \Psi_H^{ni}(x)\{\int_{\partial D}(T\Psi_H^{-ni}(y)\cdot u(y) - \Psi_H^{-ni}(y)\cdot t(y))dS\}. \tag{17}$$

The solvability of equation (17) (which has at least one solution, i.e., the boundary values u, t of the exterior field to be sought) gives

$$0 = \sum_{n=-N}^{N} \sum_{i=1}^{2} (-1)^n I^{-ni} \int_{\partial D} \Psi_H^{ni}(x)\cdot Tu^*(x)dS$$

$$= -\sum_{n=-N}^{N} \sum_{i=1}^{2} (-1)^n I^{-ni} \int_{\partial D}(T\Psi_H^{ni}(x)\cdot u^*(x) - \Psi_H^{ni}(x)\cdot Tu^*(x))dS \tag{18}$$

by theorems of the Fredholm type [Kupradze (1965)]. Of course this condition is trivially satisfied when $\omega \notin \Omega(D^-, \lambda, \mu, \rho)$ since $u^* = 0$. Let u^*_p $(1 \le p \le M)$ be a maximal set of independent eigenmodes which can be orthogonalized in the following sense;

$$\sum_{n=-N}^{N} \sum_{i=1}^{2} C_p^{ni}\overline{C_q^{ni}} = \delta_{pq}, \quad 1 \le p, q \le M, \tag{19}$$

where C_p^{ni} is the coefficient of $u^*_p(x)$ in the series of $\Psi_J^{ni}(x)$ near the origin, i.e.,

$$u^*_p(x) = \sum_{n=-\infty}^{\infty} \sum_{i=1}^{2} C_p^{ni}\Psi_J^{ni}(x).$$

Since $u^*_p(x)$ can be chosen real we demand

$$C_p^{-ni} = (-1)^{n+1}\overline{C_p^{ni}}. \qquad (20)$$

It can be shown that we can extend the definition of C_p^{ni} for $M+1 \leq p \leq 4N+2$ preserving the relations given by equations (19) and (20). Equation (18) can be rewritten as

$$\sum_{n=-N}^{N} \sum_{i=1}^{2} I^{ni} C_p^{ni} = 0, \quad 0 \leq p \leq M, \qquad (21)$$

where we have used equation (12).

We next define

$$\Phi_p(x) = \sum_{n=-N}^{N} \sum_{i=1}^{2} C_p^{ni} \Psi_H^{ni}(x), \quad 1 \leq p \leq 4N+2.$$

Then the summation term of equation (17) is rewritten as

$$\varepsilon \sum_{p=M+1}^{4N+2} \Phi_p(x) \sum_{n=-N}^{N} \sum_{i=1}^{2} C_p^{ni} I^{ni}$$

where we have used equations (19), (20) and (21). On the other hand we can readily establish the relation

$$\int_{\partial D} \mathrm{T}u^*_p(x) \cdot \Phi_q(x) dS = 0, \; 1 \leq p \leq M, \; M+1 \leq q \leq 4N+2,$$

which implies the existence of interior fields $\hat{\Phi}_q(x) \in U(D^-, \lambda, \mu, \rho, \omega)$ such that

$$\hat{\Phi}_q(x) = \Phi_q(x), \qquad x \in \partial D \;\; \text{for } M+1 \leq q \leq 4N+2.$$

Therefore the general solution of equation (17) is

$$u(x) - \varepsilon \sum_{p=M+1}^{4N+2} \hat{\Phi}_p(x) A_p = u^{e+}(x) - u^{*-}(x),$$

$$t(x) - \varepsilon \sum_{p=M+1}^{4N+2} \mathrm{T}\hat{\Phi}_p(x) A_p = \mathrm{T}u^{e+}(x) - \mathrm{T}u^{*-}(x), \; x \in \partial D, \qquad (22)$$

where $u^e(x) \in U(D^+, \lambda, \mu, \rho, \omega)$ and $A_p = \sum_{n=-N}^{N} \sum_{i=1}^{2} C_p^{ni} I^{ni}$.
From equation (22) we have

$$A_q - \varepsilon \sum_{p=M+1}^{N} \int_{\partial D} (\mathrm{T}\hat{\Phi}_p(x) \cdot \Phi_q(x) - \hat{\Phi}_p(x) \cdot \mathrm{T}\Phi_q(x)) dS \; A_p$$

$$= A_q - \varepsilon \sum_{p=M+1}^{N} \int_{\partial D} (\mathrm{T}\hat{\Phi}_p(x) - \mathrm{T}\Phi_p(x)) \cdot \Phi_q(x) dS \; A_p = 0,$$
$$M+1 \leq q \leq 4M+2.$$

The last integral is symmetric in p and q because

$$\Phi_q(x) = \int_{\partial D} \Gamma(x, y)(\mathrm{T}\hat{\Phi}_q(y) - \mathrm{T}\Phi_q(y)) dS, \; x \in D^+.$$

Also, since we can write

$$\hat{\Phi}_p(x) = \sum_{n=-N}^{N} \sum_{i=1}^{2} C_p^{ni} \Psi_J^{ni}(x) + \sum_{n=-\infty}^{\infty} \sum_{i=1}^{2} B_p^{ni} \Psi_J^{ni}(x)$$

where $B_p^{-ni} = (-1)^n \overline{B_p^{ni}}$, we have the following result;

$$A_q + 16\mu i \varepsilon \sum_{p=M+1}^{4M+2} (\delta_{pq} + \sum_{n=-N}^{N} \sum_{i=1}^{2} B_p^{ni} \overline{C_q^{ni}}) A_p = 0,$$
$$M+1 \leq q \leq 4M+2.$$

We can easily see that $\sum\sum_{ni} B_p^{ni} \overline{C_q^{ni}}$ is pure imaginary. The rest of analysis goes as Jones showed which, together with equation (21), yields

$I^{ni} = 0, -N \leq n \leq N, 1 \leq i \leq 2.$

This shows the equivalence of the first and second formulations.

We thus have shown that Jones' methods are valid also for mixed problems of elastodynamics.

4. NUMERICAL EXAMPLES

In this section we test the methods discussed in 3 by a simple example.

We calculated the hoop stress around a circular hole having a radius of a subject to plane time-harmonic P wave. Poisson's ratio is 1/4. The fictitious eigenfrequencies for this problem are the zeros of the following expressions;

$$J_{n+1}(k_T a)J_{n-1}(k_L a) + J_{n+1}(k_L a)J_{n-1}(k_T a), \; n = 0, 1, \ldots \quad (23)$$

[e.g. Vekua (1967), or readily from the expressions of $\Psi_J^{ni}(x)$]. As can be easily checked $k_L a = 3.906$ ($n = 3$) is one of the zeros of equation (23).

The numerical analysis is carried out as follows;

1) 24 quadratic isoparametric elements using 8 point Gaussian quadrature are employed.
2) Boundary stress is calculated by differentiating boundary quantities directly [Kobayashi & Nishimura (1982 a)].
3) $N = 3$ for both first and second methods.
4) We set $\varepsilon = -i / 32\mu$.
5) The first method requires the use of a least-square solver. We used Householder transformations. For the second method, we used the Crout method.
6) We used M-200 of Kyoto University Data Processing Center.

Fig. 2 shows the improvement of the result when $k_L a = 3.906$. The real (imaginary) part indicates the stress when the peak (node) of the incident stress wave arrives at the center of the hole. The results of the both remedies agree completely. The imaginary parts of the conventional and modified BIE coincide by chance. This coincidence reflects the fact that the eigenmode is essentially real. Table 1 compares the CPU time of

Table 1. CPU time (sec.)

	conventional BIE	first method	second method
make matrix	6.5	7.6	8.7
solver	0.5	1.7	0.5
total	7.0	9.3	9.2

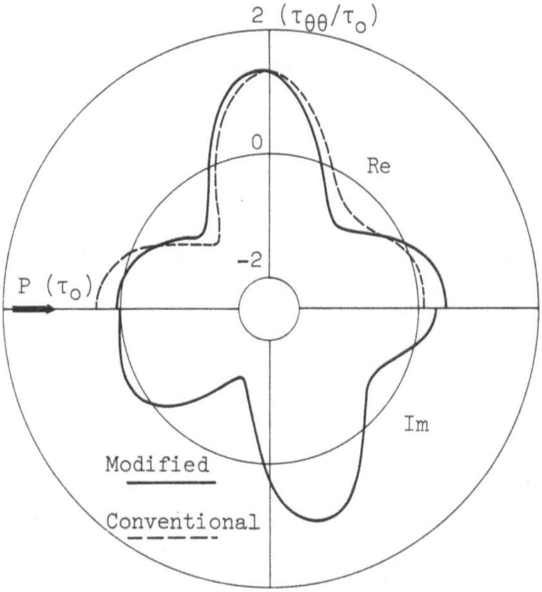

Figure 2. Hoop stress around a circular hole subject to plane time-harmonic P wave ($k_L a$ = 3.906, ν = 1/4).

these methods This result shows that the modified methods requires about 30 % more CPU time than the conventional method. It seems that the CPU time for the first and second methods are almost the same.

5. DISCUSSIONS

One of the difficulties concerning the present formulations is the choice of the constants N and ε. Jones [Jones (1974)] proposed a method of estimation of N in his paper, the elastodynamic counterpart of which is not available at present. Practically one may use some known formulae, such as equation (23), of eigenfrequencies for simple domains which are close in shape and dimension to the actual body.

For transient problems using Fourier transform approach, one may use conventional BIE and interpolation techniques to calculate steady states [Kobayashi & Nishimura (1982 a)]. It is because the fictitious eigenfrequencies are distributed rather sparse in the frequency ranges of practical importance for civil engineers. In addition, the conventional method gives fairly good results even when the frequency is near one of the fictitious eigenfrequencies provided one uses a carefully coded computer program. Fig. 3 shows an example of transient analysis using conventional BIE (not the elastodynamic version of Jones' method) and FFT. This figure shows the deformation around a circular hole in a half-plane subject to plane S wave of critical incidence. Poisson's ratio (ν) = 1/4. The parenthesized number indicates the corresponding location of wave fronts (Fig. 3(b)). The half-plane boundary is truncated and the usual fundamental solution (not Green's function) is used [for the detail see Kobayashi & Nishimura (1982 a, b)]. Of course it is essential to use a BIE which gives a unique solution when one is interested in a time-harmonic analysis for a particular frequency ω.

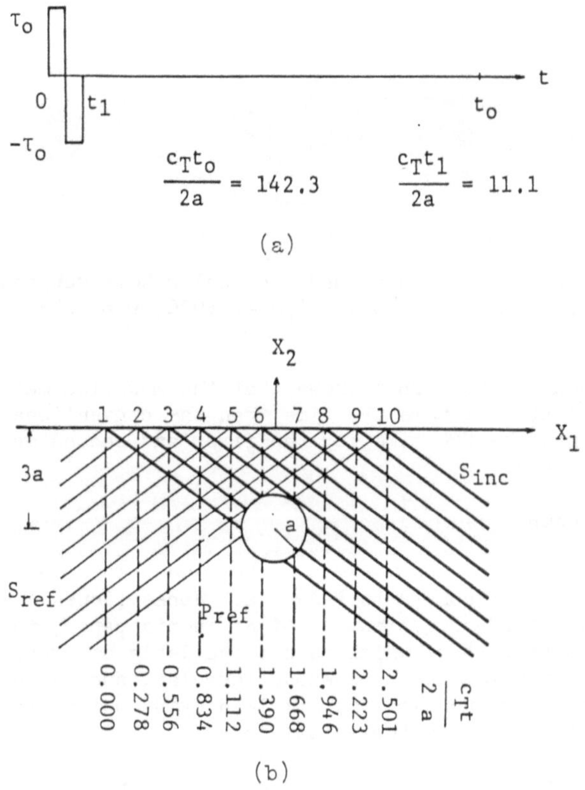

Figure 3. Response of a circular hole in a half-plane to incident plane SV wave of critical incidence (ν = 1/4). (a) Incident stress history (t_o : duration time, $c_{L,T}$: wave velocities) (b) Wave fronts of incident and reflected waves (c) Displacement (E : Young's modulus).

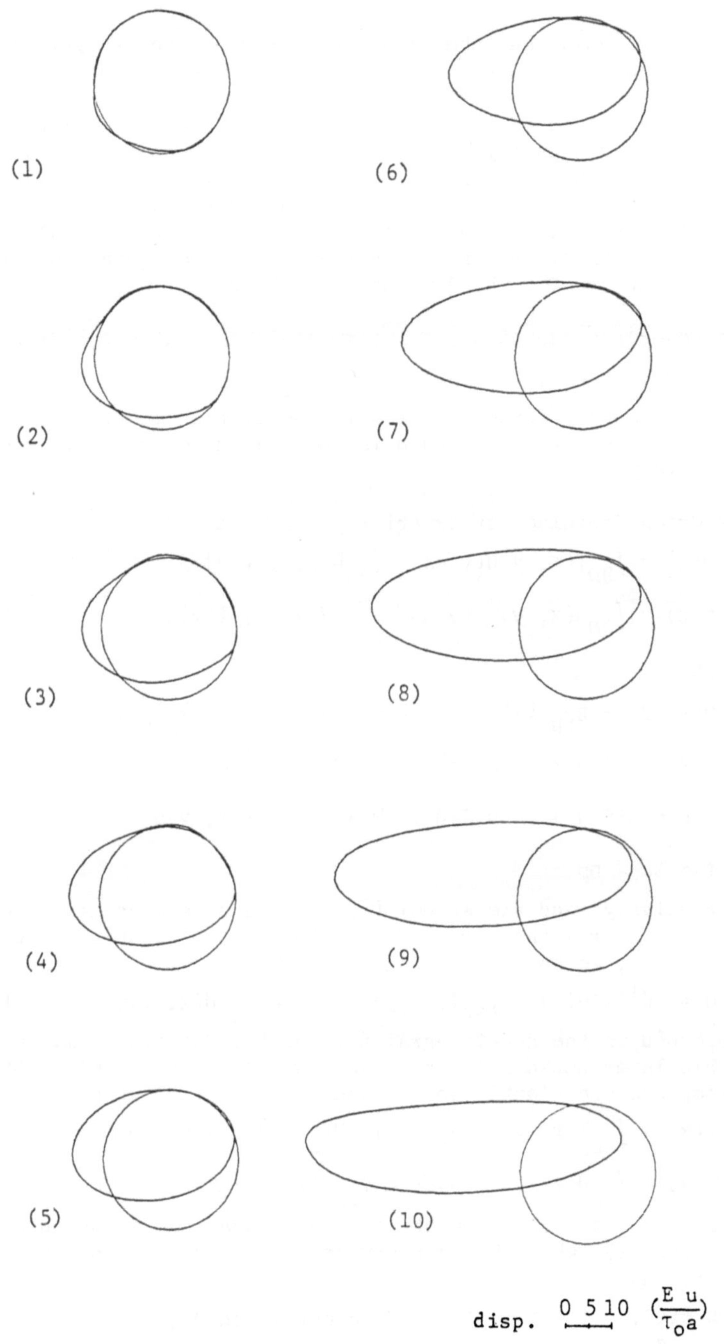

Figure 3 (c) Contd.

6. INCOMPRESSIBLE ELASTOSTATICS

It can be shown that the same phenomenon as the fictitious eigenfrequencies occurs to BIE for the incompressible solids (or incompressible slow viscous fluid). The indeterminacy of the BIE solution of the exterior problem of Stokes' flow under the velocity boundary conditions is known in connection with the question of existence of the solution [Odqvist (1930)]. In this paper we discuss this problem and show that the methods described in 3 can be applied to obtain a BIE which yields a unique solution. We investigate 3-dimensional problems using the same notations for the domains as in 2 and 3.

The governing equations of incompressible elastic solids are

$$-\mu\Delta \mathrm{u} + \nabla p = 0, \qquad \nabla\cdot\mathrm{u} = 0$$

neglecting body force, where u is the displacement, p is the indeterminate pressure and μ is the modulus of rigidity, respectively.

The Green formulae for exterior problems are

$$\tilde{\mathrm{u}}(\mathrm{x}) = \int_{\partial D}\mathrm{T}(\mathrm{x}, \mathrm{y})\mathrm{u}(\mathrm{y})dS - \int_{\partial D}\mathrm{U}(\mathrm{x}, \mathrm{y})\mathrm{t}(\mathrm{y})dS$$

$$\tilde{p}(\mathrm{x}) = \int_{\partial D}\mathrm{W}(\mathrm{x}, \mathrm{y})\cdot\mathrm{u}(\mathrm{y})dS - \int_{\partial D}\mathrm{V}(\mathrm{x}, \mathrm{y})\cdot\mathrm{t}(\mathrm{y})dS \tag{24}$$

where

$$\mathrm{U}(\mathrm{x}, \mathrm{y}) = \frac{1}{8\pi\mu}(1\Delta - \nabla\nabla)|\mathrm{x}-\mathrm{y}|, \quad \mathrm{V}(\mathrm{x}, \mathrm{y}) = \nabla_{\mathrm{y}}\frac{1}{4\pi}|\mathrm{x}-\mathrm{y}|^{-1}$$

$$\mathrm{T}(\mathrm{x}, \mathrm{y}) = \mathrm{U}(\mathrm{x}, \mathrm{y})\mathrm{S}_{\mathrm{y}} + \nabla_{\mathrm{y}}\frac{1}{4\pi}|\mathrm{x}-\mathrm{y}|^{-1}\otimes\mathrm{n}_{\mathrm{y}},$$

$$\mathrm{S}_{\mathrm{y}}\mathrm{u} = \mu\{\nabla_{\mathrm{y}}\mathrm{u} + (\nabla_{\mathrm{y}}\mathrm{u})^{T}\}\mathrm{n}_{\mathrm{y}}, \quad \mathrm{W}(\mathrm{x}, \mathrm{y}) = \mathrm{V}(\mathrm{x}, \mathrm{y})\mathrm{S}_{\mathrm{y}},$$

$$\mathrm{t} = \mathrm{S}\mathrm{u} - \mathrm{n}p,$$

respectively, and the symbol ($\tilde{\ }$) has the same meaning as equation (4). From the first of equation (24) by letting x approach ∂D from D^-, we have a BIE for $\mathrm{x} \in \partial D$

$$0 = \mathrm{C}^{i}(\mathrm{x})\mathrm{u}(\mathrm{x}) + \int_{\partial D}\mathrm{T}(\mathrm{x}, \mathrm{y})\mathrm{u}(\mathrm{y})dS - \int_{\partial D}\mathrm{U}(\mathrm{x}, \mathrm{y})\mathrm{t}(\mathrm{y})dS, \tag{25}$$

where $\mathrm{C}^{i}\mathrm{u}$ is the non-integral term of the interior limit of the double layer potential. From the uniqueness theorems of the incompressible elastic solids and equation (25) we have

$$\mathrm{v}(\mathrm{x}) = \int_{\partial D}\mathrm{T}(\mathrm{x}, \mathrm{y})\mathrm{u}(\mathrm{y})dS - \int_{\partial D}\mathrm{U}(\mathrm{x}, \mathrm{y})\mathrm{t}(\mathrm{y})dS = 0 \quad \text{in } D^-,$$

$$q(\mathrm{x}) = \int_{\partial D}\mathrm{W}(\mathrm{x}, \mathrm{y})\cdot\mathrm{u}(\mathrm{y})dS - \int_{\partial D}\mathrm{V}(\mathrm{x}, \mathrm{y})\cdot\mathrm{t}(\mathrm{y})dS = \text{const. in } D^-.$$

Therefore the same argument as 2 shows that the solution of equation (25) when the boundary condition is homogeneous can be written as

$$\mathrm{u} = \mathrm{u}^{e}, \quad \mathrm{t} = \mathrm{t}^{e} - \mathrm{n}C \quad (C : \text{const.}) \quad \text{on } \partial D,$$

where u^{e} is an exterior displacement field, t^{e} being its traction on ∂D. The boundary condition for u^e is

$$u^e = 0 \text{ on } \partial D_1, \quad t^e = nC \text{ on } \partial D_2.$$

We thus conclude that the uniqueness of the solution does not hold in this case.

We can apply analogous remedies as those for elastodynamics in order to assure the uniqueness of solutions. We here give only the results since the derivation is simpler by far than the elastodynamic case.

1) The following method gives a unique solution; Solve equation (25) subject to a condition $q(x) = 0$, where x is an arbitrary point in D^-.
2) The integral equation obtained from equation (25) by replacing $U(x, y)$ and $T(x, y)$ with

$$U(x, y) + c_o r(x) \otimes r(y) \text{ and } T(x, y) + c_o r(x) \otimes s(y)$$

renders the solution unique where c_o is a nonzero constant,

$$r(x) = \nabla \frac{1}{4\pi|x - x_o|}, \quad s(x) = 2\mu \frac{\partial}{\partial n} r(x),$$

and x_o is a point in D^-.

7. CONCLUSIONS

We discussed the non-uniqueness of the solution of exterior BIE.

Specifically, we established elastodynamic counterparts of Jones' methods for acoustics and showed its validity for general mixed problems. We also showed the method's applicability by numerical examples. We finally remarked that the same phenomenon occurs in incompressible elastostatics and it can be avoided easily.

REFERENCES

Burton, A.J. & Miller, G.F. (1971) The application of integral equation methods to the numerical solution of some exterior boundary value problems. Proc. Roy. Soc. London (A), 323:201-210.

Dubois, M & Lachat, J.C. (1973) The integral formulation of boundary value problems. In; Variational Methods in Engineering, Vol. II, Chapt. 9:89-108, Southampton Univ. Press.

Jones, D.S. (1974) Integral equations for the exterior acoustic problem. Q. J. Mech. Appl. Math., 27:129-142.

Kleinman, R.E. & Roach, G.F. (1974) Boundary integral equations for the three dimensional Helmholtz equation. SIAM Review, 16: 214-326.

Kobayashi, S. & Nishimura, N. (1982 a) Transient stress analysis of tunnels and caverns of arbitrary shape due to travelling waves. In; Developments in Boundary Element Methods-II (Eds. P.K. Banerjee & R.P. Shaw):177-210, Appl. Sci. Publ.

Kobayashi, S. & Nishimura, N. (1982 b) Analysis of dynamic soil--structure interactions by boundary integral equation method. Proc. Int. Conf. FEM, Shanghai.

Kupradze, V.D. (1965) Potential Methods in the Theory of Elasticity. Israel program for scientific translations, Jerusalem.

Meyer, W.L., BELL, W.A. & Zinn, B.T. (1978) Boundary integral solutions of three dimensional acoustic radiation problems. J. Sound Vibration, 59 (2):245-262.

Odqvist, F.K.G. (1930) Uber die Randwertaufgaben der Hydrodynamik zaher Flussigkeiten. Math. Zeit., 32:329-375.

Schenck, H.A. (1968) Improved integral formulation for acoustic radiation problems. J. Acoust. Soc. Am., 44:41-58.

Shaw, R.P. (1979) Boundary integral equation methods applied to wave problems. In; Developments in boundary element methods -I (Eds. P.K. Banerjee & R. Butterfield):121-153, Appl. Sci. Publ.

Terai, T. (1980) On calculation of sound fields around three dimensional objects by integral equation methods. J. Sound Vibration, 69 (1):71-100.

Ursell, F. (1973) On the exterior problems of acoustics. Proc. Camb. Phil. Soc., 74:117-125.

Vekua, I.N. (1967) New methods for solving elliptic equations. North-Holland Publ.

APPLICATIONS OF THE BOUNDARY INTEGRAL EQUATION METHOD TO EIGENVALUE PROBLEMS OF ELASTODYNAMICS

Y. Niwa*, S. Kobayashi* and M. Kitahara**

* Dept. of Civil Engineering, Kyoto University, Kyoto, Japan
** Dept. of Ocean Civil Engineering, Tokai University, Shimizu, Shizuoka, Japan

ABSTRACT

Applicability of the boundary integral equation (BIE) method to the analysis of eigenvalue problems in elastodynamics is demonstrated by several examples, in which eigenvalues and eigenmodes are analyzed for a disc, annulus and dam on a rigid foundation. The eigenvalues thus obtained by the BIE method show good agreement with analytical ones. The BIE method is also applied to the analysis of dominant mode of a dam constructed on a semi-infinite foundation. Lastly, some remarks for the eigenvalue analysis in elastodynamics by the BIE method are given.

INTRODUCTION

Hitherto, the BIE methods applied to the analysis of eigenvalue problems have been mainly concerned with the problems of membrane (Helmholtz eq.) and plate vibrations. Complete review of work done for these problems is found in the work by Shaw(1979) and Niwa *et al.*(1982 a) except the Wong and Hutchinson's recent study (1981) of plate vibration.

In our recent work(Niwa *et al.*, 1982 a), we formulated various types of boundary integral equations to determine eigenvalues of elastodynamics and plate vibrations in terms of layer potentials and Green's(Somigliana's) formula and clarified mutual relations among those integral equations including interior equations and exterior ones.

The principal aim of this paper is to show the further development of numerical results concerning to eigenvalues, eigendensities, and eigenmodes of elastodynamic problems and to show the applicability of the BIE method to the dominant mode analysis of a dam on a semi-infinite foundation. After the brief recapitulation, for completeness, of the boundary integral equations to determine eigenvalues of the third(mixed) boundary value problem in the steady state elastodynamics, some example problems such as a disc and an annulus to demonstrate the efficiency of the BIE method to eigenvalue problems in elastodynamics are shown and these results are compared with the exact solutions. Then the BIE method is applied to the dominant mode analysis of a dam structure on a semi-infinite foundation and these results are compared with the eigenmodes of a dam structure on a rigid foundation. Lastly, some remarks particular to the eigenvalue analysis of elastodynamics by the BIE method are given.

FORMULATION OF EIGENVALUE PROBLEMS IN ELASTODYNAMICS

In the following, we restrict our attention to the boundary integral equations in terms of Green's formula and briefly summarize four types of formulations based on this formula, which are used in the computational stage. As for more detailed version of this formulation, see Niwa *et al.* (1982 b). Numerical accuracy of eigenvalues by means of these four types of equations is also discussed in this section.

Governing equations

It is convenient to express the governing (Navier-Cauchy) equation in the steady state elastodynamics as follows (see, for example, Eringen & Suhubi, 1975).

$$\mathbf{Lu}(X,\omega) = \mu[(k_T/k_L)^2\boldsymbol{\nabla}\otimes\boldsymbol{\nabla} - \boldsymbol{\nabla}\mathrm{x}\boldsymbol{\nabla}\mathrm{x} + k_T^2\mathbf{1}]\mathbf{u}(X,\omega) = -\rho\mathbf{b}(X,\omega) \qquad (1)$$

where $\mathbf{u}$ and $\mathbf{b}$ are displacement and body force vectors respectively; μ,ρ, and ω are the shear modulus, the mass density, and the angular frequency; $k_T = \omega/c_T$ and $k_L = \omega/c_L$ are the transverse and the longitudinal wave numbers respectively, where c_T and c_L are corresponding wave velocities, of course,

$$(k_T/k_L)^2 = (c_L/c_T)^2 = \{2(1-\nu)\}/(1-2\nu), \quad (\nu\text{:Poisson's ratio}); \qquad (2)$$

tensor product $\boldsymbol{\nabla}\otimes\boldsymbol{\nabla}$ of two gradient operators is defined as

$$(\boldsymbol{\nabla}\otimes\boldsymbol{\nabla})(\mathbf{v}) = \boldsymbol{\nabla}(\boldsymbol{\nabla}\cdot\mathbf{v}) \qquad \text{for every vector } \mathbf{v},$$

also, $\mathbf{1}$ is the unit tensor; and X is a point in the two or three dimensional Euclidian space $\mathbf{R}^m$ (m=2 or 3). In equation (1) the time factor exp($-i\omega t$) is suppressed and this convention is used throughout this paper. Also, it is convenient to suppress the angular frequency ω in the quantities $\mathbf{u}$ and $\mathbf{b}$, that is, we write, for example, $\mathbf{u}(X)$ instead of $\mathbf{u}(X,\omega)$ for simplicity, in what follows.

If a surface element has a unit outward normal vector $\mathbf{n}$, the traction vector $\mathbf{t}$ on the surface is expressed by means of the displacement vector $\mathbf{u}$ as follows

$$\mathbf{t} = \overset{n}{\mathbf{T}}\mathbf{u} = \mu[\{(k_T/k_L)^2 - 2\}\mathbf{n}\otimes\boldsymbol{\nabla} + 2(\mathbf{n}\cdot\boldsymbol{\nabla})\mathbf{1} + \mathbf{n}\mathrm{x}\boldsymbol{\nabla}\mathrm{x}\ ]\mathbf{u} \qquad (3)$$

where $(k_T/k_L)^2 - 2 = 2\nu/(1-2\nu)$.

Radiation conditions and fundamental solutions

If the domain under consideration contains an unbounded region, certain physical conditions which restrict the behavior at infinity should be satisfied when $r = |X| \to \infty$ (see, for example, Kupradze, 1965)

$$\mathbf{u}_L = o(\, r^{(3-m)/2}\,), \qquad \mathbf{u}_T = o(\, r^{(3-m)/2}\,),$$
$$\partial_r \mathbf{u}_L - ik_L \mathbf{u}_L = o(r^{(1-m)/2}), \quad \partial_r \mathbf{u}_T - ik_T \mathbf{u}_T = o(r^{(1-m)/2}) \tag{4}$$

where $\mathbf{u}_L$ and $\mathbf{u}_T$ are the irrotational and solenoidal parts of $\mathbf{u}$, respectively. These conditions are known as regularity and radiation conditions, which physically mean that no wave propagates toward the interior from infinity.

The fundamental solution $\mathbf{U}$ of equation (1) for a unit harmonic body force is defined as the solution of the following equation

$$\mathbf{LU} = -\delta \mathbf{1} \qquad (\ \delta : \text{Dirac measure}\ ). \tag{5}$$

In the two dimensional plane strain problem, the fundamental solution which satisfies the radiation condition (4) has the following form

$$\mathbf{U}(X,Y;\omega) = \frac{i}{4\mu}[H_0^{(1)}(k_T r)\mathbf{1} + \frac{1}{k_T^2}\nabla\otimes\nabla\{H_0^{(1)}(k_T r) - H_0^{(1)}(k_L r)\}] \tag{6}$$

where $H_0^{(1)}(\cdot)$ is the zero-order Hankel function of the first kind and $r = |X-Y|$.

Green's formula (Somigliana's formula)

Let D_- be an interior domain enclosed by its boundary ∂D and D_+ be the exterior domain in the Euclidian space $\mathbf{R}^m$, as shown in Figure 1. For example, in the third(mixed) boundary value problem, the displacement $\mathbf{u} = \mathbf{f}$ and the traction $\mathbf{t} = \mathbf{g}$ are given on ∂D_1 and ∂D_2 ($\partial D = \partial D_1 \cup \partial D_2$), respectively. The unit outward normal vector $\mathbf{n}_y$ is defined at a point y on the boundary. Unless otherwise stated, X or Y stands for a point in the interior or exterior domain, and x or y stands for a point on the boundary.

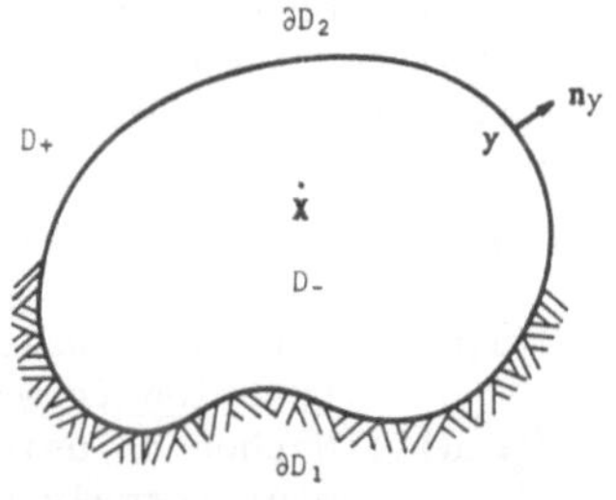

Figure 1 Domain, boundary, and normal vector

The most direct method to derive Green's third formula(Somigliana's formula) is to consider the formal adjoint of the operator $\mathbf{L}$. Then, taking into account the equations (1),(5), and the properties of potentials (Kupradze, 1965), we have the following Green's displacement formula when the body force is zero, i.e., $\mathbf{b} = \mathbf{0}$

$$\int_{\partial D}[\mathbf{U}(X,y;\omega)\cdot\{\overset{n_y}{\mathbf{T}}\mathbf{u}(y)\} - \{\overset{n_y}{\mathbf{T}}\mathbf{U}(X,y;\omega)\}\cdot\mathbf{u}(y)]\, ds_y$$
$$= (\mathbf{St})(X) - (\mathbf{Du})(X)$$
$$= \begin{cases} \mathbf{u}(X) & X\varepsilon D_- \quad (7)\\ \mathbf{C}_d^+\,\mathbf{u}(x) & x\varepsilon \partial D \quad (8)\\ \mathbf{0} & X\varepsilon D_+ \quad (9)\end{cases}$$

where $\mathbf{t} = \overset{n}{\mathbf{T}}\mathbf{u}$ as defined in equation (3). $\mathbf{St}$ and $\mathbf{Du}$ are the simple and double layer potentials, respectively. $\mathbf{C}_d^+ \mathbf{u}(x)$ is the free term of the exterior limit of the double layer potential. If the tangent plane at x is continuous, $\mathbf{C}_d^+ = (1/2)\mathbf{1}$.

If we operate $\overset{n_X}{\mathbf{T}}$ on equation (7) in the interior domain(D_-) and use the properties of potentials, the following Green's traction formula is obtained

$$\overset{n_X}{\mathbf{T}}(\mathbf{St})(X) - \overset{n_X}{\mathbf{T}}(\mathbf{Du})(X)$$

$$= \begin{cases} \mathbf{t}(X) & X \in D_- \quad (10) \\ -\mathbf{C}_s^+\mathbf{t}(x) & x \in \partial D \quad (11) \\ \mathbf{0}(X) & X \in D_+ \quad (12) \end{cases}$$

where $\mathbf{C}_s^+ \mathbf{t}(x)$ is the free term of the exterior limit of the traction expression of the simple layer potential. If the tangent plane at x is continuous, $\mathbf{C}_s^+ = -(1/2)\mathbf{1}$. So far, we considered interior formulae. Exterior formulae can be formulated in the same way.

Boundary integral equations

As an example, we consider the third(mixed) boundary value problem stated in the previous section. The formulation of the boundary integral equations by the use of Green's formula is the most suitable for the mixed problem. In this case, the following four types of formulations can be considered:

- $G(\mathbf{u})$: Method of using displacement formula both on ∂D_1 and ∂D_2.
- $G(\mathbf{u},\mathbf{t})$: Method of using displacement formula on ∂D_1 and traction formula on ∂D_2.
- $G(\mathbf{t},\mathbf{u})$: Method of using traction formula on ∂D_1 and displacement formula on ∂D_2.
- $G(\mathbf{t})$: Method of using traction formula both on ∂D_1 and ∂D_2.

Table 1 shows the integral equations for the third interior and exterior problems by the use of these four types of formulations under the assumption that the boundary is smooth. In this table, $(\cdot)_1$ and $(\cdot)_2$ stand for the integration on ∂D_1 and ∂D_2, respectively; $\mathbf{I}$ means the identity operator; and the boundary integral operators $\bar{\mathbf{K}}^*$ and $\mathbf{K}$ are defined as

$$(\bar{\mathbf{K}}^*\mathbf{u})(x) = \int_{\partial D} \{\overset{n_y}{\mathbf{T}}\mathbf{U}(x,y;\omega)\}\cdot\mathbf{u}(y)ds_y,$$

$$(\mathbf{K}\,\mathbf{u})(x) = \int_{\partial D} \{\overset{n_x}{\mathbf{T}}\mathbf{U}(x,y;\omega)\}\cdot\mathbf{t}(y)ds_y.$$

Eigenvalue problems and numerical procedures

Among the boundary integral equations listed in Table 1, Boundary integral equations corresponding to the formulations enclosed with the dotted line determine the eigenvalues of the third interior boundary value problem. These homogeneous boundery integral equations are generally expressed as

$$\mathbf{B}(r,\mathbf{n};\omega)\boldsymbol{\mu} = \mathbf{0} \qquad \text{on } \partial D \qquad (13)$$

which are obtained by just putting the right-hand side of the equations in Table 1 to zero.

After the discretization and the (numerical and/or analytical) integration process, equation (13) is finally approximated by a system of homogeneous linear equations with respect to the unknown density $\boldsymbol{\mu}$, that is,

$$[A(\omega)]\{\mu\} = \{0\} \tag{14}$$

in the matrix form. The matrix [A] contains only the parameter ω in this stage. The necessary and sufficient condition for which equation (14) has a non-trivial solution is well-known as

$$D = \det[A(\omega)] = 0. \tag{15}$$

The eigenvalues are characterized as roots of this determinant. In the numerical calculation, however, eigenvalues can be determined as parameters which attain local minima of the absolute value of the complex determinant D. As for the detailed version of this process to determine eigenvalues, see Niwa *et al.* (1982 a). After having determined eigenvalues, we can obtain boundary densities from equation (14) by the use of the least-square method or the usual Crout method with a weighted matrix element. Eigenmodes in the interior domain are obtained by the simple quadrature afrer substituting these eigendensities into the representation of the eigenfunction, that is, equation (7) with a suitable homogeneous boundary condition.

In advance of the actual eigenvalue analysis by the BIE method, some remarks should be given here. First, it is noted that the essential material constant in the eigenvalue analysis of elastodynamics is Poisson's ratio ν only. As is easily understood from equations (1), (3), (6), and Green's formulae (7)-(12), if we consider $\mu\mathbf{u}$ as the unknown displacement instead of $\mathbf{u}$ and $\mu\mathbf{U}$ as the fundamental solution, the resulting matrix element to be evaluated has no dependence to the shear modulus μ. If we remember the property of the eigenvalue problem, that is, the arbitrariness of a constant factor, it is understood that μ is immaterial in the case where homogeneity is assumed. Secondly, it is convenient to use the appropriately nondimensionalized shear wave number Lk_T (L is a characteristic length, for example, the radius of a circle) as the eigenparameter instead of the angular frequency ω in the numerical stage, as is easily understood from the fundamental solution in equation (6). Of course, it is convenient to put this characteristic length to unity in the actual numerical calculation. In this case the other longitudinal wave number Lk_L is given by the relation (2). After the eigenvalue Lk_T has been determined by the aforementioned direct search method, if we want to know the corresponding angular frequency ω, this ω is immediately obtained from the relation $Lk_T = L(\omega/c_T)$ by giving the other material constant μ and the mass density ρ in this stage.

To check the accuracy of eigenvalues by the use of four types of boundary integral equations listed in Table 1, the third(mixed)

Table 1 Boundary integral equations for the third(mixed) problem

	Formulation of Problem	Integral Equation	I.P.	U
Third (Mixed) Interior Problem, M.I.	Representation of Solution	$u = (St)_1 + (Sg)_2 - (Df)_1 - (Du)_2$ in D_-		
	G(u)	$(St)_1 - (\bar{K}^*u)_2 = [(\frac{1}{2}I + \bar{K}^*)f]_1 - (Sg)_2$ on ∂D_1 $(St)_1 - [(\frac{1}{2}I + \bar{K}^*)u]_2 = (\bar{K}^*f)_1 - (Sg)_2$ on ∂D_2		
	G(u,t)	$(St)_1 - (\bar{K}^*u)_2 = [(\frac{1}{2}I + \bar{K}^*)f]_1 - (Sg)_2$ on ∂D_1 $(Kt)_1 - (D_nu)_2 = (D_nf)_1 + [(\frac{1}{2}I - K)g]_2$ on ∂D_2	I_1	u
	G(t,u)	$[(\frac{1}{2}I - K)t]_1 + (D_nu)_2 = -(D_nf)_1 + (Kg)_2$ on ∂D_1 $(St)_1 - [(\frac{1}{2}I + \bar{K}^*)u]_2 = (\bar{K}^*f)_1 - (Sg)_2$ on ∂D_2		t
	G(t)	$[(\frac{1}{2}I - K)t]_1 + (D_nu)_2 = -(D_nf)_1 + (Kg)_2$ on ∂D_1 $(Kt)_1 - (D_nu)_2 = (D_nf)_1 + [(\frac{1}{2}I - K)g]_2$ on ∂D_2		
Third (Mixed) Exterior Problem, M.E.	Representation of Solution	$u = (Df)_1 + (Du)_2 - (St)_1 - (Sg)_2$ in D_+		
	G(u)	$(St)_1 - (\bar{K}^*u)_2 = -[(\frac{1}{2}I - \bar{K}^*)f]_1 - (Sg)_2$ on ∂D_1 $(St)_1 + [(\frac{1}{2}I - \bar{K}^*)u]_2 = (\bar{K}^*f)_1 - (Sg)_2$ on ∂D_2		
	G(u,t)	$(St)_1 - (\bar{K}^*u)_2 = -[(\frac{1}{2}I - \bar{K}^*)f]_1 - (Sg)_2$ on ∂D_1 $(Kt)_1 - (D_nu)_2 = (D_nf)_1 - [(\frac{1}{2}I + K)g]_2$ on ∂D_2	I_1	u
	G(t,u)	$[(\frac{1}{2}I + K)t]_1 - (D_nu)_2 = (D_nf)_1 - (Kg)_2$ on ∂D_1 $(St)_1 + [(\frac{1}{2}I - \bar{K}^*)u]_2 = (\bar{K}^*f)_1 - (Sg)_2$ on ∂D_2		t
	G(t)	$[(\frac{1}{2}I + K)t]_1 - (D_nu)_2 = (D_nf)_1 - (Kg)_2$ on ∂D_1 $(Kt)_1 - (D_nu)_2 = (D_nf)_1 - [(\frac{1}{2}I + K)g]_2$ on ∂D_2		

I.P.: Identity Pair, U: Unknown

Table 2 Eigenvalues of the third problem (rectangular domain, a/b = 2, N = 54)

m \ n	1				2				3				4			
0	6.283				12.566				18.850				25.133			
	6.3	6.3	6.3	6.3	12.6	12.6	12 6	12.6	18.9	18.9	18.9	18.9	25.1	25.1	25.1	25.1
	0.27	0.27	0.27	0.27	0.27	0.27	0.27	0.27	0.27	0.27	0.27	0.27	0.13	0.13	0.13	0.13
1	7.025				12.953				19.110				25.328			
	7.0	7.0	7.0	7.0	13.0	13.0	13.0	12.9	19.1	19.1	19.1	19.1	25.3	25.3	25.3	25.3
	0.36	0.36	0.36	0.36	0.36	0.36	0.36	0.41	0.05	0.05	0.05	0.05	0.11	0.11	0.11	0.11
2	8.886				14.050				19.869				25.906			
	8.9	8.9	8.9	8.9	14.1	14.0	14.1	14.1	19.9	19.9	19.9	19.9	25.9	25.9	25.9	25.9
	0.16	0.16	0.16	0.16	0.36	0.36	0.36	0.36	0.16	0.16	0.16	0.16	0.02	0.02	0.02	0.02
3	11.327				15.708				21.074				26.842			
	11.3	11.3	11.3	11.3	15.7	15.7	15.7	15.7	21.1	21.1	21.1	21.1	26.8	26.8	26.8	26.8
	0.24	0.24	0.24	0.24	0.05	0.05	0.05	0.05	0.12	0.12	0.12	0.12	0.16	0.16	0.16	0.16
4	14.050				17.772				22.654				28.099			
	14.1	14.0	14.1	14.0	17.8	17.8	17.8	17.8	22.6	22.6	22.7	22.7	28.1	28.1	28.1	28.1
	0.36	0.36	0.36	0.36	0.16	0.16	0.16	0.16	0.24	0.24	0.20	0.20	0.004	0.004	0.004	0.004
5	16.918				20.116				24.537				29.638			
	16.9	16.9	16.9	16.9	20.1	20.1	20.1	20.1	24.5	24.5	24.5	24.5	29.6	29.6	29.6	29.6
	0.11	0.11	0.11	0.11	0.08	0.08	0.08	0.08	0.15	0.15	0.15	0.15	0.13	0.13	0.13	0.13
6	19.869				22.654				26.657				31.416			
	19.9	19.9	19.9	19.9	22.6	22.6	22.7	22.7	26.6	26.6	26.6	26.6	31.4	31.4	31.4	31.4
	0.16	0.16	0.16	0.16	0.24	0.24	0.20	0.20	0.21	0.21	0.21	0.21	0.05	0.05	0.05	0.05

Exact($a\bar{k}_T$)			
G(u)	G(u,t)	G(t,u)	G(t)
e	e	e	e

$e = |a\bar{k}_T - ak_T| / a\bar{k}_T$ (%)

problem of antiplane shear in a rectangular domain with the adjacent side length ratio a/b = 2 (fixed along the long edge a (=1.0) and free along the short edge b) is chosen. Table 2 shows the eigenvalues (ak_T) obtained from these four types of formulations G(**u**), G(**u**,**t**), G(**t**,**u**), and G(**t**) with the number of boundary element N = 54. The values in each block denote, from the top to the bottom, the exact value, the value obtained by the BIE method, and the relative error in percentage. Also, m and n denote the number of nodal lines on the long and short edges, respectively. It is found that all eigenvalues obtained are within 1% of the relative error in each formulation. Although the eigenvalues are accurately obtained from any formulation, it should be noted that the boundary integral equation based on the formulation G(**u**) is preferable from the computational point of view, since the interior eigenmodes are directly calculated from this formulation.

EXAMPLE PROBLEMS IN ELASTODYNAMICS

The efficiency and the versatility of the BIE method in the analysis of eigenvalue problems are demonstrated by some example problems in elastodynamics with the assumption that the field is plane strain state. In the examples of the disc and annulus, exact eigenvalues and the properties of their distribution are also shown. In the last example, the BIE method is applied to the dominant mode analysis of a dam on a semi-infinite foundation.

Disc

We consider a fixed disc with radius a. In this case, exact (analytical) eigenvalues are calculated from the following frequency equation

$$J_{n+1}(ak_T)J_{n-1}(ak_L) + J_{n-1}(ak_T)J_{n+1}(ak_L) = 0 \qquad (16)$$

where $J_n(\cdot)$ is the Bessel function of the first kind. This frequency equation can be directly obtained by integrating the integral equation analytically (As for different treatment, see, Vekua, 1967). Table 3 shows these exact eigenvalues ($a\bar{k}_T$) for each n in the case that Poisson's ratio ν is 0.25. In this table, s merely indicates the order of eigenvalues for each n. Figure 2 shows the trend of the distribution of these eigenvalues. Table 4 shows the eigenvalues obtained by the BIE method from the first to the 10th in an increasing order of values. The values in each block denote the exact eigenvalue($a\bar{k}_T$), the value(ak_T) obtained by the BIE method, and the relative error in percentage. Also, ak_T^{sn} denotes the relation with the exact values listed in Table 3. All eigenvalues obtained by the BIE method have good accuracy, i.e., within 1% of the relative error. Figure 3 shows the eigenmodes obtained by the BIE method. It should be remarked that the eigendensities on the boundary which are obtained by solving the homogeneous system (14) are tractions for this fixed disc. These distribution shapes of eigendensities are shown on the periphery by the lines with arrows.

Table 3 Exact eigenvalues of a fixed disc ($\nu = 0.25$)

s \ n*	0	1	2	3	4
1	3.832	3.365	5.222	6.765	8.129
2	6.637	5.379	6.936	8.557	10.201
3	7.016	8.484	9.918	11.325	12.727
4	10.173	9.277	11.616	13.735	15.534
5	12.151	11.709	13.208	14.750	16.486
6	13.324	14.700	16.304	17.761	19.195
7	16.471	14.948	17.299	19.634	
8	17.621	18.014	19.520		
9	19.616	20.270			
s \ n*	**5**	**6**	**7**	**8**	**9**
1	9.398	10.613	11.796	12.957	14.102
2	11.814	13.362	14.833	16.231	17.567
3	14.144	15.586	17.057	18.548	20.042
4	17.051	18.465	19.847		
5	18.402	20.324			
6	20.624				

* $J_{n+1}(ak_T)J_{n-1}(ak_L) + J_{n-1}(ak_T)J_{n+1}(ak_L) = 0$

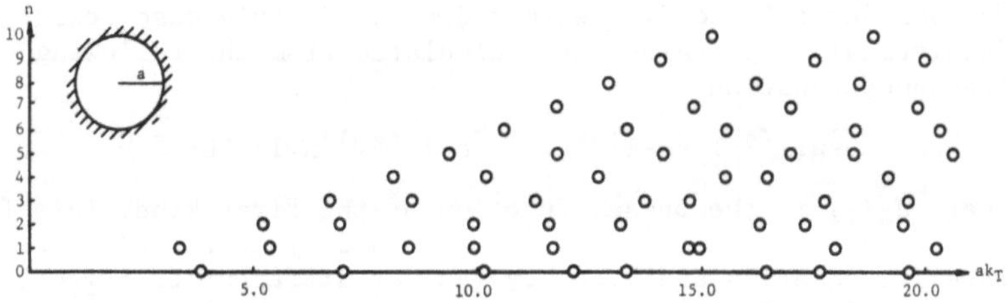

Figure 2 Trend of exact eigenvalues of a fixed disc ($\nu = 0.25$)

Table 4 Eigenvalues of a fixed disc (N = 36, $\nu = 0.25$)

	1	2	3	4	5	6	7	8	9	10
ak_T^{sn}	ak_T^{11}	ak_T^{10}	ak_T^{12}	ak_T^{21}	ak_T^{20}	ak_T^{13}	ak_T^{22}	ak_T^{30}	ak_T^{14}	ak_T^{31}
$\overline{ak}_T$	3.365	3.832	5.222	5.379	6.637	6.765	6.936	7.016	8.129	8.484
ak_T	3.38	3.84	5.24	5.40	6.66	6.78	6.96	7.04	8.14	8.51
e	0.45	0.21	0.34	0.39	0.35	0.22	0.35	0.34	0.14	0.31

$e = |\overline{ak}_T - ak_T| / \overline{ak}_T$ (%)

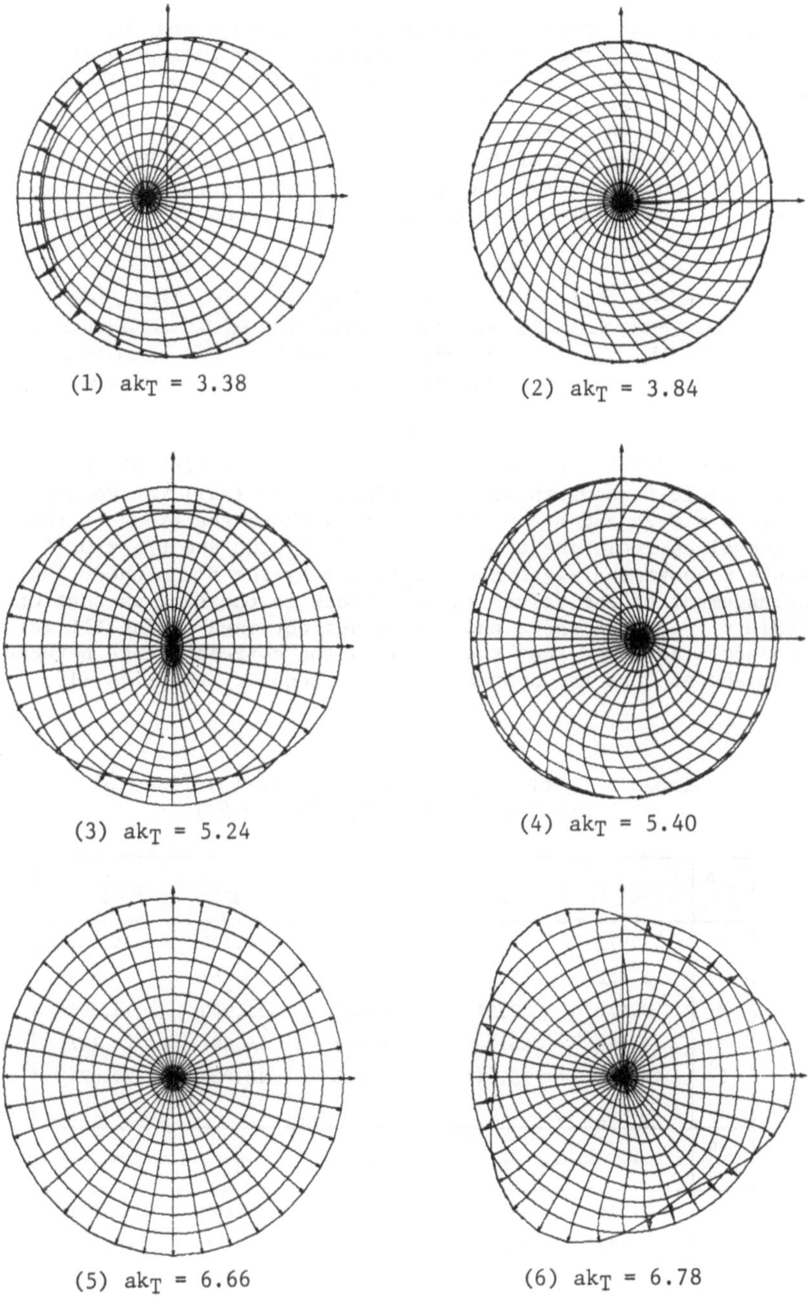

Figure 3 Eigenmodes of a fixed disc ($N = 36$, $\nu = 0.25$)

Annulus

We consider an annulus with the outer radius a and the inner radius b. As for the boundary conditions, the following four types of combinations can be considered: (i) fixed(a)-fixed(b), (ii) free(a)-free(b), (iii) free(a)-fixed(b), (iv) fixed(a)-free (b). As an example, the results for free(a)-fixed(b) annulus, that is, the annulus of free on the outer boundary and fixed on the inner boundary, are shown here. In this case, exact eigenvalues are calculated from the following frequency equation

$$\det\begin{bmatrix} -k_T J_{n+2}(ak_T) & -k_L\{\kappa J_n(ak_L)-J_{n+2}(ak_L)\} & -k_T Y_{n+2}(ak_T) & -k_L\{\kappa Y_n(ak_L)-Y_{n+2}(ak_L)\} \\ k_T J_{n-2}(ak_T) & -k_L\{\kappa J_n(ak_L)-J_{n-2}(ak_L)\} & k_T Y_{n-2}(ak_T) & -k_L\{\kappa Y_n(ak_L)-Y_{n-2}(ak_L)\} \\ J_{n+1}(bk_T) & -J_{n+1}(bk_L) & Y_{n+1}(bk_T) & -Y_{n+1}(bk_L) \\ J_{n-1}(bk_T) & J_{n-1}(bk_L) & Y_{n-1}(bk_T) & Y_{n-1}(bk_L) \end{bmatrix} = 0 \tag{17}$$

where $\kappa = 1/(1-2\nu)$ and $Y_n(\cdot)$ is the Bessel function of the second kind. Table 5 shows these exact eigenvalues ($a\bar{k}_T$) for Poisson's ratio $\nu = 0.25$ and $b/a = 0.5$. Figure 4 shows the trend of these eigenvalues including the cases of $b/a = 0.25$ and $b/a = 0.75$. Table 6 shows the eigenvalues obtained by the BIE method. The arrangement of the values in each block is the same as that of Table 4. Figure 5 shows the eigenmodes obtained by the BIE method. It should be noted that there exists the correspondence between mode shapes and the number n in the frequency equation (17).

Table 5 Exact eigenvalues of a free(a)-fixed(b) annulus ($\nu = 0.25$, $b/a = 0.5$)

s \ n	0	1	2	3	4
1	1.973	2.616	3.813	4.736	5.389
2	5.552	5.511	5.649	6.467	7.855
3	9.177	9.579	10.600	11.980	13.556
4	15.563	15.244	14.878	14.697	14.737
5	16.394	16.899	17.818	18.900	20.073

s \ n	5	6	7	8	9
1	6.099	6.907	7.781	8.691	9.619
2	9.379	10.896	12.359	13.746	15.053
3	14.952	15.459	16.154	17.045	18.104
4	15.285	16.973	18.648	20.218	

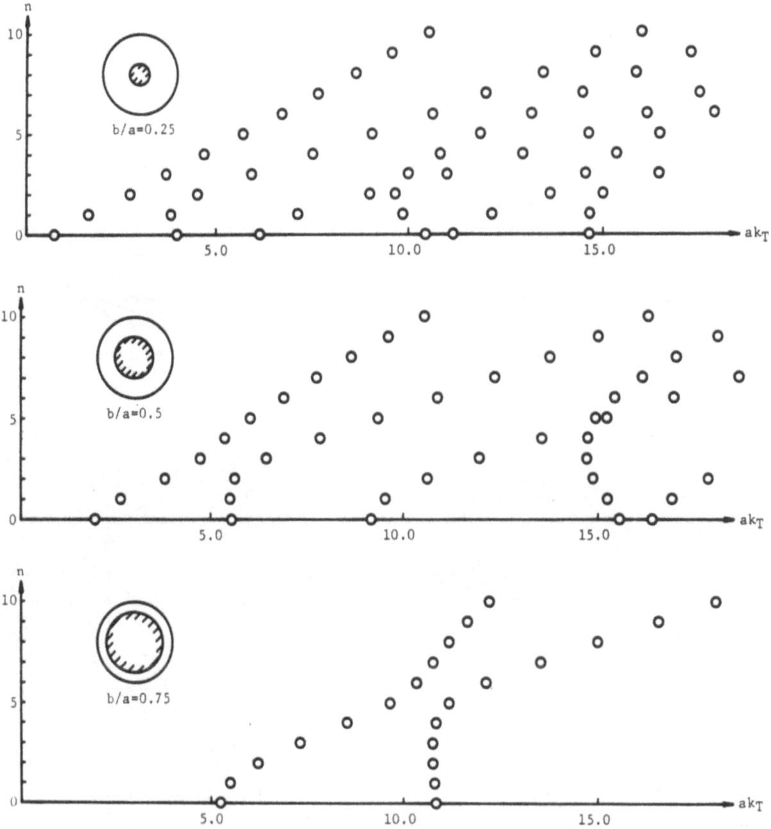

Figure 4 Trend of exact eigenvalues of a free(a)-fixed(b) annulus (ν = 0.25)

Table 6 Eigenvalues of a free(a)-fixed(b) annulus (N=48, ν=0.25)

ak_T^{sn}	1 ak_T^{10}	2 ak_T^{11}	3 ak_T^{12}	4 ak_T^{13}	5 ak_T^{14}	6 ak_T^{21}
$a\bar{k}_T$	1.973	2.616	3.813	4.736	5.389	5.511
ak_T	2.02	2.66	3.85	4.77	5.49	5.55
e	2.38	1.68	0.97	0.72	1.87	0.71

$e = |a\bar{k}_T - ak_T| / a\bar{k}_T$ (%)

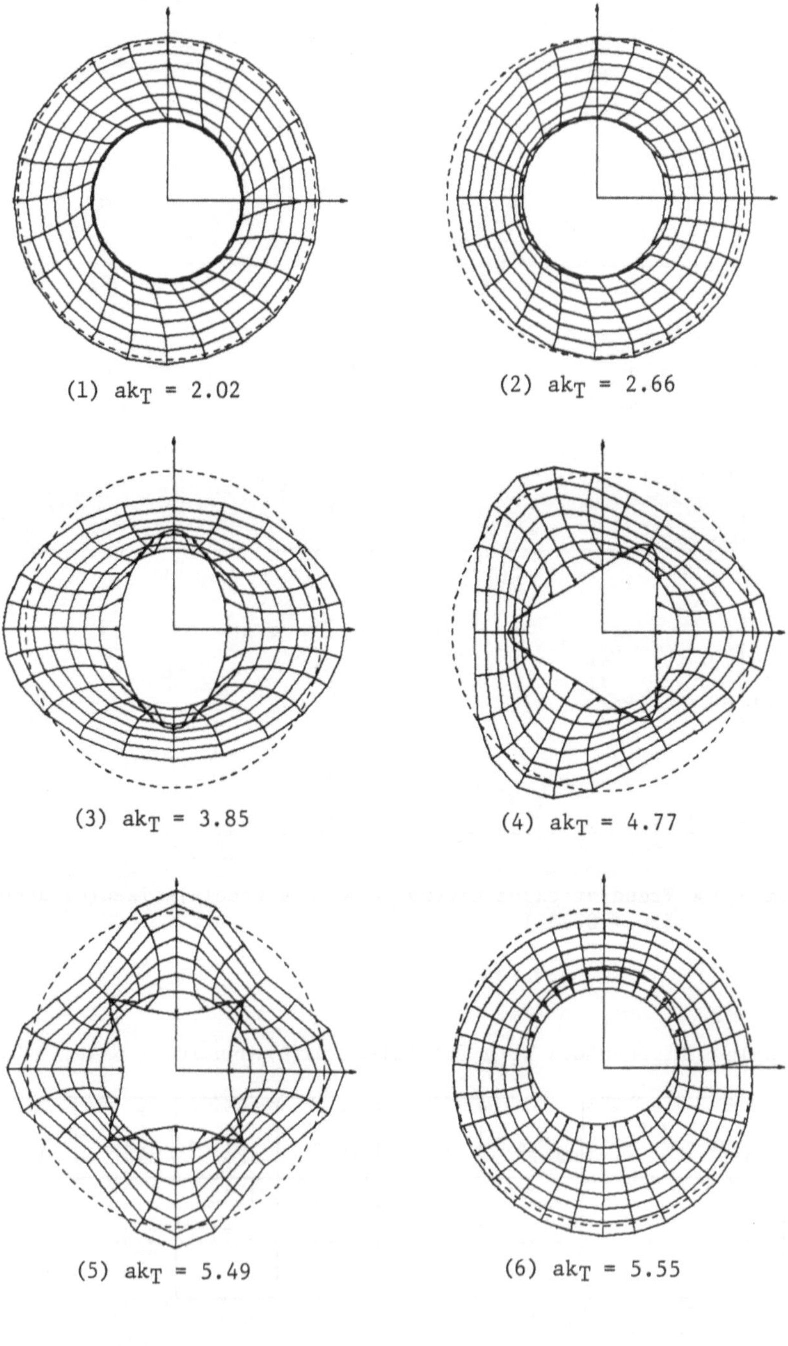

(1) $ak_T = 2.02$

(2) $ak_T = 2.66$

(3) $ak_T = 3.85$

(4) $ak_T = 4.77$

(5) $ak_T = 5.49$

(6) $ak_T = 5.55$

Figure 5 Eigenmodes of a free(a)-fixed(b) annulus (N=48, ν=0.25)

Dam structure on a rigid foundation

As the geometry of the dam structure, we consider the isosceles triangular shaped dam with the height h and the base length of three times of the height, that is, $b = 3h$, and Poisson's ratio ν is assumed to be 0.45. As the boundary condition, the dam base is assumed to be fixed and the other peripheries to be stress free. Figure 6 shows the eigenmodes obtained by the BIE method and corresponding eigenvalues are shown at the bottom of each mode shape as the nondimensionalized shear wave number (hk_T). As for the more detailed discussion of this problem, which includes the comparison with the results by the finite element method, see Niwa *et al.* (1982 b).

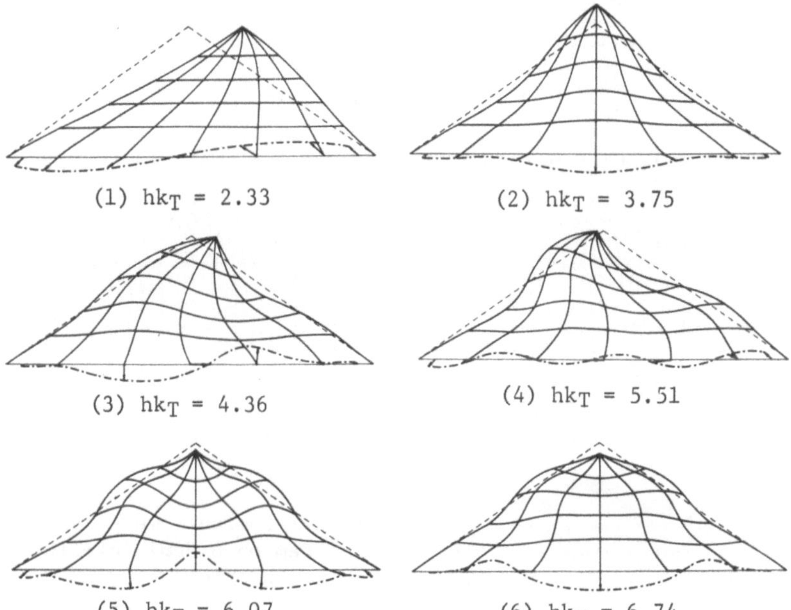

(1) $hk_T = 2.33$ (2) $hk_T = 3.75$

(3) $hk_T = 4.36$ (4) $hk_T = 5.51$

(5) $hk_T = 6.07$ (6) $hk_T = 6.74$

Figure 6 Eigenmodes of a dam on a rigid foundation

Dam structure on a semi-infinite foundation

First, it should be remarked that the dam base was assumed to be fixed in the preceding analysis. Therefore the frequencies thus obtained seem to be somewhat higher than those of the real dam structure on a semi-infinite foundation. Thus, our next step is to apply the present BIE method to the dominant frequency analysis of a dam-type structure on a semi-infinite foundation. In this case, unlike the domain type methods the BIE method requires no artificial wave transmitting boundary in the foundation and infinite domain is easily taken into account by choosing the fundamental solution so as to satisfy the radiation conditions at infinity. Figure 7 shows the dominant modes obtained by the BIE method and the corresponding dominant frequencies. Here, the geometrical configuration of the dam and material properties of the dam-foundation system are the same as those of the previous example on a rigid foundation. It is observed that the mode shapes of the dam are quite the same as those of the previous example except that the corresponding frequencies are fairly lower than those of the dam on the rigid foundation.

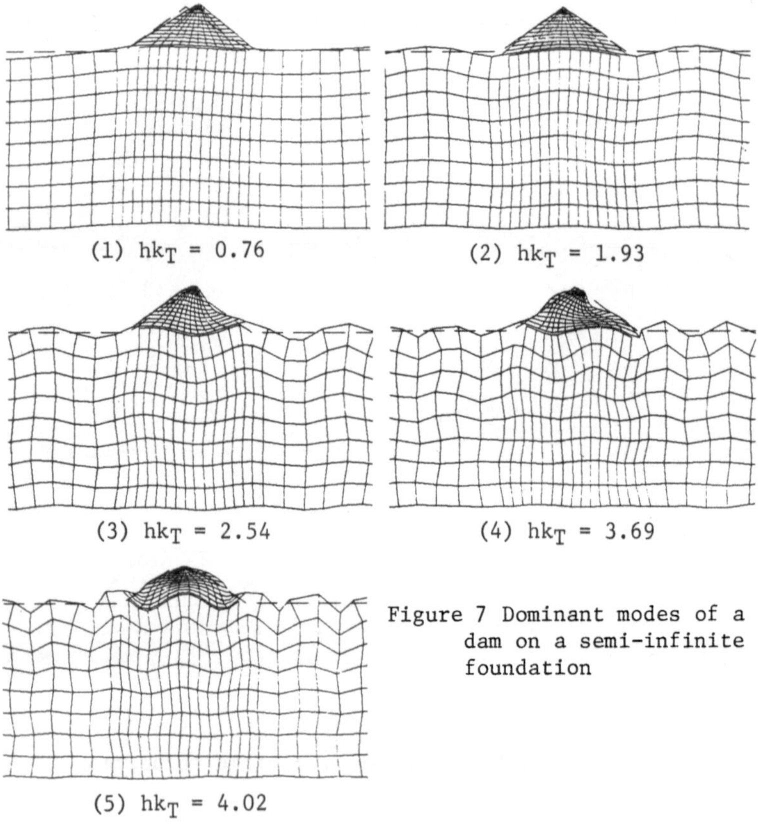

Figure 7 Dominant modes of a dam on a semi-infinite foundation

CONCLUDING REMARKS

In the present paper, the efficiency and the versatility of the BIE method applied to the eigenvalue problems of elastodynamics have been demonstrated by some example problems such as a disc, annulus, dam on a rigid foundation, and dam on a semi-infinite foundation.

Some remarks on the eigenvalue analysis of elastodynamics by the BIE method are summarized here:

1) To find the eigenvalues correctly, it is necessary to evaluate the minima of the absolute values of the complex determinant.
2) It is convenient to use the nondimensionalized shear wave number(Lk_T) as the eigenparameter.
3) Essential material constant in the stage to calculate the eigenvalues(Lk_T), eigendensities, and eigenmodes is Poisson's ratio only (of course, homogeneity and isotropy are assumed). The other material constant and the mass density are needed only when Lk_T is converted into the dimension of frequency.
4) The traction mode shapes on the fixed boundary as well as the displacement mode shapes on the free boundary are directly obtained as the eigendensities by solving the homogeneous system.

Lastly, it should be remarked that the dam models analyzed in this paper are the most simple ones and further research is necessary. In particular, more practical dominant mode analysis of the dam which has the different material properties from those of the semi-infinite stratum will be advantageously carried out by the BIE method.

REFERENCES

Eringen, A. C. and Suhubi, E. S. (1975) Elastodynamics, Volume II: Linear Theory, Academic Press, New York.

Kupradze, V. D. (1965) Potential Methods in the Theory of Elasticity, Israel Program for Scientific Translations, Jerusalem.

Niwa, Y., Kobayashi, S. and Kitahara, M. (1982 a) Determination of Eigenvalues by Boundary Element Methods, Developments in Boundary Element Methods-2, Eds. P. K. Banerjee and R. P. Shaw, Applied Science Pub., London, Ch.6, pp.143-176.

Niwa, Y., Kobayashi, S. and Kitahara, M. (1982 b) Eigenfrequency Analysis of a Dam by the Boundary Integral Equation Method, Proc. of the 4th International Conference on Numerical Methods in Geomechanics, Edmonton, Canada.

Shaw, R. P. (1979) Boundary Integral Equation Methods Applied to Wave Problems, Developments in Boundary Element Methods-1, Eds. P. K. Banerjee and R. Butterfield, Applied Science Pub., London, Ch.6, pp.121-153.

Vekua, I. N. (1967) New Methods for Solving Elliptic Equations, North-Holland Pub., Amsterdam, Ch.6.

Wong, G. K. K. and Hutchinson, J. R. (1981) An Improved Boundary Element Method for Plate Vibrations, Boundary Element Methods (Proc. 3rd International Seminor, Irvine, California, 1981), Ed. C. A. Brebbia, Springer-Verlag, Berlin, pp.272-289.

A NEW APPROACH TO FREE VIBRATION ANALYSIS USING BOUNDARY ELEMENTS

D. Nardini and C.A. Brebbia

University of Zagreb, Yugoslavia and University of Southampton, England.

1. INTRODUCTION

The boundary element method has now become a well established technique of structural analysis [1],[2], and is starting to be widely used to solve elastostatics and general steady state problems, for which it is in many cases much more economical and accurate than the classical domain methods. The boundary element technique has also been successfully applied to forced vibration analysis of structures [3] and for determination of the dynamic stiffness of foundations [4]. These applications require sweeping the frequency range of interest to obtain the dynamic response of the system. Free vibration analysis however, consists of finding only the natural frequencies and modes of the system. Up to now this analysis has been carried out in the same fashion as for forced vibration, i.e. by successively applying different forcing frequencies to the undamped system until resonance occurs. This type of solution has been applied to the Helmholz equation [5] as well as to the equations of elastodynamics [6].

The reason why this search is needed is because the fundamental solution of the governing differential equations is itself frequency dependent, producing therefore a non-algebraic eigenvalue problem. In addition the fundamental solution for elastodynamics is expressed in terms of exponential or Hankel functions with the eigenvalue as a parameter. The integration of these functions over the boundary has to be carried out for each trial frequency, which makes the procedure extremely uneconomic for many practical applications.

In this paper a new procedure is described which reduces the problem of free vibrations to an algebraic eigenvalue problem, solution of which is straightforward. The main advantage of this original approach is that the boundary integrals need to be computed only once as they are frequency independent. Hence the procedure is extremely economic when compared to the ones previously presented.

2. GOVERNING EQUATIONS

The governing differential equations for free vibrations of an isotropic homogeneous elastic body can be written as:

$$\sigma_{ij,j} + \omega^2 \rho\, u_i = 0 \tag{1}$$

where u_i = components of displacement amplitudes

σ_{ij} = stress tensor components

ω = natural circular frequency

ρ = mass density

The domain Ω is bounded by the surface Γ, which is subject to the following boundary conditions:

$$\begin{aligned} u_i &= 0 \quad \text{on } \Gamma_1 \\ t_i &= 0 \quad \text{on } \Gamma_2 \end{aligned} \tag{2}$$

where the total boundary is $\Gamma = \Gamma_1 + \Gamma_2$. The body traction components are

$$t_i = \sigma_{ij}\, n_j \tag{3}$$

with n_j = components of outward normal on the boundary.

3. INTEGRAL FORMULATION

In this analysis we will propose a new free vibrations formulation using the static - i.e. frequency independent - fundamental solution rather than frequency dependent functions.

The fundamental solution resulting from a static unit force acting at a point A in the direction k in an infinite medium is employed. The resulting displacement at any point X in direction i of the medium will be denoted by $U_{ki}(A,X)$, and is the well known Kelvin solution.

The corresponding stress field will be denoted by S_{kij}, and the resulting tractions at the boundary by $T_{ki} = S_{kij} n_j$.

Using U_{ki} as a weighting function [1], eq. (1) is expressed in its equivalent integral form as:

$$\int_\Omega \sigma_{ij,j}\, U_{ki}\, d\Omega + \omega^2 \rho \int_\Omega u_i\, U_{ki}\, d\Omega = 0 \tag{4}$$

Applying the divergence theorem, the first integral in eq. (4) is expressed as:

$$\int_\Omega \sigma_{ij,j}\, U_{ki}\, d\Omega = \int_\Omega u_i(X)\, S_{kij,j}(A,X)d\Omega + \int_\Gamma U_{ki}(A,B)t_i(B)d\Gamma - \\ - \int_\Gamma T_{ki}(A,B)\, u_i(B)d\Gamma \tag{5}$$

where B defines a boundary point and X a point in the domain. $S_{kij,j}(A,X)$ is zero everywhere except at the point A in direction k, so the domain integral is simply

$$\int_\Omega u_i(X)\, S_{kij,j}(A,X)d\Omega = -\, \delta_{ki}\, u_i(A) \tag{6}$$

when A is within the domain (δ_{ki} is the Kronecker delta).

If the point A is on the boundary, the last integral of eq. (5) has a finite value at that point which can be expressed as:

$$(c_{ki}-\delta_{ki})u_i(A) = \lim_{\varepsilon \to o} \int_{\Gamma_\varepsilon} T_{ki}(A,B)\, u_i(B)d\Gamma \tag{7}$$

Using eqs (6) and (7), eq. (5) yields

$$\int_\Omega \sigma_{ij,j}\, U_{ki}\, d\Omega = -\, c_{ki}u_i(A) + \int_\Gamma U_{ki}(A,B)\, t_i(B)d\Gamma - \\ - \int_\Gamma T_{ki}(A,B)\, u_i(B)d\Gamma \tag{8}$$

with the final integral taken in the Cauchy principal-value sense. Using this result, eq. (4) is transformed into

$$c_{ki}u_i(A) = \int_\Gamma U_{ki}(A,B)\, t_i(B)d\Gamma - \int_\Gamma T_{ki}(A,B)\, u_i(B)d\Gamma + \\ + \omega^2\rho \int_\Omega u_i(X)\, U_{ki}(A,X)d\Omega \tag{9}$$

This expression contains the unknown displacement $u_i(B)$ and the tractions $t_i(B)$ on the boundary, but also the unknown displacements $u_i(X)$ within the domain, appearing in the "inertial term". In order to formulate the problem in terms of the boundary unknowns only, a suitable approximation to the displacements inside the domain needs to be employed.

4. APPROXIMATION TO THE DOMAIN INTEGRAL

A class of functions denoted by $f^j(X)$ (superscript j denoting the member of the class) will be chosen, and then the displacements $u_i(X)$ will be approximated using a set of unknown coefficients α_{ij} such that,

$$u_i(X) = \alpha_{ij}\; f^j(X) \tag{10}$$

With this substitution, the domain integral in eq. (8) becomes,

$$\int_\Omega u_i(X)\; U_{ki}(A,X)d\Omega = \alpha_{ij} \int_\Omega f^j(X)\; U_{ki}(A,X)d\Omega \tag{11}$$

Notice that the resulting integral now contains only known functions. Our final objective is to transform it into an equivalent boundary integral. This can be achieved by a suitable choice of functions $f^j(X)$, which will be described in the following.

The similarity in the form of the domain integral $\int_\Omega \sigma_{ij,j}\; U_{ki}\; d\Omega$ from eq. (8) which has been transformed into boundary integrals and the integral $\int_\Omega f^j U_{ki} d\Omega$ from eq. (10) suggests that if one can find a displacement field $\psi^j_{\ell i}$ with the corresponding stress tensor $\tau^j_{\ell im}$ such that

$$\tau^j_{\ell im,m} = \delta_{\ell i}\; f^j \tag{12}$$

the transformation of the later domain integral would also be possible. This transformation can be done as follows,

$$\alpha_{ij} \int_\Omega f^j(X)\, U_{ki}(A,X)\, d\Omega = \alpha_{\ell j} \int_\Omega \delta_{\ell i}\, f^j(X)\, U_{ki}(A,X)\, d\Omega =$$

$$= \alpha_{\ell j}\{ \int_\Omega \psi^j_{\ell i}(X)\, S_{kim,m}\, d\Omega + \int_\Gamma U_{ki}(A,B)\, p^i_{\ell i}(B)\, d\Gamma =$$

$$- \int_\Gamma T_{ki}(A,B)\, \psi^j_{\ell i}(B)\, d\Gamma\} =$$

$$= \alpha_{\ell j}\{ -c_{ki}\, \psi^j_{\ell i}(A) + \int_\Gamma U_{ki}(A,B)\, p^i_{\ell i}(B)\, d\Gamma$$

$$- \int_\Gamma T_{ki}(A,B)\psi^j_{\ell i}(B)\, d\Gamma\} \qquad (13)$$

where $p^j_{\ell i} = \tau^j_{\ell im}\, n_m$ is the traction vector at the boundary corresponding to the displacement field $\psi^j_{\ell i}$.

Substituting this result into eq. (9) we obtain,

$$c_{ki}\, u_i(A) - \int_\Gamma U_{ki}(A,B)\, t_i(B)\, d\Gamma + \int_\Gamma T_{ki}(A,B)\, u_i(B)\, d\Gamma =$$

$$= \omega^2 \rho\{ -c_{ki}\, \psi^j{}_i(A) + \int_\Gamma U_{ki}(A,B)\, p^j_{\ell i}(B)\, d\Gamma$$

$$- \int_\Gamma T_{ki}(A,B)\psi^j_{\ell i}(B)\, d\Gamma\}\, \alpha_{\ell j} \qquad (14)$$

From this expression, a boundary element procedure which does not require domain integration, can be established.

5. BOUNDARY ELEMENT FORMULATION

In what follows, matrix rather than the indicial notation will be used for simplicity.

Eq. (14) contains the unknown tractions and displacements t_i and u_i on boundary, as well as the unknown coefficients $\alpha_{\ell j}$. The boundary will be divided into elements of a given type and the tractions and displacements will be approximated in terms of the nodal values using element shape functions Φ, i.e.

$$\int_{\Gamma} U_{ki} t_i d\Gamma = \sum_{e} \int_{\Gamma_e} U_{ki} t_i d\Gamma \tag{15}$$

$$\{t\} = [\Phi] \, \{t\}^e \tag{16}$$

$$\int_{\Gamma_e} [U]\{t\}d\Gamma = \int_{\Gamma_e} [U][\Phi]d\Gamma \, \{t\}^e = [g]^e \{t\}^e \tag{17}$$

where: $[\Phi]$ = matrix of element shape functions

$\{t\}^e$ = traction vector at local element nodes

$$[g]^e = \int_{\Gamma_e} [U][\Phi]d\Gamma$$

Equally for the boundary displacements,

$$\int_{\Gamma} T_{ki} \, u_i \, d\Gamma = \sum_{e} \int_{\Gamma_e} T_{ki} \, u_i \, d\Gamma \tag{18}$$

$$\{u\} = [\Phi] \, \{u\}^e \tag{19}$$

$$\int_{\Gamma_e} [T]\{u\}d\Gamma = \int_{\Gamma_e} [T][\Phi]d\Gamma \, \{u\}^e = [h]^e \, \{u\}^e \tag{20}$$

where: $\{u\}^e$ = displacement vector at local element nodes

$$[h]^e = \int_{\Gamma_e} [T][\Phi]d\Gamma$$

The integrals in eq. (14) associated with the inertial term do not contain unknown functions and could be computed as they are, but such a computation would represent a substantial effort because it requires an integration over the whole boundary for every coefficient.

However, these integrals are identical in form to the integrals containing t and u, suggesting the following procedure. If the same shape functions which approximate the variation of u and t over the boundary elements are used to approximate the variation of ψ and p, no additional integrals need to be computed at all! Therefore,

$$\int_{\Gamma} U_{ki}\, p^{j}_{\ell i}\, d\Gamma = \sum_{e} \int_{\Gamma_e} U_{ki}\, p^{j}_{\ell i}\, d\Gamma \tag{21}$$

$$[p]_j = [\Phi]\,[p]^e_j \tag{22}$$

$$\int_{\Gamma_e} [U][p]_j\, d\Gamma = \int_{\Gamma_e} [U][\Phi]d\Gamma\, [p]^e_j = [g]^e\, [p]^e_j \tag{23}$$

where $[p]^e_j$ = the matrix of values $p^j_{\ell i}$ at element nodes.

Equally,

$$\int_{\Gamma} T_{ki}\, \psi^{j}_{\ell i}\, d\Gamma = \sum_{e} \int_{\Gamma_e} T_{ki}\, \psi^{j}_{\ell i}\, d\Gamma \tag{24}$$

$$[\psi]_j = [\Phi][\psi]^e_j \tag{25}$$

$$\int_{\Gamma_e} [T][\psi]_j\, d\Gamma = \int_{\Gamma_e} [T][\Phi]d\Gamma\, [\psi]^e_j = [h]^e\, [\psi]^e_j \tag{26}$$

where $[\psi]^e_j$ = matrix of values $\psi^j_{\ell i}$ at element nodes.

Assembling all element contributions into the global system yields:

$$[H]\{u\} - [G]\,\{t\} = \omega^2\rho\, (-\,[H][\psi] + [G][p])\{\alpha\} \tag{27}$$

where the diagonal submatrices c_{ki} have been included into [H].

The relationship between {u} and {α} can be established using eq. (10), formally being:

$$\{u\} = [F]\{\alpha\} \tag{28}$$

where elements of matrix [F] are simply the values of the functions $f^j(X)$ at the nodal points. If the functions f^j are linearly independent, and their number if chosen to be equal to the number of nodes, the matrix [F] is square and regular and possesses an inverse $[E] = [F]^{-1}$, therefore,

$$\{\alpha\} = [E]\ \{u\} \tag{29}$$

Substituting (27) into (25) we obtain,

$$[H]\ \{u\} - [G]\ \{t\} = \omega^2\ [M]\ \{u\} \tag{30}$$

where: $[M] = \rho(-\ [H][\psi] + [G][p])[E]$ (31)

So far we have not employed the homogeneous boundary conditions which state that on any part of the boundary either u or t are zero. If we denote $u(\Gamma_1) = u_1$, $u(\Gamma_2) = u_2$, $t(\Gamma_1) = t_1$ and $t(\Gamma_2) = t_2$, eq. (28) can be rewritten using submatrices as:

$$\begin{bmatrix} H_{11} & H_{12} \\ H_{21} & H_{22} \end{bmatrix} \begin{bmatrix} u_1 \\ u_2 \end{bmatrix} - \begin{bmatrix} G_{11} & G_{12} \\ G_{21} & G_{22} \end{bmatrix} \begin{bmatrix} t_1 \\ t_2 \end{bmatrix} = \omega^2 \begin{bmatrix} M_{11} & M_{12} \\ M_{21} & M_{22} \end{bmatrix} \begin{bmatrix} u_1 \\ u_2 \end{bmatrix} \tag{32}$$

Taking into account that $u_1 = 0$ and $t_2 = 0$:

$$[H_{12}]\{u_2\} - [G_{11}]\{t_1\} = \omega^2 [M_{12}]\{u_2\} \tag{33}$$

$$[H_{22}]\{u_2\} - [G_{21}]\{t_1\} = \omega^2 [M_{22}]\{u_2\} \tag{34}$$

From these two sets of equations, $\{t_1\}$ can be eliminated resulting in:

$$[\hat{H}]\ \{u_2\} = \omega^2\ [\hat{M}]\ \{u_2\} \tag{35}$$

where:

$$[\hat{H}] = [H_{22}] - [G_{21}][G_{11}]^{-1}[H_{12}] \tag{36}$$

$$[\hat{M}] = [M_{22}] - [G_{21}][G_{11}]^{-1}[M_{12}] \tag{37}$$

which represents the generalised algebraic eigenvalue problem.

6. TWO-DIMENSIONAL FORMULATION

For the case of plane strain the fundamental solution has the form:

$$U_{ki} = \frac{1}{8\pi G(1-\nu)} [\delta_{ki}(4\nu-3)\ln R + r_k r_i] \tag{38}$$

where R = distance from the point of application of unity force to the observed point

r_k = direction cosines of vector $\vec{R}$.

G ,ν = shear modulus and Poisson's ratio

The corresponding tractions are:

$$T_{ki} = \frac{1}{4\pi(1-\nu)} \{(1-2\nu)(n_i r_k - r_i n_k) - [\delta_{ki}(1-2\nu)+2r_k r_i] r_m n_m\} \tag{39}$$

As it has been previously indicated, a set of functions f^j needs to be defined in order to approximate the displacements within the domain. For computational reasons they should be easy to generate and to compute, and should also meet the requirement given by eq. (12), i.e. enabling us to transform the domain integral into equivalent boundary integrals.

A simple class of functions that are associated with particular points in the plane, meeting these requirements are

$$f^j(X) = C - R(A_j,X) \tag{40}$$

where: $R(A_j,X)$ = distance from the point A_j where the function is applied to a point X.

C = a suitably chosen constant

The corresponding function ψ, satisfying eq. (12) is found to be:

$$\psi_{\ell i} = \left[\frac{1-2\nu}{5-4\nu} C + \frac{R}{30(1-\nu)}\right] x_\ell x_i - \frac{9-10\nu}{90(1-\nu)} \delta_{\ell i} R^3 \tag{41}$$

where x_ℓ = coordinate components from A_j to X.

In order to improve the approximation, a number of other functions can be added to the proposed class:

a) A constant: $f(X) = 1$ (42)

in which case $\psi_{\ell i} = \frac{1-2\nu}{5-4\nu} x_\ell x_i$ (43)

b) A linear function in terms of the coordinate x_k, i.e.

$$f(X) = x_k \quad (44)$$

in which case $\psi_{\ell i} = (a_1 \delta_{ik} x_\ell + a_2 \delta x_k x_i + a_3 \delta x_i x_k)$

(45)

where: $a_1 = - \frac{1}{16(3-4\nu)}$

$$a_2 = \frac{1-2\nu}{2(1-\nu)} a_1$$

$$a_3 = - (5-8\nu) a_1$$

Other, more complicated functions can also be added, but it is not certain whether the additional programming and computational effort would justify the improvement in the approximation. Besides, increase in the accuracy can also be achieved by the procedure described in the following section.

7. INTERNAL DEGREES OF FREEDOM

The class of functions presented, which are associated with particular points, are well suited for computer implementation, since they can use the same nodes as those defined for the boundary element model. Therefore each node will carry its own shape function, the resulting displacement being the linear combination of all contributions, as expressed by eq. (10).

In order to further improve the accuracy, internal nodes can be introduced, which will also mean additional dynamic degrees of freedom. The same equations derived so far apply to internal points as well, with the submatrix c_{ki} being a unity matrix. The internal degrees of freedom will be useful in determining accurately higher modes of free vibrations.

8. ON THE ACCURACY AND EFFICIENCY OF THE METHOD

It is important to note that the approximation to the displacements in the domain by the chosen class of functions has only been used in the integral corresponding to the inertial term itself. The fact that the shape functions used to obtain inertial terms can be considerably simpler than those used in the formulation of stiffness terms has been realized quite early in numerical methods (the widespread use of lumped mass matrix in FEM being a typical example). This is due to the absence of any derivatives in the mass integral, thus imposing fewer restrictions on the choice of the shape functions.

It is therefore to be expected, that the proposed approximation to the displacement field for computation of the mass coefficients will be sufficient for determining at least the lower modes of vibration accurately (the inevitable consequence of any numerical method with a finite number of dynamic degrees of freedom is the loss of accuracy for the higher modes).

From the computational point of view, the method presented here is very efficient for the following two reasons:

a) The technique requires the same amount of integration as for the elastostatics. The matrices $[\psi]$ and $[p]$ need only the evaluation of simple functions at the boundary nodes.

b) The resulting eigenvalue problem is an algebraic one which can be easily solved by a number of the existing eigenvalue routines.

9. COMPUTER IMPLEMENTATION

A computer program for plane elasticity based on the outlined method has been developed. The implemented boundary elements are straight, but with quadratic interpolation functions. Although the computation involves five different global matrices: $[H]$, $[G]$, $[\psi]$, $[p]$ and $[F]$ (eqs (27) and (28)), it is possible to arrange the operations in such a way as to accommodate all the required data at any one time in only two matrices. Therefore, the working area required by the program is much the same as for the case of elastostatics.

The resulting generalised algebraic eigenvalue problem $[\hat{H}]\{u_2\} = \omega^2[\hat{M}]\{u_2\}$ could be solved directly by a variant of the subspace iteration. However, in order to simplify the procedure, in the present version of the program the matrix $[\hat{H}]$ is inverted and postmultiplied by $[\hat{M}]$ thus forming the standard eigenvalue problem $[A]\{u_2\} = \lambda\{u_2\}$ ($[A] = [\hat{H}]^{-1}[\hat{M}]$, $\lambda = 1/\omega^2$), which can then be solved by a computer library routine. It is worth noting that since the dimensions of the matrices involved are an order of magnitude less than in the domain methods, the choice of the appropriate eigenvalue algorithm is not as critical.

10. APPLICATIONS

Several applications of in-plane free vibrations are presented here in order to demonstrate the feasibility and accuracy of this original formulation. The boundary element results obtained here were compared to the results produced by the two dimensional finite elements of SAPIV. The material constants in all examples are: $E/\rho = 10^4$, $\nu = 0.2$.

Example 1 - Simple Geometrical Shapes

As a first verification the in-plane free vibrations of two simple geometrical shapes, a square (a = 6.0) and a triangle (b = 10.0, h = 8.0) were studied. The problems were solved using several different boundary element discretizations. In fig. 1 the first four periods of free vibrations are plotted versus the number of elements employed. As a comparison, eigenvalue results using the SAP finite element program are given in the same diagrams as dotted lines. The agreement is very good even with a small number of boundary elements, and the results show convergence.

Example 2 - Cantilever Beam

The dynamic properties of a deep cantilever beam (h = 6, ℓ = 24) were studied. The structure shown in fig. 2 was analysed using several boundary element meshes, and the results compared to the finite element ones. The third and the fifth modes are the longitudinal ones for which the results are very accurate even for the smaller discretization. For the transverse modes (first, second and fourth) it is to be expected that more elements will be needed to obtain the same degree of accuracy. In all cases however the solutions show very good agreement with the finite element model.

Example 3 - Shear Wall

Fig. 3 shows a more realistic application, i.e. a shear wall with four openings. The boundary element model consists of 29 quadratic elements with 58 nodes, as shown on the figure. The finite element mesh comprising of 559 nodes and used for comparison purposes is shown in the same figure. The results for the free vibration periods for the first eight natural modes are given in table 1. In spite of the complicated geometry and the rather small number of boundary elements used, the agreement of the results is surprisingly good.

Mode	1	2	3	4	5	6	7	8
BEM	3.022	0.875	0.822	0.531	0.394	0.337	0.310	0.276
FEM	3.029	0.885	0.824	0.526	0.409	0.342	0.316	0.283

Table 1 - Periods of free vibrations for the two models

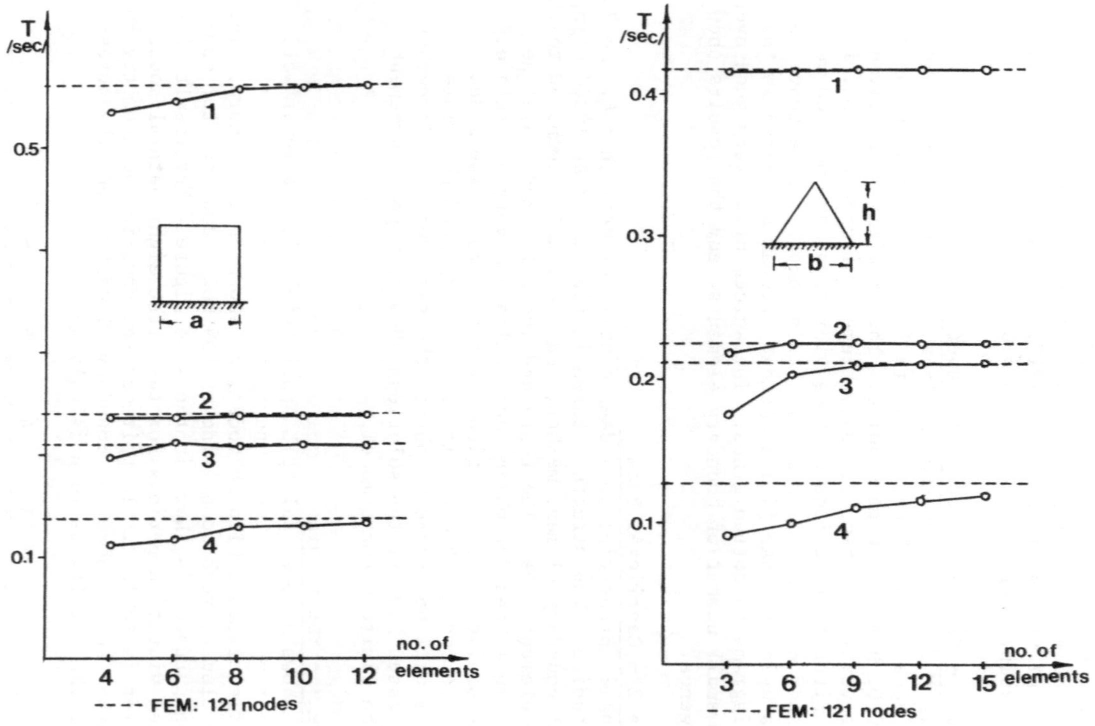

Fig. 1 Periods of in-plane free vibrations for two simple geometrical shapes

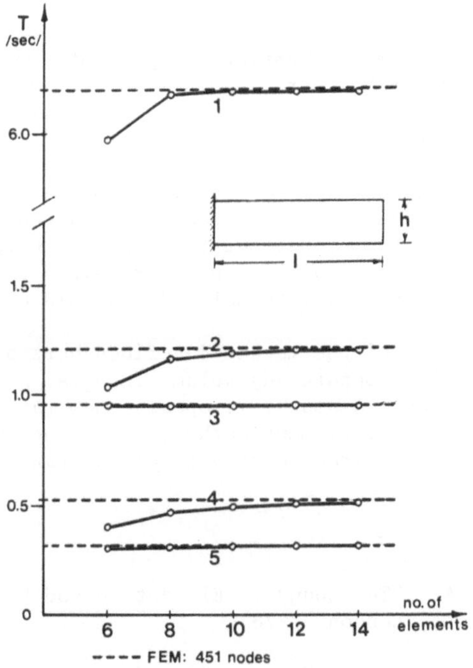

Fig. 2 Periods of free vibrations of a deep cantilever beam

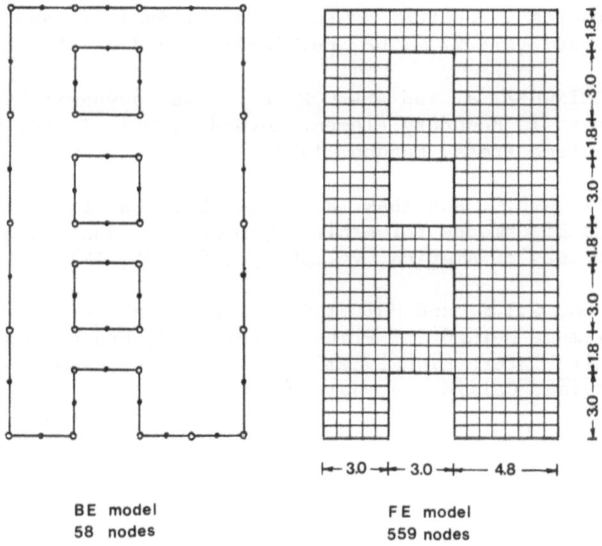

Fig. 3 The two numerical models of a shear wall

11. CONCLUSION

The new method for free vibration analysis of homogeneous elastic bodies described here is shown to be very efficient and accurate. The amount of work needed to evaluate boundary integrals is the same as that required for the solution of an elastostatic problem. Contrary to previous free vibration solutions using boundary elements,this paper describes in detail how the problem can be reduced to an algebraic eigenvalue problem. It is pointed out that special care should be taken if higher modes of vibration are required to be computed accurately, in which case internal nodes should be employed.

The advantages of the method described here are that it requires neither to compute any volume integral nor to perform eigenvalue searches. As such, it opens a new range of applications for the boundary element method, including the extension of the described technique to many types of transient dynamic problems.

REFERENCES

1. BREBBIA, C.A. "The Boundary Element Method for Engineers", Pentech Press, London, 1978.

2. BREBBIA, C.A. and WALKER, S. "Boundary Element Techniques in Engineering", Newnes-Butterworths, London, 1980.

3. CRUSE, T.A. and RIZZO, F.J. "A Direct Formulation and Numerical Solution of the General Transient Elastodynamic Problem", Journal of Mathematical Analysis and Applications, Vol. 22, 1968, pp.244-259 and 341-355.

4. DOMINGUEZ, J. and ALARCON, E. "Elastodynamics" in "Progress in Boundary Element Methods", Vol. 1, Ed. C.A. Brebbia Pentech Press, London, 1981.

5. TAI, G.R.C. and SHAW, R.P. "Helmholz-Equation Eigenvalues and Eigenmodes for Arbitrary Domains" Journal of Acoustical Society of America, Vol.56, pp.796-804, 1974.

6. WONG, G.I.K. and HUTCHINSON, J.R. "An Improved Boundary Element Method for Plate Vibrations", Proceedings of the Third International Seminar on Boundary Element Methods, Irvine, California, July 1981.

VISCOPLASTIC ANALYSIS

J. C. F. Telles
Department of Civil Engineering
COPPE-Federal University of
Rio de Janeiro - Cx.P. 68506
Cep 21944 - BRAZIL

C. A. Brebbia
Department of Civil
Engineering
University of Southampton
Southampton SO9 5NH
ENGLAND

INTRODUCTION

This paper extends the boundary element method to solve elastic/viscoplastic problems.

One of the first finite element formulations for viscoplasticity appears to be due to Zienkiewicz and Cormeau [7], where the viscoplastic model due to Perzyna [2] was used. Many other finite element papers have been written afterwards. Boundary elements were first applied to viscoplasticity by Chaudonneret [6] and Mukherjee [8]. In these references different constitutive relations were used to model the time dependent material behaviour. Other boundary element applications have only concentrated on the solution of linear viscoelastic problems [26] [1].

In the present paper a complete formulation for viscoplasticity using boundary elements is presented. The procedure can be used for creep problems as well. The Perzyna's approach has been adopted since it is appropriate for computer applications and can be used to simulate pure elastoplastic solutions. The time-dependent solution is obtained by a simple Euler one step procedure and some guide lines for the selection of the time step length are discussed.

By using the proper integral equation the inelastic strains are obtained at a series of points (internal points and boundary nodes) and cast into an initial stress form to be interpolated in linear piecewise fashion. This gives improved accuracy when compared to other boundary element formulations [6] [8] [9] which use constant strain approximation over each internal cell.

The examples presented and disucssed in the end of the paper point out the accuracy of the boundary element solution and illustrate the potentialities of the technique for these sort of nonlinear problems.

GOVERNING EQUATIONS

Within the context of small strain theory, the equilibrium equations for three-dimensional bodies in terms of displacement rates ($\dot{u}_i$) and inelastic strain rates ($\dot{\varepsilon}^a_{ij}$) are

$$\dot{u}_{j,\ell\ell} + \frac{1}{1-2\nu}\dot{u}_{\ell,\ell j} = 2\left(\dot{\varepsilon}^a_{ij,i} + \frac{\nu}{1-2\nu}\dot{\theta}_{,j}\right) - \dot{b}_j/G \tag{1}$$

where $\dot{\theta} = \dot{\varepsilon}^a_{kk}$, $\dot{b}_j$ represents the body force rate components, ν is Poisson's ratio and G is the shear modulus. Time derivatives are indicated by a dot and space derivatives by a comma.

Here, by inelastic strains one means any kind of strain field which can be considered as initial strains, i.e.

$$\dot{\varepsilon}^a_{ij} = \dot{\varepsilon}^p_{ij} + \dot{\varepsilon}^c_{ij} + \dot{\varepsilon}^T_{ij} \tag{2}$$

in which

$\dot{\varepsilon}^p_{ij}$ - plastic strain rate

$\dot{\varepsilon}^c_{ij}$ - creep strain rate

$\dot{\varepsilon}^T_{ij}$ - thermal strain rate

The traction boundary conditions for equation (1) can be written as

$$\dot{p}_i + 2G\left(\dot{\varepsilon}^a_{ij}\, n_j + \frac{\nu}{1-2\nu}\dot{\theta} n_i\right) = \frac{2G\nu}{1-2\nu}\dot{u}_{\ell,\ell}\, n_i + G(\dot{u}_{i,j} + \dot{u}_{j,i})\, n_j \tag{3}$$

where $\dot{p}_i$ represents the traction rate components and n_j stands for the direction cosines of the outward normal to the boundary of the body.

Alternatively, one can write equations (1) and (3) in terms of "initial stresses" as follows

$$\dot{u}_{j,\ell\ell} + \frac{1}{1-2\nu}\dot{u}_{\ell,\ell j} = \dot{\bar{b}}_j/G \tag{4}$$

and

$$\dot{\bar{p}}_i = \frac{2G\nu}{1-2\nu}\dot{u}_{\ell,\ell}\, n_i + G(\dot{u}_{i,j} + \dot{u}_{j,i})\, n_j \tag{5}$$

where $\dot{\bar{b}}_j$ and $\dot{\bar{p}}_i$ are pseudobody forces and pseudotractions given by

$$\dot{\bar{b}}_j = \dot{b}_j - \dot{\sigma}^a_{ij,i} \tag{6}$$

and

$$\bar{\dot{p}}_i = \dot{p}_i + \dot{\sigma}^a_{ij} n_j \tag{7}$$

in which the "initial stresses" are of the following form

$$\dot{\sigma}^a_{ij} = 2G \dot{\varepsilon}^a_{ij} + \frac{2G\nu}{1-2\nu} \dot{\theta} \delta_{ij} \tag{8}$$

(δ_{ij} is the Kronecker delta)

Different procedures (i.e. initial strain, initial stress and fictitious tractions and body forces) for the solution of equations (1) or (4) using boundary elements have been presented by the present authors [3-5] for plasticity problems. These included not only the fundamental solution due to Kelvin, but also the complete fundamental solution for half-plane problems (Melan's problem) presented in [25]. Herein the initial stress equations have been adopted since they present the advantage of handling compressible or incompressible inelastic strains in plane strain or plane stress problems with minor alterations.

The starting integral equation for the boundary element method applied to the initial stress problem is as follows

$$c_{ij} \dot{u}_j = \int_\Gamma u^*_{ij} \dot{p}_j \, d\Gamma - \int_\Gamma p^*_{ij} \dot{u}_j \, d\Gamma + \int_\Omega u^*_{ij} \dot{b}_j \, d\Omega$$

$$+ \int_\Omega \varepsilon^*_{jki} \dot{\sigma}^a_{jk} \, d\Omega \tag{9}$$

where Γ represents the boundary of the body and Ω its interior. The tensors u^*_{ij}, p^*_{ij} and ε^*_{jki} stand for the displacements, tractions and strains due to a unit point load applied in i direction (fundamental solution).

Equation (9) is valid for any location of the load point (interior or boundary points) provided that c_{ij} and the second boundary integral on the right-hand-side are properly interpreted as known from the elastic application of the technique [1].

For plane problems, equations (1) - (9) can be applied (i, j, k, ℓ = 1, 2) with $\dot{\theta} = \dot{\varepsilon}^a_{11} + \dot{\varepsilon}^a_{22} + \dot{\varepsilon}^a_{33}$ in plane strain and ν replaced by $\bar{\nu} = \nu/(1+\nu)$ with $\dot{\theta} = \dot{\varepsilon}^a_{11} + \dot{\varepsilon}^a_{22}$ in plane stress. Here it is interesting to note that if the half-plane fundamental solution is used instead of Kelvin's, integration over the traction-free surface of the semi-plane becomes unnecessary, producing an accurate and efficient formulation for half-plane type inelastic problems [5].

Of fundamental importance for the stepwise solution of nonlinear material problems is the calculation of stresses at internal points. In order to combine both, accuracy and computational efficiency, it has been demonstrated [5] that the proper integral equation should be used in preference to computing displacements at internal points and differentiating them numerically as it is done in finite differences or finite elements.

The correct integral equation for stresses at internal points has been presented in previous papers by the present authors. Since its derivation requires the derivative of the singular integral of the inelastic term and this had often led to incorrect expressions in the past [6, 10, 11], a proper procedure for obtaining this equation has been recently presented by the authors in reference [14].

From the application of Hooke's law to the elastic part of the total strain rate tensor comes the following expression for the stress rates

$$\dot{\sigma}_{ij} = G\left(\frac{\partial \dot{u}_i}{\partial X_j} + \frac{\partial \dot{u}_j}{\partial X_i}\right) + \frac{2G\nu}{1-2\times}\frac{\partial \dot{u}_k}{\partial X_k}\delta_{ij} - \dot{\sigma}^a_{ij} \tag{10}$$

Stresses at points located within the body can be computed by substituting equation (9) ($c_{ij} = \delta_{ij}$) in equation (10) on condition that the space derivatives present in (10) be taken with reference to the coordinates of the load point. Computation of the derivatives of the last integral of equation (9) requires special care. The complete demonstration has been thoroughly discussed in reference [14] and involves a generalization of Leibnitz rule in the presence of singularities. This leads to the final expression for stresses at internal points

$$\dot{\sigma}_{ij} = \int_\Gamma u^*_{ijk}\,\dot{p}_k\,d\Gamma - \int_\Gamma p^*_{ijk}\,\dot{u}_k\,d\Gamma + \int_\Omega u^*_{ijk}\,\dot{b}_k\,d\Omega$$

$$+ \int_\Omega \varepsilon^*_{ijk\ell}\,\dot{\sigma}^a_{k\ell}\,d\Omega + g_{ij}(\dot{\sigma}^a_{k\ell}) \tag{11}$$

where the integral of the initial stress term is in the principal value sense and the tensors corresponding to the fundamental solution have been presented in previous publications.

The expressions for g_{ij} (Kelvin fundamental solution) are of the following form:

$$g_{ij} = -\frac{1}{15(1-\nu)}\left[(7-5\nu)\dot{\sigma}^a_{ij} + (1-5\nu)\dot{\sigma}^a_{\ell\ell}\,\delta_{ij}\right] \text{ for 3-D} \tag{12}$$

and

$$g_{ij} = -\frac{1}{8(1-\nu)}\left[2\dot{\sigma}^a_{ij} + (1-4\nu)\dot{\sigma}^a_{\ell\ell}\,\delta_{ij}\right] \text{ for 2-D plane strain} \tag{13}$$

for plane stress ν is replaced by $\bar{\nu}$.

The appropriate expressions for half-plane problems were given elsewhere [5].

RATE DEPENDENT CONSTITUTIVE RELATIONS

In the present work we will restrict ourselves to the solution of either creep or elastic/viscoplastic problems in the sense described by Perzyna [2].

Viscoplastic Equations: the static yield criterion for isotropic hardening can be written in general form as

$$F(\sigma_{ij}, k) = 0 \tag{14}$$

where k represents a hardening parameter which dictates the position of this static yield surface in the nine-dimensional stress space. This condition can be better visualized in the following form

$$f(\sigma_{ij}) = \psi(k) \tag{15}$$

where $F = f - \psi$ and if the work hardening hypothesis is being adopted

$$k = W^p = \int_0^{\varepsilon^p_{ij}} \sigma_{ij}\, d\varepsilon^p_{ij} \tag{16}$$

One can notice that the condition expressed in (14) or (15) does not differ from the corresponding yield condition for the so-called inviscid theory of plasticity. Therefore, the different expressions for F (such as those due to Tresca, von Mises, Drucker-Prager and Mohr-Coulomb) present in the literature [16, 3] can still be used. This encourages a further interpretation; let us designate the scalar function $f(\sigma_{ij})$ by σ_e which plays the role of an equivalent or effective stress. Such designation allows for the definition of an equivalent plastic strain ε^p_e whose increment produces an increment in the plastic strain energy of the form

$$\sigma_e\, d\varepsilon^p_e = \sigma_{ij}\, d\varepsilon^p_{ij} = dk \tag{17}$$

Following the generalized normality principle due to Perzyna [2], the viscoplastic strain rates are given by

$$\dot{\varepsilon}^{p}_{ij} = \gamma < \Phi \left(\frac{F}{\psi}\right) > \frac{\partial F}{\partial \sigma_{ij}} \tag{18}$$

where γ denotes a viscosity parameter of the material which can be function of time, temperature, hardening, etc., and

$$< \Phi \left(\frac{F}{\psi}\right) > = \begin{cases} 0 \text{ for } F < 0 \\ \\ \Phi \left(\frac{F}{\psi}\right) \text{ for } F > 0 \end{cases} \tag{19}$$

The function Φ is selected from experimental results.

Equation (18) can be further written as

$$\dot{\varepsilon}^{p}_{ij} = \gamma < \Phi \left(\frac{F}{\psi}\right) > \frac{\partial f}{\partial \sigma_{ij}} \tag{20}$$

which after multiplying both sides by σ_{ij} gives

$$\sigma_{ij}\, \dot{\varepsilon}^{p}_{ij} = \gamma < \Phi \left(\frac{F}{\psi}\right) > \sigma_{ij} \frac{\partial f}{\partial \sigma_{ij}} \tag{21}$$

Assuming that $f(\sigma_{ij})$ is homogeneous of degree one (a requirement satisfied by the yield criteria adopted here) and applying Euler's theorem comes

$$\sigma_{ij}\, \dot{\varepsilon}^{p}_{ij} = \gamma < \Phi\left(\frac{F}{\psi}\right) > f(\sigma_{ij}) \tag{22}$$

Recalling definition (17), expression (22) can be finally represented by

$$\dot{\varepsilon}^{p}_{e} = \gamma < \Phi\left(\frac{F}{\psi}\right) > \tag{23}$$

a relation which for $F > 0$ leads to

$$f(\sigma_{ij}) = \psi(k) \left[1 + \Phi^{-1}\left(\frac{\dot{\varepsilon}^{p}_{e}}{\gamma}\right)\right] \tag{24}$$

Equation (24) when compared to (15) clearly demonstrates the explicit dependence of the flow surface on the equivalent plastic strain rate.

As further illustration, consider the following definition:

$$\dot{\varepsilon}_e = \frac{\dot{\sigma}_e}{E} + \dot{\varepsilon}_e^p \qquad (25)$$

in which E is the Young's modulus and $\dot{\varepsilon}_e$ stands for an equivalent measure of the total strain rate. Note that in uniaxial problems $\dot{\varepsilon}_e$, $\dot{\sigma}_e$ and $\dot{\varepsilon}_e^p$ become the actual total strain, stress and plastic strain rates if ψ is defined as the uniaxial yield stress.

The flow surface can now be written as

$$f(\sigma_{ij}) = \psi(k)\left[1 + \Phi^{-1}\left(\frac{\dot{\varepsilon}_e - \dot{\sigma}_e/E}{\gamma}\right)\right] \qquad (26)$$

indicating the dependence of $f(\sigma_{ij})$ on the rate of induced strains/stresses.

<u>Creep Equations</u>: for creep problems the equivalent version of expression (23) is assumed to be

$$\dot{\varepsilon}_e^c = K\sigma_e^m \, t^n \qquad (27)$$

where K is a material parameter and $\sigma_e = f(\sigma_{ij})$ represents the von Mises equivalent stress.

It is interesting to note that the time-hardening function t^n can be removed from (27) by the following transformation [17]

$$\bar{t} = \int_0^t p^n \, dp \qquad (28)$$

where $\bar{t}$ denotes a transformed time leading to

$$\frac{d\varepsilon_e^c}{d\bar{t}} = K\,\sigma_e^m \qquad (29)$$

This means that the problem can be solved in terms of a fictitious time which relates to the true time t by means of the inverse relation

$$t = \left[\bar{t}(n+1)\right]^{\frac{1}{n+1}} \qquad (30)$$

Different time hardening functions can be equally transformed by the above procedure assuming that (27) is taken from experimental analysis under constant stress.

The creep strain rates can therefore be written as

$$\dot{\varepsilon}^{c}_{ij} = K\,\sigma^{m}_{e}\,\frac{\partial f}{\partial \sigma_{ij}} \tag{31}$$

where the dot indicates derivative with respect to $\bar{t}$ if $n \neq 0$.

Equation (31) corresponds to the Prandtl-Reuss relations and can be cast into the form of equation (20). In both cases the initial stress rates are computed by the simple relation

$$\dot{\sigma}^{a}_{ij} = \gamma < \Phi > d_{ij} \tag{32}$$

where

$$d_{ij} = C_{ijk\ell}\,\frac{\partial f}{\partial \sigma_{k\ell}} \tag{33}$$

and $C_{ijk\ell}$ is the isotropic tensor of elastic constants.

SPATIAL DISCRETIZATION

The spatial discretization of equations (9) and (11) has been carried out for two-dimensional problems. The boundary of the body is assumed to be represented by surface elements and the part of the domain in which non-zero inelastic strains are expected to develop is discretized using internal cells for integration . Herein; tractions, displacements and "initial stresses" have been interpolated in linear piecewise form, the two former over boundary elements and the latter over the internal cells. In addition, the body force term, though not causing any difficulty for implementation, was not considered for simplicity. The problem of accurately integrating over the internal cells has been solved by adopting the semi-analytical integration scheme fully described in another reference [4]. This procedure has proved to be efficient and sufficiently general, being also applied to computing the principal values of equation (11).

The application of equation (9) in discretized form to all boundary nodes generates the following matrix relationship:

$$\underset{\sim}{H}\,\dot{\underset{\sim}{u}} = \underset{\sim}{G}\,\dot{\underset{\sim}{p}} + \underset{\sim}{Q}\,\dot{\underset{\sim}{\sigma}}^{a} \tag{34}$$

in which matrix $\underset{\sim}{Q}$ stands for the initial stress integral.

Similarly, the stress rates at the internal points are given by equation (11) in discretized form

$$\dot{\underset{\sim}{\sigma}} = \underset{\sim}{G}'\,\dot{\underset{\sim}{p}} - \underset{\sim}{H}'\,\dot{\underset{\sim}{u}} + \overline{\underset{\sim}{Q}}\,\dot{\underset{\sim}{\sigma}}^{a} \tag{35}$$

where matrix $\overline{Q}$ represents the initial stress integral plus the independent terms g_{ij}. Matrices $\underset{\sim}{H}$, $\underset{\sim}{G}$, $\underset{\sim}{H}'$ and $\underset{\sim}{G}'$ are the same as those obtained for pure elastic analysis [1].

In order to extend the validity of equation (35) for stress rates at selected boundary nodes, expressions other than (11) have to be used. These expressions do not require any integration and can be taken from the rate version of the interpolated boundary displacements and tractions [3, 4].

Taking into consideration that for a well-posed problem a sufficient number of tractions and boundary displacements is prescribed, equation (34) and (35) can be further written as

$$\underset{\sim}{A}\,\dot{\underset{\sim}{x}} = \dot{\underset{\sim}{f}} + \underset{\sim}{Q}\,\dot{\underset{\sim}{\sigma}}^{a} \tag{36}$$

and

$$\dot{\underset{\sim}{\sigma}} = -\,\underset{\sim}{A}'\,\dot{\underset{\sim}{x}} + \dot{\underset{\sim}{f}}' + \overline{\underset{\sim}{Q}}\,\dot{\underset{\sim}{\sigma}}^{a} \tag{37}$$

where vector $\dot{\underset{\sim}{x}}$ is formed by the unknown tractions and boundary displacements and the contribution of the prescribed values is included in vectors $\dot{\underset{\sim}{f}}$ and $\dot{\underset{\sim}{f}}'$.

From the multiplication of equation (36) by $\underset{\sim}{A}^{-1}$ comes

$$\dot{\underset{\sim}{x}} = \underset{\sim}{R}\,\dot{\underset{\sim}{\sigma}}^{a} + \dot{\underset{\sim}{m}} \tag{38}$$

where

$$\underset{\sim}{R} = \underset{\sim}{A}^{-1}\,\underset{\sim}{Q} \tag{39}$$

and

$$\dot{\underset{\sim}{m}} = \underset{\sim}{A}^{-1}\,\dot{\underset{\sim}{f}} \tag{40}$$

Substituting (38) in (37) yields

$$\dot{\underset{\sim}{\sigma}} = \underset{\sim}{V}\,\dot{\underset{\sim}{\sigma}}^{a} + \dot{\underset{\sim}{n}} \tag{41}$$

in which

$$\underset{\sim}{V} = \overline{\underset{\sim}{Q}} - \underset{\sim}{A}'\,\underset{\sim}{R} \tag{42}$$

and

$$\dot{\underset{\sim}{n}} = \dot{\underset{\sim}{f}}' - \underset{\sim}{A}'\,\dot{\underset{\sim}{m}} \tag{43}$$

Note that the elastic solution to the rate problem is given by the vectors $\dot{\underset{\sim}{m}}$ and $\dot{\underset{\sim}{n}}$.

From the above it is seen that (see expression (32)) equation (41) represents a system of ordinary differential equations for stresses at selected boundary nodes and internal points which can be solved by standard methods (provided it satisfies the Lipschitz condition), producing a unique solution for the rate dependent problem. A simple and efficient solution procedure for this matrix equation is the subject of the next section.

SOLUTION TECHNIQUE

For the solution of the examples presented in this paper, the simple Euler one-step procedure has been adopted in the following fashion; let us assume a load factor $\lambda(t)$ which is considered to be a known function of time. Equations (38) and (41) can be integrated on time to give

$$\underset{\sim}{x} = \underset{\sim}{R}\,\underset{\sim}{\sigma}^a + \lambda(t)\,\underset{\sim}{m} \qquad (44)$$

and

$$\underset{\sim}{\sigma} = \underset{\sim}{V}\,\underset{\sim}{\sigma}^a + \lambda(t)\,\underset{\sim}{n} \qquad (45)$$

where vectors $\underset{\sim}{m}$ and $\underset{\sim}{n}$ correspond to the elastic solution at some reference load level.

For the time marching procedure, equation (45) is applied after each discrete time step ($\Delta t = {}^{k+1}t - {}^{k}t$) with the value of the initial stresses being computed at the selected boundary nodes and internal points by the Euler's formula

$${}^{k+1}\sigma^a_{ij} = {}^{k}\sigma^a_{ij} + \Delta t\;\gamma <{}^{k}\Phi>\,{}^{k}d_{ij} \qquad (46)$$

During this process one may have that $\lambda(t)$ is left constant for some time, creating a situation in which after a sufficient number of time steps has been applied, the values of $\Delta t\,\dot{\varepsilon}^p_e$ or ${}^{k+1}\sigma_e - {}^{k}\sigma_e$ become vanishingly small everywhere. In such cases, a stationary condition is deemed to have occured and the time marching scheme can be stopped.

It is interesting to note that the time integration procedure does not require computation of the boundary unknowns. Consequently, equation (44) need only be used to print the boundary unknowns at some requested specific time/load values.

The success of this simple time integration scheme is dependent on the proper selection of the time step lengths. It has been known for quite some time [17] that ideally small time steps should be applied in the early stages of the computation (i.e. after the application of the load or load increment) and that these can be increased in size as stationary or steady state is approached.

Following the experience of many authors [17, 18, 7, 19] with different spatial discretization techniques (mainly finite elements), the time step size should be controlled by a relation between accumulated and rate value of some variables to produce the above described automatic lengthening as asymptotic state is achieved. This can be considered at each node or point as follows:

$$\Delta t \leqslant \eta \frac{\varepsilon_e}{\dot{\varepsilon}_e^a} = \frac{\eta}{\dot{\varepsilon}_e^a}\left(\frac{\sigma_e}{E} + \varepsilon_e^a\right) \tag{47}$$

under the condition that

$$^{k+1}\Delta t \leqslant \eta_o \, ^{k}\Delta t \tag{48}$$

where η and η_o are problem dependent parameters that should be chosen to compromise between computer time and accuracy. Normally, $0.01 \leqslant \eta \leqslant 0.15$ and $1.2 \leqslant \eta_o \leqslant 2$.

A drawback of relations (47) and (48) is that they do not guarantee complete stability of the explicit time integration scheme, particularly near to the steady state which produces large time step values. Useful bounds for the maximum time step length have been presented by Cormeau [20] for perfectly viscoplastic materials only. Herein, a simpler, yet general, limiting value for the time step has been adopted [14]. Its expression for work hardening viscoplastic materials is as follows,

$$\Delta t \leqslant \Delta t_{crit.} = \frac{2\psi^2}{\gamma\, \Phi'(E\psi + H'\sigma_e)} \tag{49}$$

where $\Phi' = d\Phi/d(F/\psi)$ and $H' = d\psi/d\varepsilon_e^p$ can be interpreted as the slope of the static uniaxial curve plotted as stress versus plastic strains.

For creep problems the equivalent expression is

$$\Delta t_{crit.} = \frac{2}{K\, Em\, \sigma_e^{m-1}} \tag{50}$$

and if equation (27) is used instead, the term t^n should appear in the denominator producing the same critical time step obtained by Cormeau [20] and Irons [21] when $\nu \rightarrow 0.5$.

Expression (49) has proved to work well not only for ideal viscoplastic materials, but also for hardening or softening material behaviour [14].

APPLICATIONS

An interesting feature of the elastic/viscoplastic theory is that if the load is applied in small increments, allowing for stationary conditions to be achieved after each load step, a pure elastoplastic solution is obtained. The question of how small these increments should be taken, remains an open question and is, in fact, problem dependent. In the first example presented here, this feature is fully explored for solving a current elastoplastic problem. But in the second and third applications, the total load is applied in one step and two problems of the type power law creep and quasilinear viscoplastic are analysed.

i) Deep beam: in the first example the elastoplastic behaviour of a simply supported deep beam under uniform load is studied by the viscoplastic boundary element technique. The discretization employed is shown in figure 1 and the material parameters are as follows:

$e = 30 \times 10^6$ psi

$\sigma_0 = 36 \times 10^3$ psi (uniaxial yield stress)

$\nu = 0.3$

$H' = 0, \left(\text{Tresca yield criterion} - \Phi\left(\frac{F}{\psi}\right) = \frac{F}{\psi}\ ;\ \gamma = 1.\text{sec}^{-1}\right)$

This problem has been analysed by Anand et al. [22] by using a mesh of 272 linear displacement triangular finite elements, which corresponds to 33% more elements on the boundary than the discretization used here.

A comparison of results is depicted in figure 2 where the load-midspan displacement curves, for both numerical techniques, are plotted together with the beam theory solution. As can be seen, the boundary element solution asymptotically approaches the limit load obtained by the beam theory, whereas the finite element results slightly exceed this load level. A vanishing small difference is already noticed in the elastic results, with the BE technique predicting a lower load value for initial yield and larger displacements for the same load level. The plastic zones produced by both techniques were in good agreement with the beam theory and therefore are not shown here.

A further confirmation of the BE results was obtained by solving the same example using the pure elastoplastic BE implementation [3]. Remarkable agreement was then achieved in every aspect of the solution (differences within prescribed tolerance for convergence/stationarity).

ii) Thin Disc: accurate bounds for the creep problem of a thin disc with a central rigid insert under constant external edge loads were produced by Sim [23]. These were obtained by direct time integration of the analytical solution and presented in dimensionless form using the so-called "reference stress" technique. In order to test the boundary element performance in the same problem, the following material parameters were chosen:

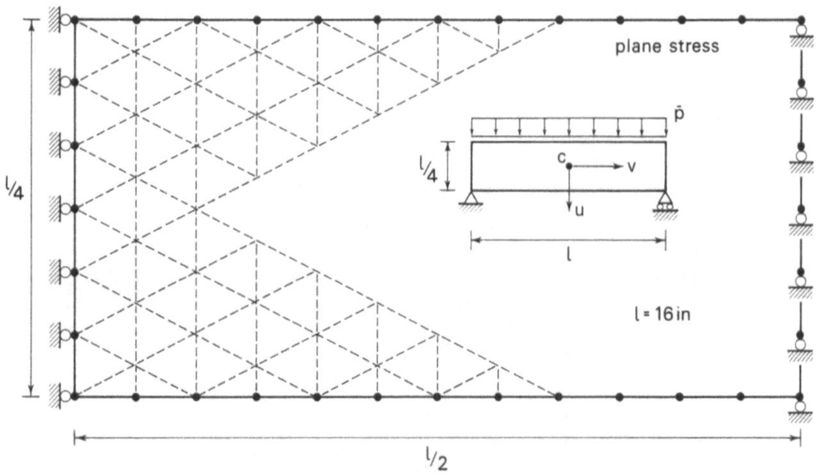

Figure 1. Deep beam elastoplastic problem. Geometry and discretization used for B.E. results.

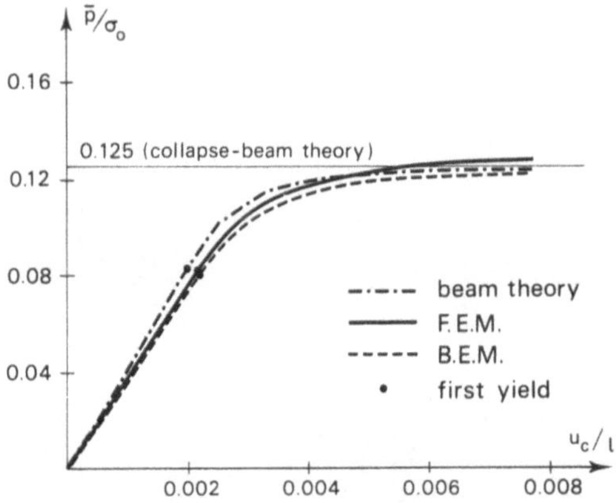

Figure 2. Load-Midspan displacement curves for deep beam problem.

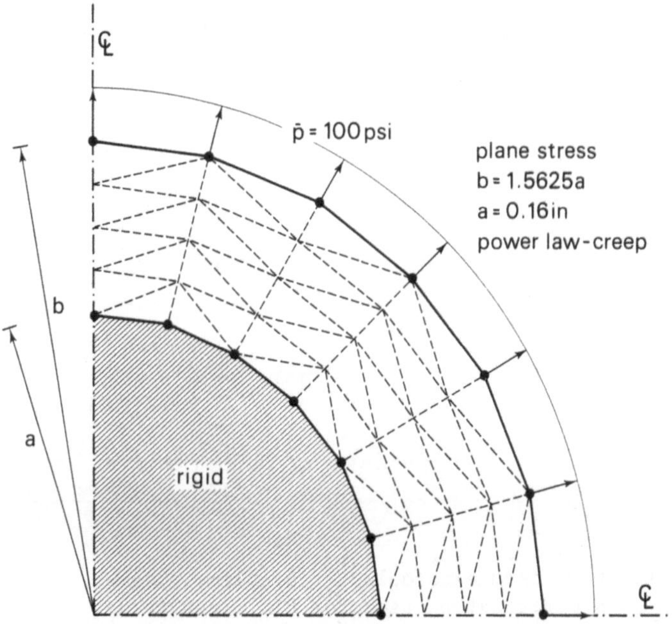

Figure 3. Geometry of thin disc creep problem including boundary element and internal cell discretization.

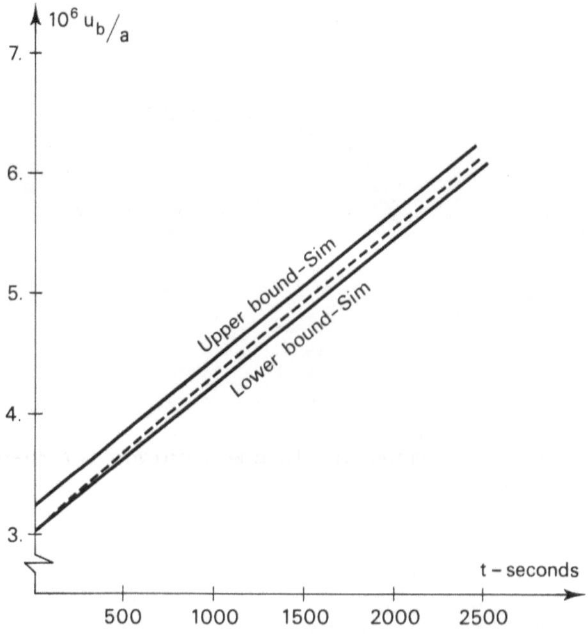

Figure 4. Variation of outer boundary radial displacement with time for thin disc creep problem.

$E = 17 \times 10^6$ psi

$\nu = 0.33$

$\dot{\varepsilon}_e^c = 5.8 \times 10^{-18} \sigma_e^{4.4}$ (units : lb, in and sec.)

The geometry and load value are given in figure 3 where the boundary element and internal cell discretization is also shown. Notice that improved axial symmetry was obtained by avoiding boundary discretization of the symmetry axes.

Radial displacements computed over the outer boundary are plotted against time for comparison with the solution bounds in figure 4. As expected, the boundary element technique produces a flat curve which lies within the narrow space between the two limiting lines taken from the reference. It is interesting to note that the slope of these parallel lines was calculated for an approximate stationary condition in which the variation of the displacement rates was 1%. Consequently, the same stationarity criterion was adopted here, generating the final straight part of the curve.

iii) Plate under thermal shrinkage: in this example the analysis of a rectangular plate, bonded on one edge to a rigid support and subjected to a sudden uniform temperature drop is presented. The thermal shrinkage was assumed to be such that $\varepsilon_{ij}^T = -0.01\ \delta_{ij}$. The problem can be properly solved by prescribing tangential displacements corresponding to $\varepsilon_{ij} = -\varepsilon_{ij}^T$ over the fixed edge and computing the final displacements by simple superposition.

The material was assumed to be quasilinear ($\Phi(F/\psi) = F/\psi$) ideal viscoplastic, obeying the von Mises criterion.

Due to symmetry, only half the plate was discretized using 26 boundary elements and 17 internal points located in the region near to the restrained edge as shown in figure 5. Also included is the plastic zone produced by the instantaneous cooling process.

Finite element results for this problem have been presented by Zienkiewicz and Cormeau [7]. They used a mesh of 96 quadrilateral elements which was equivalent in size to the boundary element discretization over the bonded edge (A-B), but presented more refinement over the opposite edge (C-D).

An interesting comparison of results is depicted in figures 6, 7 and 8 where the stresses computed at the fixed edge are shown for times $t = 0$ (elastic) and $t \to \infty$ when stationary condition is achieved. These include not only the FE and BE results, but also the sufficiently refined elastic finite difference solution produced by Bauer and Reiss [24], which provides a useful reference result for $t = 0$.

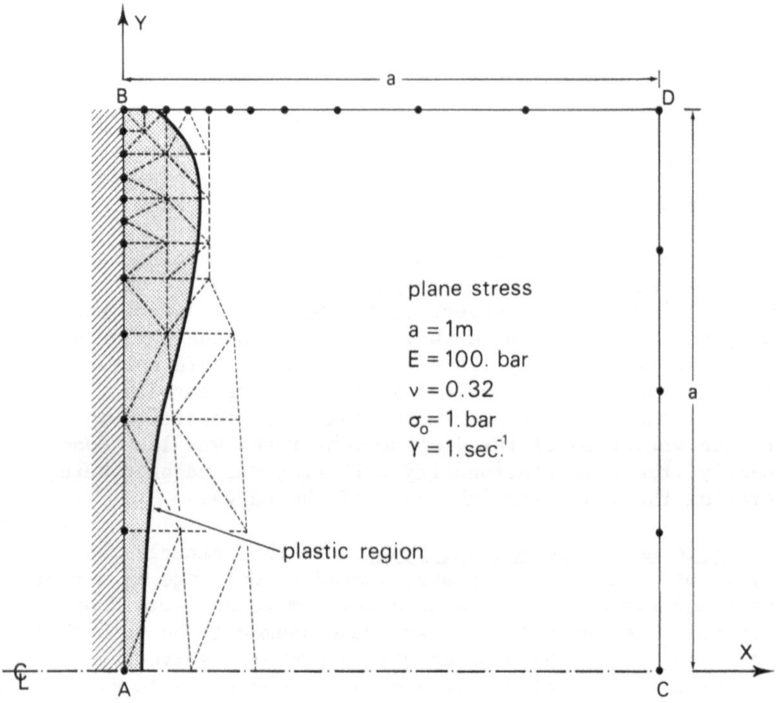

Figure 5. Discretization for rectangular plate under thermal shrinkage and total extent of plastic region.

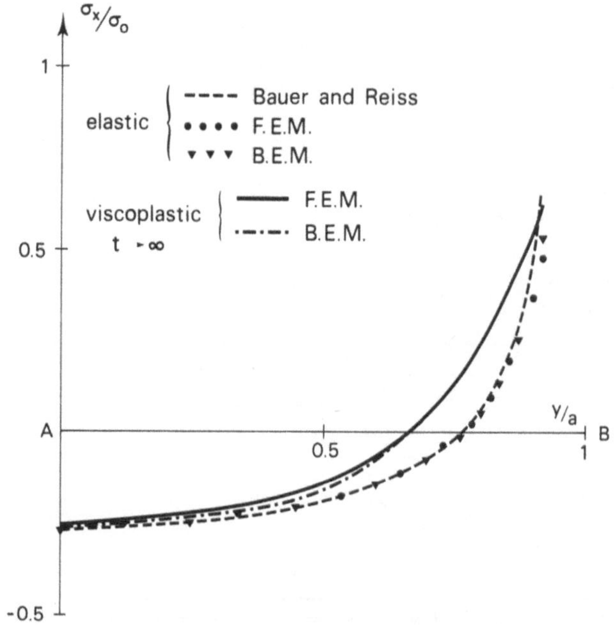

Figure 6. Variation of σ_x over fixed edge for t = 0 and asymptotic state.

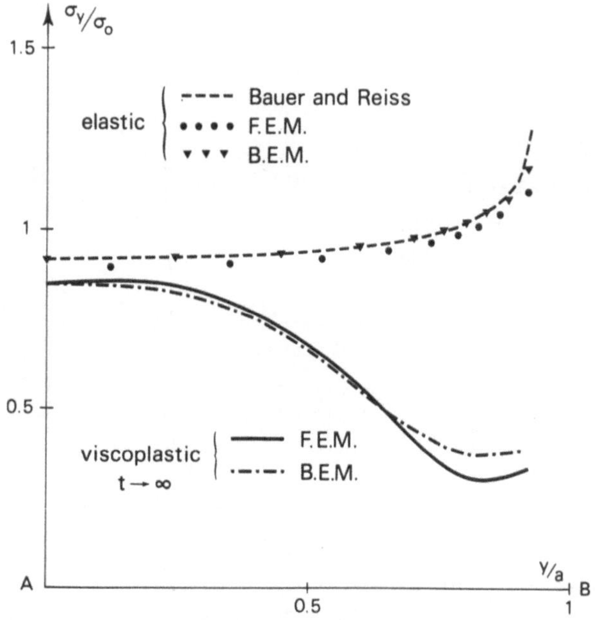

Figure 7. Variation of σ_y over fixed edge for t = 0 and asymptotic state.

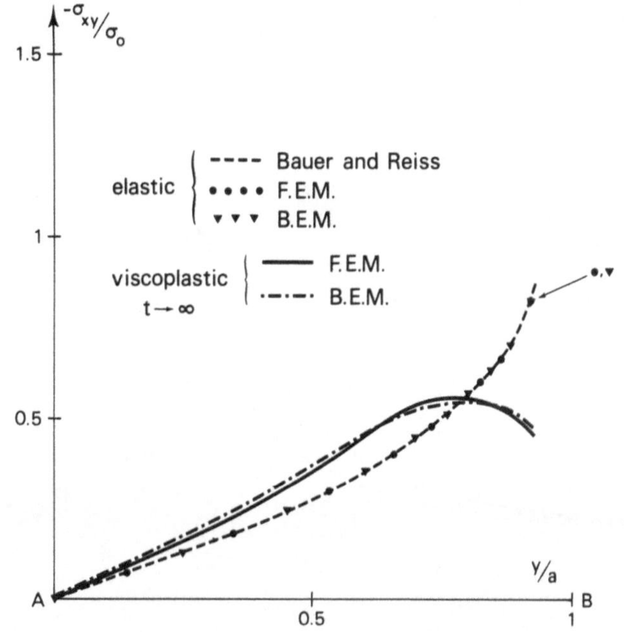

Figure 8. Variation of σ_{xy} over fixed edge for t = 0 and asymptotic state.

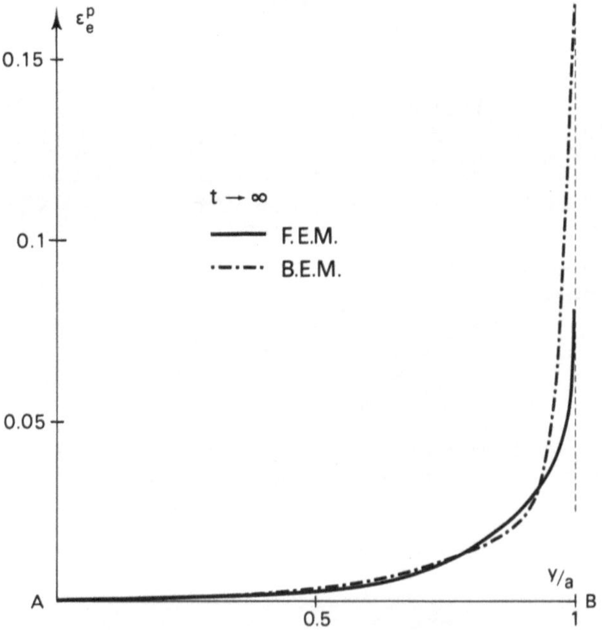

Figure 9. Equivalent plastic strain distribution over the bonded edge at stationary state.

It is worth mentioning that neither method can predict the infinite value of the elastic stresses at the corners of the fixed edge. Consequently, a localized perturbation in the solutions is expected in the vicinity of corner B. Nevertheless, even though we neglect the results near to the singular node, one can notice that the boundary elements tend to produce a better representation of the singular behaviour than the finite elements. This difference may be partly explained by the fact that the FE stresses were calculated at the Gauss points (2 x 2 integration) and is particularly apparent in figure 7 where the σ_y stresses are noticeably unequal over a large range.

Another comparison is presented in figure 9 in which the equivalent plastic strains computed by the boundary element technique are indicating a more severe concentration of plasticity near the corner than the finite element results.

The computer time for the FE solution of this example has been given in reference [27], where a CDC 7600 was used. Herein, all the computer codes were implemented in an IBM 360/195. Thus a comparison of the overall solution times is given in the table below.

Method	Computer	CPU time (sec.)
FE	CDC 7600	20.
BE	IBM 360/195	10.9

It is worth mentioning that for these type of applications the CDC computer is faster than the IBM machine, which makes the difference in equivalent CPU time between the two methods even more pronounced.

CONCLUSIONS

This paper presents a complete boundary element formulation for the solution of viscoplastic problems using the constitutive equations due to Perzyna. The formulation is simple to implement numerically and is capable of handling pure creep problems in the same fashion. In addition, elastoplastic solutions can be simulated through the application of long term load increments, followed by stationary conditions.

The examples discussed illustrate the potentialities of the formulation and demonstrate that efficient solutions for these classes of problems are now possible within the context of the boundary element technique.

REFERENCES

1. C. A. Brebbia and S. Walker, The Boundary Element Techniques in Engineering. Butterworths, London (1980).

2. P. Perzyna, Fundamental Problems in Viscoplasticity. Advan. Appl. Mech. 9, 243 (1966).

3. J.C.F. Telles and C. A. Brebbia, Plasticity, Chapter 5 in Progress in Boundary Element Methods. V. 1, (Edited by C. A. Brebbia), Pentech Press, London (1981).

4. J.C.F. Telles and C.A. Brebbia, The Boundary Element Method in Plasticity. Proc. 2nd Int. Seminar on Recent Advances in Boundary Element Methods, (Edited by C.A. Brebbia), p. 295, University of Southampton, U.K. (1980).

5. J.C.F. Telles and C.A. Brebbia, New Developments in Elastoplastic Analysis. Proc. 3rd Int. Seminar on Recent Advances in Boundary Element Methods, (Edited by C.A. Brebbia), p. 350, Irvine, California (1981).

6. M. Chaudonneret, Méthode des Équations Intégrales Appliquées a la Résolution de Problèmes de Viscoplasticité. J. Méchanique Appliquée, 1, 113 (1977).

7. O.C. Zienkiewicz and I.C. Cormeau, Viscoplasticity and Plasticity, an Alternative for Finite Element Solution of Material Nonlinearities. Proc. Colloque Méthodes Calcul. Sci. Tech., p. 171, IRIA, Paris (1973).

8. S. Mukherjee and V. Kumar, Numerical Analysis of Time-Dependent Inelastic Deformation in Metallic Media using Boundary Integral Equation Method. Trans. ASME, J. Appl. Mech., 45, 785 (1978).

9. M. Morjaria and S. Mukherjee, Improved Boundary Integral Equation Method for Time-Dependent Inelastic Deformation in Metals. Int. J. Num. Meth. Engng., 15, 97 (1980).

10. S. Mukherjee, Corrected Boundary Integral Equation in Planar Thermoelastoplasticity. Int. J. Solids Structures, 13, 331 (1977).

11. A Mendelson, Boundary Integral Equation Methods in Elasticity and Plasticity. Report. No. NASA-TN-D-7418 (1973).

12. H.D. Bui, Some Remarks about the Formulation of Three-Dimensional Thermoelastoplastic Problems by Integral Equations. Int. J. Solids Structures, 14, 935 (1978).

13. J.C.F. Telles and C.A. Brebbia, On the Application of the Boundary Element Method to Plasticity. Appl. Math. Modelling, 3, 466 (1979).

14. J.C.F. Telles and C.A. Brebbia, Elastic/Viscoplastic Problems Using Boundary Elements. Int. J. Mech. Sci., 24, (1982).

15. S.G. Mikhlin, Singular Integral Equations. Amer. Math. Soc. Trans., Series 1, 10, 84 (1962).

16. G.C. Nayak and O.C. Zienkiewicz, Convenient Form of Stress Invariants for Plasticity. Proc. Am Soc. Civ. Engns., J. Struct. div., 98, 949 (1972).

17. R.K. Penny, The Creep of Spherical Shells Containing Discontinuities. Int. J. Mech. Sci., 9, 373 (1967).

18. R.K. Penny and D. R. Hayhurst, The Deformations and Stresses in a Stretched Thin Plate Containing a Hole During Stress Redistribution Caused by Creep. Int. J. Mech. Sci., 11, 23 (1969).

19. W.H. Sutherland, AXICRIP - Finite Element Computer Code for Creep Analysis of Plane Stress, Plane Strain and Axisymmetric Bodies. Nucl. Eng. and Design, 11, 269 (1970).

20. I. Cormeau, Numerical Stability in Quasi-Static Elasto/Viscoplasticity. Int. J. Num. Meth. Engng., 9, 109 (1975).

21. B. Irons and G. Treharne, A Bound Theorem in Eigenvalues and its Practical Applications. Proc. 3rd Conf. Matrix Meth. Struct. Mech. p. 245, Wright-Patterson A.F.B., Ohio (1971).

22. S.C. Anand, S.L. Lee and E.C. Rossow, Finite Element Analysis based upon Tresca Yield Criterion. Ingenieur-Archiv, 39, 73 (1970).

23. R.G. Sim, Reference Results for Plane Stress Creep Behaviour. J. Mech. Engng. Sci., 14, 404 (1972).

24. F. Bauer and E.L. Reiss, On the Numerical Determination of Shrinkage Stresses. Trans. ASME, J. Appl. Mech., March, 123 (1970).

25. J.C.F. Telles and C.A. Brebbia, Boundary Element Solution for Half-Plane Problems. Int. J. Solids Structures, 17, 1149 (1981).

26. F.J. Rizzo and D.J. Shippy, An Application of the Correspondence Principle of Linear Viscoelasticity Theory. SIAM, J. Appl. Math., 21, 321 (1971).

27. I.C. Cormeau, Viscoplasticity and Plasticity in the Finite Element Method. Ph.D. Thesis, University College of Swansea, (1976).

NUMERICAL ANALYSIS OF VISCOPLASTICITY USING THE BOUNDARY ELEMENT METHOD.

M. BRUNET

Laboratoire de Mécanique des Solides - INSA Lyon - Bât. 304
69621 VILLEURBANNE CEDEX - France

INTRODUCTION

If the F.E.M. has been commonly used to solve various problems in plasticity and viscoplasticity, the B.E.M. is a possible alternative to F.E.M. for this class of problems in the two dimensional case and can present some advantages.

In this paper, the proper formulation for two-dimensional elasto-viscoplasticity is presented giving the relations between the initial stress and initial strain case . We can employ the linear boundary element with constant internal cells and the quadratic isoparametric boundary element with linear internal cells which greatly improve the accuracy for the same number of boundary elements. The time integration schemes with a time step control are formulated in matrix form and the type can be explicit or implicit. An implicit time integration scheme is developed using a simple iterative relationship on stresses during a time step.

Several examples are presented to illustrate the method using the well-known general model of elasto-viscoplasticity developed by Zienkiewicz and Cormeau [1] in the context of a finite element discretization and a more sophisticated viscoplastic constitutive law for the description of cyclic and anisotropic behavior of metal developed by Chaboche [2].

BOUNDARY ELEMENT FORMULATION

We start with an elasto-viscoplastic formulation in which traction and body force rates are fictitious (depend of the viscoplastic strain rates). The displacement rate at a boundary point P takes the form

$$C_{ij}\dot{u}_j(P) = \dot{u}_i^o(P) + \int_V U_{ij}(P,q)\dot{f}_j^P(q)dV + \int_{\partial V} U_{ij}(P,Q)\dot{t}_j^P(Q)dS \quad (1)$$

where $\dot{u}_i^o(P)$ stands for the classical elastic solution (see [3]) and U_{ij} is the well known Kelvin'singular solution due to a point load in an infinite elastic solid.

Equation(1) is valid for interior point p with $C_{ij} = \delta_{ij}$. The equilibrium equations for the pseudo body force rates and traction rates are :

$$\dot{\sigma}^P_{ij,i} + \dot{f}^P_j = 0 \qquad (2)$$

$$\dot{\sigma}^P_{ij} n_j = \dot{t}^P_i \qquad (3)$$

due to the fact that the total strain rate is assumed to be represented by :

$$\dot{\varepsilon}_{ij} = \frac{1}{2}(\dot{u}_{i,j} + \dot{u}_{j,i}) = \dot{\varepsilon}^e_{ij} + \dot{\varepsilon}^P_{ij} \qquad (4)$$

and
$$\dot{\sigma}^e_{ij} = C_{ijkl}\dot{\varepsilon}_{kl} = \dot{\sigma}_{ij} + \dot{\sigma}^P_{ij} \qquad (5)$$

where C_{ijkl} is the isotropic tensor of elastic constants which can be written as :

$$C_{ijkl} = \frac{2\,G\nu}{1-2\nu}\,\delta_{ij}\delta_{kl} + G(\delta_{ik}\delta_{jl} + \delta_{il}\delta_{jk}) \qquad (6)$$

As usual, time derivatives are indicated by a dot and space derivatives by a comma.

The substitution of equations (2) and (3) into (1) leads to :

$$C_{ij}\dot{u}_j(P) = \dot{u}_i^o(P) + \int_V U_{ij,k}(P,q)\dot{\sigma}^P_{jk}(q)dV \qquad (7)$$

For plane strain and plane stress we have :

$$U_{ij,k} = \frac{-1}{8\pi(1-\nu)Gr}\left[(3-4\nu)\delta_{ij}r_{,k} - \delta_{ik}r_{,j} - \delta_{jk}r_{,i} + 2r_{,i}r_{,j}r_{,k}\right] \qquad (8)$$

where ν is replaced by $\bar{\nu} = \frac{\nu}{1+\nu}$ for plane stress.

The last term in equation (7) can be rewritten in the following form :

$$\int_V E_{jki}(P,q)\dot{\sigma}^P_{jk}(q)dV \quad (9) \quad \text{where } E_{jki} = \frac{1}{2}\left[U_{ij,k} + U_{ik,j}\right] \qquad (10)$$

due to the symetry of the stress tensor.

The equations (7) and (9) are referred to the initial stress formulation and the initial strain formulation can be obtained as follow :

For the plane stress case (9) is replaced by

$$\int_V \Sigma_{rsi}(P,q)\dot{\varepsilon}^P_{rs}(q)dV \tag{11}$$

where $\Sigma_{rsi} = E_{jki}C_{jkrs}$ (12)

with ν replaced by $\bar{\nu}$ everywhere.

However, for the plane strain case, the initial strain formulation is particularly simplified by considering the viscoplastic strain to be incompressible

$$\dot{\varepsilon}^P_{ll} = 0 \tag{13}$$

thus, expression (9) can be replaced by

$$\int_V \Sigma^*_{jki}(P,q)\dot{\varepsilon}^P_{jk}(q)dV \tag{14}$$

where $\Sigma^*_{jki} = 2\ G\ \ E_{jki}$ (15)

The above expression (1) together with (4) and (5) allow for the determination of the internal stresses in terms of the fictitious traction and body force rates

$$\dot{\sigma}_{ij}(p) = \dot{\sigma}^o_{ij}(p) - \int_V \Sigma_{ijk}(p,q)\dot{f}^P_k(q)dV - \int_{\partial V}\Sigma_{ijk}(p,Q)\dot{t}^P_k(Q)ds$$
$$- \dot{\sigma}^P_{ij}(p) \tag{16}$$

where $\dot{\sigma}^o_{ij}(p)$ stands for the classical elastic solution [3] . Introducing the equations (2) and (3) in (16) and applying Stoke's theorem carefully, we follow the same procedure as H.D. Bui and K. Dang Van in [4] with an initial strain formulation :

$$\dot{\sigma}_{ij}(p) = \dot{\sigma}^o_{ij}(p) - \lim_{\varepsilon\to 0}\int_{\partial B(p,\varepsilon)} \Sigma_{ijk}\dot{\sigma}^P_{kl}\ n_l dS - \lim_{\varepsilon\to 0}\int_{V-B(p,\varepsilon)}\Sigma_{ijk,l}\ \dot{\sigma}^P_{kl}dV$$
$$- \dot{\sigma}^P_{ij}(p) \tag{17}$$

where $B(p,\varepsilon)$ is the sphere of ε radius centred at the singular point p.

Due to the fact that $\int_{\partial B(p,1)}\Sigma_{ijk,l}\ dS = 0$ (18)

the last integral in (17) can be interpreted in the sens of the Cauchy principal value and we obtain for the two dimensional case :

$$\lim_{\varepsilon\to 0} - \int_{\partial B(p,\varepsilon)} \Sigma_{ijk}\, \dot{\sigma}^{P}_{kl}\, n_l dS - \dot{\sigma}^{P}_{ij}(p) = - \frac{1}{8(1-\nu)}\left[2\dot{\sigma}^{P}_{ij} + (1-4\nu)\dot{\sigma}^{P}_{ll}\delta_{ij}\right] \tag{19}$$

$$\text{Noticing : } E_{ijkl} = -\frac{1}{2}\left[\Sigma_{ijk,l} + \Sigma_{ijl,k}\right] \tag{20}$$

we get the initial stress formulation for the internal stress rate.

$$\dot{\sigma}_{ij}(p) = \dot{\sigma}^{o}_{ij}(p) + \int_V E_{ijkl}\dot{\sigma}^{P}_{kl} dV - \frac{1}{8(1-\nu)}\left[2\dot{\sigma}^{P}_{ij} + (1-4\nu)\dot{\sigma}^{P}_{ll}\delta_{ij}\right] \tag{21}$$

which is valid for plane strain and plan stress where ν is replaced by $\bar{\nu}$ everywhere.

The initial strain formulation can be applied with equal ease, the equivalent expression for the plane stress case is

$$\dot{\sigma}_{ij}(p) = \dot{\sigma}^{o}_{ij}(p) + \int_V \Sigma_{ijkl}\, \dot{\varepsilon}^{P}_{kl} dV - \frac{G}{4(1-\bar{\nu})}\left[2\dot{\varepsilon}^{P}_{ij} + \dot{\varepsilon}^{P}_{ll}\delta_{ij}\right] \tag{22}$$

$$\text{where } \Sigma_{ijkl} = E_{ijrs}C_{rskl} \tag{23}$$

and for the plane strain case with the hypothesis of incompressibility

$$\dot{\sigma}_{ij}(p) = \dot{\sigma}^{o}_{ij}(p) + \int_V \Sigma^{*}_{ijkl}\dot{\varepsilon}^{P}_{kl} dV - \frac{G}{4(1-\nu)}\left[2\dot{\varepsilon}^{P}_{ij} + (1-4\nu)\dot{\varepsilon}^{P}_{ll}\delta_{ij}\right] \tag{24}$$

$$\text{where } \Sigma^{*}_{ijkl} = 2\, G\, E_{ijkl} \tag{25}$$

VISCOPLASTICITY RELATIONS

It is now necessary to introduce one or more specific laws defining the viscoplastic strains.

One explicit model which has wide applicability for isotropic material is offered by the following visco-plastic flow rule [1], [5]

$$\{\dot{\varepsilon}^{P}\} = \gamma < \Phi\,(F) > \frac{\partial Q}{\partial\{\sigma\}} \tag{26}$$

We restrict ourselves to associated plasticity situations, in which case $F \equiv Q$ and a sufficiently general expression for viscoplastic strain rate is given by a power law for $\Phi(F)$ such as :

$$\Phi(F) = \left[\frac{F-F_o}{F_o}\right]^N \tag{27}$$

The expression (27) is often used with an exponent N equal to unity and γ is a fluidity parameter. The onset of viscoplastic behaviour is governed by a scalar yield condition of the form

$$F(\sigma) - F_o = 0 \tag{28}$$

in which F_o can be the uniaxial yield stress for Von Mises material for instance. F_o may itself be a function of a hardening parameter $\bar{\varepsilon}^P$ and for a linear strain hardening response, we have :

$$F_o = \sigma_y + H'\bar{\varepsilon}^P \tag{29}$$

where σ_y is the uniaxial yield stress, H'is the slope of the strain hardening portion of the uniaxial stress-strain curve after removal of the elastic strain and $\bar{\varepsilon}^P$ is the current effective viscoplastic strain. The expression (27) can be used to model the Norton power law of creep by assigning the threshold uniaxial yield value F_o to zero.

In order to take account of the complex history effets, as well as for the Bauschinger effect and for the description of cyclic behaviour, J.L. Chaboche [2] has developped a set of constitutive equations which can describe most of the experimentally observed phenomena during creep or cyclic sollicitations. One of these constitutive equations can be written :

$$\dot{\varepsilon}^P_{ij} = \left\langle \frac{\sqrt{S_{ij}S_{ij}} - R(p)}{K} \right\rangle^n \frac{S_{ij}}{\sqrt{S_{ij}S_{ij}}} \tag{30}$$

$$\dot{X}_{ij} = C(A\dot{\varepsilon}^P_{ij} - X_{ij}\,\dot{P}) \quad \text{with} \quad \dot{P} = \sqrt{\dot{\varepsilon}^P_{ij}\dot{\varepsilon}^P_{ij}} \tag{31}$$

where the stress deviator s_{ij} is divided into the pseudo stress deviator X_{ij} (internal state parameter) and the active stress deviator S_{ij}.

$$S_{ij} = s_{ij} - X_{ij} \tag{32}$$

The material is viscoplastically incompressible and the anisotropic hardening is only described by the additional stress X_{ij} in a kinematical hardening form as those used in plasticity.

The one dimensional form of the constitutive equations is :

$$\dot{\varepsilon}^P = \left\langle \frac{|\sigma - x| - R_o}{K_o} \right\rangle^n \frac{(\sigma - x)}{|\sigma - x|} \tag{33}$$

$$\dot{x} = C_o(a_o \dot{\varepsilon}^P - x|\dot{\varepsilon}^P|) \tag{34}$$

where n, K_o, R_o, C_o and a_o are five parameters characterizing each material obtained by trial and error from uniaxial experimental data.

Comparison with the one-dimensional case shows that

$$R = \sqrt{\frac{2}{3}}\, R_o \;;\; K = K_o \sqrt{\frac{2}{3}}^{\,1+1/n} \;;\; A = \sqrt{\frac{2}{3}}\, a_o \quad \text{and} \quad C = \sqrt{\frac{2}{3}}\, C_o \tag{35}$$

BOUNDARY ELEMENT DISCRETIZATION

The boundary of the region is discretized into a series of elements which can be linear or parabolic (iso-parametric) in the program developed. The boundary integral are carried out using a numerical integration scheme and the principal values are calculated by applying rigid body movements to the system of equations, assuming zero body forces and zero viscoplastic strains.

The interior of the region where viscoplasticity is likely to occur is discretized into a number of quadrilateral cells assuming constant or linear interpolation functions. Non singular cases are evaluated accurately using normal integration rules. The volume integration of the displacement equation in the singular case can be done by introducing a new variable of integration such that the Jacobian is O(r) and cancels the 1/r singularity in the kernel. As it was shown in [6], the principal value of the volume integral in the stress equation can be computed by applying constant viscoplastic strain fields. We give the following matrix expressions for the initial strain formulation which is more suitable than the initial stress formulation for viscoplastic analysis

$$[H]\{\dot{u}\} = [G]\{\dot{t}\} + [D]\{\dot{\varepsilon}^P\} \tag{36}$$

where matrices [H] and [G] are the same obtained for elastic analysis and matrix [D] is due to the viscoplastic strain rate integral of equations (11) or (14). After applying the boundary conditions the system (36) can be reordered and we obtain :

$$[A]\{\dot{x}\} = [D]\{\dot{\varepsilon}^P\} + \{\dot{f}\} \tag{37}$$

Premultiplying (37) by $[A]^{-1}$ gives

$$\{\dot{x}\} = [K]\{\dot{\varepsilon}^P\} + \{\dot{b}\} \tag{38}$$

where $\{\dot{x}\}$ contains the unknown displacement and traction rates and the vector $\{\dot{b}\}$ represents the elastic solution to the boundary problem rate.

In the same way, we obtain the internal stress rate matrix relationship

$$\{\dot{\sigma}\} = [G']\{\dot{t}\} - [H']\{\dot{u}\} + [D' + C']\{\dot{\varepsilon}^P\} \tag{39}$$

or $$\{\dot{\sigma}\} = - [A']\{\dot{x}\} + \{\dot{f}'\} + [D'+C']\{\dot{\varepsilon}^P\} \tag{40}$$

in which $[G']$ and $[H']$ are the same as those obtained for elastic analysis. $[C']$ represents the independant terms in the equations (22) or (24) and $[D']$ stands for the viscoplastic strain rate integral.

Substituting (38) into (40) gives

$$\{\dot{\sigma}\} = [B]\{\dot{\varepsilon}^P\} + \{\dot{d}\} \tag{41}$$

EXPLICIT AND IMPLICIT ALGORITHMS

The initial distribution of internal stresses and boundary tractions and displacements is obtained from the solution of the corresponding elastic problem using the equations (38) and (40) where $\{\dot{\varepsilon}^P\} = 0$. The load increments are applied as discrete steps and thus $\{\Delta f_n\} = 0$ for all time steps other than the first within an increment.

a) Starting from known values of $\{x_n\}$, $\{\sigma_n\}$, $\{\varepsilon_n^P\}$ at a time instant t_n compute the rate of viscoplastic strain $^n\{\dot{\varepsilon}_n^P\}$ by the constitutive relations (26) or (30).

b) We get the change in $\{\varepsilon^P\}$ as

$$\{\Delta\varepsilon_n^P\} = \{\dot{\varepsilon}_n^P\}\ \Delta t_n \tag{42}$$

and $$\{\varepsilon_{n+1}^P\} = \{\varepsilon_n^P\} + \{\Delta\varepsilon_n^P\} \tag{43}$$

c) Compute the value of the boundary unknowns

$$\{\Delta x_n\} = [K]\{\Delta\varepsilon_n^P\} \tag{44}$$

$$\{x_{n+1}\} = \{x_n\} + \{\Delta x_n\} \tag{45}$$

d) Determine the internal stresses

$$\{\Delta\sigma_n\} = [B\ ]\{\Delta\varepsilon_n^P\} \tag{46}$$

$$\{\sigma_{n+1}\} = \{\sigma_n\} + \{\Delta\sigma_n\}$$

We arrive at all the starting values of the next step for which the same process can be started in the next time interval. In order to improve the accuracy of this explicit time stepping algorithm which is equivalent to the Euler extrapolation procedure, an inner iterative loop can be imposed on the steps (a) to (d) where step (b) is replaced by calculating :

$$\{\Delta\varepsilon_n^P\} = (\dot{\varepsilon}_n^P + \dot{\varepsilon}_{n+1}^P)\Delta t_n/2 \tag{48}$$

taking initially

$$\{\dot{\varepsilon}_{n+1}^P\} = \{\dot{\varepsilon}_n^P\} \tag{49}$$

and repeating the process to (d) until some convergence criteria is reached using the second viscoplastic strain invariant for instance.

An other point of vue is to define $\{\dot{\varepsilon}_{n+1}^P\}$ by using a limited Taylor series expansion such as : [5]

$$\{\dot{\varepsilon}_{n+1}^P\} = \{\dot{\varepsilon}_n^P\} + [H_n]\{\Delta\sigma_n\} \tag{50}$$

where

$$[H_n] = \frac{\partial\{\dot{\varepsilon}_n^P\}^T}{\partial\{\sigma_n\}} = [H_n(\sigma_n)] \tag{51}$$

The matrices $[H_n]$ depend on the stress level $\{\sigma_n\}$ and can be easily evaluated with the constitutive laws (26) or (30). The viscoplastic strain increment can be written as

$$\{\Delta\varepsilon_n^P\} = \Delta t_n[(1-\theta)\{\dot{\varepsilon}_n^P\} + \theta\{\dot{\varepsilon}_{n+1}^P\}] \tag{52}$$

For $\theta = 0$ we obtain the Euler time integration scheme, the case $\theta = 1/2$ results in the so-called implicit trapezoïdal scheme and $\theta = 1$ gives the fully implicit scheme, we get :

$$\{\Delta\varepsilon_n^P\} = \{\varepsilon_n^P\}\Delta t_n + \theta[H_n]\Delta t_n\{\Delta\sigma_n\} \tag{53}$$

Using the incremental form (46) and substituting for $\{\Delta\varepsilon_n^P\}$ from (53), then (46) becomes

$$\{\Delta\sigma_n\} = [B]\{\dot{\varepsilon}_n^P\}\Delta t_n + \theta[B][H_n]\Delta t_n\{\Delta\sigma_n\} \tag{54}$$

Equation (54) can be solved iteratively for the increment $\{\Delta\sigma_n\}$ and the iterations go on until some convergence is achieved using the ratio test

$$\left|\frac{||\Delta\sigma_n||_\infty^k - ||\Delta\sigma_n||_\infty^{k-1}}{||\Delta\sigma_n||_\infty^{k-1}}\right| < 10^{-3} \tag{55}$$

As a result, it has been found from numerical experience that this implicit scheme is convergent and stable for even large size time steps. However, some form of instability can occur for $\theta = 1$ with a strain hardening slope parameter $H' = 0$.

Following the method used by O.C. Zienkiewicz and I.C. Cormeau the magnitude of the time step is controlled by a factor τ which limits the maximum effective viscoplastic strain increment $\Delta\bar{\varepsilon}_n^p$ as a fraction of the total effective strain $\bar{\varepsilon}_n$, so that :

$$\Delta\bar{\varepsilon}_n^P = \sqrt{\frac{2}{3}\dot{\varepsilon}_{ij}^P\dot{\varepsilon}_{ij}^P}\,\Delta t_n \leqslant \tau\,\bar{\varepsilon}_n \tag{56}$$

where Δt_n must be computed to satisfy (56) at each internal point and the least value taken for computation. The above time step limiting method is basically empirical and τ in the range 0.01 to 0.15 is found to be effective for explicit algorithms and for implicit schemes values of τ up to 2 have been found to be stable with the B.E.M. A further limit is generally imposed where the change in the time step length between any two intervals is limited according to :

$$\Delta t_{n+1} \leqslant 1.5\,\Delta t_n \tag{57}$$

NUMERICAL EXAMPLES

The first example is the elasto-viscoplastic deformation of a thick tube under the action of internal pressure loading with the exterior surface remaining free with plane strain condition being assumed in the axial direction. We use 30 linear boundary elements with 27 constant quadrilateral cells in order to show that even for this quite coarse internal cell distribution, the steady state displacement is seen to be in good agreement with the F.E.M. solution given by Owen and Hinton [5] using a mesh of 102 degrees of freedom instead of 68 degrees of freedom in the B.E.M. computation. As shown, the implicit trapezoïdal time stepping scheme $\theta = 1/2$ with the stress iteration algorithm is more accurate than the explicit scheme but the solution by the implicit method increases the computation time by a factor of 2.8 in comparison with the explicit approach for the same tolerance factor $\tau = 0.05$. However, we can take greater time step lenghts in the implicit method but with a strain hardening slope parameter $H' = 0$, divergence in the iterations can occur.

The second example shows that viscoplastic B.E.M. is very suitable when a severe spatial stress gradient occurs such that at a notch root of an elliptic cutout where the ratio of axes of the ellipse is 4 which leads to an elastic stress concentration factor of about 10 (see ref. [7]). Very accurate results are obtained by using 23 isoparametric parabolic boundary elements and 32 linear quadrilateral cells concentrated at the notch root

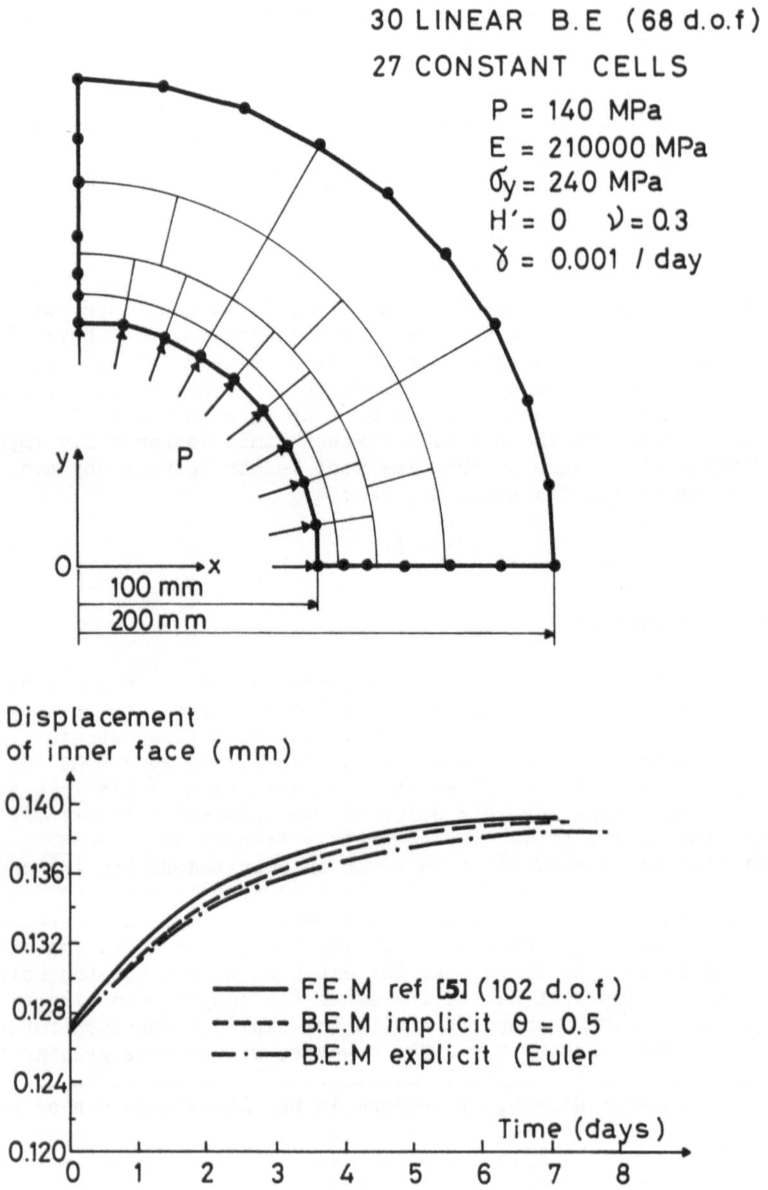

FIGURE 1 INTERNALLY PRESSURISED CYLINDER

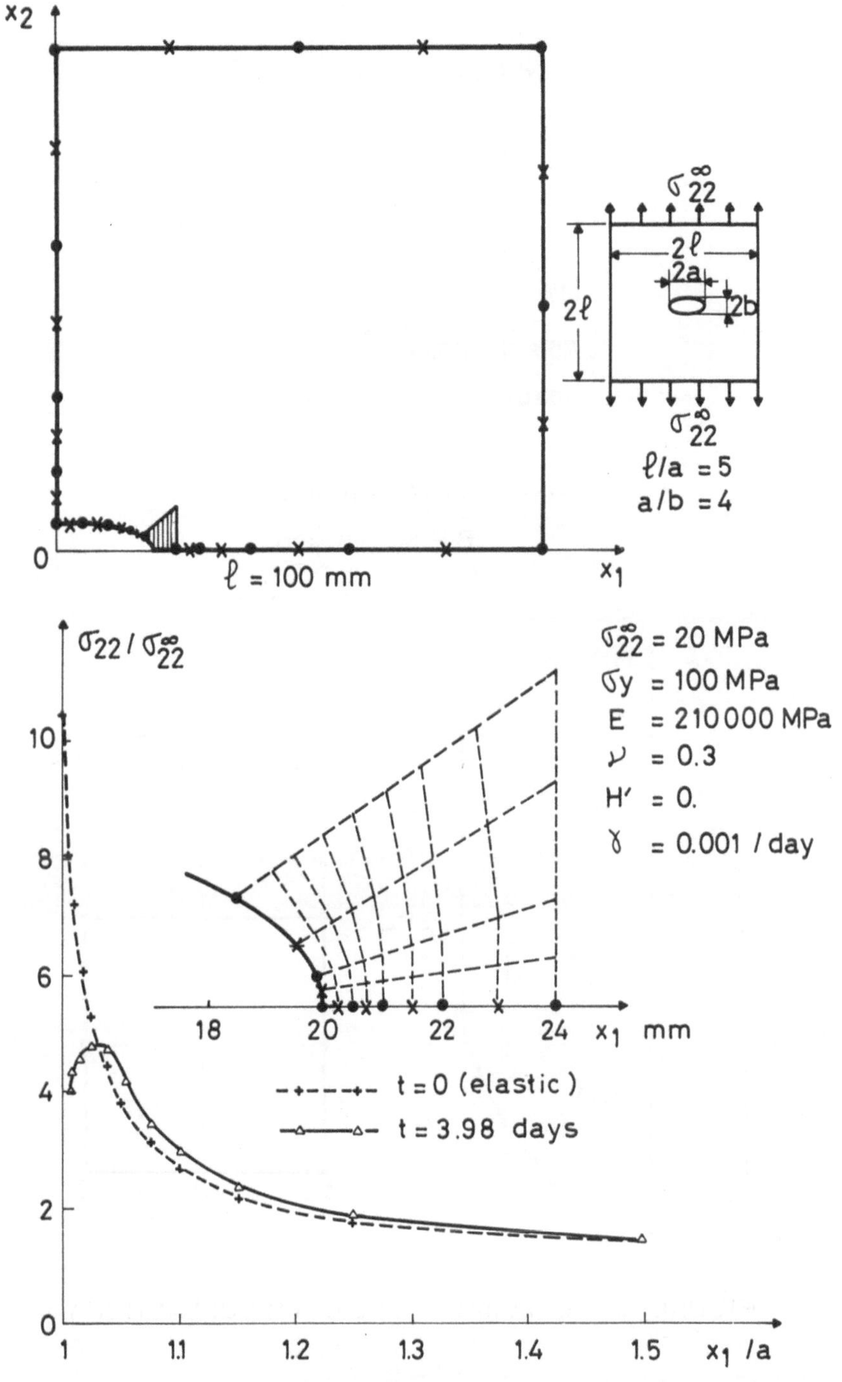

FIGURE 2 STRESS REDISTRIBUTION (PLANE STRESS)

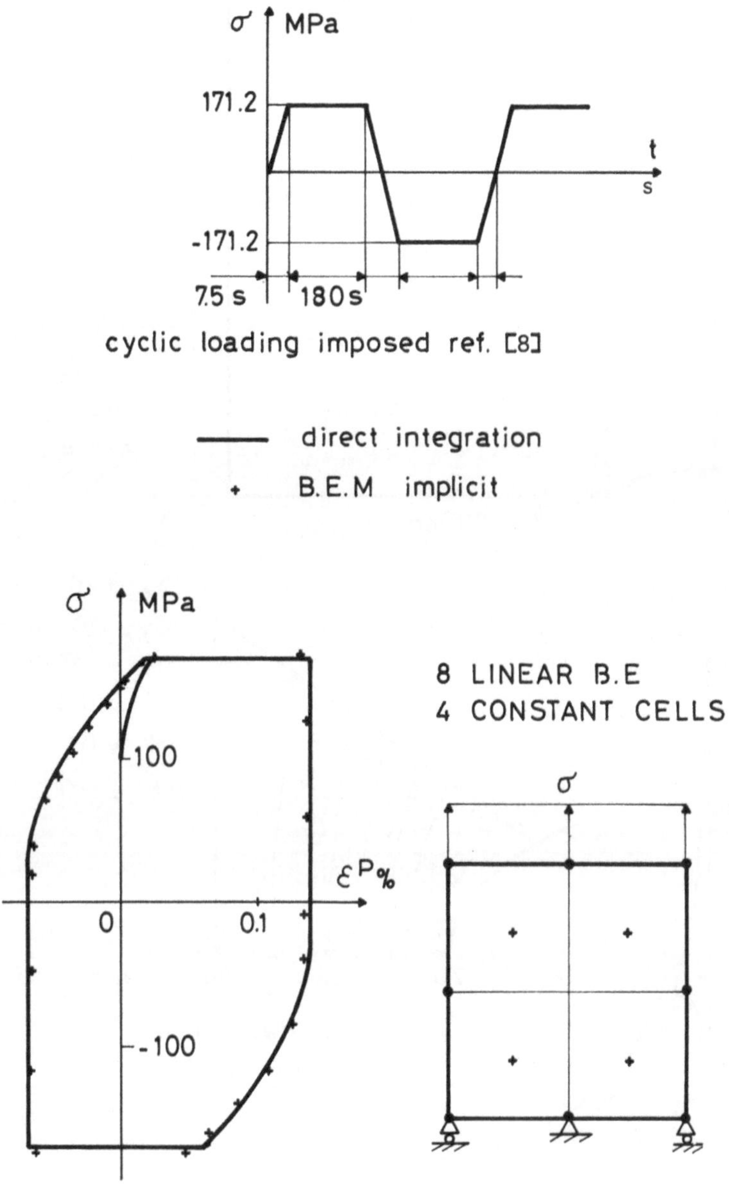

FIGURE 3 UNIAXIAL CYCLIC VISCOPLASTICITY

in a small region where viscoplastic strains are expected to develop. Using the same constitutive law as the previous example but in a plane stress case, the redistribution of stress with time on the axe Ox_1 is shown in figure 2 using an implicit time marching scheme.

Results obtained by the B.E.M. method for one dimensional cyclic viscoplastic uniaxial tests are compared with those from direct numerical integration of equations (33) and (34). The parameters of the constitutive law used are those of IN 100 alloy at 1000°C (ref [8]) (Unities : daN,mm,s) $R_o = 0$; $K_o = 45.5$ $C_o = 690$; $a_o = 14$; $n=8$.

The comparison with 8 linear boundary elements and 4 constant internal cells are excellent using the complete iterative semi-implicit scheme.

CONCLUSIONS

Efficient solution of viscoplasticity problems with material behaviour modelled by several realistic constitutive laws has been presented using either explicit time integration scheme or two kinds of iterative implicit algorithms. An improve computational approach is proposed with quadratic boundary elements and linear internal cells particularly for hight stress gradient problems. It appears from the numerical results of this paper that the B.E.M. is a competitive computational procedure for the efficient solutions of such problems.

REFERENCES

1 O.C. Zienkiewicz and I.C. Cormeau : Visco-plasticity, plasticity and creep in elastic solids. A unified numerical solution approach. Int. J. Num. Meth. vol. 8, 821-845 (1974)

2 J.L. Chaboche : Viscoplastic constitutive equations for the description of cyclic and anisotropic behaviour of metals Bull. Acad. Polonaise des Sciences, vol. 25, n° 1 (1977)

3 J.C.F. Telles and C.A. Brebbia : Elasto-plastic Boundary element analysis. U.S. Europe Workshop Ruhr University Bo chum, West Germany, July 28-31 (1980)

4 H.D. Bui and K. Dang Van : sur le problème aux limites en vitesse des contraintes du solide elasto-plastique - Int. J. Solids Struct. vol. 6, p. 183-193 (1970)

5 D.R.J. Owen and E. Hinton : Finite element in Plasticity Theory and Practice. Pineridge Press (1980)

6 J.C.F. Telles and C.A. Brebbia : On the application of the boundary element method to plasticity. Appl. Math. Modelling, vol. 3, December (1979)

7 M. Morjaria and S. Mukherjee : Improved Boundary Integral Equation Method for time dependent Inelastic deformation in Metals. Int. J. Num. Meth. Engng - vol. 15 - 97-111 (1980)

8 M. Chaudonneret : calcul des concentrations de contraintes en elastoviscoplasticité
Publication ONERA n° 1 (1978).

CONSOLIDATION ANALYSIS BY BOUNDARY ELEMENT METHOD

G. ARAMAKI, T. KUROKI, and K. ONISHI

Civil Engineering Dept., Saga University, Honjo, Saga 840, Japan: Civil Engineering Dept., Fukuoka University, Fukuoka 814-01, Japan: Applied Mathematics Dept., Fukuoka University.

INTRODUCTION

From the engineering point of view, it is often necessary to predict the settlement of an embankment over layered soils. In boundary element concept, following questions arise: (1) Which should be applied, coupled scheme or uncoupled scheme between seepage flow and soil deformation? (2) Can we evaluate the effect of layered soils, especially of very thin layer? (3) How can we characterize material properties in numerical models by using data available from experimental observation?

Predeleanu[1981] discussed the boundary integral method for Biot's consolidation theory. He suggested that important simplifications may be introduced if supplementary hypothesis is adopted concerning the uncoupled interaction between liquid and solid phases.

One of the advantages of boundary element method over finite element method is that the method of subregions takes automatically the continuity of both pore pressure and mass flux between adjacent layers into consideration. Thin layer may be introduced with no further information except for the permeability and thickness.

The purpose of this paper is to develope the

uncoupled boundary element method for Biot's linear theory with layerd soils. We shall show the validity of the scheme by specific examples.

Boundary integral equations are formulated by using a time dependent fundamental solution. The flow equation is expressed in terms of pore pressure and the rate of total volumetric stress. Integral equations are descretized by linear boundary elements for pore water pressure and by constant boundary elements for both displacement and stress. Initial pore pressure distribution is evaluated from the force equilibrium equations using iteration technique.

Numerical examples in two dimensions include (1) comparison of uncoupled boundary element solutions with coupled finite element solutions for zoned inhomogeneous layers, (2) effect of horizontal thin layer on the fluid flow, and (3) boundary element solutions versus an experimental observation.

BOUNDARY ELEMENT METHOD

Boundary integral equations

Let Ω denote an isotropic homogeneous porous medium enclosed by the boundary Γ . We assume that the medium is two-dimensional, elastic, and it is completely saturated with compressible pore fluid. We also assume that the flow obeys Darcy's law.

From the discussion in Kuroki et al. [1982], we can obtain the boundary integral equations for the flow problem

$$c(P_i)v(P_i,t_n) - \int_\Gamma d\Gamma \int_{t_{n-1}}^{t_n} vq^*dt = - \int_\Gamma d\Gamma \int_{t_{n-1}}^{t_n} qv^*dt + \int_\Omega [vv^*]_{t_{n-1}} d\Omega + \int_\Omega d\Omega \int_{t_{n-1}}^{t_n} v^*\gamma\dot{\sigma}_{\mathrm{vol}}dt \qquad (1)$$

and for the elastic problem

$$c(P_i)u_k(P_i,t_n) + \int_\Gamma p_j^k\, u_j\, d\Gamma \qquad (2)$$

$$= \int_{\Gamma} p_j \, u_j^k \, d\Gamma - \int_{\Omega} v\delta_{ij} \, \varepsilon_{ij}^k \, d\Omega \quad .$$

Here v is the pressure (Pascals), q is the Darcy flux (in meters per second) in the outward normal direction with its components n_i (i=1,2), $\dot{\sigma}_{vol}$ is the rate of total volumetic stress (Pascals per second), u_k is the component of displacement (in meters), p_j is the component of surface traction (Pascals), P_i is an arbitrary spatial point, and t_n are break points in time (n=1,2,3,..., in seconds). The weight c is given by the formula $c = \Theta/2\pi$ for the solid angle Θ according to the geometry around the point P_i. Functions q*, p_j^k, and ε_{ij}^k, are derived from wellknown fundamental solutions, v* and u_j^k. The constants are defined as $\gamma = 1/(1+n\beta K)$ for the porosity n, the compressibility β of the pore fluid (in square meters per Newton), and for the compression modulus K (Pascals), and δ_{ij} is the Kronecker's delta.

The principle of effective stresses is written in the form

$$\sigma_{ij} = \sigma'_{ij} + v\delta_{ij} \tag{3}$$

The internal effective stresses can be expressed in terms of displacements and tractions along the boundary in the form

$$\begin{aligned} \sigma'_{kl} = & - \int_{\Gamma} u_j S_{jkl} d\Gamma + \int_{\Gamma} p_j D_{jkl} d\Gamma - \\ & - \frac{1}{2} \int_{\Gamma} v\delta_{ij}(D_{jkl}n_i + D_{ikl}n_j) \, d\Gamma + \\ & + \frac{1}{2} \int_{\Omega} \delta_{ij}(v_{,i}D_{jkl} + v_{,j}D_{ikl}) \, d\Omega \end{aligned} \tag{4}$$

in which kernels, S_{jkl} and D_{jkl}, were given in Brebbia[1978].

Boundary element approximation

We divide the boundary Γ into boundary elements. We also divide the domain Ω into finite elements or cells. The rate of volumetric total stress in Equation (1) is assumed to be constant over each cell and is simply approximated by the backward difference

$$\dot{\sigma}^{n-1}_{vol} = \frac{1}{\Delta t} \left(\sigma^{n-1}_{vol} - \sigma^{n-2}_{vol} \right) \quad . \tag{5}$$

By interpolating the pressure and flux linearly in terms of nodal values, and by assuming constant values for displacement and traction on each boundary element, Equations (1) and (2) can be discretized respectively in matrix forms

$$[H]\{v^n\}_\Gamma = [G]\{q^n\}_\Gamma + \{b(v^{n-1}_\Omega, \dot{\sigma}^{n-1}_{vol})\} \tag{6}$$

and

$$[\hat{H}]\{u^n\}_\Gamma = [\hat{G}]\{p^n\}_\Gamma + \{\hat{b}(v^n_\Omega)\} \quad . \tag{7}$$

The b-column vectors depend on the arguments indicated.

Computational procedures

Uncoupled computational scheme is devised between flow problem and elastic problem. Time marching scheme for the solution to Equations (6),(7) is divided into two parts: estimation of the initial pore pressure distribution and step by step integration. The complete algorithm is described as follows.

i. Initial Calculations:

Set $v^0 = 0$ in Equation (7).
Iterate the next until satisfied:
| Solve Equation (7) to find
| unknown displacements and
| tractions on the boundary.
| Calculate stresses σ_{ij} at
| internal points by Equation (4).
| Set new $v^0 = (\sigma_{11} + \sigma_{22})/2$.
Now the initial pore pressure v^0 is

obtained.
Set $\dot{\sigma}^0_{vol} = 0$ in Equation (6).

ii. Time Marching Algorithm:

For n = 1,2,3,..., until satisfied, do:
| Solve Equation (6) to find
| the boundary pressure v^n.
| Calculate pressures at internal
| points using Equation (1).
| Solve Equation (7) to find
| displacements u^n and tractions p^n
| on the boundary.
| Calculate internal stresses using
| Equation (4).
Next n.

It was ascertained from numerical experiments that the iteration in the above initial calculations converges rapidly for the typical two-dimensional examples. In practice two iterations are satisfactory for the initial setting of the pressure distribution.

EXAMPLES

Inhomogeneous layered soil

An example was considered to show the validity of the present algorithm applied to zoned inhomogeneous media. Figure 1 shows the geometry of two layers under an embankment together with element mesh. By taking the advantage of symmetry, the right half of the solution domain was illustrated. The domain was chosen large enough to make the side effect negligible due to truncation of the domain extending to infinity. The embankment was assumed to be flexible and its weight is transmitted to the soil layer. The time increment was taken as Δt = 10 days to retain the accuracy.

Water is drained from the top and bottom surface. The water is undrained along the line of symmetry and the right side surface. The displacement is fixed in the y direction on the bottom surface and fixed in the x direction on both sides. The rest remains free. The trapezoidal embankment of 5 (m) in height and 1.76E04 (N/m^3) in weight is settled.

Coefficients chosen are summarized in Table 1. We denote the elastic modulus by E (Pascals),

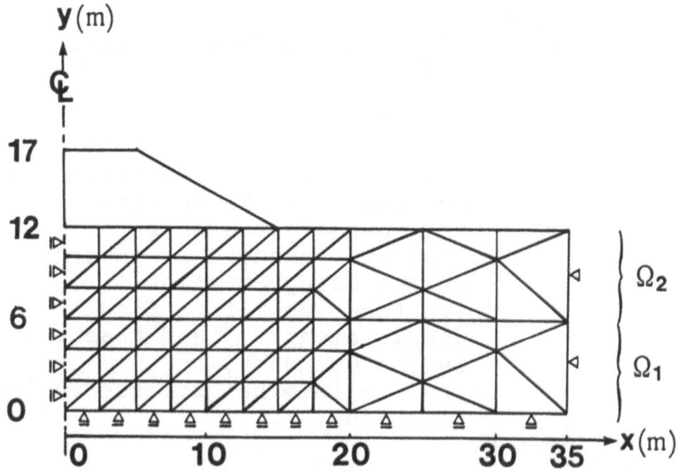

Figure 1. Two - layered soil

Poisson's ratio by ν, the permeability by $\varkappa$ (in square meters), and the dynamic viscosity by μ (in kilograms per meter, second).

Table 1. Material properties

Coefficients	Layers 1	Layers 2
E (Pa)	0.490 E+06	0.980 E+06
K (Pa)	0.480 E+06	0.961 E+06
ν	0.330	0.330
$\varkappa/\mu$ (m^4/Ns)	0.353 E-05	0.265 E-06
$n\beta$ (m^2/N)	0.501 E-09	0.125 E-09

Calculated results were compared with finite element solutions obtained by using Verruijt's FORTRAN codes [1977] for plane strain consolidation with triangular linear finite elements.

Figure 2 shows calculated subsidences of the top

surface and the interface after 100 days. Good agreement is achieved between uncoupled boundary element solutions and coupled finite element solutions. The same is true for solutions of pore water pressure distribution at small times as shown in Figure 3.

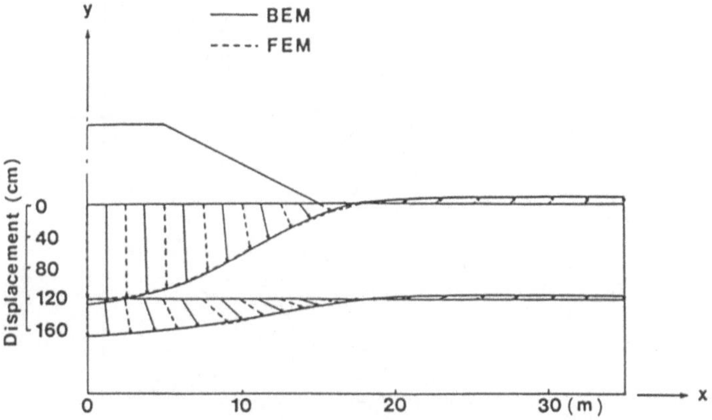

Figure 2. Calculated subsidences after 100 days

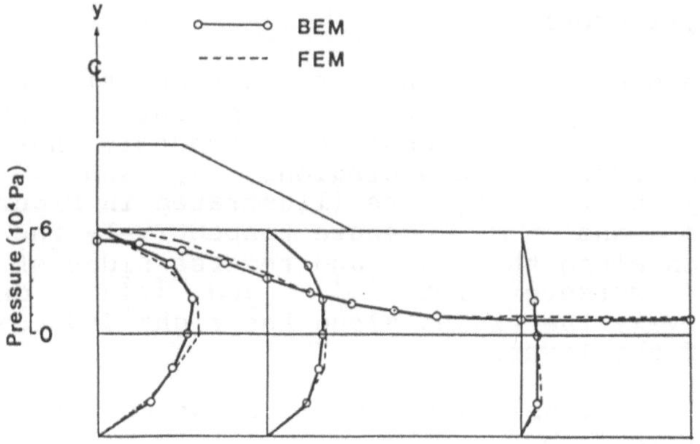

Figure 3. Pore water pressure distribution after 10 days

Figure 4 shows the decreasing water pressure on the

left vertical boundary. The calculated pore water pressure by the boundary element method tends to reduce faster than the pressure by the finite element method does.

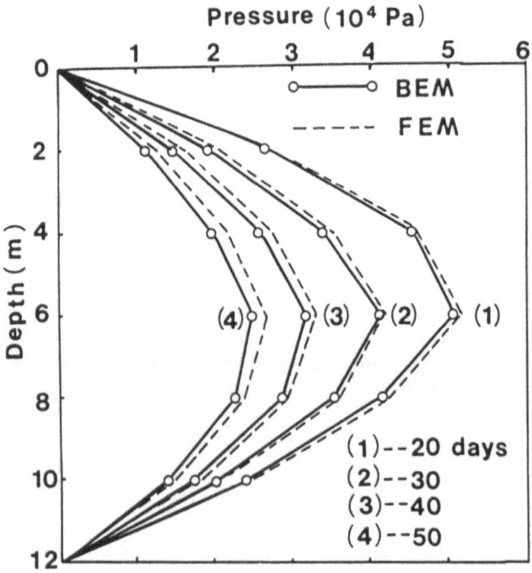

Figure 4. Calculated pressure profiles

Thin layer model

The method of subregions is particularly useful when we consider very thin layers contained in the zoned inhomogeneous foundation. Suppose that the interface between two subregions, Ω_1 and Ω_2, is actually the thin layer as illustrated in Figure 5. Let $\{u_I^1\}$ and $\{u_I^2\}$ denote respectively the pore pressures along the right and reverse sides of the layer. Moreover let $\{q_I^1\}$ and $\{q_I^2\}$ denote respectively the fluxes along the right and reverse sides of the layer.

For the flow problem, we consider the continuity of fluxes written by

$$\{ q_I^1 \} + \{ q_I^2 \} = \{ 0 \} \tag{8}$$

and the Darcy's law expressed by

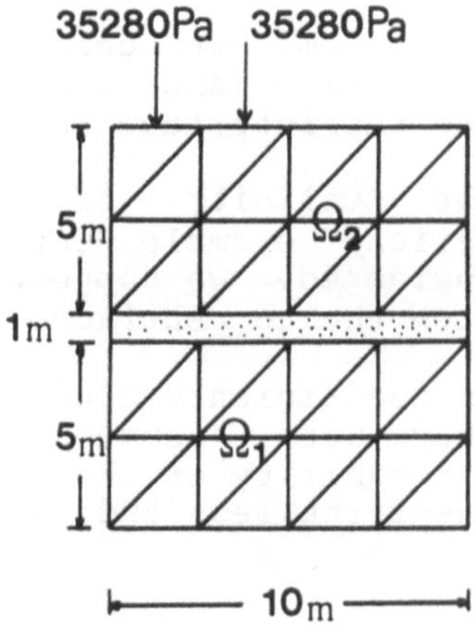

Figure 5. Thin layer model

$$\{ q_I^2 \} = - \frac{\lambda}{h}(\{ u_I^1\} - \{ u_I^2 \}) \tag{9}$$

in which $\lambda = \varkappa/(n\beta\mu)/(1+1/(n\beta k))$, and h is the thickness (in meters) of the thin layer.

If we combine two boundary element equations derived in Ω_1 and Ω_2 by taking Equations (8) and (9) into account, we can obtain the linear system

$$\begin{vmatrix} H^1 & H_I^1 & 0 \\ 0 & H_I^2 & H^2 \end{vmatrix} \begin{vmatrix} u_1 \\ u_I^2 \\ u_2 \end{vmatrix}$$

$$= \begin{vmatrix} G^1 & G_I^1 - \frac{h}{\lambda} H_I^1 & 0 \\ 0 & - G_I^2 & G^2 \end{vmatrix} \begin{vmatrix} q_1 \\ q_I^1 \\ q_2 \end{vmatrix} + \begin{vmatrix} b_1 \\ b_2 \end{vmatrix} . \tag{10}$$

As the permeability $\varkappa$ tends to infinity, we know that λ also becomes infinity, and that Equation (10) will become identical with two-region equations. In the limit the continuity of pore pressure will be recovered.

As to the elastic problem, we consider both the condition of equilibrium and the condition of compatibility along the interface as usual.

To demonstrate the validity of the present algorithm, a numerical example for the model in Figure 5 was reconsidered. We suppose that water is drained only from the top surface.

Table 2 summarizes coefficients used for regions, Ω_1 and Ω_2 . They were assumed to be the same materials. The permeability of the thin layer is supposed to be one-fifth less and the thichness is 1 (m).

Table 2. Material properties

E	0.980 E+06	(Pa)
K	0.961 E+06	(Pa)
ν	0.330	
$\varkappa/\mu$	0.102 E-10	(m^4/Ns)
$n\beta$	0.250 E-09	(m^2/N)

Figure 6 shows calculated pore water pressures. We can observe the discontinuities in the slope at interfaces reflecting the inhomogeneous thin layer. If equivalent material properties are chosen as in Table 2 for the thin layer, then we can obtain smooth curves for calculated pressures.

Boundary element solution v.s. experiment

Figure 7 shows the geometry of soil layer under an experimental embankment conducted by Sakamaki and Mochizuki [1980], where 1.81E04 (N/m^3) in weight is settled. The layer was loaded gradually as follows: The rate of piling the embankment was 0.1 (m/day) until the embankment reaches the hight of 3.9 (m). There was no additional embankment for next 9 days. Then the piling was restarted with the rate 0.18 (m/day) until the embankment reaches the final hight of 5 (m).

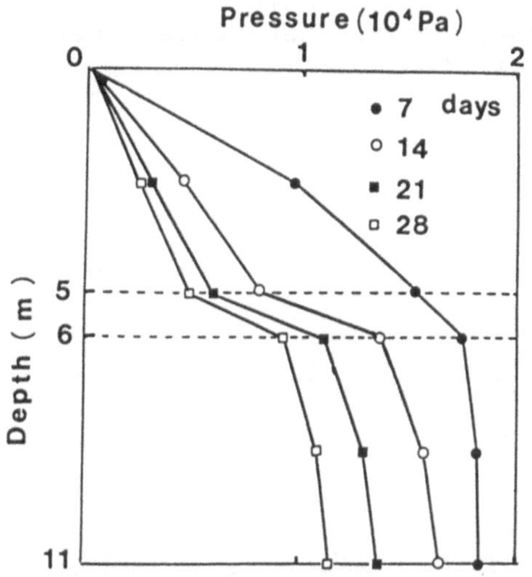

Figure 6. Calculated pressure profiles

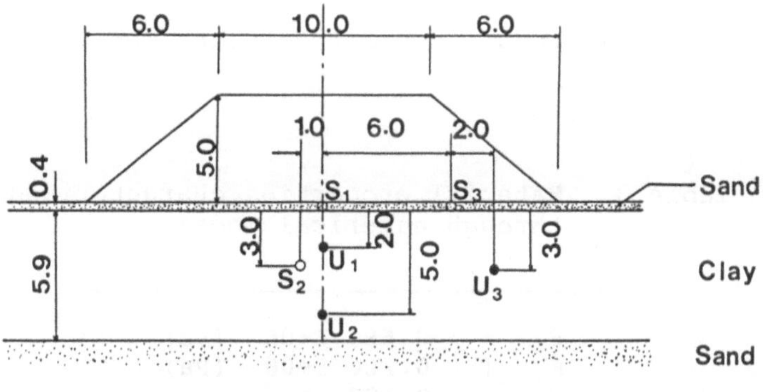

Figure 7. Experimental prototype (m)
Pore pressures are observed in U_1 - U_3 , displacements in S_1 - S_3

A boundary element model for Figure 7 is shown in Figure 8. Coefficients in the numerical model were determined by trial and error among values obtained from laboratory test. Suitable coefficients are listed in Table 3.

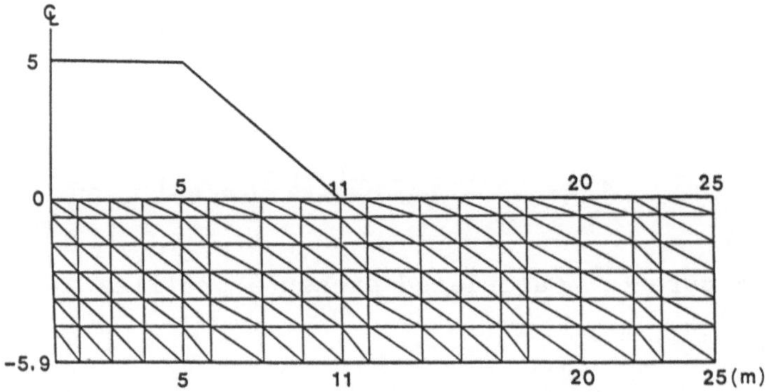

Figure 8. Boundary element model

Table 3. Material properties characterized through numerical model

E	0.686 E+06 (Pa)
K	0.286 E+06 (Pa)
ν	0.100
$\varkappa/\mu$	0.209 E-15 (m^4/Ns)
$n\beta$	0.350 E-09 (m^2/N)

Figure 9 shows transient subsidences. Experimental results are reproduced only for the range of interest. Good agreement is achieved between boundary element solution and observations. The same is true for solutions of pore water pressures as shown in Figure 10.

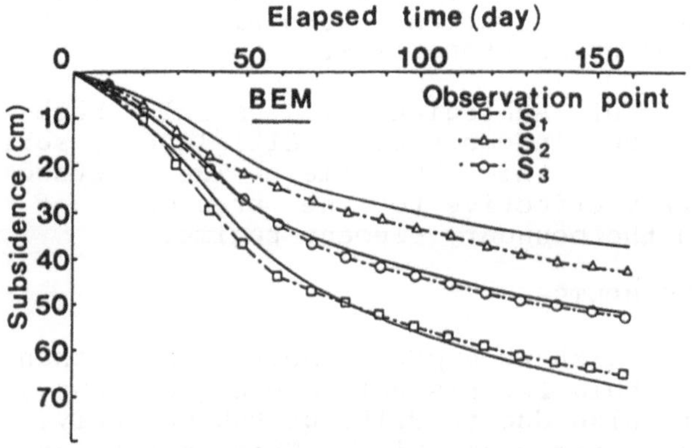

Figure 9. Transient subsidence

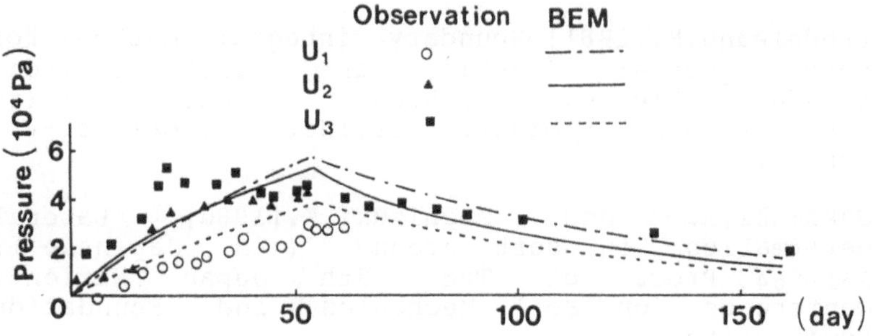

Figure 10. Pore water pressures

CONCLUSION

An uncoupled boundary element method was compared favorably with a coupled finite element method. It was found that the present method is applicable to practical consolidation problems by choosing suitable material properties.

The method of subregions is useful both for treating zoned inhomogeneous multi-layered soil and for reducing the CPU time. The method is particularly effective to the treatment of thin layers in the boundary element regime.

ACKNOWLEDGEMENTS

The authors wish to express their cordial thanks to Dr. C.A.Brebbia for his valuable suggestions. Much thanks are also due to T.Ito of Fukuoka Univ. for the computer implementation. This investigation is supported by part by The Japanese Ministry of Education, Grant in Aid for Scientific Research.

REFERENCES

Brebbia,C.A.[1978] The Boundary Element Method for Engineers. Pentech Press, London.

Kuroki,T., Ito,T., and Onishi,K.[1982] Boundary element method in Biot's linear consolidation. Appl. Math. Modelling (in press, April)

Predeleanu,M.[1981] Boundary integral method for porous media: 325-334, in C.A.Brebbia(Ed.) Boundary Element Methods. Proc. Third International Seminar, Irvine. Springer-V., Berlin.

Sakamaki,A. and Mochizuki,K.[1980] Lateral deformation in soft ground (in Japanese): 861-864. Proc. of The 15th Japan National Conference on Soil Mechanics and Foundation Engineering.

Verruijt,A.[1977] Generation and dissipation of pore water pressure: 293-316, in G.Gudehus(Ed.) Finite Elements in Geomechanics. John Wiley & Sons., London.

CONSOLIDATION PROBLEMS

C.García-Suárez, E.Alarcón

Polytechnical University Madrid

Introduction

The analysis of deformation in soils is of paramount importance in geotechnical engineering. For a long time the complex behaviour of natural deposits defied the ingenuity of engineers. The time has come that, with the aid of computers, numerical methods will allow the solution of every problem if the material law can be specified with a certain accuracy.

Boundary techniques (B.E.) have recently exploded in a splendid flowering of methods and applications that compare advantageously with other well-established procedures like the finite element method (F.E.). Its application to soil mechanics problems (Brebbia 1981) has started and will grow in the future.

This paper tries to present a simple formulation to a classical problem. In fact there is already a large amount of applications of B.E. to diffusion problems (Rizzo et al, Shaw, Chang et al, Combescure et al, Wrobel et al, Roures et al, Onishi et al) and very recently the first specific application to consolidation problems has been published by Onishi et al.

Here we develop an alternative formulation to that presented in the last reference. Fundamentally the idea is to introduce a finite difference discretization in the time domain in order to use the fundamental solution of a Helmholtz type equation governing the neutral pressure distribution.

Although this procedure seems to have been unappreciated in the previous technical literature it is nevertheless effective and straightforward to implement. Indeed for the special problem in study it is perfectly suited, because a step by step interaction between the elastic and the flow problems is needed. It allows also the introduction of non-linear elastic properties and time dependent conditions very easily as will be shown and compares well with performances of other approaches.

Basic equations

As is well known the classical consolidation problems describe the coupling of fluid and solid phases of saturated soils.

If the compressibility of solid particles is not taken into account the law of mass conservation can be written as

$$\dot{\theta} - \frac{n}{k_w}\dot{\phi} + \operatorname{div} \underset{\sim}{\upsilon} = 0 \tag{1}$$

where the dots indicate time derivative and

$\theta = u_{i,i}$ is the volumetric deformation

u_i : is the displacement vector

n : is the porosity

k_w : is the bulk modulus of water

ϕ: is the excess pore pressure

$\underset{\sim}{\upsilon}$: is the apparent velocity of the fluid

Taking ϕ as negative, DARCY's law is

$$\upsilon_i = \frac{k}{\gamma_w} \phi_{,i} \qquad \phi < 0 \tag{2}$$

where

k : is the permeability (m/s)

γ_w: is the specific weight of water (N/m^3)

The first field equation is then obtained by introducing (2) in (1) as follows

$$\frac{\partial \theta}{\partial t} - \frac{n}{k_w}\frac{\partial \phi}{\partial t} + \frac{k}{\gamma_w}\phi_{,ii} = 0 \tag{3}$$

The other coupled phenomen is the deformation of the soil skeleton.

Equilibrium equations are

$$\begin{aligned} \underset{\sim}{\nabla}\, \underset{\sim}{\sigma} + X &= 0 \qquad \text{in } \Omega \\ \underset{\sim}{\sigma}\, \underset{\sim}{\upsilon}\,. &= \underset{\sim}{t} \qquad \partial\Omega \end{aligned} \tag{4}$$

where

$\underset{\sim}{\sigma}$ is the total stress tensor
$\underset{\sim}{X}$ are the body forces
$\underset{\sim}{\nu}$ is the exterior normal to $\partial\Omega$
$\underset{\sim}{t}$ are the boundary tractions

Equations (4) can also be written, following the classical assumption of TERZAGHI, as the sum of effective $\underset{\sim}{\sigma}$ and neutral $\phi\underset{\sim}{\partial}$ pressures

$$\underset{\sim}{\sigma} = \underset{\sim}{\sigma} + \phi\underset{\sim}{\partial} \tag{5}$$

which produces

$$\begin{aligned} &\underset{\sim}{\nabla}\,\underset{\sim}{\sigma} + \underset{\sim}{X} + \underset{\sim}{\nabla}\,\phi = 0 && \text{in } \Omega \\ &\underset{\sim}{\sigma}\,\underset{\sim}{\nu} + \phi\,\underset{\sim}{\nu} = \underset{\sim}{t} && \text{in } \partial\Omega \end{aligned} \tag{6}$$

(5) can also be introduced into (3) through the use of a material law as (which necessarily needs to be expressed in effective stresses)

$$\dot{\theta} = \frac{1}{k_s}\,\dot{\sigma}'_{oct} = \frac{1}{k_s}\,(\dot{\sigma}_{oct} - \dot{\phi}) \tag{7}$$

where σ_{oct} is the so-called octahedral stress

$$\sigma_{oct} = \frac{1}{3}\,\sigma_{ii} \tag{8}$$

In this way equation (3) is

$$\frac{1}{k_s}\,\dot{\sigma}_{oct} = \left(\frac{n}{k_w} + \frac{1}{k}\right)\dot{\phi} - \frac{k}{w}\,\phi_{,ii} \tag{9}$$

or, using the same notation as VERRUIJT

$$\begin{aligned} &\gamma\,\nabla^2\phi = \dot{\phi} - \mu\,\dot{\sigma}_{oct} \\ &\alpha = \frac{kK_s}{\gamma_w} \\ &\beta = \frac{Kn}{\gamma_w} + 1 \\ &\gamma = \frac{\alpha}{\beta}\,;\quad \mu = \frac{1}{\beta} \end{aligned} \tag{10}$$

With the appropriate boundary conditions (6) and (10) defined the consolidation problem.

Boundary element discretization

In order to produce the boundary discretization it is necessary to introduce fundamental solutions of the field equations and to establish inner products in the whole domain. For equations (6) this can be done through Kelvin type solution $\underset{\sim}{u}^*$ in the following way

$$(\nabla \underset{\sim}{\sigma}', \underset{\sim}{u}^*)_\Omega = -(\underset{\sim}{\nabla} \phi, u^*)_\Omega - (\underset{\sim}{X}, \underset{\sim}{u}^*)_\Omega \tag{11}$$

But

$$(\underset{\sim}{\nabla} \underset{\sim}{\sigma}', \underset{\sim}{u}^*)_\Omega = \underset{\sim}{\nabla}(\underset{\sim}{\sigma}', \underset{\sim}{u}^*)_\Omega - (\underset{\sim}{\sigma}', \underset{\sim}{\nabla} \underset{\sim}{u}^*)_\Omega = (\underset{\sim}{\sigma}' \underset{\sim}{\nu}, \underset{\sim}{u}^*)_{\partial\Omega} - (\underset{\sim}{\sigma}', \underset{\partial}{\varepsilon}^*)_\Omega =$$

$$= (\underset{\sim}{\sigma}' \underset{\sim}{\nu}, \underset{\sim}{u}^*)_{\partial\Omega} - (\underset{\sim}{\varepsilon}, \underset{\sim}{\sigma}^*)_\Omega = (\underset{\sim}{\sigma}' \underset{\sim}{\nu}, \underset{\sim}{u}^*)_{\partial\Omega} - (\underset{\sim}{u}, \underset{\sim}{t}^*)_{\partial\Omega} \qquad -(\underset{\sim}{u}, \underset{\sim}{X}^*)_\Omega \tag{12}$$

In addition

$$-(\underset{\sim}{\nabla} \phi, u^*)_\Omega = -\nabla(\phi, \underset{\sim}{u}^*)_\Omega + (\phi, \underset{\sim}{\nabla} \underset{\sim}{u}^*)_\Omega = -(\phi \underset{\sim}{\nu}, \underset{\sim}{u}^*)_{\partial\Omega} + (\phi, \underset{\sim}{\nabla} \underset{\sim}{u}^*)_\Omega \tag{13}$$

So that

$$(\underset{\sim}{\sigma}' \underset{\sim}{\nu}, \underset{\sim}{u}^*)_{\partial\Omega} - (\underset{\sim}{u}, \underset{\sim}{t}^*)_{\partial\Omega} - (\underset{\sim}{u}, \underset{\sim}{X}^*)_\Omega = -(\phi \underset{\sim}{\nu}, \underset{\sim}{u}^*)_{\partial\Omega} + (\phi, \underset{\sim}{\nabla} \underset{\sim}{u}^*)_\Omega - (\underset{\sim}{X}, \underset{\sim}{u}^*)_\Omega \tag{14}$$

and

$$-(\underset{\sim}{u}, \underset{\sim}{X}^*)_\Omega - (\underset{\sim}{u}, \underset{\sim}{t}^*)_{\partial\Omega} = -((\underset{\sim}{\sigma}' + \phi\partial) \underset{\sim}{\nu}, \underset{\sim}{u}^*)_{\partial\Omega} - (\underset{\sim}{X}, \underset{\sim}{u}^*)_\Omega + (\phi, \underset{\sim}{\nabla} \underset{\sim}{u}^*)_\Omega \tag{15}$$

that produces the desired relationship

$$(\underset{\sim}{u}, \underset{\sim}{X}^*)_\Omega + (\underset{\sim}{u}, \underset{\sim}{t}^*)_{\partial\Omega} = (\underset{\sim}{t}, \underset{\sim}{u}^*)_{\partial\Omega} + (\underset{\sim}{X}, \underset{\sim}{u}^*)_\Omega - (\phi, \underset{\sim}{\nabla} \underset{\sim}{u}^*)_\Omega \tag{16}$$

When $\underset{\sim}{X}^*$ is chosen as the fundamental Kelvin solutionthe classical B.E.M. are obtained plus the additional term $(\phi, \underset{\sim}{\nabla} \underset{\sim}{u}^*)$ which marks the coupling with the diffusion phenomena. For the usual cases $(\underset{\sim}{X}, \underset{\sim}{u})$ can be reduced to the boundary, but the coupling term has to be evaluated in volume cells.

For equation (10) and previous to any spatial discretization we introduce a time stepping by putting

$$\dot{\phi}_i \simeq \frac{1}{\Delta t}(\phi_i - \phi_{i-1}) \tag{17}$$

or any other convenient rule. In this way it is possible to write

$$\gamma \nabla^2 \phi - \frac{1}{\Delta t} \phi_i = - \frac{\phi_{i-1}}{\Delta t} - \mu \dot{\sigma}_{oct} \tag{18}$$

The operator on the right hand side is the Helmholtz one, so that using its fundamental solution ϕ^* we can establish

$$(B \phi_i, \phi^*)_\Omega = (g, \phi^*)_\Omega \tag{19}$$

with

$$B = \gamma \nabla^2 - \frac{1}{\Delta t}$$

$$g = - \frac{\phi_{i-1}}{\Delta t} - \mu \dot{\sigma}_{oct} \tag{20}$$

arriving at the final equation

$$(\phi, g^*)_\Omega + \gamma (\phi, q^*)_{\partial\Omega} = \gamma (\phi^*, q)_{\partial\Omega} + (g, \phi^*)_\Omega \tag{21}$$

where q and q^* represent the fluxes associated with respectively the actual and the fundamental solution.

Computational scheme

As was indicated the solution of the problem follows a marching process progressing with the following scheme: (BANERJEE & BUTTERFIELD)

a) Assuming undrained conditions. Working on total stresses the elastic problem is solved, what allows the computation of σ_{oct}

b) Initial conditions for pore water excess are evaluated as

$$\phi^\circ = \sigma^\circ_{oct}$$

Observe that the method has no problem with management of values $\nu = \frac{1}{2}$.

c) The initial slope of the (σ_{oct}, t) curve is assumed zero, i.e.:

$$\dot{\sigma}^\circ_{oct} = 0$$

d) With $\dot{\sigma}^\circ_{oct} = 0$ (21) is solved producing the field of neutral pressures ϕ^1.

e) Knowing ϕ^1 it is an easy matter to solve (16), obtaining then the displacements $\underset{\sim}{u}^1$, the stress state $\underset{\sim}{\sigma}^1$ and the octahedral stresses σ^1_{oct}.

f) Now it is assumed that

$$\dot{\sigma}^1_{oct} = \frac{\sigma^1_{oct} - \sigma^\circ_{oct}}{\Delta t}$$

i.e.: the initial step is extrapolated from the previous computed values.

g) With σ^1_{oct} steps from d) on, can be repeated for every time interval.

The previous ideas have been implemented in a computer program whose flow-chart is described in tables 1 2 and 3. The basic support for them is the two previous programs described by PARIS et al and ROURES et al with several computational improvements. Both use linear interpolation in the boundary while the potential ϕ is assumed as having constant values inside volume cells.

As usual the most expensive part of the method is the numerical integration that in the boundary is done using a standard Gauss rule with 4 points except in certain cases where closed-form expressiones have been developed. The domain integrals are always evaluated numericals with a rule of 16 points inside each quadrilateral cell.

The integration coefficients have to be computed for each different Δt. When a unique Δt is chosen they are computed only once during the first iteration and then stored for subsequent use.

The problem is the election of the interval size in order to maintain an acceptable accuracy without increasing the cost by too much.

Roures et al suggest a three step range for decoupled problems (RENDULIC's assumption of $\dot{\sigma}_{oct} = 0 \ \forall t$) as follows

$$S = kK_s / \gamma_w$$

$$0 < t < 0.025 \frac{L^2}{S} \quad \text{take} \quad \Delta t = 0.005 \frac{A}{S}$$

$$0.025 < t < 0.20 \frac{L^2}{S} \quad \text{"} \quad \Delta t = 0.0125 \frac{A}{S}$$

$$t > 0.20 \frac{L^2}{S} \quad \text{"} \quad \Delta t = 0.06 \frac{A}{K}$$

Nevertheless these rules have been tested only in a limited number of cases and the matter is open to further study.

Non-linear behaviour

Several improvements are easily implemented in the previous scheme at the price of growing computation time.

DAVIS and RAYMOND proposed for instance the use of variable values for the oe dometric modulus which can be done without difficulty but at the cost of computing the constants at every time step.

A better improvement is to use a non-linear elastic law to model the behaviour of the soil skeleton. (see for instance Naylor et al.)

In addition to the well-known hyperbolic laws the most interesting models are the so-called K – G, that is a

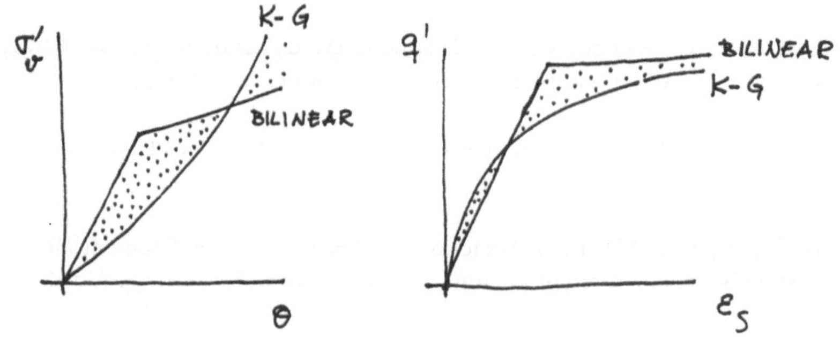

Figure 1.

continuous model, and the bilinear one.

The deviatoric and volumetric parts are separated in both models and a plasticity-type criterion is added to simulate collapse.

The K – G model allows a good representation of the increase in bulk stiffness while the bilinear model defines more neatly the collapse load.

To simulate unloading the equivalent stress inside each cell of the domain is improved through the plasticity criterion and when the limit values are reached the shear modulus is reduced to zero.

This produces good qualitative results when more accurate (but more expensive also!) analyses are not necessary.

References

Banerjee & Butterfield (1981) The Boundary Element Method for Engineers. McGraw.

Brebbia, C. (1981) B.E.M. in Geotechnical Problems, in Numerical Methods in Geomechanics. NATO Seminar, Vimeiro.

Combescure, A. & Lachat, J.C. (1977) Laplace Transform and B.I.E. Application to Transient Heat Conduction Problems. 1st Int.Conf. on Innov.Num.Anal. in App.Eng. Versailles.

Davis and Raymond (1965) A Nonlinear Theory of Consolidation. Geotechnique, 15: 161-171.

Naylor, P.J., Pande, G.N., Simpson, B. & Tabb, R. (1981) Finite Elements in Geotechnical Engineering. Pineridge Press.

Onishi, K. (1981) Convergence in the B.E.M. for Heat Equation. TRV Mathematics, 17-2.

Onishi, K, Kusoki, T. & Tomoko I. (1982) Boundary Element Method in BIOT's Linear Consolidation. Applied Math. Modelling, Vol 2. Nº2.

Rizzo, F.J. and Shippy, D.J. (1970) A Method of Solution of Certain Problems of Transient Heat Conduction. AIAA Journal, 11, Vol 8.

Roures, V. and Alarcon, E. (1982) Transient Heat Conduction Problems using B.I.E.M. To be published in Computers and Structures.

Shaw, R.P. (1974) An Integral Equation Approach to Diffusion. Int. J. Heat Mass Transfer, 17, 693-699.

Verruijt, A. (1977) Generation and Dissipation of Pore-Water Pressures, in Finite Elements in Geomechanics, ed by Gudehus. J.Wiley.

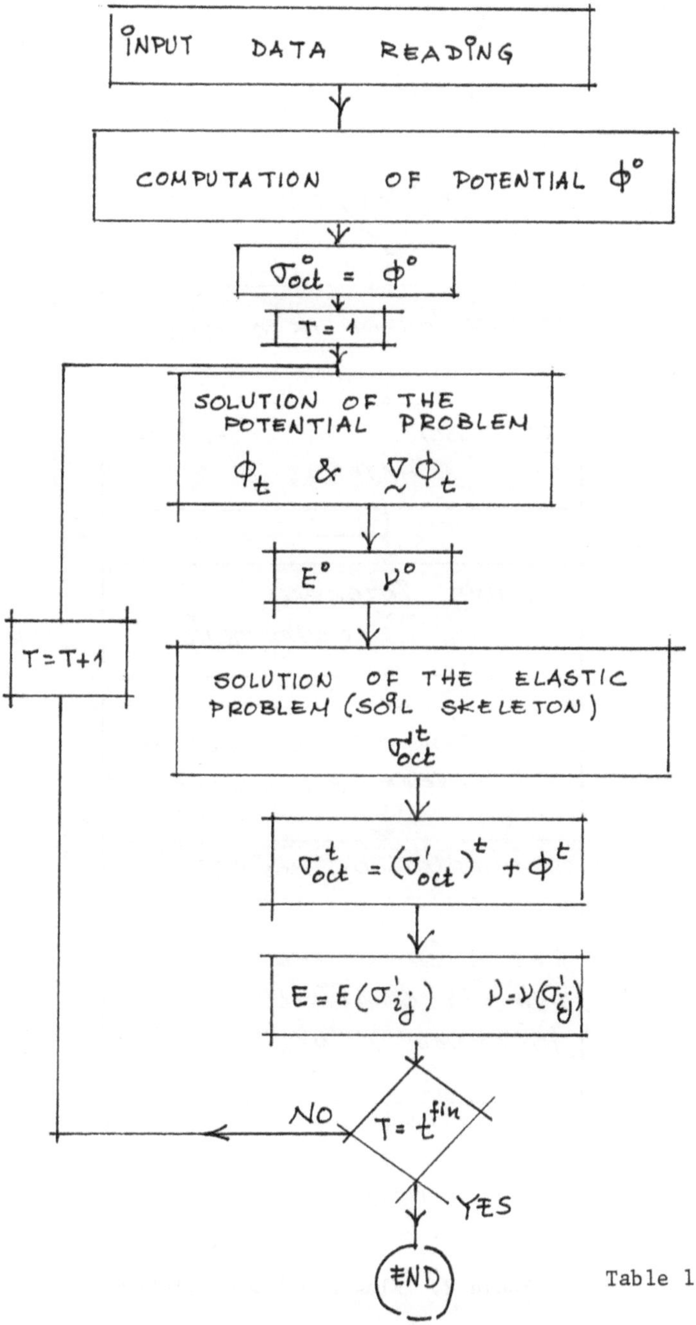
INPUT DATA READING
COMPUTATION OF POTENTIAL Φ^0
$\sigma^0_{oct} = \Phi^0$
T = 1
SOLUTION OF THE POTENTIAL PROBLEM
Φ_t & $\nabla \Phi_t$
E^0 ν^0
T=T+1
SOLUTION OF THE ELASTIC PROBLEM (SOIL SKELETON)
σ'^t_{oct}
$\sigma^t_{oct} = (\sigma'_{oct})^t + \Phi^t$
$E = E(\sigma'_{ij})$ $\nu = \nu(\sigma'_{ij})$
NO
$T = t^{fin}$
YES
END

Table 1.

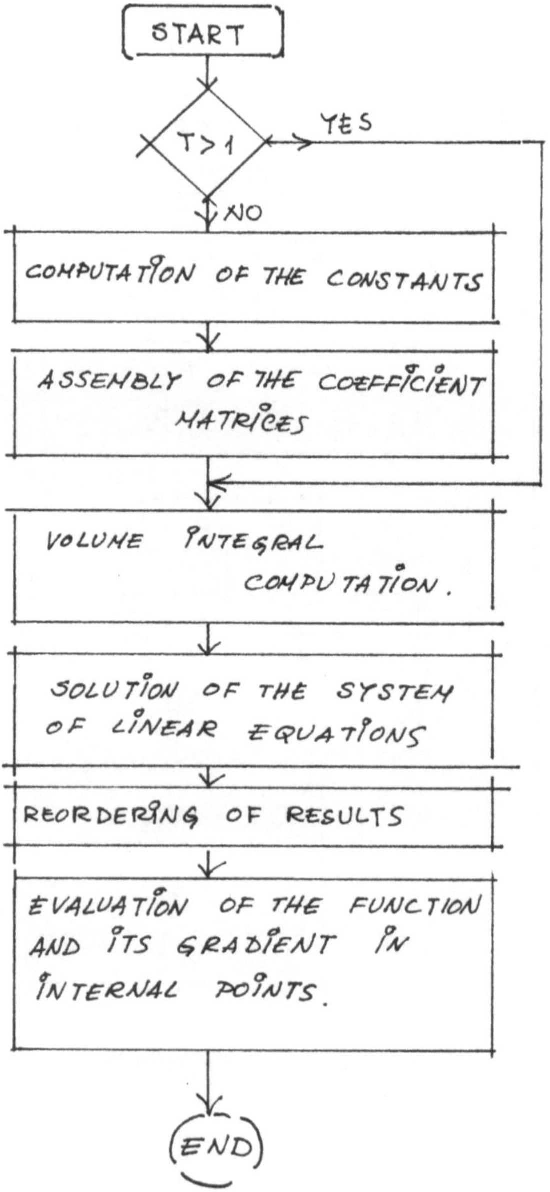

Table 2. Flow Problem Resolution

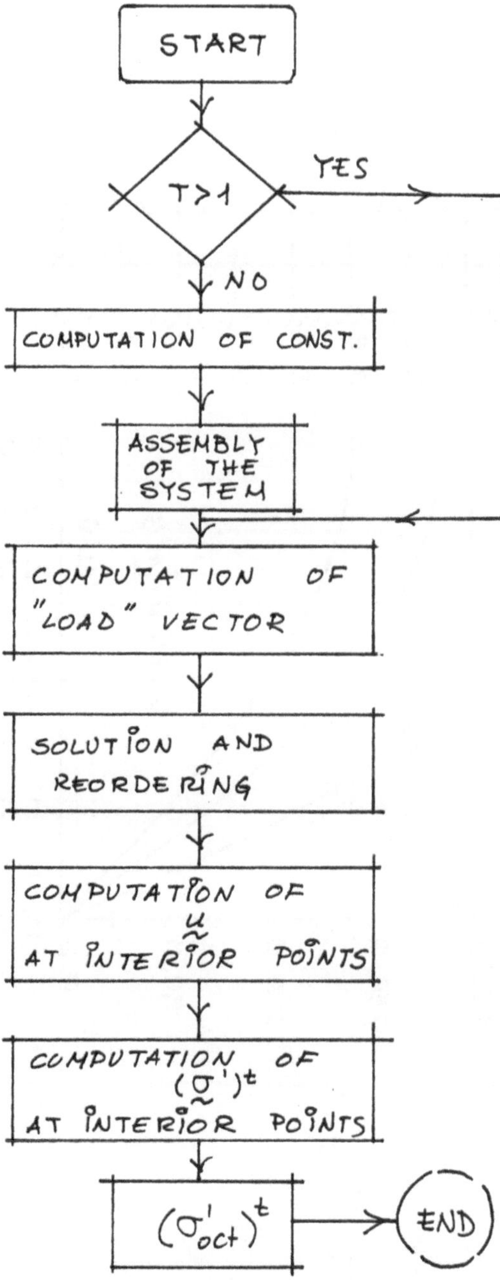

Table 3. Elastic Problem Resolution

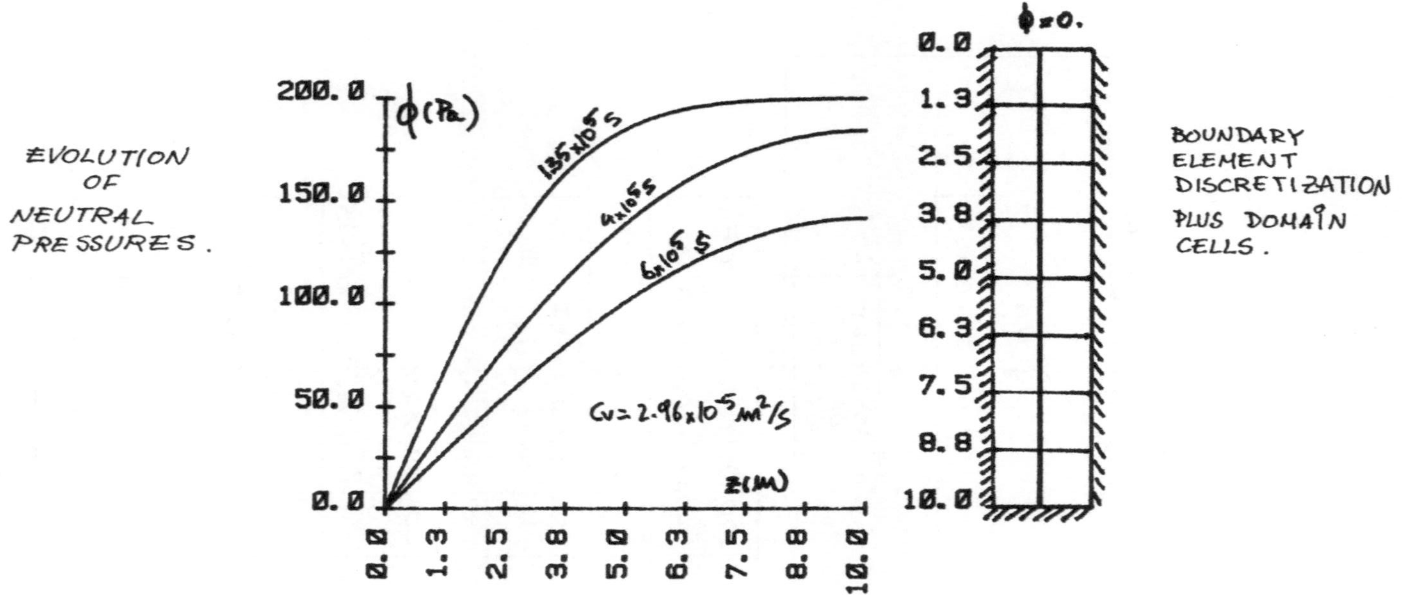

Figure 2. Uncoupled Theory Applied to One Dimensional Consolidation

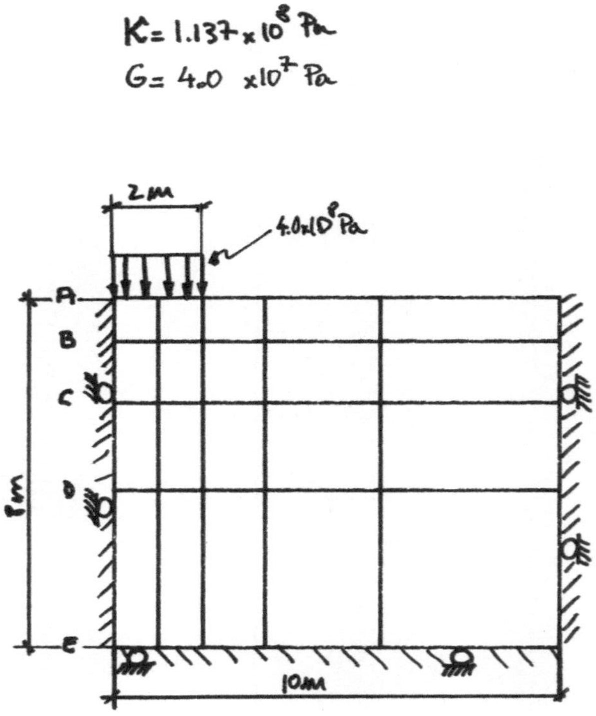

Figure 3a. Boundary Element Mesh for the Consolidation of a Layer Under a Footing Loading

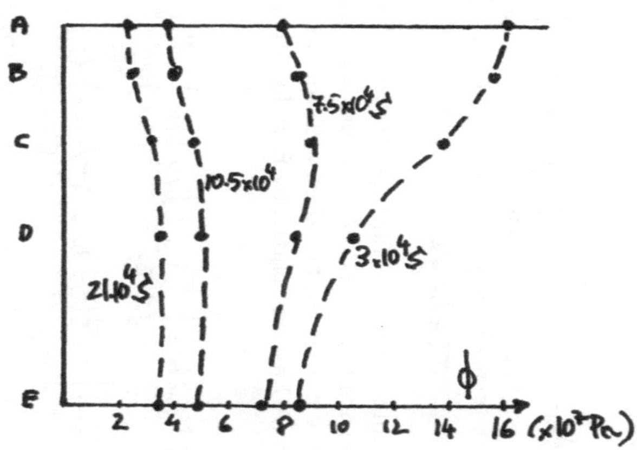

Figure 3b. Evolution of Neutral Pressure Under the Footing

APENDIX

In the solution of the elastic problem when only self weight and external applied forces are considered the following expresion is obtained for the internal displacements

$$u_j + \int_{\partial\Omega} T_{ji} u_i \, ds = \int_{\partial\Omega} U_{ji} \cdot t_i \, ds - \int_{\partial\Omega} gz U_{ji} \cdot n_i \, ds - \int_{\Omega} \zeta U_{ji,i} \, dv \qquad \text{(A1)}$$

where all the terms have the usual meaning an ζ is defined as

$$\zeta = \phi - gz \qquad \text{(A2)}$$

For the calculation of the internal effetive stress is necessary to derive the expresion (A1). All the derivatives can be easily performed but some care must be taken with the one corresponding to the last term in (A1) as was shown by BUI for the plasticity problem.

Let

$$I_j = \int_{\Omega} \zeta U_{ji,i} \, dv = \int_{\Omega - B_\varepsilon} \zeta U_{ji,i} \, dv + \lim_{\varepsilon \to 0} \int_{B_\varepsilon} \zeta U_{ji,i} \, dv \qquad \text{(A3)}$$

$$\frac{\partial I_j}{\partial x_k} = \lim_{\varepsilon \to 0} \int_{\Omega - B_\varepsilon} \zeta \frac{\partial U_{ji,i}}{\partial x_k} \, dv - \zeta \lim_{\varepsilon \to 0} \int_{\partial B_\varepsilon} U_{ji,i} \cdot r_{,k} \, ds \qquad \text{(A4)}$$

The first integral of (A4) can be easily computed numerically and for the last term an analytical expresion can be found.

THE APPLICATION OF BOUNDARY ELEMENTS IN OFFSHORE ENGINEERING

L. A. Wood and S.G. Creed

Lecturer, Research Assistant, Department of Civil Engineering, Queen Mary College, London.

SUMMARY

A simple boundary element formulation for the modelling of heterogeneous soil deposits is discussed. It is shown that the coupling of this element with standard structural finite elements has lead to the development of a powerful tool for the economic analysis of soil-structure interaction problems. The use of the method is illustrated with respect to two off-shore structures; namely a gravity platform and a jacket structure.

INTRODUCTION

When dealing with soils, which unlike structures rarely exhibit totally defined geometric boundaries and tend therefore to form large aspect ratio problems, the use of a boundary element model is particularly inviting. The use of such an approach invariably reduces the amount of effort involved in data preparation and may be more efficient in terms of accuracy for a given computational effort than the more conventional techniques utilising finite element or finite difference solutions.

In the context of foundations, the soil represents the boundary to the structure and although it is important that the soil behaviour is modelled correctly, the Engineering interest is, in the main, concentrated upon the design of the structure. Hence it is inefficient to have to solve for displacements within the whole continuum as with most finite element solutions. On the other hand in a boundary element model the action may be concentrated in the area of concern.

The coupling of structural finite elements with soil boundary elements represents a most useful and efficient tool for the investigation of the soil-structure interaction problems associated with Offshore Engineering. At the present time the major structural forms of production platforms are large gravity or jacket structures. The former, usually constructed of reinforced concrete, relies upon its self-weight for stability and rests upon the sea-bed. Whilst the latter constructed in the form of a tubular steel lattice is tied to the sea-bed by an array of pipe piles situated at the base of each leg of the structure. In the future it may be expected, as oil recovery in deeper water and exploitation of oceanic resources increase that the development and design of anchoring systems for tethered platforms will grow in importance.

In the context of a wide ranging research programme into the behaviour of offshore structures attention is focused below on results obtained from the analysis of the Frigg TP-1 gravity platform and the Heather jacket structure, both located within the North Sea.

BOUNDARY ELEMENT FORMULATION

The particular boundary element formulation used has been developed over a number of years (Wood, 1977, 1978, 1979) and takes into account the natural heterogeneity of soil deposits. The formulation is based upon the use of the Boussinesq and Mindlin analytical solutions for the stresses and displacements occurring within a homogeneous, elastic half-space due to the application of both surface and embedded forces. An approximate extension to non-homogeneous, layered elastic continua has been developed in a similar manner to that suggested by Poulos (1971). In addition non-linear soil behaviour has been incorporated by the utilisation of the simple concept of limiting the developed stresses at the structure-soil interface. The limits being computed from consideration of plastic failure of the soil. In this manner a realistic and economic, albeit approximate, soil boundary element has been developed which when coupled to a standard finite element model of the structure provides the engineer with a most powerful tool for use in the design of offshore structures in particular, and in other branches of geotechnical engineering in general.

GRAVITY PLATFORM

Foss and Warming (1979) have published some results obtained from the monitoring of three gravity platforms of the SEA TANK design installed in the North Sea. In order to illustrate briefly the use of the boundary element technique in modelling the settlement performance of such a structure reference has been made to the Frigg TP-1 installation shown in Fig. 1. The platform base approximately 70m square in plan has been taken to act as a rigid body in the analysis. The available soils data is shown in Fig. 2 and the derived variation in vertical, drained Young's modulus, E' is given in Table 1.

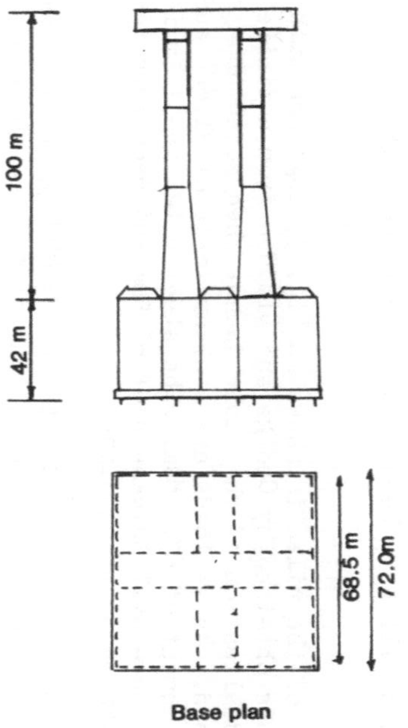

Figure 1 Frigg TP-1 Gravity Platform

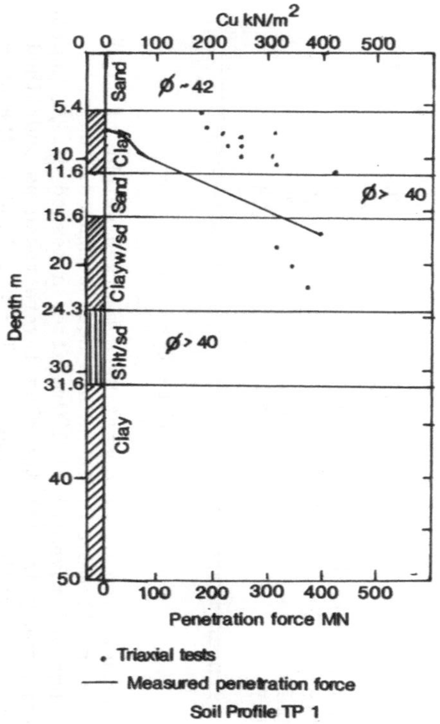

Figure 2 Soil Data Frigg TP-1

For all layers an effective Poisson's Ratio, ν' of 0.1 has been assumed.

Description	Thickness (m)	E'(MN/m²)
SAND	5.4	15 → 45*
CLAY	6.2	36 → 55*
SAND	4.0	45
CLAY with sand	8.7	58
Silty SAND	7.3	45
CLAY	6.4	72
CLAY	40.0	108

* linear increase

Table 1

From consideration of an operating submerged weight of 1150MN a uniform settlement of 140mm has been computed; compared to the observed movement of 100mm after a period of $2\frac{1}{2}$ years. Subsequent analysis with the clay layers taken as undrained with Poisson's ratio = 0.5 and E_u = 1.36E', lead to the computation of an immediate settlement of 96mm and a differential movement under the design environmental moment of 7500MNm of 30mm between the centre and the edge. The former is somewhat in excess of the recorded movement at one month of 40mm.

JACKET STRUCTURE

George and Sladden (1978) have presented in detail the results of the analyses undertaken for Lloyds certification of the piled foundation to the Heather platform. The present analysis has been concentrated on a detailed study of the computed performance of the piles under lateral loading only. It is anticipated that the results obtained from further work in which the behaviour of the piles under both vertical and lateral load will have been studied, will be published in the near future.

The layout of the piles is shown in Fig. 3, together with the design loads obtained by George and Sladden from a finite element idealisation of the whole jacket when acted upon by a storm load. These loads are a shear force of 21.8MN and a restoring moment of 63.4 MNm. The steel piles are 1.5m in diameter with a wall thickness of 63.5mm and penetrate to a depth of 45m below the mudline.

George and Sladden utilised both the PGROUP and PLAYER computer programs in order to predict the performance of the structure. The results they obtained from both programs were similar under working loads. Where possible the results of the present analysis and those obtained previously are compared.

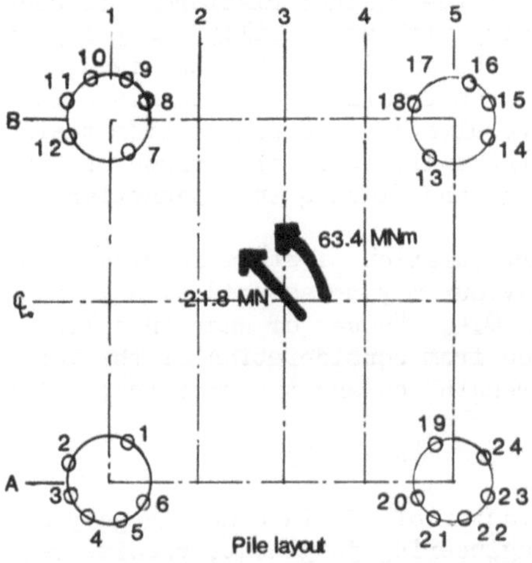

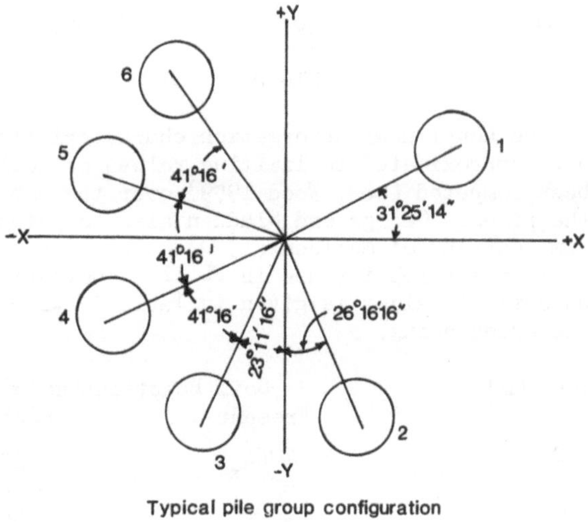

Figure 3 Pile Layout for Heather Platform

Soil properties

Foundation soils at the Heather platform site consist of hard silty and sandy clay with dense silty sandy layers below a depth of 40m from the mudline. The undrained shear strength profile obtained from triaxial tests is given in Fig. 4, together with the results of cone penetration tests. The profile assumed in the analysis is also shown and is the same as that used by George and Sladden in conjunction with PLAYER.

The relative short duration of storm loading is such that undrained soil behaviour may be assumed with Poisson's ratio taken as equal to 0.5. Values of undrained Young's modulus, E_u may be obtained from consideration of the variation in undrained shear strength, c_u and the empirical relationship,

$$E_u = 250 c_u . \qquad \text{... (1)}$$

The choice of a factor of 250 is based largely on experience and engineering judgement, precise reasons as to its selection lie outside the scope of this paper. The assumed soil parameters used in the present analysis are given in Table 2.

Thickness (m)	$c_u (kN/m^2)$	$E_u (MN/m^2)$
3	300	75
8	500	125
39	350	87.5

Table 2

In order for the non-linear deformation characteristics of the soil to be approximated to limiting values of soil reaction have been computed (see, Wood 1979) over the embedded length of the piles. George and Sladden also computed, based on the recommendations of Matlock (1970), a corresponding set of maximum soil pressures for use in PLAYER. Comparison between the two sets of values is given in Table 3, where the similarity is encouraging.

Depth (m)	Soil Reactions (kN/m) Present	PLAYER
0	1064	1220
5	6240*	2410
10	4600	3560
15	4360	4310
20 and below	4400	4310

* within stiff layer

Table 3 Limiting soil reactions

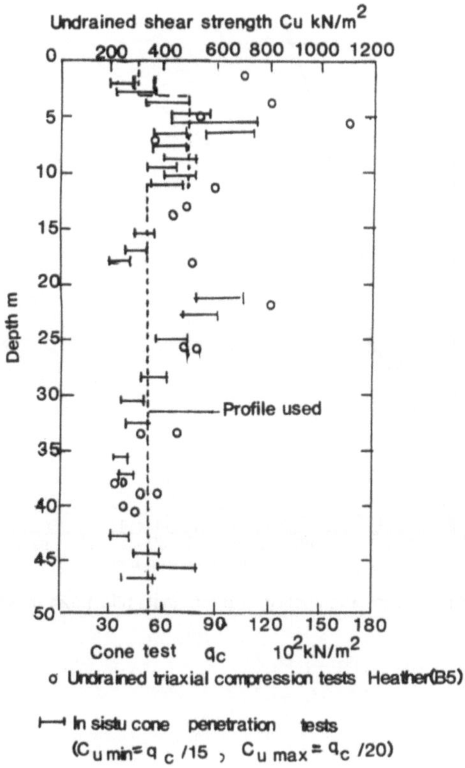

Figure 4 Soil Data for Heather Platform

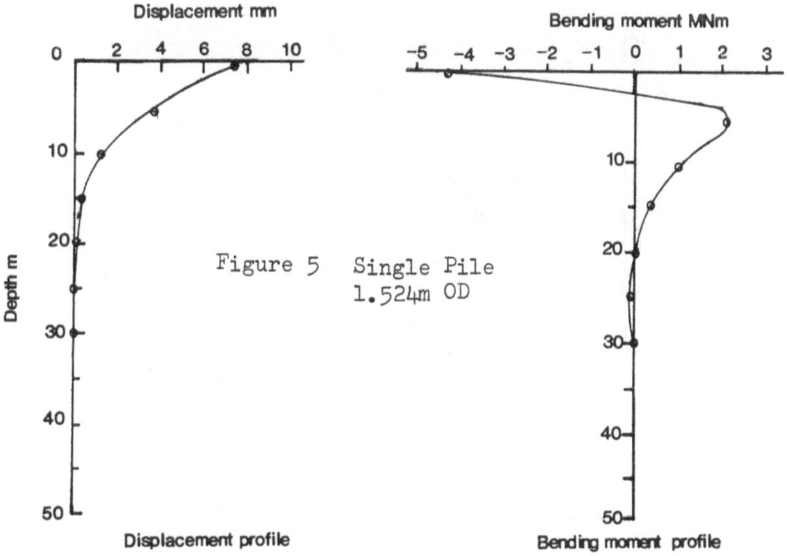

Figure 5 Single Pile 1.524m OD

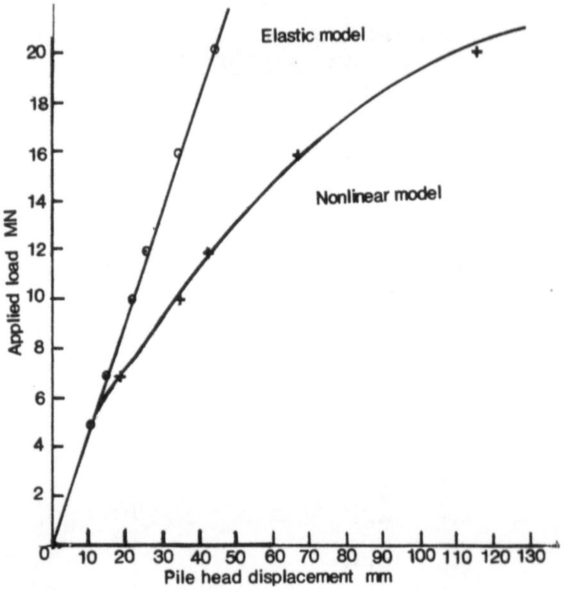

Pile head displacement vs applied load for single pile 1.524 m OD

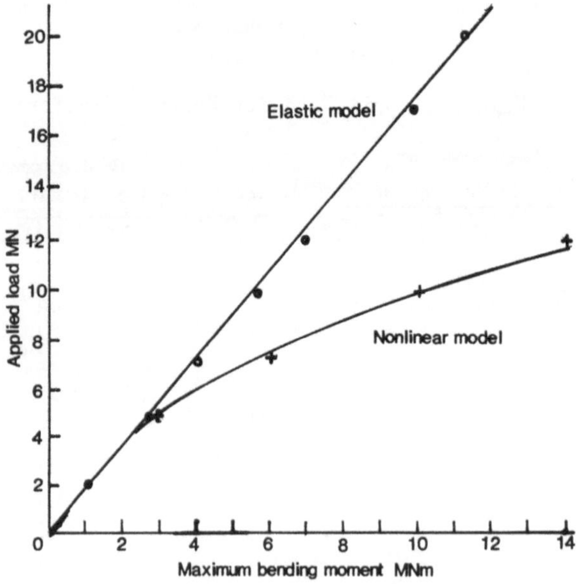

Maximum bending moment vs applied load for single pile 1.524 m OD

Figure 6 Single Pile 1.524m OD

Results of analysis

In order to investigate the interaction of the piles within one of the pile groups and gain an insight into the overall stability of the foundation several situations have been given consideration. Namely,

(a) the behaviour of a single pile carrying one-sixth of the load carried by the group,
(b) the behaviour of the pile group,
and (c) the behaviour of the group represented as a single pile of the same diameter as the group.

Single pile In order to simulate the presence of the rotational constraint provided by the jacket leg at the head of the piles in the group; the single pile has been analysed with its head fixed against rotation. The displacement and bending moment profiles for the pile computed under an applied load equal to one-sixth of the design load for the group are shown in Fig. 5. It is of interest to note that the maximum positive moment is 2MNm, the displacement of the pile head is 7.5mm and, the moment developed at the pile head is - 4.4MNm.

The variation of pile head displacement and maximum positive moment with applied load are shown in Fig. 6. The pile behaves in an elastic manner up to an applied load of 5MN and then becomes non-linear as progressive soil failure occurs from the surface down. The results show that the elastic results are valid for about one-fifth of the "failure" load of the pile.

Pile group The connection between the piles in the group afforded by the jacket leg has been simulated by connecting the pile heads together through a series of rigid beams. The results indicate that all of the piles in the group respond to load in a similar manner and that the load was shared almost equally between the six piles. The displacement and bending moment profiles obtained for the piles in the group under the design loadings are shown in Fig. 7 together with the bending moments computed by George and Sladden using PLAYER. It is of interest to note that the pile head displacement of the pile group is 20mm, 2.6 times that computed for the single pile, whereas the maximum positive moment induced is 1.8MNm compared with 2.0MNm for the single pile. Load-displacement and load-maximum positive bending moment relations for a typical pile (pile No. 1) are shown in Fig. 8 where it is apparent that elastic behaviour is maintained up to an applied load of 50MN (cf. 5MN for the single pile).

Group as a single pile In order to further access the behaviour of the group it has been represented as a single 7.4m diameter pile exhibiting a bending stiffness, EI equal to six times that of a single pile; again the head of the pile has

Pile Head
Displacement
PLAYER 19mm
PGROUP 23mm

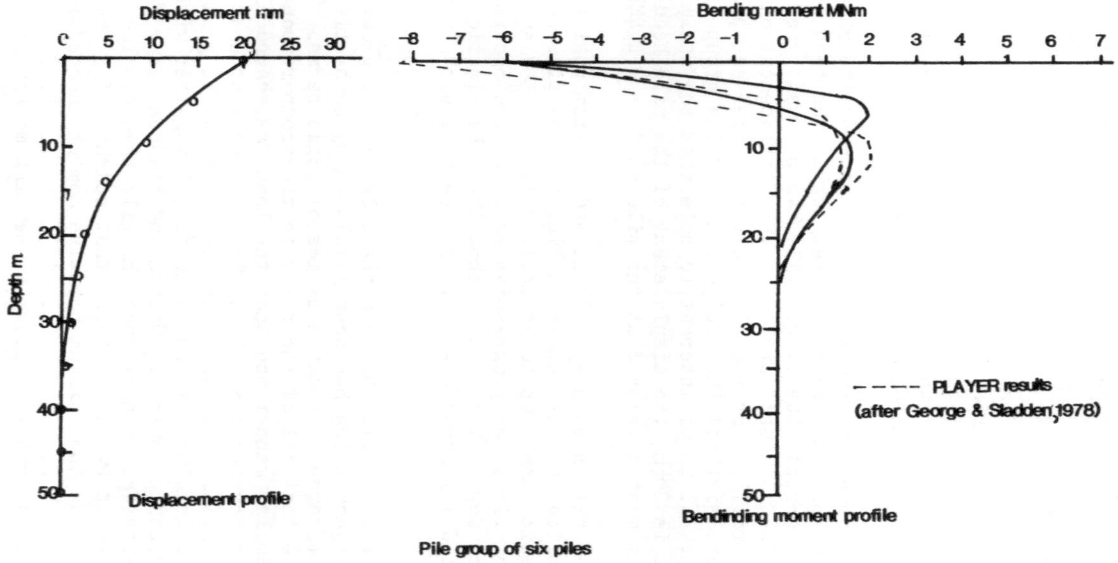

Figure 7 Pile Group

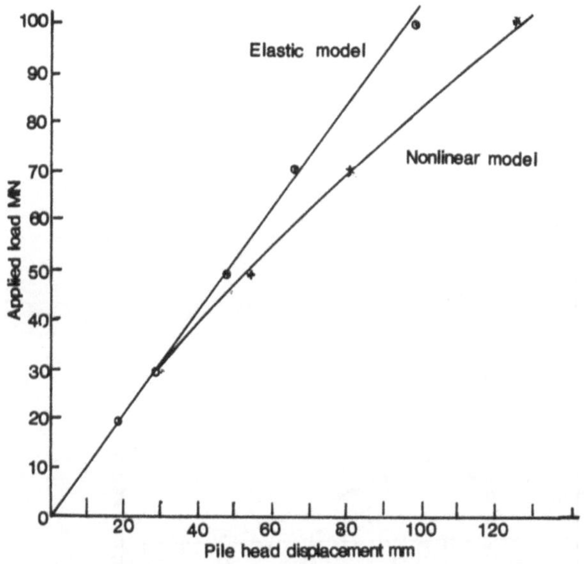

Applied load vs pile head displacement for six piles in a group

Pile 1 - taken as typical

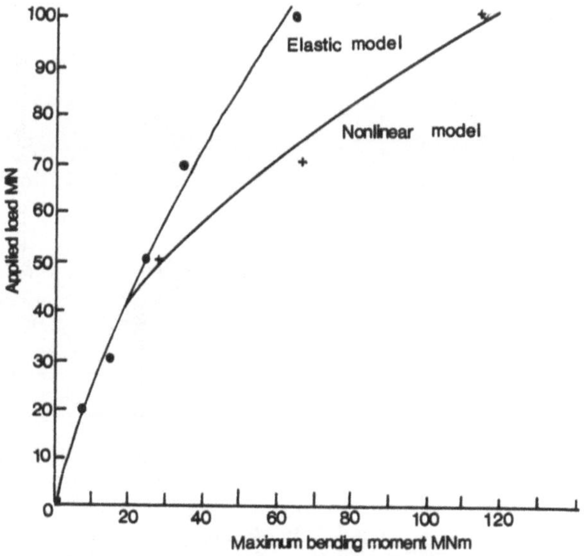

Applied load vs sum maximum bending moment for six piles in a group

Figure 8 Pile Group

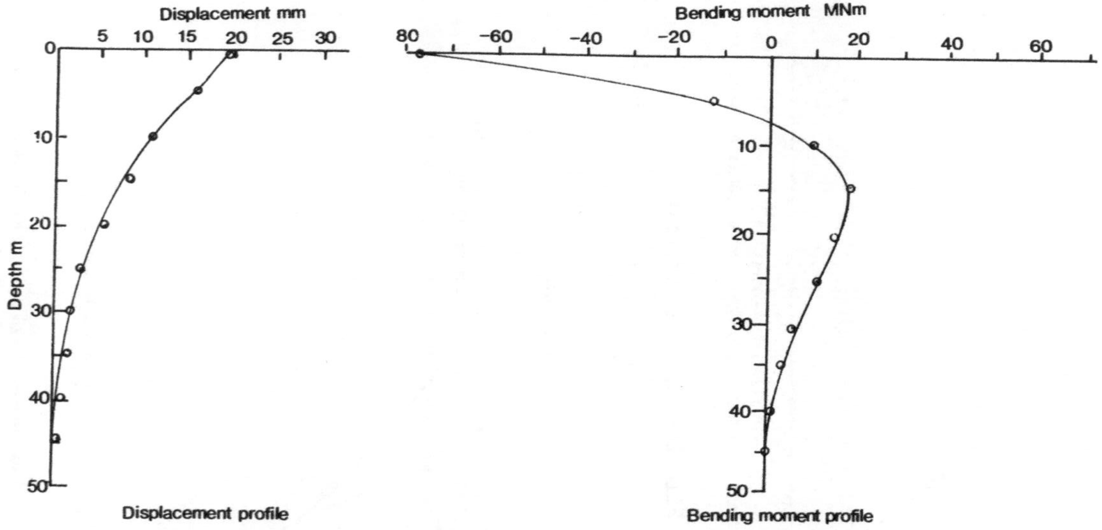

Figure 9 Group as a Single Pile

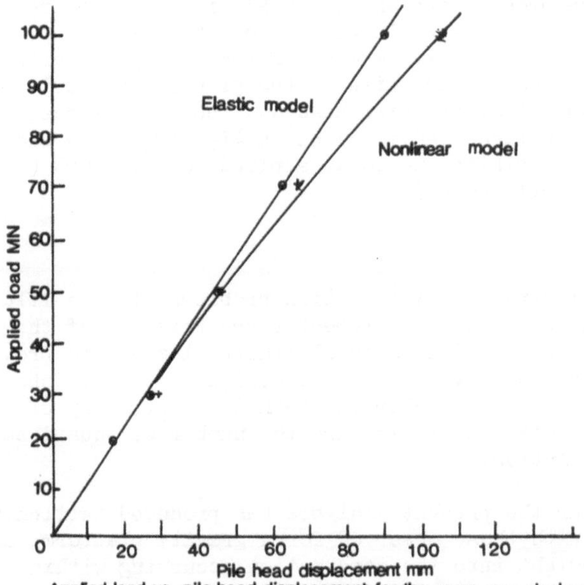

Applied load vs pile head displacement for the group as a single pile

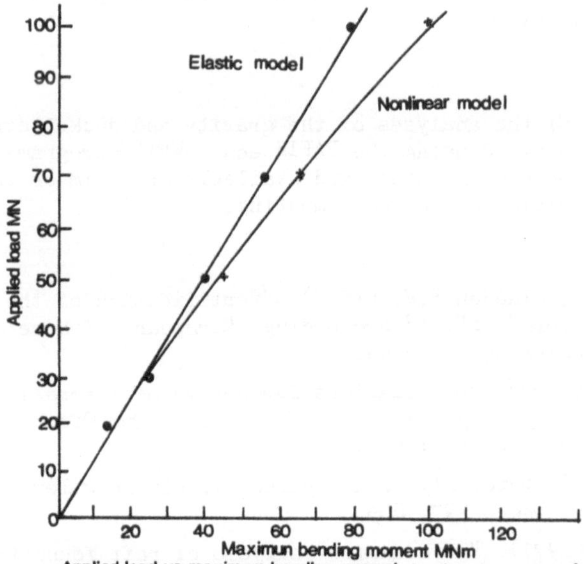

Applied load vs maximum bending moment for the group as a single pile

Figure 10 Group as a Single Pile

been constrained against rotation. The displacement and moment profiles obtained at the design loads are shown in Fig.9, where the deformed shape and the head displacement are almost identical to those given in Fig. 7 for the pile group. The maximum computed positive moment is 18MNm compared with the 1.8MNm developed in each pile of the group. The variation of the displacement of the pile head and the maximum positive moment with applied load is given in Fig. 10, where elastic behaviour is exhibited up to an applied load of 20MN (cf. 50MN for the pile group).

CONCLUSIONS

The two soil-structure interaction problems studied illustrate the ability of a boundary element representation of the soil coupled with standard structural finite elements to provide the engineer with a very powerful design tool in offshore engineering. In particular the method is extremely efficient in terms of data preparation and the number of equations requiring solution.

In particular the present analysis has produced settlements comparably with those observed for a gravity platform, and given an insight into the interaction occurring within the pile group forming the foundation of a jacket structure. For the latter not only have displacements and bending moments at design load been evaluated, but also the behaviour at much higher applied loads has been studied. The results of the latter suggest that the load factor with respect to ultimate is at least 5.

ACKNOWLEDGEMENTS

The results of the analyses of the gravity and jacket structures were obtained using the RAFTS and LAWPILE programs written by the senior author and available from him or through the bureau service of United Computing.

REFERENCES

George, P.J., Sladden P.R. (1978) "Certification of the Heather P atform", EUR 48 Proceedings European Offshore Petroleum Conference, London.

Matlock, H.(1970) "Correlations for design of laterally loaded piles in soft clay" OTC 1204 Proceedings 2nd Offshore Technology Conference, Texas.

Poulos (1971) Laterally loaded piles, I. Single Piles, J. Am. Soc. Civ. Engrs. 97, SM5

Wood, L.A. (1977) The economic analysis of raft foundations. Int. J. Num. Anal. Meths. Geomechanics. 1.

Wood, L.A. (1978) RAFTS: a program for the analysis soil-structure interaction. Advances Eng. Soft., 1, 11-17.

Wood, L.A. (1979) LAWPILE : a program for the analysis of laterally loaded pile groups and propped sheetpile and diaphragm walls. Advances Eng. Soft., 1, 173-179.

Foss, I and Warming, J. (1979) Three gavity platform foundations. Second Int. Conf. Behaviour of Offshore Structures, Imperial College, London.

Wong, [illegible] (1976) [illegible] programs for the [illegible] with [illegible] international [illegible] Proc. Soc. [illegible]

[illegible] program [illegible]
[illegible] and [illegible] pass [illegible]
[illegible]

[illegible] and Kramer, [illegible] (1974) [illegible]
Foundations [illegible]
[illegible]

Session IV
Stress Concentration and Fracture Mechanics

THE SECOND GENERATION BOUNDARY ELEMENT CONTACT PROGRAM

Torbjörn Andersson

Linköping Institute of Technology, Dept of Mech Eng,
S-581 83 Linköping, Sweden

INTRODUCTION

The problem of the contact of elastic bodies was first studied by H Hertz (1895). His theory is valid under the assumptions that the contact area is small compared to the dimensions of the bodies in contact and that the frictional forces in the contact area can be neglected. Typical engineering applications of Hertz contact are ball and roller bearings. Formulas for stress and deformations in Hertz contact, computed by elastic potential theory, are given by Lundberg and Sjövall (1958). A contact with large contact area, compared with the dimensions of the bodies, is also a typical engineering problem. Rivet and bolted joints or connections are typical examples of such cases. This later problem has been solved by Persson (1964) under the assumption that the shear stresses vanish along the entire contour of the hole.

An interesting survey on theoretical treatments of contact problems is given by Gladwell (1980) and Kalker (1977). The later one also gives an account of the variational formulation of contact problems. The variational formulation of elasticity problems has been used especially for numerical calculations, one of them well known as the Finite Element Method, FEM. The contact problem has been studied using FEM technique by Fredriksson (1976). Using a numerical technique with incremental load steps complex contact problems with friction, can be solved. Two-dimensional contact problems have been analysed using the Boundary Element Method, BEM, by Andersson et al (1980) and Andersson (1981). Constant boundary elements were used in those two papers. A comprehensive discussion on BEM contact solutions in two-dimensional elasticity with constant, linear and parabolic elements is given in Andersson and Persson (1982).

This paper describes a technique for applying BEM to two-dimensional contact problems with friction. Linear and parabolic (second generation) elements are used. By use of a conventional BEM program the relations between displacement and traction increments at the boundaries of the bodies are determined. Considering the friction properties inside the contact area the two systems of equations are linked together and solved numerically. An incremental algorithm is presented which follows the load history as closely as the element model allows. The pressure and shear stress distributions between the two bodies in contact are calculated in each load step. The boundary solution for each load step is easy to use as input to a conventional BEM routine for calculating the resulting stress and strain fields in each of the two bodies in contact. Applications are made to some practical problems and the BEM solutions are compared to solutions from other methods.

BASIC RELATIONS

Boundary Integral Formulation

Consider the linear elastic body Ω, Figure 1, bounded by a cylindrical surface Γ. The geometry and all variable quantities are independent of the co-ordinate perpendicular to the paper. The problem is thus considered as a plane problem in the (x_1,x_2)-system.

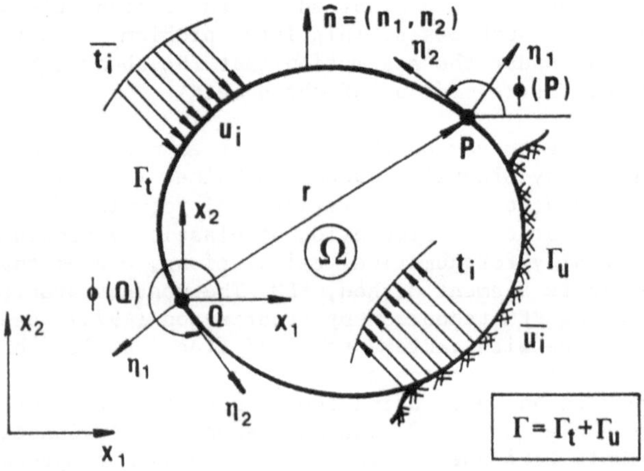

Figure 1 Problem definition

The tractions acting on the boundary are denoted t_i and the displacements u_i. The problem is governed by the Navier equation, obtained by substituting the displacement derivatives into the equilibrium equation by use of the constitutive equations.

Using the fundamental solution, i e a solution of Navier's equation with disturbance term, a boundary integral relation of the Somigliana type can be established, Rizzo (1967). Assume P and Q to be two points inside or on the boundary of the body. The relation is thus written

$$cu_i(P) + \int_\Gamma T'_{ij}(P,Q)u_j(Q)d\Gamma = \int_\Gamma U_{ij}(P,Q)t_j(Q)d\Gamma(Q) \quad (1)$$

The fundamental tensors $T'_{ij}(P,Q)$ and $U_{ij}(P,Q)$ can be found from Rizzo (1967) or Brebbia and Walker (1980). The constant c is equal to 1.0 if P is within Ω and equal to 0.5 if P is on a smooth part of the boundary. It is in the following practical to introduce the Dirac delta function $\delta_{ij}(P,Q)$ for the definition of new integral kermels,

$$T_{ij}(P,Q) = \delta_{ij}(P,Q) + T'_{ij}(P,Q) \quad (2)$$

$$\int_{\Omega+\Gamma} \delta_{ij}(P,Q)u_j(Q)d\Omega = \begin{cases} u_i(P) & P \in \Omega \\ cu_i(P) & P \in \Gamma \\ 0 & P \notin \Omega+\Gamma \end{cases} \quad (3)$$

Equation (1) is thus written in a more compact form as

$$\int_\Gamma T_{ij}(P,Q)u_j(Q)d\Gamma(Q) = \int_\Gamma U_{ij}(P,Q)t_j(Q)d\Gamma(Q) \quad (4)$$

Allow P and Q to be located on the boundary and define local co-ordinate systems (η_1,η_2) in these points. The η_1-axis is perpendicular to the boundary and the η_2-axis tangent to the boundary, Figure 1. The transformation between the two systems is made by the transformation matrix

$$\beta_{ij}(P) = \begin{bmatrix} \cos[\phi(P)] & -\sin[\phi(P)] \\ \sin[\phi(P)] & \cos[\phi(P)] \end{bmatrix} \quad (5)$$

Inside the contact area the normal direction is calculated with a special technique due to geometry and contact conditions, described later in the text. The local co-ordinate systems and equation (4) in local co-ordinates are used throughout in the rest of the text.

Incremental technique

When solving non-linear problems with linear methods the specific problem usually has to be considered in increments. The total quantities are then calculated as the sum of increments, i e for displacements and tractions

$$\begin{aligned} t_i^n &= \Delta t_i^n + t_i^{n-1} = \sum_{k=1}^{n} \Delta t_i^k \\ u_i^n &= \Delta u_i^n = u_i^{n-1} = \sum_{k=1}^{n} \Delta u_i^k \end{aligned} \quad (6)$$

If the increments are sufficiently small the total quantities are sufficiently good approximations to the exact values. When applicating this technique equation (4) is used in each increment and thus takes the form

$$\int_\Gamma T_{ij}(P,Q)\Delta u_j(Q)d\Gamma(Q) = \int_\Gamma U_{ij}(P,Q)\Delta t_j(Q)d\Gamma(Q) \quad (7)$$

However, when solving elastic contact problems, the non-linearities arise in the boundary conditions and not in the material. Thus, from the uniqueness theorem, see Fung (1965), the stresses and strains do not have to be calculated by increments but can be calculated directly from the total quantities.

Contact conditions

Consider two linearly elastic bodies Ω^A and Ω^B, Figure 2, bounded by cylindrical surfaces Γ^A and Γ^B respectively and contacting each other on the contact surface Γ_c.

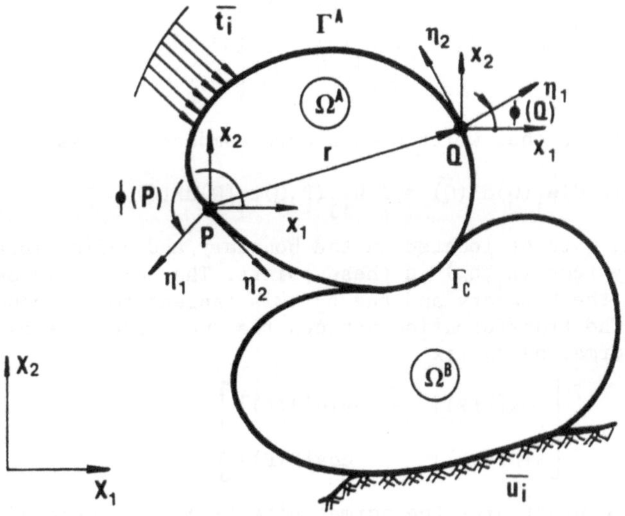

Figure 2 Two elastic bodies in contact along the boundary Γ_c

The boundaries of the bodies are divided into three parts corresponding to the boundary conditions and the contact zone,

$$\Gamma^A = \Gamma^A_t + \Gamma^A_u + \Gamma^A_c$$
$$\Gamma^B = \Gamma^B_t + \Gamma^B_u + \Gamma^B_c \qquad (8)$$

In Equation (8) Γ^A_c and Γ^B_c are the contact boundaries in the undeformed state. According to small displacement theory they have approximately the same length as the contact boundary Γ_c.

From the initial state or a previous load step the normal gap u^o_1 between the two contact boundaries are calculated, see Figure 3. The distance from Γ^A_c to Γ_c and from Γ^B_c to Γ_c are αu^o_1 and $(1-\alpha)u^o_1$ respectively.

Introducing two new co-ordinate systems, located on Γ_c, four new displacement components are defined as

$$v_1^A = u_1^A - \alpha u_1^o \qquad v_1^B = u_1^B - (1-\alpha)u_1^o$$
$$v_2^A = u_2^A \qquad v_2^B = u_2^B \tag{9}$$

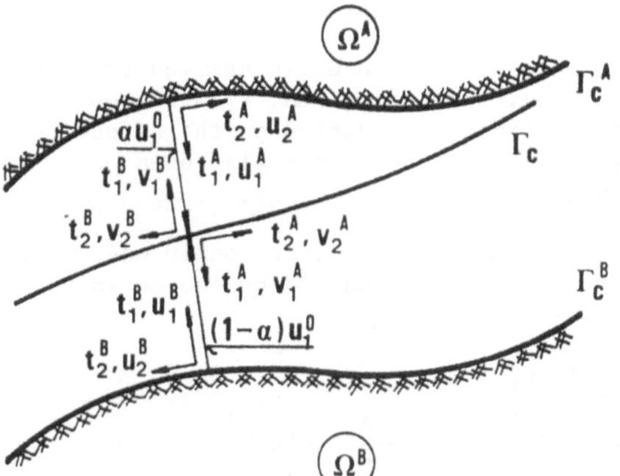

Figure 3 Definition of the local co-ordinate system, the gap u_1^o and the parameter α

By use of the new displacement components and in the absence of friction the boundary conditions are written

$$v_1^A + v_1^B = 0 \qquad t_1^A - t_1^B = 0$$
$$t_2^A - t_2^B = 0 \qquad t_1^A < 0,\ t_1^B < 0 \tag{10}$$

In equation (10) the quantities are referred to each of the local co-ordinate systems. It is worth noting that the components v_1^A and v_1^B represent the misfit of the contact location Γ_c and should be equal to zero. However, the location of the contact zone is a priori unknown and the correct value of the parameter α is impossible to guess. The solution is in practice done by giving trail values to α, for example $\alpha = 0.5$, and solving for the misfit in location of the contact zone.

When friction is considered, the relation between normal and tangential tractions is defined by the friction coefficient μ. The contact zone is divided into two parts corresponding to the friction state, slip or adhesion, for which the relations are

$$t_2 = \pm\mu t_1 \text{ on } \Gamma_{cs} \qquad \text{"slip"} \tag{11a}$$

$$|t_2| < \mu|t_1| \text{ on } \Gamma_{ca} \qquad \text{"adhesion"} \tag{11b}$$

In the case of adhesion the increment of relative tangential displacements $(\Delta v_2^A + \Delta v_2^B)$ equals zero while in the case of a slip it differs from zero. In this latter case the sign of the friction coefficient in Equation (11a) is chosen so that energy is dissipating i e

$$\mathrm{sign}(t_2) \neq \mathrm{sign}\,(\Delta v_2^A + \Delta v_2^B) \tag{12}$$

The friction coefficient is usually assumed to be a constant (Coulomb friction). When using FEM, Fredriksson (1976) assumed the friction coefficient to depend on the effective slip. The same effective slip dependence has also been used in BEM, Andersson (1981).

For each load increment the contact conditions, where the sign of the friction coefficient should be chosen in accordance to equation (12), are written

$$\begin{array}{lll}
\Gamma_{ca}: & \Delta v_1^A + \Delta v_1^B = 0 & \Delta v_2^A + \Delta v_2^B = 0 \\
 & \Delta t_1^A - \Delta t_1^B = 0 & \Delta t_2^A - \Delta t_2^B = 0 \\
 & t_1^A < 0 & t_1^B < 0 \\
\Gamma_{cs}: & \Delta v_1^A + \Delta v_1^B = 0 & \\
 & \Delta t_1^A - \Delta t_1^B = 0 & \Delta t_2^A - \Delta t_2^B = 0 \\
 & t_2^A = \pm\mu t_1^A & t_2^B = \pm\mu t_1^B \\
 & t_1^A < 0 & t_1^B < 0
\end{array} \tag{13}$$

BEM-FORMULATION OF THE TWO-DIMENSIONAL CONTACT PROBLEM WITH FRICTION

Forming the systems of equations

Consider the two elastic bodies of Figure 2. For the solving of the contact problem we assume the length of the contact zone to be Γ_c. Establishing equation (7) for both the bodies and introducing the contact displacements from equation (9) results in two uncoupled integral equations. To couple these integral equations, the contact conditions of equation (13) are used to eliminate the body B contact variables. If the superscript for the remaining body A contact variables are dropped the coupled system of integral equations is formed. The last integral in both equations comes from the change of displacement variables and represents the work needed to obtain the assumed total contact area together when no external load is applied. For simplicity the kernels are written T_{ij} instead of $T_{ij}(P,Q)$ etc.

$$\int_{\Gamma^A-\Gamma^A_c} T^A_{ij}\,\Delta u^A_j d\Gamma + \int_{\Gamma^A_{ca}} T^A_{ij}\Delta v_j d\Gamma + \int_{\Gamma^A_{cs}} (T^A_{i1}\Delta v_1 + T^A_{i2}\Delta v^A_2) d\Gamma =$$

$$= \int_{\Gamma^A-\Gamma^A_c} U^A_{ij}\Delta t^A_j d\Gamma + \int_{\Gamma^A_{ca}} U^A_{ij}\Delta t_j d\Gamma + \int_{\Gamma^A_{cs}} (U^A_{i1} \pm \mu U^A_{i2})\Delta t_1 d\Gamma +$$

$$+ \int_{\Gamma^A_c} \alpha T^A_{ij} u^o_1 d\Gamma \qquad (14a)$$

$$\int_{\Gamma^B-\Gamma^B_c} T^B_{ij}\,\Delta u^B_j d\Gamma + \int_{\Gamma^B_{ca}} - T^B_{ij}\,\Delta v_j d\Gamma + \int_{\Gamma^B_{cs}} (-T^B_{i1}\,\Delta v_1 + T^B_{i2}\Delta v^B_2) d\Gamma =$$

$$= \int_{\Gamma^B-\Gamma^B_c} U^B_{ij}\,\Delta t^B_j d\Gamma + \int_{\Gamma^B_{ca}} U^B_{ij}\,\Delta t_j d\Gamma + \int_{\Gamma^B_{cs}} (U^B_{i1} \pm \mu U^B_{i2})\,\Delta t_1 d\Gamma +$$

$$- \int_{\Gamma^B_c} (1-\alpha) T^B_{i1}\, u^o_1\, d\Gamma \qquad (14b)$$

Treating the integrals containing the applied load and the gap u^o_1 as two load systems they can be solved separately. The total increment is thus calculated as their sum. The contribution from the applied load is linear and easily scaled to form the desired load increment.

In the absence of friction the load history does not has to be followed and a start guess of the size of the contact region can be made. In this case $u^o_1 \neq 0$ if there is a gap between the two bodies over the assumed contact area. When friction is considered the load history has to be followed. Thus, when the contact area is increasing $u^o_1 \equiv 0$.

The resulting system of integral equations gives one relation for each dimension and boundary point P located outside the contact zone. When P is on the contact surface, two relations for each dimension and contact point are established. Thus, there is always one relation for each unknown quantity.

The assumptions to be done concern the location and the length of the contact zone. The location of the contact zone, the choice of α, is not a real assumption but rather a change of variables. When using equation (9) to calculate the boundary displacements they should be independent of the choice of α. The independence of α has been verified by numerical calculations, but can probably be shown by consideration of work equations and linear elasticity.

Boundary element modelling

The boundary elements used in the paper are the (second generation) two node lienar and the three node parabolic elements, Figure 4.

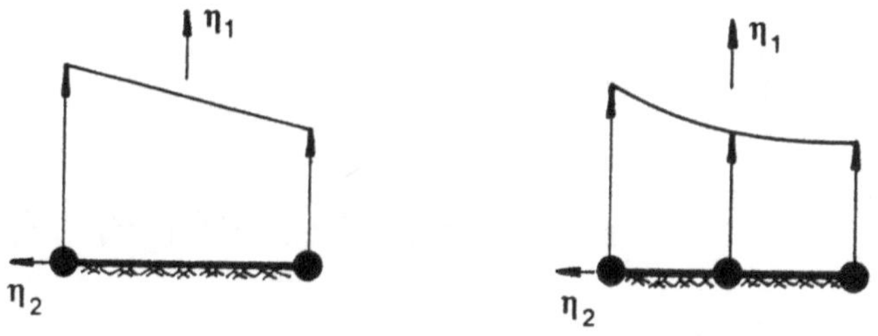

Figure 4 Boundary elements used in the paper
a) linear
b) parabolic

The conventional BEM numerical system is formed, Lachat (1975), taking care of discontinuous traction distribution, i e the traction in the node can take two values depending on to which element it is referred. On the contact surface the contacting nodes are coupled node by node. The normal gap is calculated as the dot product between two vectors. The first vector is the vector between the nodes $\underset{\sim}{u}^{o}$ and the second an associated normal vector $\underset{\sim}{n}$ for these nodes. The associated normal vector is calculated as the "mean" value of the "mean" normal vectors of the contacting elements in the node, Figure 5a.

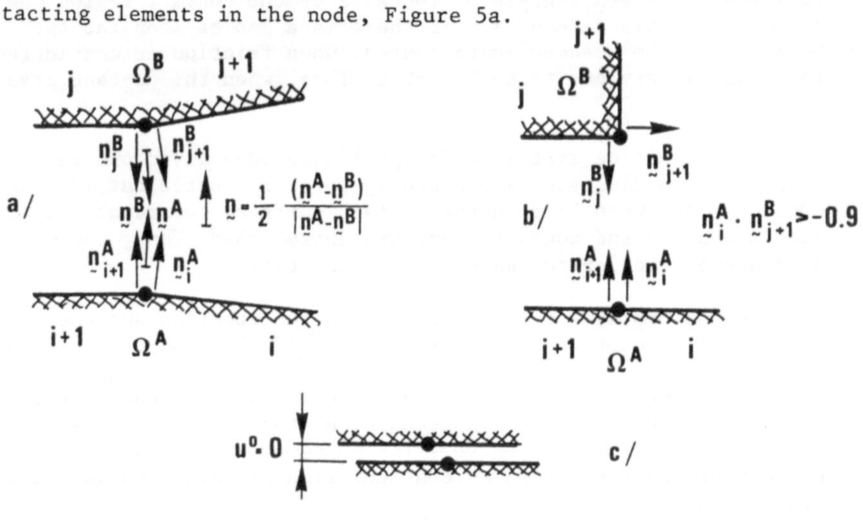

Figure 5 Calculation of associated normal vector and the gap u_1^o

If the node is connected to two contact elements the normal vector of the node is calculated as the "mean" normal vectors of the contact elements in the node. Whether the elements connected to the node are possible contact elements or not is desided be calculating the dot products of the corresponding normal vectors, $\underset{\sim}{n}_i^A$ and $\underset{\sim}{n}_{i+1}^B$ or $\underset{\sim}{n}_{i+1}^A$ and $\underset{\sim}{n}_j^B$. If the product is greater than -0.9 (a value which is set in the program) they are not possible contact elements, Figure 5b. When discontinuous tractions occur the traction of the non-contacting elements has to be known in the node, usually taken as zero. The normal vector also defines the local co-ordinate systems.

This technique for the calculation of the gap u_1^o is necessary since the length of the vector between the nodes does not give the correct information, for example in the case of a zero normal gap, Figure 5c. In this case $u_1^o = 0$ while the nodes may differ from each other as a result of the numerical modelling of the element mesh or relative tangential displacements before contact is established. The "discontinuous" technique automatically takes care of discontinues tractions in such cases where the contact area does not have physical possibility to grow.

According to the boundary and contact conditions the equation system is formed. Unknown quantities related to the u_1^o-integral are collected in $\underset{\sim}{x}^o$, unknowns from the load contribution in $\underset{\sim}{x}$ and the load vectors respectively in $\underset{\sim}{b}^o$ and $\underset{\sim}{b}$.

$$\underset{\sim}{A}(\underset{\sim}{x}^o,\underset{\sim}{x}) = (\underset{\sim}{b}^o,\underset{\sim}{b}) \tag{15}$$

The system matrix $\underset{\sim}{A}$, written out in extensio, takes the form

$$\underset{\sim}{A} = \begin{bmatrix} T^A & U^A & T^A_{1a} & T^A_{1s} & T^A_{2a} & T^A_{2s} & 0 & 0 & -U^A_{1a} & (-U_{1s} \pm \mu U^A_{2s}) & -U^A_{2a} & 0 \\ 0 & 0 & T^B_{1a} & T^B_{1s} & T^B_{2a} & 0 & T^B & U^B & -U^B_{1a} & (-U^B_{1s} \pm \mu U^B_{2s}) & -U^B_{2a} & T^B_{2s} \end{bmatrix}$$

Scaling of the load increment

After solution of equation (15) the load has to be scaled according to contact conditions. In the absence of the friction the normal pressure should be positive over the entire contact and no overlapping of elements should be present outside the contact. The load increment is scaled up to the point where the contact conditions should be changed or up to the totally applied load. In the first case the load is scaled up to the point where two new nodes are contacting each other or the node at the limit of the contact gets zero contact pressure. The contact conditions are thus changed and a new load-step is taken.

When friction is present the conditions for slip and adhesion have to be scaled too. In the case of a node in adhesion the direction of the increment vector $\Delta\underset{\sim}{t}$ of normal and tangential tractions in the node is studied. The slip surface can be expressed in the vectors $\underset{\sim}{s}^+ = [-1.0,\mu]^t$ and $\underset{\sim}{s}^- = [-1.0,-\mu]^t$, see Figure 6.

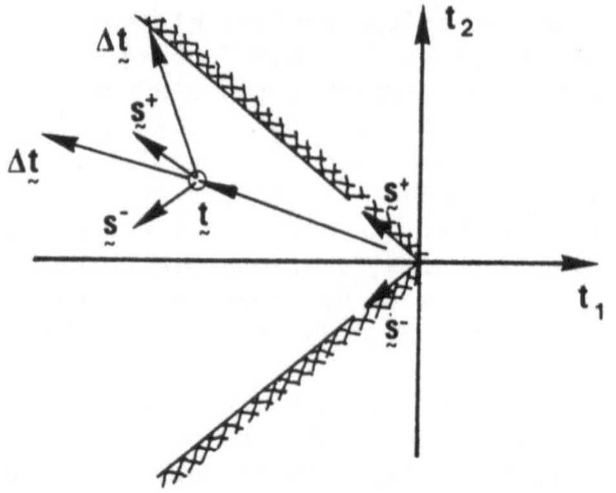

Figure 6 Slip surfaces and increment vectors

By solving the system

$$\begin{bmatrix} s^+ & s^- \end{bmatrix} \begin{bmatrix} \beta^+ \\ \beta^- \end{bmatrix} = \Delta t \tag{16}$$

the direction of Δt is found. If $\beta^+, \beta^- \geq 0$ the tangential traction will never reach the slip surface. If $\beta^+ < 0$, $\beta^- > 0$ or $\beta^+ > 0$, $\beta^- < 0$ the tangential traction can reach the slip surface. The scaling increment for the node is found by solving the system

$$\begin{bmatrix} s^\pm, -\Delta t \end{bmatrix} \begin{bmatrix} \beta^\pm \\ \beta \end{bmatrix} = t \qquad \begin{matrix} + \text{ if } \beta^+ > 0, \ \beta^- < 0 \\ - \text{ if } \beta^+ < 0, \ \beta^- > 0 \end{matrix} \tag{17}$$

The increment, i e the magnitude of Δt when it reaches the slip surface, is given by β. The same technique is used when $\beta^+ < 0$, $\beta^- < 0$.

The second case of friction conditions, a node in slip, is scaled for "forward" slip. The sign of the increment in relative tangential displacements $(\beta\Delta u_2^A + \beta\Delta u_2^B)$ is compared with the sign of total tangential traction $(t_2 + \beta\Delta t_2)$ in the node. The signs should be opposite to each other so that energy is dissipating in the slip zone. The scaling is made for changes in sign of the tangential traction, i e zero tangential traction.

Starting with a value on the scaling parameter β corresponding to the total applied load the algorithm searches for the smallest $\beta \geq 0$. The node is noted and after the scaling procedure a call to a suitable routine for change of contact status is made.

For the parabolic element not only the mid point node but the entire element is taken into the contact when the contact area increases. This technique is used to avoid geometric incompatibilities,

Figure 7. The gap between such new nodes is taken care of by the gap term (u_1^o). Unfortunately, this gap gives rise to oscillations

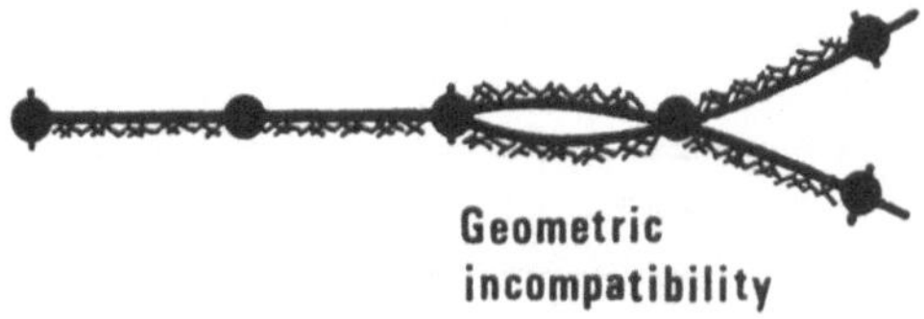

Figure 7 Geometric incompatibility in the contact element

in the tangential tractions when friction is present.This effect can probably be avoided by use of linear constraints for the mid-point node of the elements in close vicinity to the contact.

Stresses and strains

Due to the uniqueness theorem of elastic bodies the stresses and strains can be calculated from the traction and displacement results in each load step. The calculation is easily made by using a conventional BEM routine.

Computer program system

The computer program system for solving two-dimensional contact problems is developed for interactive running possibilities. Allocation of available primary memory and all data handling is made by a data management system, Glemberg and Petersson (1982). The contact program is built up of four stand alone modules. They are, see also Figure 8,

DB1 - Input module for contact bodies and contact data. Includes routines for the inspection of the database.

DB2 - Establishing of contact matrix and load vectors.

DB3 - Solver routine, blocks the matrix to suit available core storage.

DB4 - Scaling of load step. Change of contact status. Printing of results.

The data management system allocates space in primary memory only for those fields which are to be used by the module. The system is developed for FEM programs but can of coarse also be used with a BEM program. It makes it easy to include a new module or change for a better one.

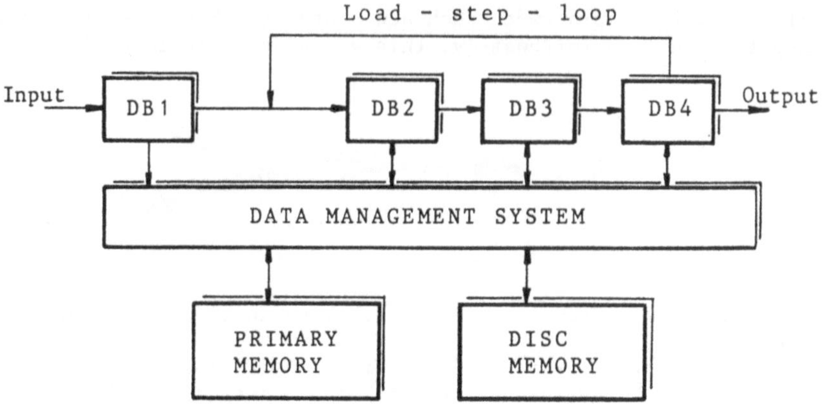

Figure 8 Computer program system

APPLICATIONS

Three examples of problems which can be studied by the program will be given. The first is the Hertz problem with friction which shows how the incremental algorithm works. The second is the problem of contact and shear stresses in a bolt-lug connection and will show the influence on the contact stress distribution of lug-thickness, clearance and friction. The third is the problem of an elastic punch on an elastic foundation. This example will examine the singular contact pressure problem at the end of the contact. It also demonstrates how well this problem is handled by a conventional BEM-modelling of the structures. In all cases of friction the tangential traction is divided by the friction coefficient so that the contact pressure and shear stress distributions takes the same curve in the slip region.

Hertz problem with friction

The element models of the two bodies, with parabolic elements, are shown in Figure 9. Symmetry properties have been used. For the linear element case twice the number of elements where used so that degrees-of-freedom nodes take the same co-ordinates for easy comparison. In Figure 10 (right half) the contact pressure and shear stress distributions are shown for different load steps. From the formulas of Hertz the increment in contact pressure and contact width should have a constant ratio. This is well verified by the calculations except for the first load step where it differs slightly. The contact starts with a point contact and adhesion at the symmetry line. In load step number 5 the new limit node has changed for slip before the load has been applied. In load step number 6 the scaling is made for slip in the next node while load step number 7 again is scaled for a new contact node. For parabolic elements the frictional forces do not give a smooth curve probably due to the gap between nodes when taking new elements into the contact. For parabolic elements the case of vanishing friction is presented, Figure 10 (left half).

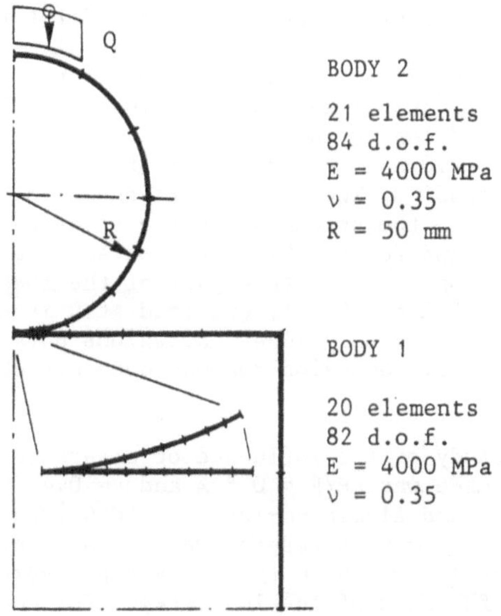

Figure 9 Elastic roller on elastic foundation BEM-models

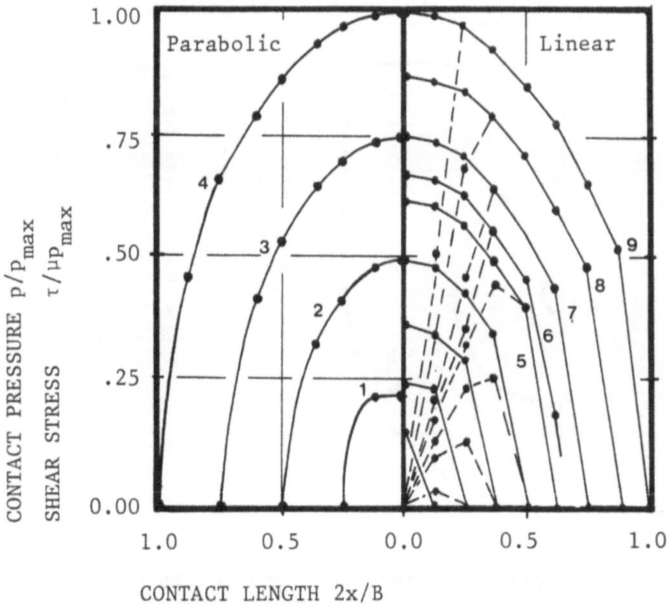

Figure 10 Contact pressure and shear stress distributions. p_{max} = 79 MPa, B_{max} = 8.0 mm, μ = 0.005. (— = contact pressure, --- = shear stress)

Bolt-lug connection

The element models, for linear elements, are shown in Figure 11. The relative clearance between bolt and hole is $\Delta R/R = 0.0$ and the bodies are made of the same material. First the influence of the lug-thickness was studied using linear elements and no friction. The results for $R_y/R = 1.5$, 2.0 and 5.0 are shown in Figure 12 (right half). For comparison the solution given by Persson (1964) for an "infinite" lug is included. Next parabolic elements were used for $R_y/R = 2.0$ and the influence of a friction coefficient equal to 0.25 studied. The results are shown in Figure 12 (left half). For $\mu = 0.0$ the agreement to the solution with linear element is good and for $\mu = 0.25$ the contact pressure decreases since the frictional forces take part of the load. Due to the linearity when $\Delta R/R = 0.0$ only one load step is necessary when the contact width and slip-adhesion regions have been found, i e the size of the slip-adhesion regions does not change during the loading.

For the study of the influence of clearance between bolt and lug the choce made was $\Delta R/R = 0.5$ % and $\mu = 0.4$. The materials were steel-bolt and Aluminium-lug E = 20000 [MPa], $\nu = 0.30$ and E = 10000 [MPa], $\nu = 0.35$ respectively. Linear elements were used. In Figure 13 are shown the contact pressure and shear stress distributions for some of the load steps. For comparison the result from $\Delta R/R = 0.00$ is shown.

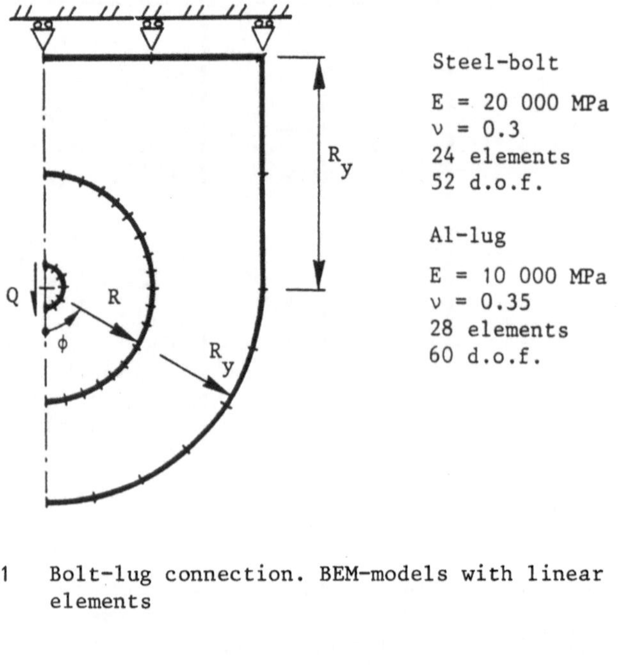

Figure 11 Bolt-lug connection. BEM-models with linear elements

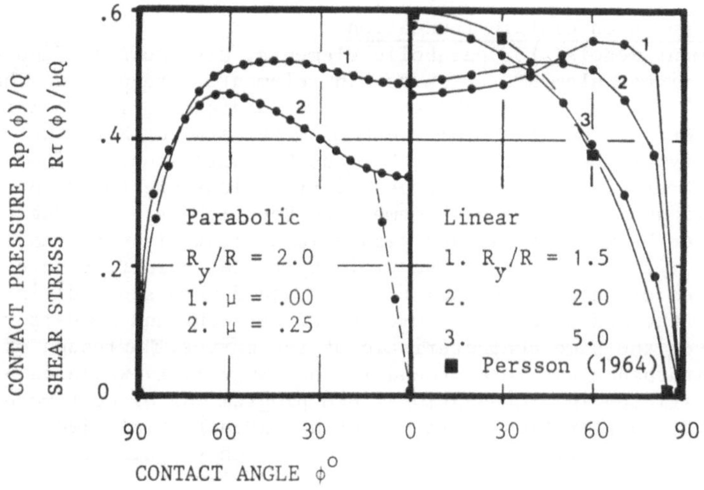

Figure 12 Contact pressure and shear stress distribution between bolt and lug (--- = shear stress)

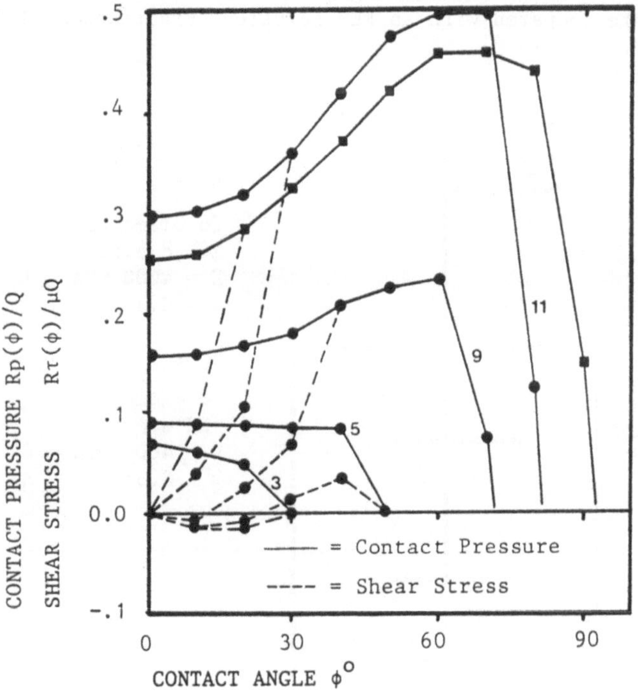

Figure 13 Contact pressure and shear stress distribution between bolt and lug. Load steps 3-5-9-11 for $\Delta R/R = 0.5$ % compared to the case $\Delta R/R = 0$ (■). $\mu = 0.4$.

Elastic punch on elastic foundation

The element models, for parabolic elements, are shown in Figure 14. For linear elements twice as many elements were used so that degree-of-freedom nodes take the same co-ordinates. The bodies are assumed to be made from the same material. For both the element types the friction μ = 0.0 and 0.2 were studied. The contact pressure and shear stress distributions are shown in Figure 15 and are in good agreement to Borowicka (1939). The difference between linear and parabolic elements is very small except for the region in close vicinity to the corner, Figure 16. The number of integration points (2, 3 and 5 were studied) influence the end defect only with a few per cent and a coarser mesh gives the same characteristic of the curves. The reason of this disturbance has to be found in the singularity which cannot be approximated in a conventional BEM program. An element modelling with singularities, pointed out by Xanthis et al (1981), has a great potential for taking care of such singularities.

In Figure 15 the value of the friction coefficient influences the contact pressure. The reason for this has to be found in the coupling between contact pressure and tangential traction. Especially the magnitude of the singular traction at the corner is highly influenced. The relative tangential displacements in the contact zone are compared with an FEM solution, Fredriksson (1976), in Figure 17.

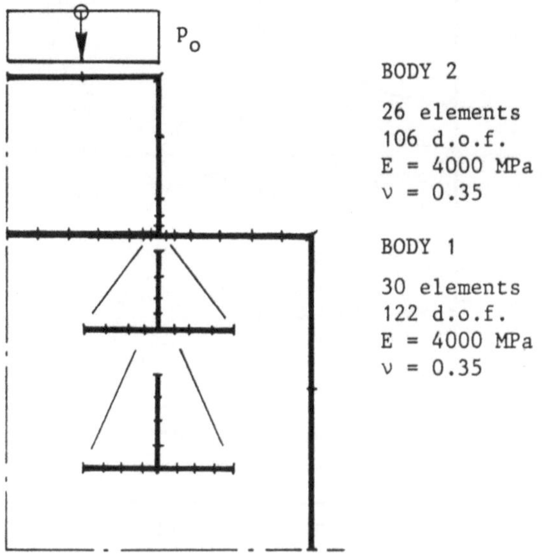

Figure 14 Elastic punch on elastic foundation, BEM-models

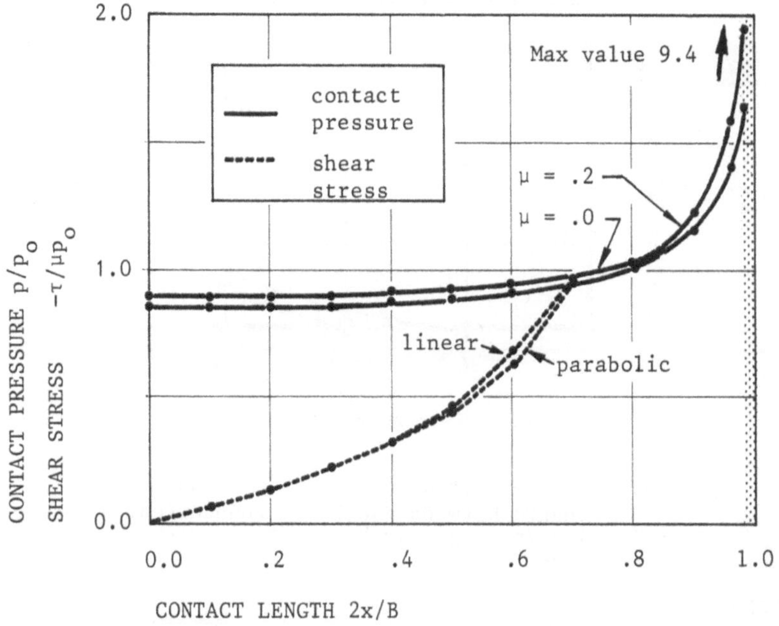

Figure 15 Contact pressure and shear stress distribution between punch and foundation

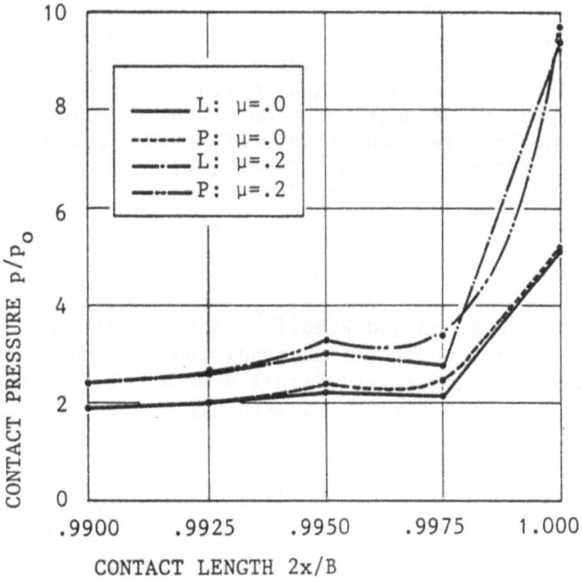

Figure 16 Contact pressure in close vicinity of the punch corner

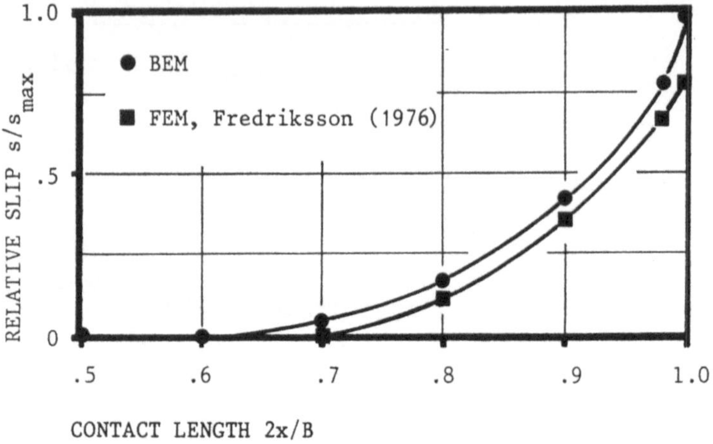

Figure 17 Relative (tangential) slip versus contact length between punch and foundation. $s_{max}/B = 28.6 \cdot 10^{-6}$

CONCLUSION

A numerical technique for solving two-dimensional contact problems including friction has been developed. The contact program couples two linear elastic bodies together by use of suitable contact and friction conditions. The bodies in contact are modelled with linear or parabolic (second generation) boundary elements. The applied load and the initial gap in the contact zone are treated as two different load systems to make it easy to scale the solution afterwards. The incrementing algorithm is scaling each load increment up to the point where new contact or friction conditions should be chosen. The algorithm thus follows the load history as close as the element model allows. Results from solutions of some engineering problems are presented for demonstrating of the algorithm and comparison with other results.

REFERENCES

Andersson T, Persson B G A and Fredriksson B (1980) "The Boundary Element Method Applied to Two-dimensional Contact Problems". Proceedings of the Second International Seminar on Recent Advances in Boundary Element Methods, Southampton, Ed C A Brebbia, CML Publications

Andersson T (1981) "The Boundary Element Method Applied to Two-dimensional Contact Problems with Friction". Proceedings of the Third International Seminar on Recent Advances in Boundary Element Methods, Irvine, California, Ed C A Brebbia, Springer Verlag, Berlin

Andersson T and Persson B G A (1982) "The Boundary Element Method Applied to Two-Dimensional Contact Problems", Chapter 5 in "Progress in Boundary Elements" Vol II, Ed C A Brebbia, Pentech Press, London (to be published)

Borowicka H (1939) Druckerverteilung under elastischen Platten, Ingenieur Archiv, Vol x, No. 2, pp 113-125

Brebbia C A and Walker S (1980) Boundary Element Technique in Engineering, Newnes-Butterworths, London

Fredriksson B (1976) Finite Element Solution of Surface Non-linearities in Structural Mechanics, Computers & Structures, 6, pp 281-290

Fredriksson B (1976) On Elastic Contact Problems with Friction, a Finite Element Analysis. Diss No. 6, Linköping Institute of Technology

Fung Y C (1965) Foundation of Solid Mechanics, Prentice-Hall

Gladwell G M L (1980) Contact Problems in the Classical Theory of Elasticity, Sijthoff and Noordhoff, Alphen aan den Rijn, The Netherlands

Glemberg R and Petersson H (1982) A Data Handling System for Computer Programs Based on the Finite Element Method. Lund Institute of Technology, Division of Structural Mechanics, Lund. Report TVFM-3001 (in Swedish)

Hertz H (1895) Gesammelte Werke, Bd 1.-Leipzig

Kalker J J (1977) A Survey of the Mechanics of Contact Between Solid Bodies, ZAMM75, T3-T17

Lundberg G and Sjövall H (1958) Stress and Deformation in Elastic Contacts. Chalmers University of Technology, Gothenburg, Sweden

Persson B G A (1964) On the Stress Distribution of Cylindrical Elastic Bodies in Contact. Diss Chalmers Institute of Technology, Gothenburg, Sweden

Rizzo F J (1967) An Integral Equation Approach to Boundary Value Problems of Classical Elastostatics, Q Appl Math, Vol 25, pp 83-95

Xanthis L S, Bernal M J M and Atkinson C (1981) The Treatment of Singularities in the Calculation of Stress Intensity Factors Using the Boundary Integral Equation Method. Comput. Meth. Appl. Mech. Engng, 26, pp 285-304

BOUNDARY ELEMENT METHOD FOR TWO-DIMENSIONAL FRACTURE MECHANICS PROBLEMS

F.G. BENITEZ , C. RUIZ
Department of Engineering Science
Oxford University,Oxford,England

SUMMARY

Methods for the determination of stress intensity factors are briefly reviewed and compared for the case of a central crack in an infinite plate subjected to uniaxial tension. A boundary integral equation method that does not require a special function for the treatment of the singularity at the crack tip is presented. The solutions are compared to the Westergaard exact and approximative results and to experimental data.

It is shown that the boundary integral equation provides a correct representation of the stress field remote from the crack tip but that it predicts a variation of the stress intensity factor with orientation as the crack tip is approached.

1. INTRODUCTION

Several techniques have been used for the numerical stress analisys of two-dimensional bodies with cracks, based on either finite elements (F.E.M) or boundary integral equations (B.I.E.M). The former are particularly well documented and established. The latter, not as well known. These techniques include:

1. The use of special interpolation functions on the element or elements near to the crack tip to model the singularity at the tip.

2. Techniques of substraction of the singularity, consisting in considering in the element that contains the crack tip

the ratio t/p^a, where t is the tension and p the distance to the tip, as the unknown and extrapolating to p=0. This technique has been used in B.I.E.M by GOMEZ-LERA[7].

3. Quarter-Point techniques. Used in F.E.M [1,8,10] and latter in B.I.E.M.. They are similar to the first technique as they establish the behaviour of the element near to the singularity by a special interpolation function.

A detailed study, from the mathematical point of view, may be found in XANTHIS [16].

The main difficulty consists in the necessity to model the singularity by preempting the behaviour of the crack tip element or elements. In this paper, a B.I.E.M method that does not imply the a priori knowledge of the type of singularity,is presented. The method is illustrated by treating the case of an infinite plate with a central crack, Fig.1.

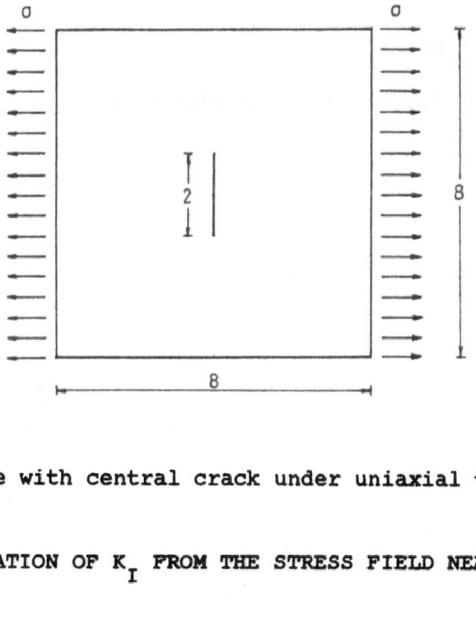

Fig.1. Plate with central crack under uniaxial tension.

2. DETERMINATION OF K_I FROM THE STRESS FIELD NEAR THE CRACK TIP.

The elastic stress field for the tensile loading of a crack, was described by WESTERGAARD [15] by the equations:

$$\sigma_x = \mathrm{Re}\, Z - y\, \mathrm{Im}\, Z'$$

$$\sigma_y = \mathrm{Re}\, Z + y\, \mathrm{Im}\, Z'$$

$$\tau_{xy} = -\, y\, \mathrm{Re}\, Z'$$

where Z is a stress function of the complex variable z. For a crack in an infinite plate under uniaxial tension σ normal to the crack axis, the equations may be modified [11,14] as follows:

$$\sigma_x = \text{Re } Z - y \text{ Im } Z' - \sigma$$

$$\sigma_y = \text{Re } Z + y \text{ Im } Z' \qquad (1)$$

$$\tau_{xy} = - y \text{ Re } Z'$$

with

$$Z = z/(z^2 - a^2)^{.5}$$

$$Z' = - a^2/(z^2 - a^2)^{1.5}$$

$$z = x + iy$$

The coordinate system is shown in Fig.2.

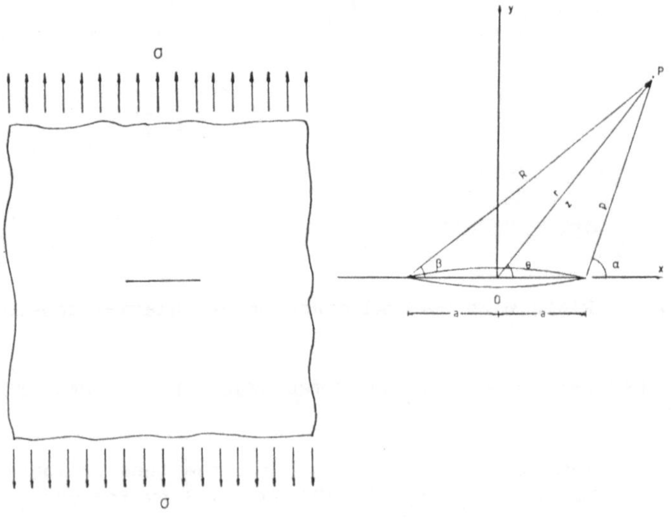

Fig.2. Co-ordinate system in the neighbourhood of the crack tip.

Equations (1) may be modified to:

$$\sigma_x = \sigma\,[S1 + S2 - 1]$$

$$\sigma_y = \sigma\,[S1 - S2] \qquad (2)$$

$$\tau_{xy} = \sigma\,S3$$

where

$$S1 = r(Rp)^{-.5}\cos(\theta - 0.5(\alpha + \beta))$$

$$S2 = ya^2(Rp)^{-1.5}\sin(-1.5(\alpha + \beta))$$

$$S3 = ya^2(Rp)^{-1.5}\cos(-1.5(\alpha + \beta))$$

For a region close to the crack tip, the following approximations can be made:

$$\theta \simeq \beta \simeq 0$$

$$r \simeq a$$

$$R \simeq 2a$$

$$S1 \simeq (a/2p)^{.5}\cos(-\alpha/2)$$

$$S2 \simeq ya^{.5}(2p)^{-1.5}\sin(-1.5\alpha)$$

$$S3 \simeq ya^{.5}(2p)^{-1.5}\cos(-1.5\alpha)$$

being the approximate form of equations (2).

A detailed comparison between both solutions can be found in EVANS and LUXMOORE [6].

It is interesting to compare the loci of constant shear stress $\sigma_1 - \sigma_2$, since these correspond to the photoelastic fringes in an actual experiment, the fringe order when the first fringe appears remote from the crack being,

$$\sigma_1 - \sigma_2 = cte = \sigma\,.N$$

photoelastic fringes have been superimposed on the corresponding loci, plotted from the exact Westergaard equations or their approximation, in figure Fig.3. It is apparent that the experimental results are in excelent agreement with the results predicted by equations (2).

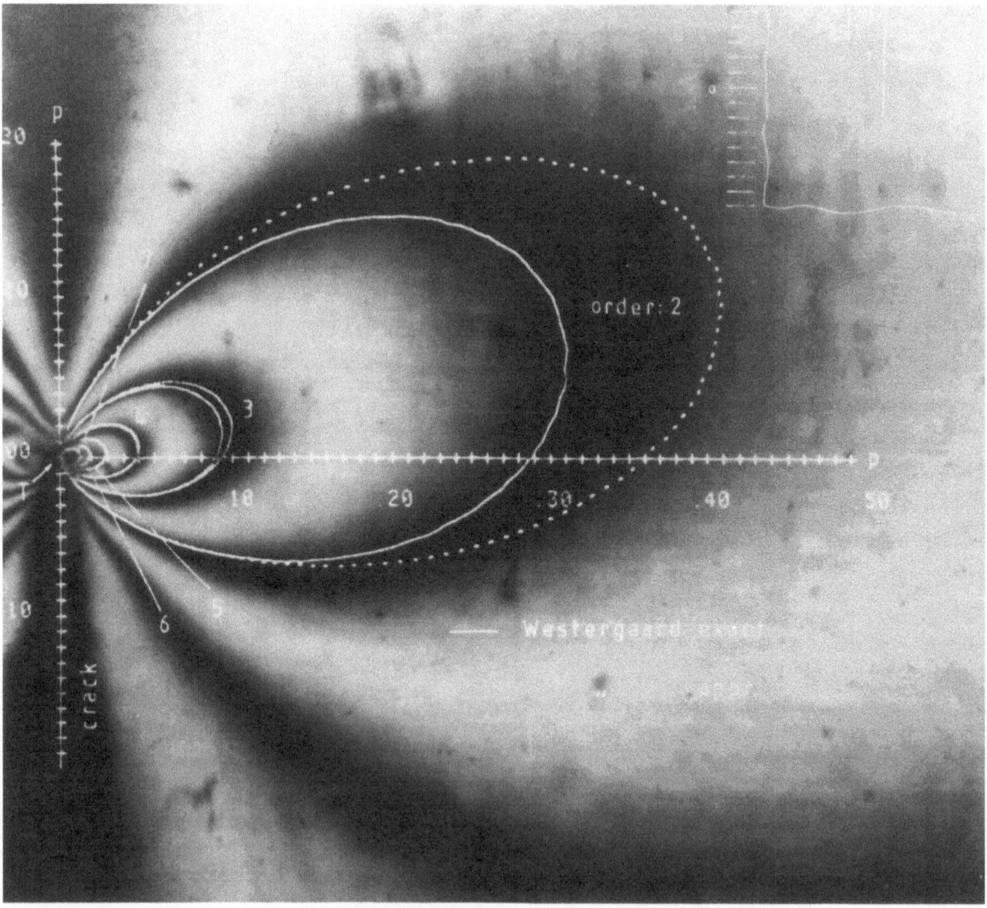

Fig.3. Photoelastic fringes and loci of constant $(\sigma_1 - \sigma_2)/\sigma$ predicted by the exact and approximate Westergaard equations.

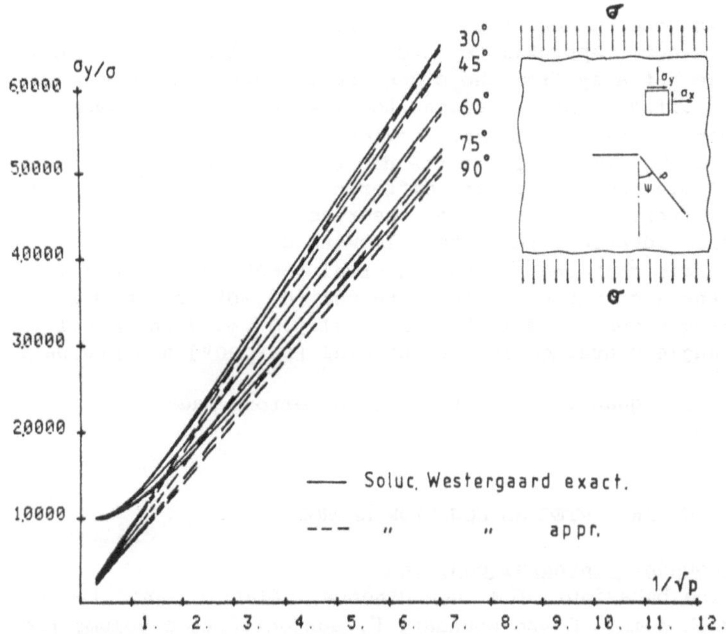

Fig.4. Plot of σ_y against $p^{-0.5}$ from the exact and approximate Westergaard solutions.

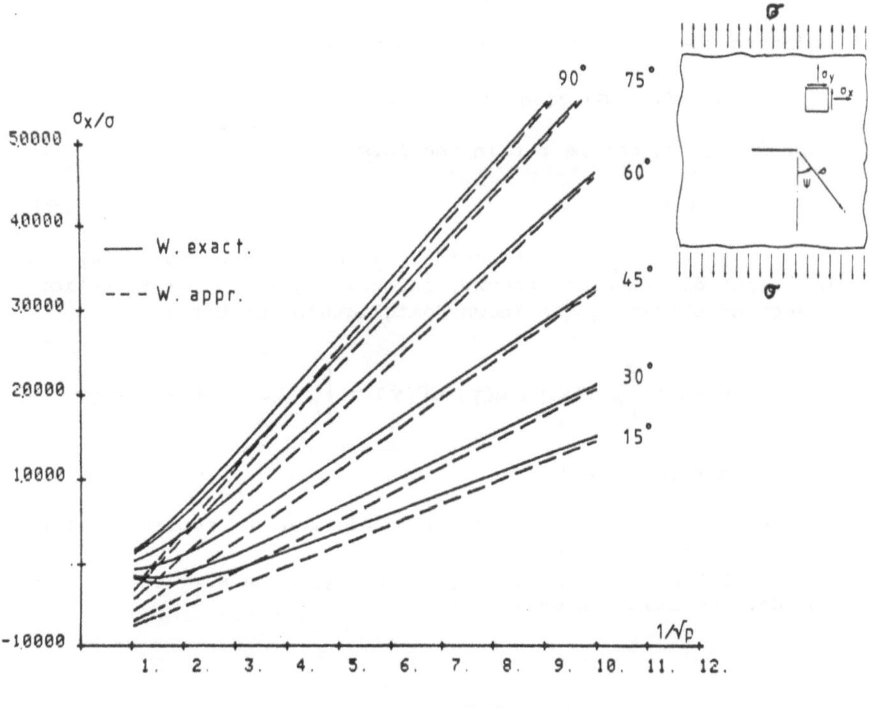

Fig.5. Plot of σ_x against $p^{-0.5}$ from the exact and approximate Westergaard solutions.

As the crack tip is approached, the differences between exact, approximate and experimental results becomes negligible but away from the crack tip the error involved in the simplified approximation to the exact equations is significant. This become more obvious when plotting σ_x and σ_y against $1/\sqrt{p}$, where p is the distance to the crack tip, in figures Fig.4 and Fig.5, for different orientations of the radius vector. It will be observed that solutions are practically identical for the following cases:

- For angle contained in the interval $[-60^0, 0^0]$ and $p \leq 1$.
- For angle contained in the interval $[0^0, 30^0]$ and $p \leq 0.25$.
- For angle contained in the interval $[30^0, 45^0]$ and $p \leq 0.1$.
- For angle contained in the interval $[60^0, 90^0]$ and $p \leq 0.06$.

see [6] for a quantitative study of the errors made.

3. THE BOUNDARY INTEGRAL EQUATION METHOD

3.1 The boundary integral equation

The behaviour of a homogeneous, elastic and linear system with domain Ω and boundary Γ, subjected to a volumetric force system f, is defined by the equation of displacement fields u of their internal points, or Navier equations:

$$(\lambda + G)\ \mathrm{grad}(\mathrm{div}\ u) + G\ \mathrm{lap}\ u + f = 0 \qquad (3)$$

Defining the linear differential operator:

$$\Delta^* = (\lambda + G)\ \mathrm{grad\ div} + G\ \Delta$$

expression (3) can be put in the form:

$$\Delta^* u = -f \qquad (4)$$

The solution of equation [4] can be done, for example, by using distribution theory, integrating in the distribution space and obtaining the known Somigliana's identity:

$$u(x) + \int_\Gamma T(x,y)\ u(y)\ d\Gamma(y) = \int_\Gamma U(x,y)\ t(y)\ d\Gamma(y)$$

$$x \in \Omega\ ;\ y \in \Gamma \qquad (5)$$

as the volumetric forces are null. T and U represent the known Kelvin's tensors.

Extending equation (5) to the system's surface Γ, the boundary equation is obtained:

$$c\, u(x) + \int_\Gamma T(x,y)\, u(y)\, d\Gamma(y) = \int_\Gamma U(x,y)\, t(y)\, d\Gamma(y)$$

$$x,y \in \Gamma \qquad (6)$$

which connects the displacement and stress fields in the surface of system $\Omega \cup \Gamma$.

The expression for the stress fields in the internal points can be obtained using equation (5) and the stress-displacement equation

$$\sigma(x) = \int_\Gamma D(x,y)\, t(y)\, d\Gamma(y) - \int_\Gamma S(x,y)\, u(y)\, d\Gamma(y)$$

$$x \in \Omega \; ; \; y \in \Gamma \qquad (7)$$

3.2 Boundary approximation

For the numerical implementation of the integral equations (5), (6) and (7) it is necessary to discretize of the surface Γ of the domain Ω in elements.

For two-dimensional domains, the boundary can be discretized in one-dimensional elements defined, in the case of parabolic interpolation for the geometry approximation, by three nodes.

The cartesian coordinates x_i^n of each node are known specifications, therefore, the cartesian co-ordinates of a point, nodal or otherwise, will be defined by the parametric equation:

$$x_i(\zeta) = N^n(\zeta)\, x_i^n \qquad \forall\, \zeta \in [-1,1]$$

$$N^1(\zeta) = 0.5\, \zeta(\zeta-1) \qquad \forall\, n = 1,2,3$$

$$N^2(\zeta) = 1 - \zeta^2 \qquad \forall\, i = 1,2 \qquad (8)$$

$$N^3(\zeta) = 0.5\, \zeta(\zeta+1)$$

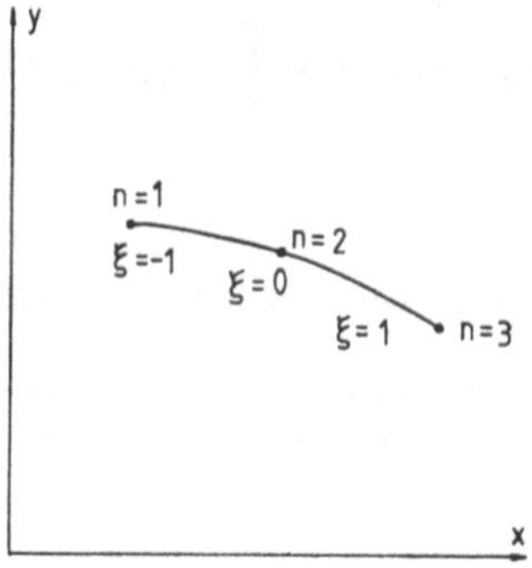

Fig.6. One-dimensional element with parabolic variation of the geometry.

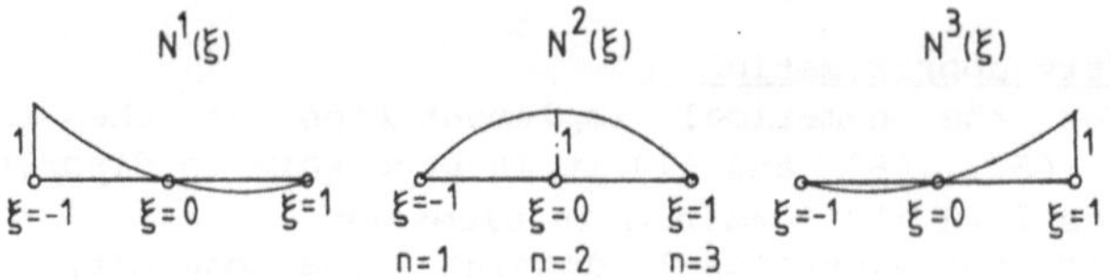

Fig.7. Parabolic interpolation shape functions.

3.3 Approximation of Variables

The functions which are involved in the boundary formulation, (6), and the domain, (5) and (7), as are the stresses and displacements in the bondary t and u, can be interpolated in each node with the discrete values that those functions have in certain points of the element.

For isoparametric elemets, where the geometric interpolation functions as the variables coincide, the value of a variable Φ in a generic point x of the boundary will be expressed by:

$$\Phi(\zeta) = N^n(\zeta)\ \Phi^n \qquad \forall\ \Phi = u,t$$

where the shape functions $N^n(\zeta)$ are defined in (8); Φ^n are the discrete values taken by the variable Φ in the nodes (n=1,2,3) of the element considered.

3.4 Discretization of the equations

The substitution of the boundary, Γ, of the system, by another discretized formed by "E" boundary elements with n (=3) nodes in each of them, causes the equation (5) to transform itself in the discretized equation:

$$c\ u(x) + \sum_{e=1}^{E}\left(\int_{\Gamma e} T\ N^{n}\ d\Gamma e\right) u_{e} = \sum_{e=1}^{E}\left(\int_{\Gamma e} U\ N^{n}\ d\Gamma e\right) t_{e}$$

where:

$$u(x)=\begin{Bmatrix} u \\ v \end{Bmatrix}$$

$$u_{e}^{T}=\{u_{1},v_{1},u_{2},v_{2},u_{3},v_{3}\}_{e} \quad ; \quad t_{e}^{T}=\{\bar{X}_{1},\bar{Y}_{1},\bar{X}_{2},\bar{Y}_{2},\bar{X}_{3},\bar{Y}_{3}\}_{e}$$

u, v : displacements of the node wherein is defined the integral equation
u_e : displacement vector of the nodes which define the element e
t_e : stress vector of the nodes which define the element e

or in compact form:

$$c\ u(x) + \sum_{e=1}^{E} H_{e}\ u_{e} = \sum_{e=1}^{E} G_{e}\ t_{e} \tag{9}$$

where H_e and G_e are shape matrices:

$$B_{e} = \int_{\Gamma e}\begin{Bmatrix} B_{11} & B_{12} \\ B_{21} & B_{22} \end{Bmatrix}\begin{Bmatrix} N_{1} & 0 & N_{2} & 0 & N_{3} & 0 \\ 0 & N_{1} & 0 & N_{2} & 0 & N_{3} \end{Bmatrix} d\Gamma_{e} \tag{10}$$

$\forall$ B = T,U

Establishing equation (9) for all the points, x, where the boundary is discretized, an algebraic system is obtained such as:

$$c\ u + \hat{H}\ u = G\ t$$

or

$$H\ u = G\ t \tag{11}$$

Similary, the equations (5) and (7) will stand discretized as:

$$u(x) = -\sum_{e=1}^{E} H_e\, u_e + \sum_{e=1}^{E} G_e\, t_e$$

$$\sigma(x) = -\sum_{e=1}^{E} S_e\, u_e + \sum_{e=1}^{E} D_e\, t_e \qquad (12)$$

3.5 Compatibility

In expression (11), displacements and stresses are expresed in global co-ordinates. In general, the practical data will be in elemental coordinates so this change should be reflected in the formulation.

In addition, the terms of matrix c can be obtained by the consideration of displacement as rigid body.

The establishement of the equations system (11) for a specified problem adequately defined, require the knowledge of the stresses and/or displacements on a determined number of boundary nodes in same number that the unknown stresses and/or displacements (Dirichlet,Neumann or mixed).

An arrangement of data and unknowns will transform equation (11) in:

$$\mathbf{A}\,\mathbf{x} = \mathbf{b} \qquad (13)$$

which will permit to solve the problem in the boundary.

3.6 Calculation of integrations

The coefficients of equations (9) are defined in (10), and equations (12) can be expressed as:

$$H_e = \int_{\Gamma e} T(\zeta)N(\zeta)\, d\Gamma_e$$

$$G_e = \int_{\Gamma e} U(\zeta)N(\zeta)\, d\Gamma_e$$

$$S_e = \int_{\Gamma e} S(\zeta)N(\zeta)\, d\Gamma_e$$

$$D_e = \int_{\Gamma e} D(\zeta)N(\zeta)\, d\Gamma_e$$

or in compact form by:

$$I = \int_{\zeta=-1}^{\zeta=1} f(\zeta)\, d\zeta \qquad f=(T,U,S,D).N.\text{Jacobian} \qquad (14)$$

The coefficient (14) represents the value provided by the integration of a boundary element from a point, which may belong to that element or not.

The numerical evaluation of (14) through a Gauss interpolation process, for example, of m points (m=6), will permit to calculate it as:

$$I = \sum_{i=1}^{m} f(\alpha_i)\omega_i$$

where α_i, ω_i correspond to ponderated coordinates and weights.

The number of points m where integral (14) is evaluated is related with the error made, and this error affects the results obtained at the boundary on solving the system of equations.

A way to decrease this error consists in maintaining the number of integration points m variable depending on an error function of the distance of the point from which it is integrated, x, to the integrated element, e; as defined by LACHAT [9]. Or, using the subelement technique [3,4], each element Γe is subdivided in subelements Γ_{Se} with local natural coordinates (ζ_1,ζ_2), (14) then becomes:

$$I = \sum_{se=1}^{SE}\int_{\zeta 1}^{\zeta 2} f(\zeta)\, d\zeta =$$

$$= \sum_{se=1}^{SE}\int_{\alpha=-1}^{\alpha=1} f[0.5\{\alpha(\zeta_2-\zeta_1) + (\zeta_2+\zeta_1)\}].(\zeta_2-\zeta_1)/2\, d\alpha =$$

$$= \sum_{se=1}^{SE} 0.5(\zeta_2-\zeta_1)\cdot\sum_{i=1}^{m} f[0.5\{\alpha(\zeta_2-\zeta_1) + (\zeta_2+\zeta_1)\}]\ \omega_i$$

where "SE" represents the number of subelements into which the original element "e" has been subdivided.

With this technique, all the integrations performed are numerical, and the error made in the evaluation of the coefficients will depend of the subelements number wherein is divided the element integrated at that moment, depending also, of some parameter as the minimum distance from the point of integration to the element, the distance to the centroid, the minimum length required for the subelement obtained and the severity parameter needed in the process, BENITEZ [5].

4. NUMERICAL RESULTS FOR THE MODEL PROBLEM

The program, which will be described in section 5, has been used to evaluate numerically the model described in section 1.

Two different models, corresponding to the problem defined in Fig.1, have been used.

4.1 Quarter plate model

Making use of the symmetry considerations, the plate with central crack shown in Fig.1, can be analyzed solving the quarter plate model presented in Fig.8:

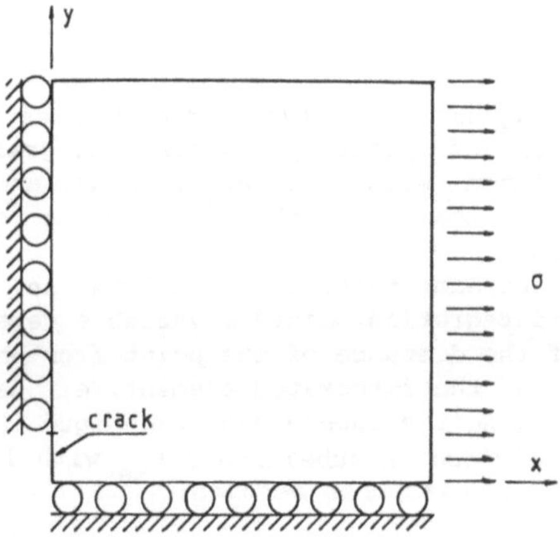

Fig.8. Quarter plate model

Two different discretizations have been used for the quantitative evaluation of the model. The first with 52 elements, and the second with 68 elements.

The elements distribution, for each side of the plate and for each discretization, is shown in Fig.9.

The results obtained for the stress distribution, reactions, in the elements x=0 is shown collected in the grafh of Fig.10, inwhich the abscissae represent distance to the crack tip, and ordinates are the stress, in the direction of tension, normal to the elements.

It will be appreciated that a well in the values provided by the element contiguous to the crack appears for both discretizations. This well appears also in the potential theory when solving problems numerically with singularities governed by Laplace equations, BENITEZ,ALARCON [2]. This behaviour proves the existence of a singularity at the crack tip.

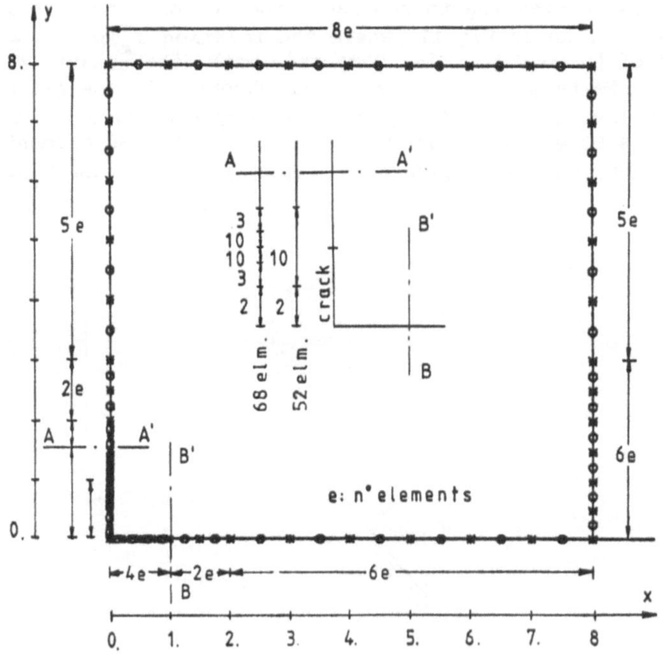

Fig.9. Quarter plate model discretization.

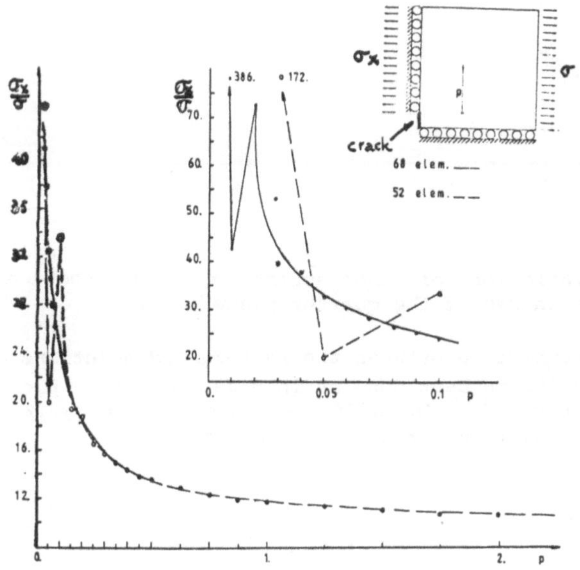

Fig.10. Stress distribution perpendicular to the elements x=0.

It is interesting to represent the graphs of Fig.10 against $p^{-0.5}$, as in Fig.11, where the relation $\sigma_x \propto p^{-0.5}$ in a range of p of [0.03,0.5], is compared with the exact and approximate Westergaard solutions. As observed in the graph, values of p less that 0.1 ($1/\sqrt{p}<3.$) are very far apart in the case of the 52 elements discretization; for the 68 elements discretization, values of $p<0.025$ ($1/\sqrt{p}<6.$) exhibit analogous behaviour.

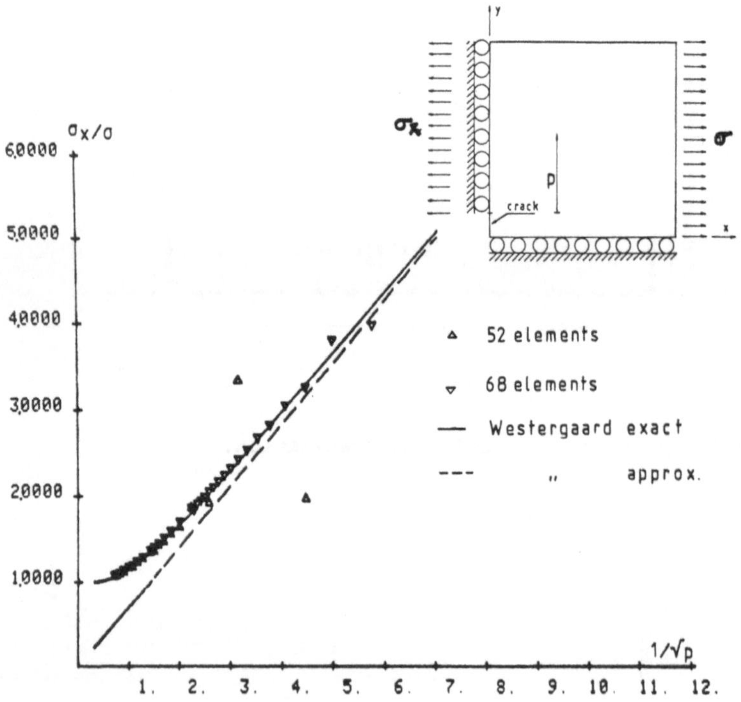

Fig.11. Variation of the stress σ_x in the elements distributed on x=0 of the quarter plate model.

A comparison between the Westergaard solutions and the B.I.E.M results is best done as in Figs.4 and 5 by plotting σ_x and σ_y against $p^{-0.5}$ for different angles. Figs.12 and 13 show this comparison for both discretizations.

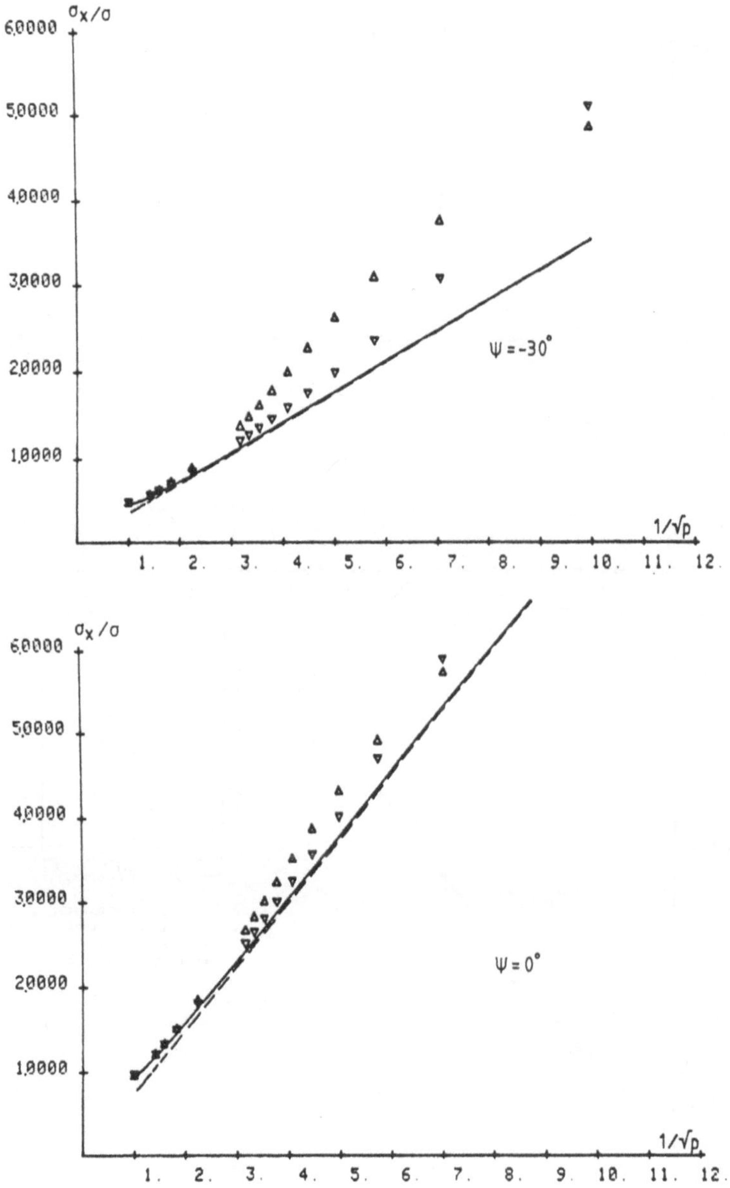

Fig.12. Variation of the stress σ_x in the elements distributed on Ψ=ct of the quarter plate model.

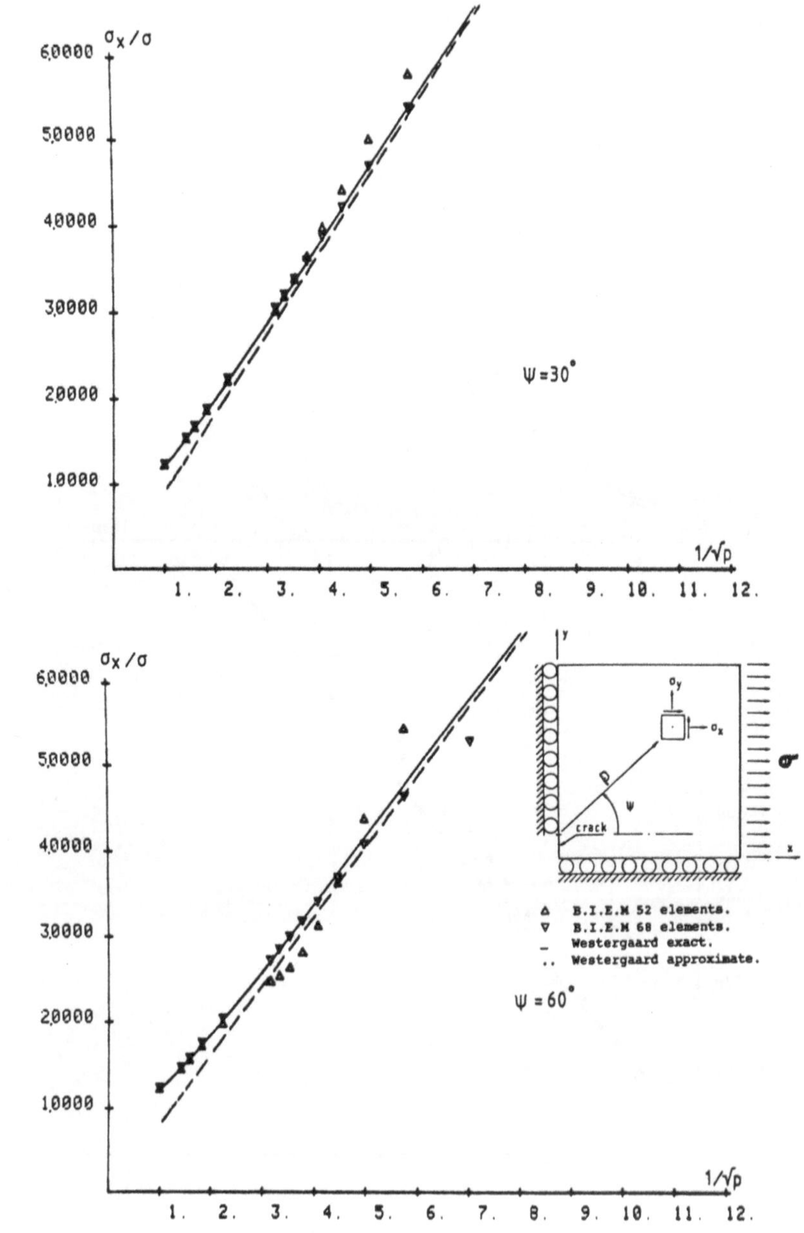

Fig.12(Cont.).Variation of the stress σ_x in the elements distributed on Ψ=ct of the quarter plate model.

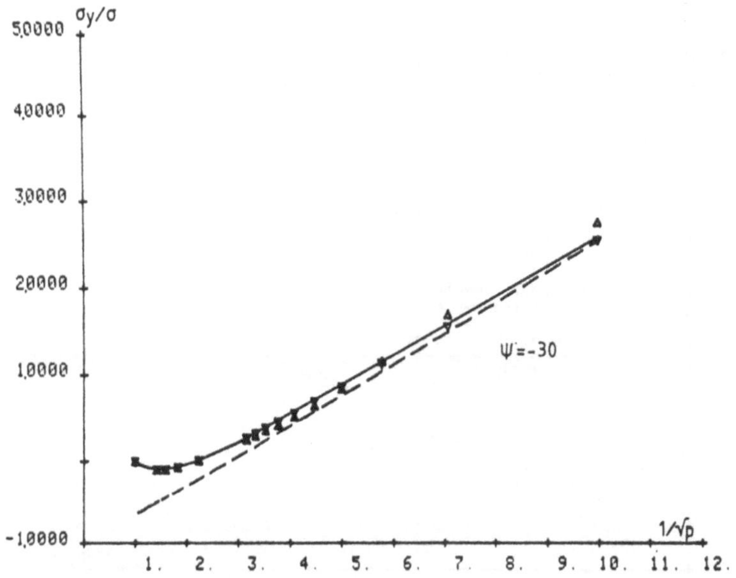

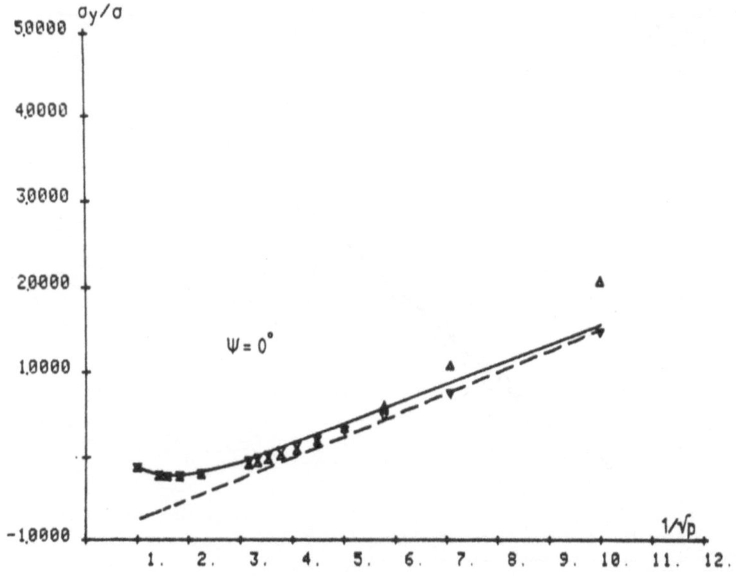

Fig.13. Variation of the stress σ_y in the elements distributed on Ψ=ct of the quarter plate model.

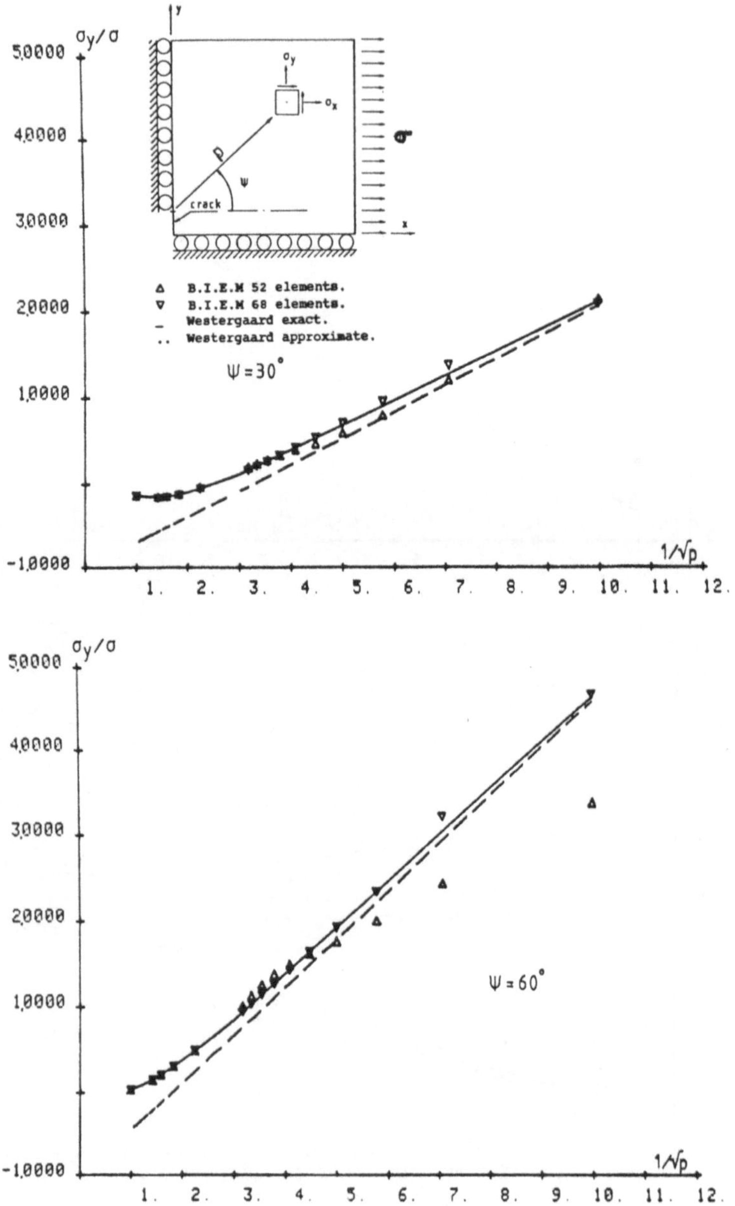

Fig.13(Cont.) Variation of the stress σ_y in the elements distributed on Ψ=ct of the quarter plate model.

4.2 Full plate model

Only one type of discretization has been used for the quantitative evaluation of the model proposed in Fig.1. As described in Fig.14, the discretization consist of 60 elements. The crack is modelled by lips separated by a gap and discretized with 24 elements.

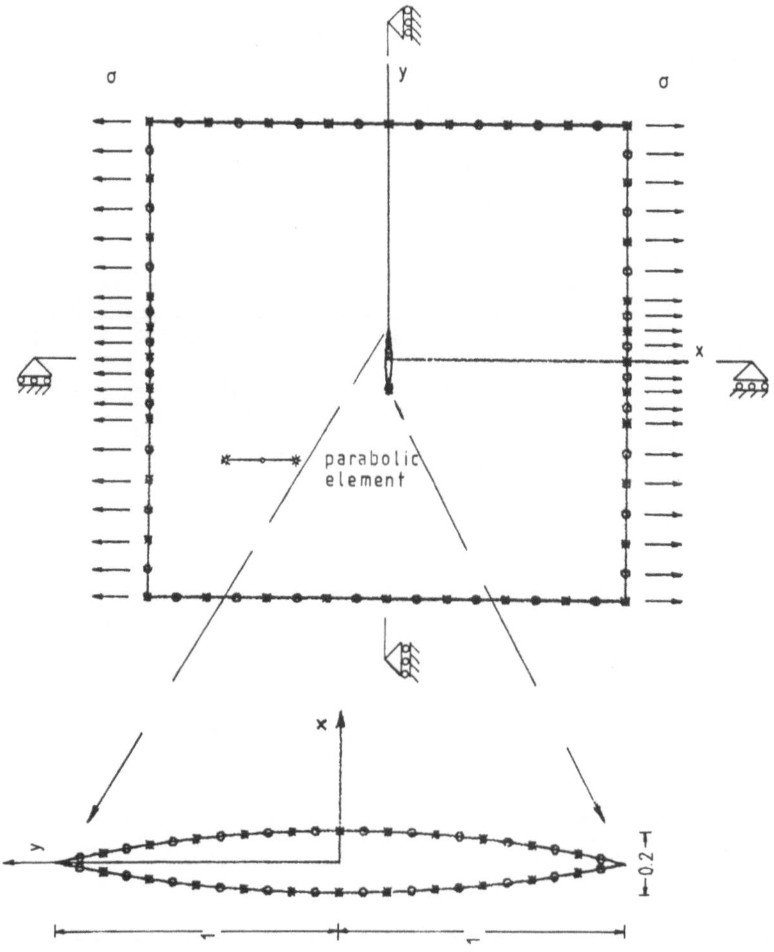

Fig.14. Plate with central crack model discretized.

In Figs.15 and 16 are plotted the stresses σ_x and σ_y respectively for internal points distributed along a straight line of given slope for values of p contained in the interval [0.1,1.]. Also in each graph are shown the theoretical solutions provided by the exact and approximate Westergaard equations.

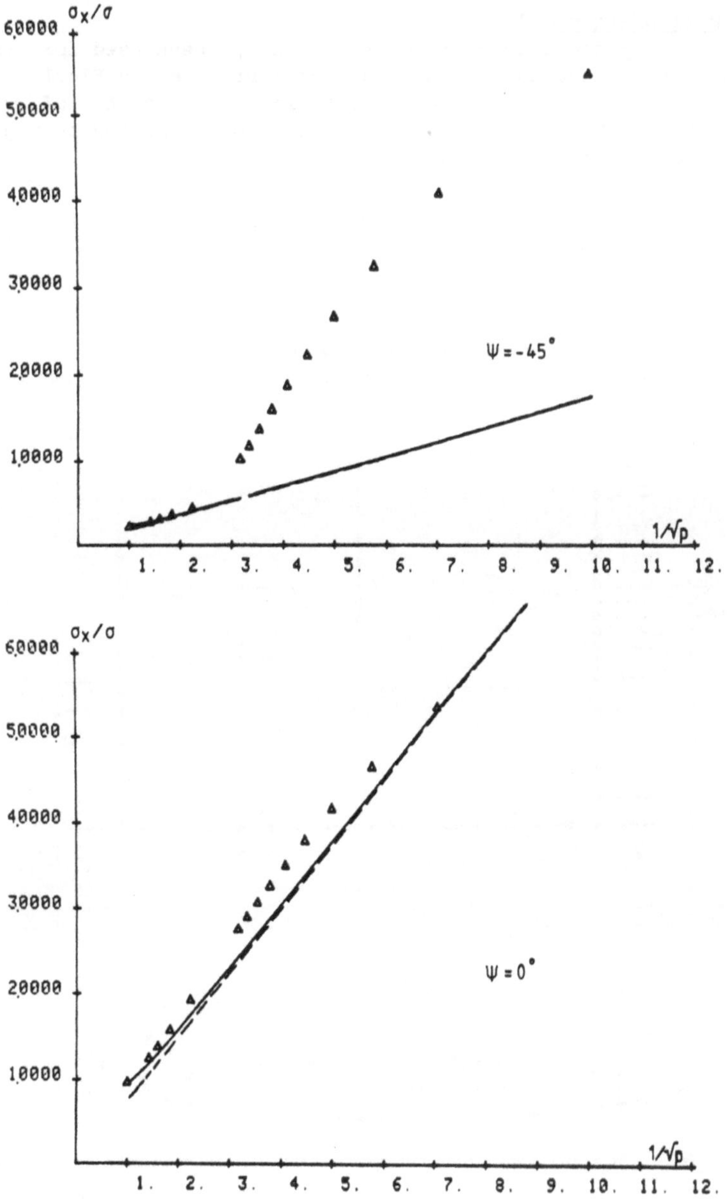

Fig.15. Variation of the stress σ_x in the elements distributed on Ψ=ct of full plate model.

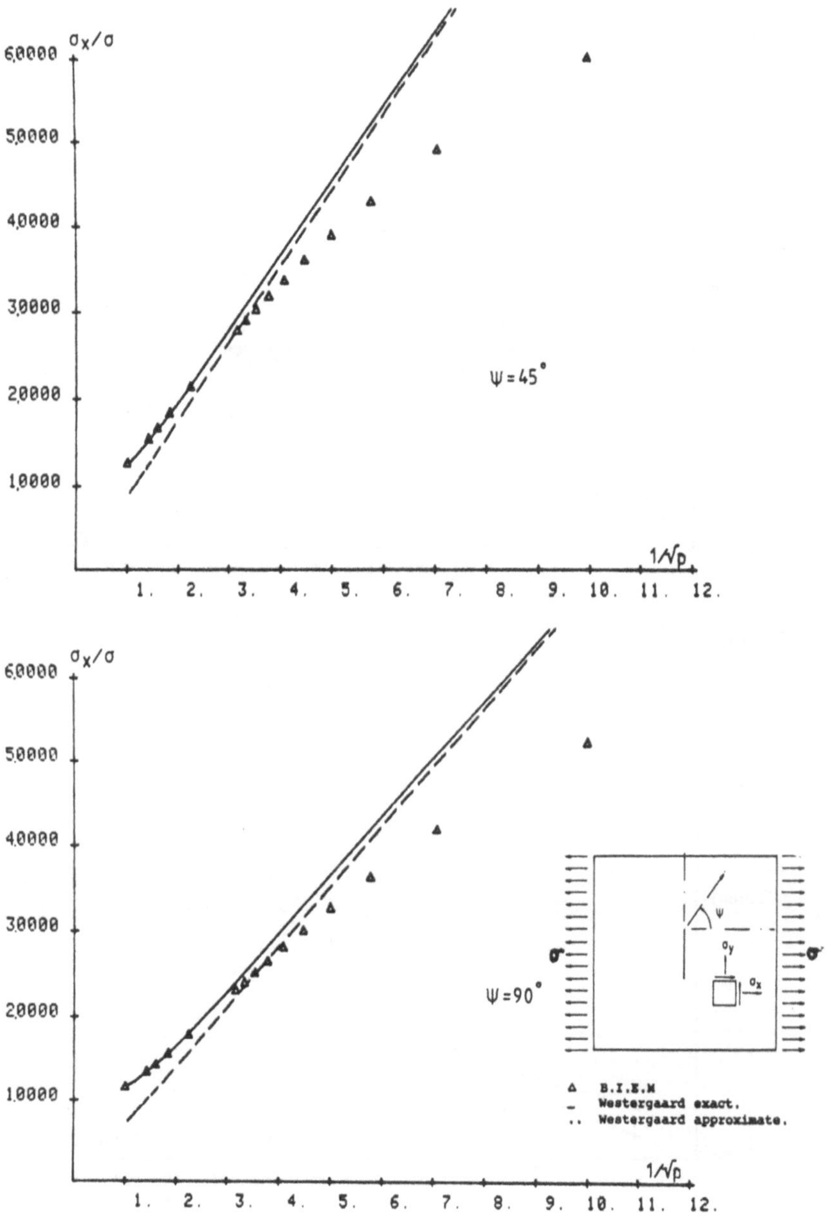

Fig.15(Cont.). Variation of the stress σ_x in the elements distributed on Ψ=ct of full plate model.

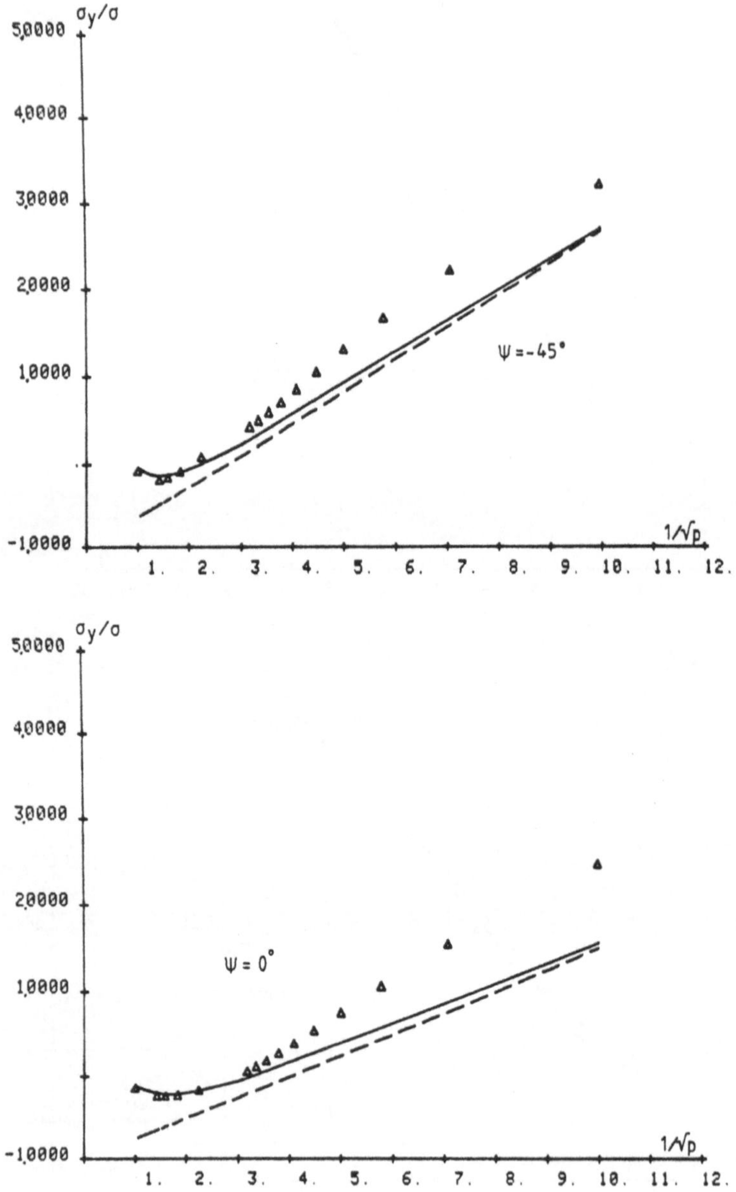

Fig.16. Variation of the stress σ_y in the elements distributed on Ψ=ct of full plate model.

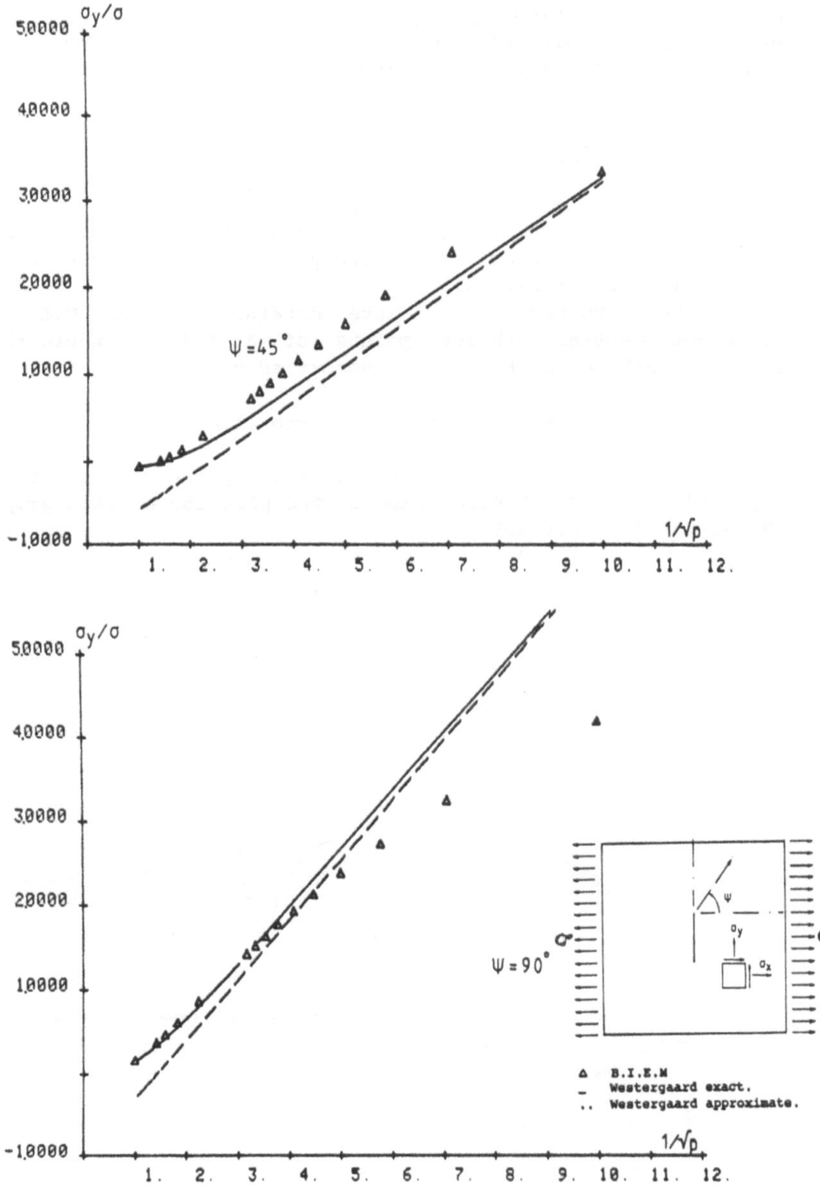

Fig.16(Cont.). Variation of the stress σ_y in the elements distributed on Ψ=ct of full plate model.

<u>Numerical calculation of fringes</u> For the full plate model a numerical analysis of the stress fields has been done, in a zone near the crack tip, plotting in a graph the points that verify the equation:

$$|\sigma_1 - \sigma_2| = cte = \sigma .order$$

where σ_1 and σ_2 represent the principal stresses.

For each order, from 2 to 5, the fringes provided by the exact and approximate Westergaard solutions have been plotted in Fig.3 [Part I].

For each order, the points obtained by the B.I.E.M represent central average points of a fringe, where the fringes width for each point is specified by:

fringe width : 2.r.Error.100

where r is the radius or distance from the point to the crack tip and "Error" the error made in the plotting of each graph is expressed in percentage.

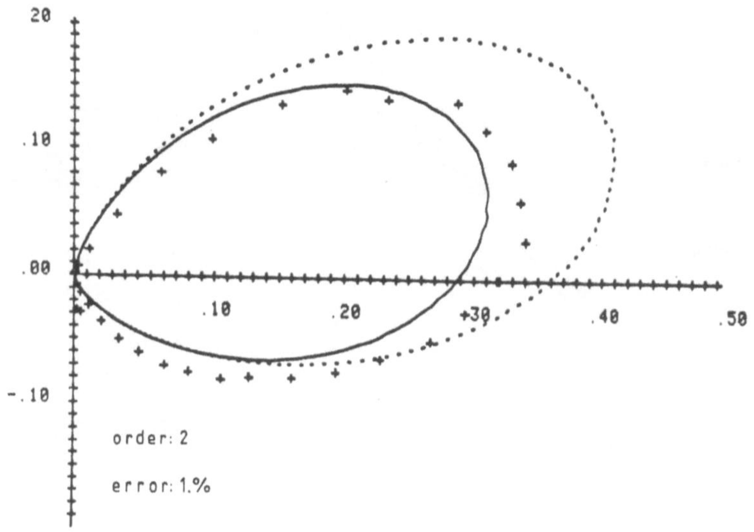

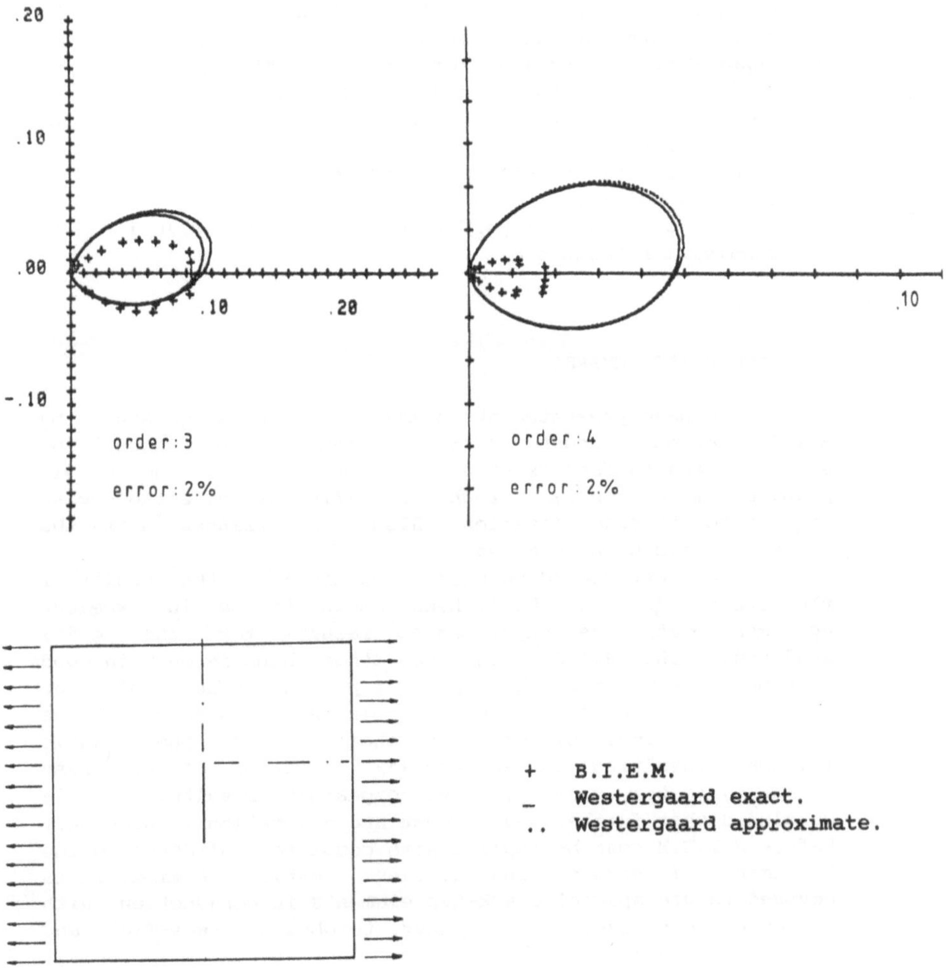

Fig.17. Fringes in the neighbourhood of the crack tip.

5. NUMERICAL PROGRAM

The numerical solution of the models presented has been completed using a program writen in FORTRAN-V language implemented in the VAX 11/780 computer of the Department of Engineering Science of Oxford University.

The program presents as main features:

1. Numerical analysis of continuum media, two-dimensional, homogeneous, isotropic, multiply connected and elastic by the Boundary Integral Equation method theory.

2. Utilisation of parabolic interpolation functions for the stresses and displacements fields.
3. Input data in global and/or local co-ordinates.
4. No limitation in the boundary conditions.
5. No limitation in the number of elements of the discretization.
6. Calculation of stresses and displacements in any required internal points.
7. All integrations are performed numerically following the subdivision technique.

6. CONCLUDING REMARKS

We have presented the numerical results obtained by B.I.E.M method, through a program of general application. The stress intensity factors have been obtained for the clasic problem of a central crack in different directions with respect to the crack position. Also, the fringes near the crack tip have been obtained.

The Westergaard solutions, particularly the results of the exact equations have been shown to be in excelent agreement with the experimental results for the model analysed. The B.I.E.M on the other hand is only in good agreement away from the crack tip, the effect of the singularity as the crack tip is approached being to modify the shape of the loci of constant shear stress (photoelastic fringes) resulting in an angular variation of the stress intensity factor, contrary to Westergaard's results. It is concluded that further development and validation is necessary before B.I.E.M ccan be applied with complete confidence to the treatment of stress singularities. Until such time, it is advised to use special crack-tip elements in conjunction with the method discussed in this paper, to obtain a near-field and far-field solution.

7. ACKNOWLEDGMENT

The authors wish to express their thanks to the Spanish Ministerio de Universidades e Investigacion. Seccion de Personal Investigador who sponsored F.G.Benitez and to the Department of Engineering Science of Oxford University for the facilities provided.

8. REFERENCES

1. BARSOUM,R.S.'On the use of isoparametric finite elements in linear fracture mechanics', Int. J. Num. Math. in Engineering,1976,10,25-37.

2. BENITEZ,F.G.;ALARCON,E.'Problemas con singularidades de contorno.Aplicacion del metodo de los elementos de contorno',III Congreso Nacional sobre la Teoria de Maquinas y Mecanismos,Sevilla,Spain,1980,291-300.

3. BENITEZ,F.G.'Formulacion del metodo de las ecuaciones integrales de contorno en elastoplasticidad tridimensional', Thesis,Escuela Superior Ingenieros Industriales,Madrid,Spain,1981.

4. BENITEZ,F.G.;ALARCON,E.;BREBBIA,C.A.;TELLES,J. 'Tridimensional plasticity using B.I.E.M.', Appl. Math. Modelling, 1981,5,6,442-447.

5. BENITEZ,F.G.'Some considerations about boundary integral equations method', Department of Engineering Science.University of Oxford,Oxford,England,1982. To appear.

6. EVANS,W.T.;LUXMOORE,A.R.'Limitations of the Westergaard equations for experimental evaluations of stress intensity factors', J. of Strain Analysis, 1976,11,3,177-185.

7. GOMEZ-LERA,S.;PARIS,F.Escuela T. S. Ingenieros Industriales.Madrid.Spain.(private communication.1981)

8. HENSHELL,R.D.;SHAW,K.G.'Crack tip finite elements are unnecessary', Int. J. Num. Meth. in Engineering,1975,9,495-507.

9. LACHAT,J.C.'A further development of the boundary integral technique for elastostatics', Ph.D. Thesis. University of Southampton. 1975.

10. LYNN,P.P.;INGRAFFEA,A.R.'Transition elements to be used with quarter-point crack-tip elements', Int. J. Num. Math. in Engineering,1978,12,6,1031-1036.

11. MacGREGOR,C.M.'The potential function method for the solution of two-dimensional stress problems', Trans. Am. Math. Soc. 1935,38.

12. RUIZ,C.;KOENIGSBERGER,F.'Design for strength and production',MacMillan and Co. 1970.

13. RUIZ,C.;PHANG,Y.P.'Stress intensity for interacting flaws',Department of Engineering Science.University of Oxford.Oxford,England,Reports 1367/81, 1368/81.

14. SIH,G.C.'On the Westergaard method of crack analysis', J. Fract. Mech. 1966, 2,628-631.

15. WESTERGAARD,H.M.'Bearing pressures and cracks', J. Appl. Mech. 1939,6,49-53.

16. XANTHIS,L.S.;BERNAL,M.J.M.;ATKINSON,C.'The treatment of singularities in the calculation of stress intensity factors using the boundary integral equation method', Computer Methods in Appl. Mech. and Eng.,1981,26,285-304.

"The Edge Function Method for Three-Dimensional Stress Analysis, Including Embedded Elliptical Cracks and Surface Flaws

P.M. Quinlan*, J.J. Grannell**, S.N. Atluri+, J.E. Fitzgerald++

* Professor, Department of Maths Physics, Univ. College, Cork.
** Lecturer, Department of Maths Physics, Univ. College, Cork.
\+ Regents'Professor of Mechanics, Georgia Instit. Tech.,Atlanta.
++ Director, School of Civil Eng., Georgia Instit. Tech., Atlanta.

INTRODUCTION

Since the publication by Shah and Kobayashi (1971) which developed the confocal potential functions, introduced by Segedin (1967) for potential problems, to determine the Stress Intensity Factor for an Elliptical Crack in an infinite body under arbitrary normal loading, the extension of that work to finite bodies has presented a major challenge. A boundary integral solution was given by Cruse (1975) which was followed by the hybrid element solution of Kathiresan (1976) involving a combination of finite and special crack tip elements.

However little progress appears to have been made in analytical methods using confocal potential functions. This is due mainly to the almost prohibitive algebraic difficulties of expressing in terms of Jacobi elliptic functions all the second and third derivatives of the stress potentials ϕ_{ij} required for the resulting boundary stresses, and corresponding difficulties in evaluation. Shah (1971) obtains but one derivative, $\delta^2\phi_{ij}/\delta z^2$, and that for only $i+j \leq 3$.

The present paper presents a straightforward calculation scheme for all the required derivatives of ϕ_{ij} without introducing any Jacobi elliptic functions and with no particular computational or algebraic difficulties for values of i and j in the range (0, 10). The resulting Elliptic Crack Functions have been included in the author's computer program PQDISK, Quinlan (1980), for "Three Dimensional Stress Analysis in Prismoidal Bodies" and numerical results are attached for both small and large cracks. In sharp contrast to other methods large cracks can be handled as readily as small cracks and the computer times required are at least an order of magnitude less. Stresses and displacements are readily calculated at any point, and each additional load case requires but a few percent of the time for one case when all cases are processed together.

CONFOCAL HARMONICS

Segedin (1967) noted that the function $V^{(n)}$ under is harmonic,

$$V^{(n)} = \int_{\lambda}^{\infty} \frac{\omega^{n}(s)ds}{Q(s)} \text{, where} \tag{1}$$

$$\omega(s) = 1 - \frac{x^2}{a^2+s} - \frac{y^2}{b^2+s} - \frac{z^2}{s} \text{ , } Q(s) = \sqrt{s(a^2+s)(b^2+s)} \text{ ,} \tag{2}$$

n being a positive integer ; a and b the semi-axes of the crack ellipse, and (λ,μ,ν) the corresponding ellipsoidal coordinates. Suitable harmonic stress functions were introduced by Shah (1971)

$$\phi_{ij} = \frac{\delta^{i+j}V^{(n)}}{\delta x^{i}\delta y^{j}} \text{ ; } n = i+j+1 \text{ ;} \tag{3}$$

which are called CRACK FUNCTIONS in this paper. We now require to present the integral ϕ_{ij} in a suitable computational form. On introducing X_{ij} to represent (3) in the form

$$\phi_{ij} = \int_{\lambda}^{\infty} \frac{X_{ij}ds}{Q(s)} \text{ ,} \tag{4}$$

and considering the case where j = 0, we obtain for X_{io}, on differentiating equation (1) successively wrt x, the functions given under in Table 1 :-

i	X_{io}
0	ω^{n}
1	$nxD_1\omega^{n-1}$
2	$nD_1\omega^{n-1} + n(n-1)x^2D_1^2\omega^{n-2}$
3	$3n(n-1)xD_1^2\omega^{n-2} + n(n-1)(n-2)x^3D_1^3\omega^{n-3}$
4	$3\,{}^{n}P_2D_1^2\omega^{n-2} + 6\,{}^{n}P_3x^2D_1^3\omega^{n-3} + {}^{n}P_4x^4D_1^4\omega^{n-4}$
5	$15\,{}^{n}P_3x\,D_1^3\omega^{n-3} + 10\,{}^{n}P_4x^3D_1^4\omega^{n-4} + {}^{n}P_5x^5D_1^5\omega^{n-5}$

$$D_1 = \frac{-2}{a^2+s} \text{ ; } D_2 = \frac{-2}{b^2+s} \text{ ; } {}^{n}P_r = n(n-1) \quad \ldots \quad (n-r+1)$$

Table 1 : Functions X_{io}

Accordingly X_{io} is seen to be of the form :

$$X_{io} = \sum_{k=1}^{i/2+1} C(i+1,\, k)x^{i-2k+2} D_1^{\;i-k+1}\; {}^{n}P_{i-k+1}\,\omega^{n-i+k-1}, \quad (5)$$

where the coefficients $C(i+1,k)$ for $i=0,8$ are given under in Table 2, together with number of terms, nots, in each expression.

i \ k	5	4	3	2	1	Nots
0					1	1
1					1	1
2				1	1	2
3				3	1	2
4			3	6	1	3
5			15	10	1	3
6		15	45	15	1	4
7		105	105	21	1	4
8	105	420	210	28	1	5

Table 2 : Coefficients C(i+1,k)

Likewise the jth derivative wrt y of $V^{(n)}$, analogous to equation (5), is of the form

$$X_{oj} = \sum_{r=1}^{j/2+1} C(j+1,r)y^{j-2r+2} D_2^{\;j-r+1}\; {}^{n}P_{j-r+1}\,\omega^{n-j+r-1} \quad (6)$$

The ith derivative wrt x of X_{oj}, denoted by X_{ij}, follows on applying operators (5) to the functions $\omega^{n'}$ in X_{oj}, where

$n' = n-j+r-1$; $n=i+j+1$; $n' = i+r$; in the form (7)

$$X_{ij} = \sum_{k=1}^{i/2+1} C(i+1,k)\, x^{i-2k+2}\, D_1^{\;i-k+1} \Big\{ \sum_{r=1}^{j/2+1} {}^{n'}P_{i-k+1} C(j+1),r)y^{j-2r+2}\, D_2^{\;j-r+1}\; {}^{n}P_{j-r+1}\,\omega^{L} \Big\}, \quad (8)$$

where $L = n-i-j+r+k-2$ and X_{ij} can be represented in a computer form consisting of nots terms :-

$$X_{ij} = \sum_{k=1}^{i/2+1} \sum_{r=1}^{j/2+1} E\, x^{i'} y^{j'} \frac{1}{(a^2+s)^{i''}} \frac{1}{(b^2+s)^{j''}}\, \omega^{L}, \quad (9)$$

where, when $n = i+j+1$:

$$E = (-2)^{L'}\, C(i+1,k) * C(j+1,r) * {}^{n'}P_{i-k+1} * {}^{n}P_{j-r+1}$$

$$L' = i+j-k-r+2\ ;\ i' = i-2k+2\ ;\ j' = j-2r+2 \quad (10)$$

$$i'' = i-k+1\ ;\ j'' = j-r+1\ ;\ L = n-i-j+r+k-2 = r+k-1$$

The series exponents obviously depend on k and r.

DISPLACEMENTS AND STRESSES FROM ϕ_{ij}

For any harmonic function ϕ the corresponding displacements and stresses, as given by Shah (1971) are

$$u = z\frac{\delta^2\phi}{\delta x\delta z} + (1-2\sigma)\frac{\delta\phi}{\delta x} \ ; \ v = z\frac{\delta^2\phi}{\delta y\delta z} + (1-2\sigma)\frac{\delta\phi}{\delta y}$$

$$w = z\frac{\delta^2\phi}{\delta z^2} - 2(1-\sigma)\frac{\delta\phi}{\delta z} \ ,$$

$$\zeta_{xx} = 2G[z\frac{\delta^3\phi}{\delta x^2\delta z} + \frac{\delta^2\phi}{\delta x^2} + 2\sigma\frac{\delta^2\phi}{\delta y^2}] \ ; \ \zeta_{yz} = 2G\ z\frac{\delta^3\phi}{\delta y\delta z^2}$$

$$\zeta_{yy} = 2G[z\frac{\delta^3\phi}{\delta y^2\delta z} + \frac{\delta^2\phi}{\delta y^2} + 2\sigma\frac{\delta^2\phi}{\delta x^2}] \ ; \ \zeta_{zx} = 2G\ z\frac{\delta^3\phi}{\delta x\delta z^2} \ ;$$

$$\zeta_{zz} = 2G[z\frac{\delta^3\phi}{\delta z^3} - \frac{\delta^2\phi}{\delta z^2}] \ ; \ \zeta_{xy} = 2G[z\frac{\delta^3\phi}{\delta x\delta y\delta z} + (1-2\sigma)\frac{\delta^2\phi}{\delta x\delta y}] \qquad (11)$$

where G is the shear modulus, and σ is Poisson's ratio.
On substituting from equation (9) in equation (4) we obtain

$$\phi_{ij} = \sum_{k=1}^{k'}\sum_{r=1}^{r'}\int_{\lambda}^{\alpha}\frac{Ex^{i'}y^{j'}\omega^{L}(s)\,ds}{\sqrt{s}(a^2+s)^{i*}(b^2+s)^{j*}} \qquad (12)$$

where $k' = i/2+1$; $r' = j/2+1$; $i* = i''+0.5$; $j* = j''+0.5$
Accordingly to evaluate the displacements and stress in equations (11) at any **point** (x,y,z), we require to compute the point values for the following derivatives of ϕ_{ij}

$$\frac{\delta\phi}{\delta x},\ \frac{\delta\phi}{\delta y},\ \frac{\delta\phi}{\delta z}$$

$$\frac{\delta^2\phi}{\delta x^2},\ \frac{\delta^2\phi}{\delta y^2},\ \frac{\delta^2\phi}{\delta z^2},\ \frac{\delta^2\phi}{\delta y\delta z},\ \frac{\delta^2\phi}{\delta z\delta x},\ \frac{\delta^2\phi}{\delta x\delta y},$$

$$\frac{\delta}{\delta z}[\frac{\delta^2\phi}{\delta x^2},\ \frac{\delta^2\phi}{\delta y^2},\ \frac{\delta^2\phi}{\delta z^2},\ \frac{\delta^2\phi}{\delta y\delta z},\ \frac{\delta^2\phi}{\delta z\delta x},\ \frac{\delta^2\phi}{\delta x\delta y}] \qquad (13)$$

We denote the above derivatives by

$$\phi^{(t)} \ ; \ t = 1, 16 \ , \ \text{where}$$

$$\phi^{(1)} = \phi_{ij} \ ; \ \phi^{(2)} = \frac{\delta\phi}{\delta x} \ ; \ \ldots \ \phi^{(16)}\ \frac{\delta^3\phi}{\delta x\delta y\delta z} \qquad (14)$$

The subscripts ij have been omitted from ϕ_{ij}, and equation (12) gives

$$\phi^{(t)} = \sum_{k=1}^{k'}\sum_{r=1}^{r'}\Psi_t^{kr} \ , \ \text{where , when } t = 1 \ ;$$

$$\Psi_1^{kr} = \int_{\lambda}^{\infty}\frac{Ex^{i'}y^{j'}\omega^{L}(s)}{\sqrt{s}\,(a^2+s)^{i*}(b^2+s)^{j*}}ds \qquad (15)$$

We shall call Ψ_1^{kr} *the basic integral*, and denote its integrand, called *the basic integrand*, by $\bar{H}^{kr}$, where

$$\bar{H}^{kr} = \frac{Ex^{i'} y^{j'} \omega^{L}(s)}{\sqrt{s}(a^2+s)^{i*}(b^2+s)^{j*}} \tag{16}$$

In subsequent work we shall drop the superscripts kr when not essential for clarity. We now proceed to obtain in a form suitable for computation the various derivatives of Ψ_1, denoted by Ψ_2 .... Ψ_{16}, as required in equations (11); the integrals arising in the process being called *primary integrals*.

Primary Integrals

On differentiating equation (15) wrt x we obtain

$$\Psi_2 = \frac{\delta\Psi_1}{\delta x} = \int_\lambda^\infty \frac{E\, y^{j'} \frac{\delta}{\delta x}[x^{i'}\ \omega^{L}(s)]\, ds}{\sqrt{s}(a^2+s)^{i*}(b^2+s)^{j*}} - [\bar{H}]_{s=\lambda} \frac{\delta\lambda}{\delta x} \tag{17}$$

The quantity $\bar{H}$ is zero at $s = \lambda$ since it contains the factor $\omega^{L}(\lambda)$ with $L > o$ as seen from Table 2, and $\omega(\lambda) = o$. Hence

$$\Psi_2 = \frac{i'}{x}\Psi_1 - 2L \int_\lambda^\infty \bar{H} \frac{x}{\omega(a^2+s)} ds \tag{18}$$

Likewise

$$\Psi_3 = \frac{j'}{y}\Psi^{(1)} - 2L \int_\lambda^\infty \bar{H} \frac{y}{\omega(b^2+s)} ds \ ; \ \Psi_4 = -2L \int_\lambda^\infty \bar{H} \frac{z}{\omega s} ds \tag{19}$$

Accordingly we require to evaluate in addition to the basic integral Ψ_1, now denoted by V_1, the following *Primary Integrals*

$$V_2 = -2L \int_\lambda^\infty \bar{H} \frac{x}{\omega(a^2+s)} ds \ ; \ V_3 = -2L \int_\lambda^\infty \bar{H} \frac{y}{\omega(b^2+s)}$$

$$V_4 = -2L \int_\lambda^\infty \bar{H} \frac{z}{\omega s} ds \ ; \ V_1 = \Psi^{(1)} \ , \tag{20}$$

On introducing the differentiation factors

$$F_x = \frac{x}{\omega(a^2+s)} \ ; \ F_y = \frac{y}{\omega(b^2+s)} \ ; \ F_z = \frac{z}{\omega s} \ , \tag{21}$$

it is seen that the primary integral corresponding to differentiation wrt x has the integrand $\bar{H}\, F_x$ — or differentiation wrt x involves a primary integral with integrand $\bar{H}\, F_x$. Similarly differentiation wrt y and z involves the primary integrals with integrand $\bar{H}\, F_y$ and $\bar{H}\, F_z$ respectively. Likewise $\frac{\delta^2}{\delta x^2}$ involves a primary integral with integrand $\bar{H}\, [F_x]^2$ etc.

Accordingly the 16 primary integrals V_k involved in evaluating the differential coefficients required in equations (11) are:

$$V_1 = \int_\lambda^\infty \bar{H}\, ds = \Psi_1 \ ;$$

$$V_2 = K_1\int_\lambda^\infty \bar{H}\, F_x ds \ ; \ V_3 = K_1\int_\lambda^\infty \bar{H}\, F_y ds \ ; \ V_4 = K_1\int_\lambda^\infty \bar{H}\, F_z ds \ ;$$

$$V_5 = K_2\int_\lambda^\infty \bar{H}\, F_x^2 ds \;;\; V_6 = K_2\int_\lambda^\infty H\, F_y^2 ds \;;\; V_7 = K_2\int_\lambda^\infty \bar{H}\, F_z^2 ds \;;$$

$$V_8 = K_2\int_\lambda^\infty \bar{H}\, F_y F_z ds \;;\; V_9 = K_2\int_\lambda^\infty \bar{H}\, F_z F_x ds \;;\; V_{10} = K_2\int_\lambda^\infty \bar{H}\, F_x F_y ds \;;$$

$$V_{11} = K_3\int_\lambda^\infty \bar{H}\, F_x^2 F_z ds \;;\; \ldots\ldots \; V_{16} = K_3\int_\lambda^\infty \bar{H}\, F_x F_y F_z ds \;; \quad (22)$$

where

$$K_1 = -2L \;;\; K_2 = -2L(-2L+2) \;;\; K_3 = K_2(-2L+4) \quad (23)$$

Derived Functions Ψ

The derivatives of Ψ in equations (18) and (19) are then given in terms of the primary integrals by :

$$\Psi_1 = V_1 \;;\; \Psi_2 = \frac{i'}{x} V_1 + V_2 \;;\; \Psi_3 = \frac{j'}{y} V_1 + V_3 \;;\; \Psi_4 = V_4 \quad (24)$$

The higher derivatives follow, on deferring to the next paragraph consideration of any variable limit contributions that may arise, as:

$$\Psi_5 = \frac{\delta}{\delta x}[\Psi_2] = \frac{i'(i'-1)}{x^2} V_1 + \frac{(2i'+1)}{x} V_2 + V_5$$

$$\Psi_6 = \frac{j'(j'-1)}{y^2} V_1 + \frac{(2j'+1)V_3}{y} + V_6 \;;\; \Psi_7 = \frac{1}{z} V_4 + V_7 \;;$$

$$\Psi_8 = \frac{j'}{y} V_4 + V_8 \;;\; \Psi_9 = \frac{i'}{x} V_4 + V_9 \;;$$

$$\Psi_{10} = \frac{i'j'}{xy} V_1 + \frac{i'}{x} V_3 + \frac{j'}{y} V_2 + V_{10} \;;$$

$$\Psi_{11} = \frac{i'(i'-1)}{x^2} V_4 + \frac{(2i'+1}{x} V_9 + V_{11} \;;$$

$$\Psi_{12} = \frac{j'(j'-1)}{y^2} V_4 + \frac{(2j'+1)}{y} V_8 + V_{12} \;;$$

$$\Psi_{13} = \frac{3}{z} V_7 + V_{13} \;;\; \Psi_{14} = \frac{j'}{yz} V_4 + \frac{j'}{y} V_7 + \frac{1}{z} V_8 + V_{14} \;;$$

$$\Psi_{15} = \frac{i'}{xz} V_4 + \frac{i'}{x} V_7 + \frac{1}{z} V_9 + V_{15} \;;$$

$$\Psi_{16} = \frac{i'y'}{xy} V_4 + \frac{i'}{x} V_8 + \frac{j'}{y} V_9 + V_{16} \quad (25)$$

Variable Limit Contribution to Ψ's

As shown in equation (17) the contributions due to the variable lower limit λ to the first derivatives in Ψ_2, Ψ_3, Ψ_4 is zero, since the integrand involves ω^L where $\omega^L(\lambda) = 0$, when $L > 0$. However, for the primary integrals involved in the second derivatives, the integrands Ψ_2, Ψ_3, and Ψ_4 being differentiated involve ω^{L-1} and if $L = 1$ then $\omega^{L-1}(\lambda) = \omega^o(\lambda) = 1$. Accordingly, if L=1, when differentiating equation (17) wrt x, an additional term,due to the effect of the variable lower limit λ, must be added to the Ψ's as given by equations (25). The adjusted expressions for Ψ under,when $K_1 = -2L = -2$, are :

$$\Psi_5 = \Psi_5 - K_1\bar{H}(\lambda)F_x(\lambda)\frac{\delta\lambda}{\delta x} \;;\; \Psi_6 = \Psi_6 - K_1\bar{H}(\lambda)F_y(\lambda)\frac{\delta\lambda}{\delta y} \;;$$

$$\Psi_7 = \Psi_7 - K_1\bar{H}(\lambda)F_z(\lambda)\,\frac{\delta\lambda}{\delta z}\ ; \quad \Psi_8 = \Psi_8 - K_1\bar{H}(\lambda)F_y(\lambda)\,\frac{\delta\lambda}{\delta z}\ ;$$

$$\Psi_9 = \Psi_9 - K_1\bar{H}(\lambda)F_z(\lambda)\,\frac{\delta\lambda}{\delta x}\ ; \quad \Psi_{10} = \Psi_{10} - K_1\bar{H}(\lambda)F_x(\lambda)\,\frac{\delta\lambda}{\delta y}\ ;$$

$$\bar{H} = \frac{x^{i'}\,y^{j'}\,\omega(\lambda)}{\sqrt{\lambda}\,(a^2+\lambda)^{i^*}(b^2+\lambda)^{j^*}}\ , \tag{26}$$

and the factor $\omega(\lambda)$ is cancelled out between $\bar{H}(\lambda)$ and the factors (21). Contributions to the third derivatives (13) will be made by the derivatives **wrt z** of the above additions together with variable limit contributions of the derivatives of the integrals V_2, V_3, and V_4 in ψ_6 .... ψ_{10} giving :- $\Psi_{11} = \Psi_{11} + A_{11}$, ... $\Psi_{16} = \Psi_{16} + A_{16}$

where

$$A_{11} = -K_1\left[\frac{\delta}{\delta z}\,\{\bar{H}(\lambda)F_x(\lambda)\,\frac{\delta\lambda}{\delta x}\}\right] - \frac{2i'+1}{x}\,A_5$$

$$A_{16} = -K_1\left[\frac{\delta}{\delta z}\,\{\bar{H}(\lambda)F_x(\lambda)\,\frac{\delta\lambda}{\delta y}\}\right] - \left[\frac{i'}{x}\,A_4 + \frac{j'}{y}\,A_5\right]\ , \tag{27}$$

and equations (26) are written as

$$\Psi_K = \Psi_K + A_K\ ; \quad K = 5,\ .....\ 10\ . \quad \text{Since} \tag{28}$$

$$1 - \frac{x^2}{a+\lambda} - \frac{y^2}{b+\lambda} - \frac{z^2}{\lambda} = 0\ ,$$

we obtain on differentiating partially wrt x :

$$\frac{\delta\lambda}{\delta x} = \frac{2x}{(a^2+\lambda)H(\lambda)}\ ; \quad H(\lambda) = \frac{x^2}{(a+\lambda)^2} + \frac{y^2}{(b^2+\lambda)^2} + \frac{z^2}{\lambda^2} \tag{29}$$

It follows that

$$\frac{\delta H}{\delta z} = \frac{2z}{\lambda^2} - 2\left[\frac{x^2}{(a^2+\lambda)^3} + \frac{y^2}{(b^2+\lambda)^3} + \frac{z^2}{\lambda^3}\right]\frac{\delta\lambda}{\delta z} \tag{30}$$

Similarly

$$\frac{\delta\lambda}{\delta y} = \frac{2y}{(b^2+\lambda)H(\lambda)}\ ; \quad \frac{\delta\lambda}{\delta z} = \frac{2z}{\lambda\,H(\lambda)} \tag{31}$$

The additional terms A_K in equations (26) and (27) are readily programmed, since on substituting when L = 1 :

$$A_5 = \frac{4x}{a^2+\lambda}\,\bar{H}(\lambda)\,\frac{\delta\lambda}{\delta x} = \frac{8x^2}{(a^2+\lambda)^2}\,\bar{H}(\lambda)/H(\lambda)$$

$$= 8E\,\frac{x^{i'+2}\,y^{j'}}{(a^2+\lambda)^{i^*+2}(b^2+\lambda)^{j^*}\sqrt{\lambda}\,H(\lambda)} \tag{32}$$

with similar expressions for A_6 to A_{10}, and A_{11} to A_{16} when L = 1.

<u>Case L = 2</u> If L = 2, ω^L reduces to ω^o after 2 differentiations and accordingly the third derivatives in equations (13) have a variable limit contribution similar to equations (26), as given by ;-

$$\Psi_{11} = \Psi_{11} - 8\bar{H}(\lambda)F_x^2(\lambda)\,\frac{\delta\lambda}{\delta z}\ ; \quad \Psi_{12} = \Psi_{12} - 8\bar{H}(\lambda)F_y^2(\lambda)\,\frac{\delta\lambda}{\delta z}$$

$$\Psi_{13} = \Psi_{13} - 8\bar{H}(\lambda)F_z^2(\lambda)\frac{\delta\lambda}{\delta z} \; ; \quad \Psi_{14} = \Psi_{14} - 8\bar{H}(\lambda)F_yF_z\frac{\delta\lambda}{\delta z}$$
$$\Psi_{15} = \Psi_{15} - 8\bar{H}(\lambda)F_zF_x\frac{\delta\lambda}{\delta z} \; ; \quad \Psi_{16} = \Psi_{16} - 8\bar{H}(\lambda)F_xF_y\frac{\delta\lambda}{\delta z} \tag{33}$$

Note if $L > 2$ then **the function** $\bar{H}(\lambda)$ arising in equations (26) and (33) involves $\omega^{L'}$ where since $L' > o$, $\omega^{L'} = o$. Consequently there are then no variable limit contributions to any of the integrals arising in equations (13).

SINGULAR INTEGRALS

The primary integrals (22) are of the type

$$I = \int_\lambda^\infty s^{-(\frac{3}{2}+\alpha)} z^{2\alpha} G(x,y,s)\omega^L(s)ds \; ; \; \alpha \geq o \text{ , where} \tag{34}$$

$$\omega(s) = \bar{\omega}(s) - z^2/s \; ; \; \bar{\omega}(s) = 1 - x^2/(a^2+s) - y^2/(b^2+s) \tag{35}$$

The integrals I are convergent for $\lambda>c>o$, but become improper as $\lambda\to o$. On expanding I in powers of $\bar{z} = z^2/s$ we obtain

$$I = \int_\lambda^\infty s^{-(\frac{3}{2}+\lambda)} z^{2\alpha}\left[F_1(s) + \bar{z}F_2(s) + \bar{z}^2F_3(s) + \bar{z}^3F_4(s)\right]ds \tag{36}$$

where the non-zero functions in the above for $L = 0,1,2,3$ .. are :

$$\begin{array}{llll} L = 0 : F_1(s) = G & L = 1 : F_1(s) = G\bar{\omega} \; ; & F_2(s) = -G & \\ L = 2 : F_1(s) = G\bar{\omega}^2 \; ; & F_2(s) = -2G\bar{\omega} \; ; & F_3(s) = G & \\ L = 3 : F_1(s) = G\bar{\omega}^3 \; ; & F_2(s) = -3G\bar{\omega}^2 \; ; & F_3(s) = 3G\bar{\omega} \; ; & F_4(s) = -G \end{array} \tag{37}$$

The singular Part $\bar{I}$ of I follows on expanding each $F_k(s)$ in s and integrating

$$\bar{I} = \int_\lambda^\infty s^{-(\frac{3}{2}+\lambda)} z^{2\alpha}\left[\bar{F}_1 + \bar{z}\bar{F}_2 + \bar{z}^2\bar{F}_3 + \bar{z}^3\bar{F}_4 + \ldots\right]ds$$
$$= 2\lambda^{-(\frac{1}{2}+\alpha)} z^{2\alpha}\left[\frac{\bar{F}_1}{1+2\alpha} + \frac{z^2}{\lambda}\frac{\bar{F}_2}{3+2\alpha} + \frac{z^4}{\lambda^2}\frac{\bar{F}_3}{5+2\alpha} + \frac{z^6}{\lambda^3}\frac{\bar{F}_4}{7+2\alpha} + \ldots\right], \tag{38}$$

where $\bar{F}_k = F_k(o)$.

Since

$\frac{z^2}{\lambda} = \bar{\omega}(\lambda) \to \bar{\omega}(o) = \bar{\omega}_o$ as $\lambda \to o$ on the crack front, the coefficient C_λ of $1/\sqrt{\lambda}$ in expression for $\bar{I}$ is :

$$C_\lambda = 2\bar{\omega}_o^{-\alpha}\left[\frac{\bar{F}_1}{1+2\alpha} + \frac{\bar{\omega}_o\bar{F}_2}{3+2\alpha} + \frac{\bar{\omega}_o^2\bar{F}_3}{5+2\alpha} \; \ldots\ldots\right] \tag{39}$$

Since $\omega_o \Rightarrow o$ approaching the crack front it follows that

$$C_\lambda = o \; ; \; \bar{I} = o, \text{ if } \alpha \neq o : \; C_\lambda = 2\bar{F}_1 \; ; \; \bar{I} = 2\bar{F}_1/\sqrt{\lambda} \text{ , if } \alpha = 0, \tag{40}$$

where on the crack front

$$\bar{F}_1 = o \text{ for } L > o \; ; \; \bar{F}_1 = G(x, y, o) \text{ for } L = 0 \; ; \; z\bar{I} = o \tag{41}$$

The Principal value of I, call it I^*, follows, on removing $\bar{I}$:

$$I^* = I - \bar{I} = \int_\lambda^\infty s^{-(\frac{3}{2}+\alpha)} \left[(F_1^* + \frac{z^2}{s} F_2^* + \ldots\ldots \right] ds \,, \quad (41)$$

where

$$F_k^* = \frac{F_k - \bar{F}_k}{s} + \frac{dF_k}{ds} \quad \text{as } s \to o \; ; \quad F_1 = G\omega^{-L} \quad (42)$$

I^* is a *convergent integral* for $z \to o$, $\lambda \to o$ since its improper part **is** shown, as above, to be of order $\sqrt{\lambda}$ and its infinite part of order $1/\sqrt{\lambda}$, $F^*_k(s)$ being of order $1/s$ as $s \to \alpha$. Hence for points on the crack front ;

$$I^* = o \text{ if } \alpha \neq o \; ; \quad I^* = \int_\lambda^\infty s^{-\frac{1}{2}} F_1^*(s)\, ds, \; \alpha = o \,, \quad (43)$$

which means that in evaluating integral (41) as $z \to o$ we formally put $z = o$ in the integrand. It can be shown that the normal stress ζ_{zz} produced by the potential field ϕ_{ij} is finite at all points within the crack ellipse as the resulting singular terms from the variable limit additions and the singular integrals cancel each other out.

Normal Stresses on Crack Ellipse $\lambda = o$

Hence to compute the normal stress ζ_{zz} on the crack ellipse $\lambda = o$

(i) take the principal value of the integrals of type (34), with $z = o$.

(ii) omit all variable limit additions

Accordingly the stress ζ_{zz} produced on $\lambda = o$ by the potential ϕ_{ij} is the sum of the stresses ζ'_{zz} resulting from Ψ_1^{kr} as given in equation (15) :

$$\zeta'_{zz} = -2G \frac{\delta^2 \Psi_1{}^{kr}}{\delta z^2} = 4GL \left[\int_\lambda^\infty \frac{G(s)\omega^{L-1}(s)\, ds}{s^{3/2}} \right]^* \; ;$$

$$G(s) \equiv G(x,y,o) = \frac{E x^{i'} y^{j'}}{(a+s)^{i^*} (b^2+s)^{j^*}} \quad (44)$$

where * denotes the principal value. On using result (43) it follows that

$$\zeta'_{zz} = 4GEL \int_\lambda^\infty s^{-\frac{1}{2}} F_1^*(s)\, ds \; ; \quad (45)$$

where the parameters for $G(x,y,s)$ are given by equations (11) with L set to L - 1, and F_1^* is given by (42). In the numerical evaluation of integral (45) the range is split into $(\lambda,1)$ and $(1, \infty)$, the appropriate form for F_1^* being used.

CONFOCAL COORDINATES ON CRACK PLANE $z = o$

Consider a confocal system with axes (a,b,o), the cracked ellipse being (a,b) corresponding to $\lambda = o$. At points outside the ellipse on the plane $z = o$, $\mu = o$, and λ is small. Hence the system through the point $P(x,y,z)$ is :

$$x^2/(a^2+\lambda \;+\; y^2/(b^2+\lambda) + z^2/\lambda = 1 \; ; \quad \text{or} \quad (46)$$

$$\lambda^3 + s_1\lambda^2 + s_2\lambda + s_3 = 0 \,, \quad \text{where} \quad (47)$$

$$s_1 = a^2 + b^2 - x^2 - y^2 - z^2$$

$$s_2 = a^2b^2 - b^2x^2 - a^2y^2 - z^2(a^2+b^2) \quad ; \quad s_3 = -z^2a^2b^2 \quad (48)$$

The cubic has 3 real roots λ, μ, ν, the confocal coordinates of P, which can be determined numerically, where

$$\infty > \lambda \geqslant o \geqslant \mu \geqslant -b^2 \geqslant \nu \geqslant -a^2 \quad (49)$$

At points P (a cos θ, b sin θ), where θ is the ellipse parameter, on the crack front λ=o, μ=o, equation (48) gives

$$\nu = \nu_o = -s_1 = -a^2-b^2+x^2+y^2 = -a^2 \sin^2\theta \; -b^2\cos^2\theta \quad (50)$$

Points Q (x+Δx, y+Δy) on the outward normal at P to the crack ellipse in the plane z=o have confocals (λ, o, ν), where $\lambda \to o$ and $\nu \to \nu_o$. On equating first order quantities in the expression for s_2 we obtain

$$s_2 = \Sigma\lambda\mu = a^2b^2 - b^2(x+\Delta x)^2 - a^2(y+\Delta y)^2$$
$$\Rightarrow \lambda\nu_o = 2b^2x\Delta x - 2a^2y\Delta y \quad (51)$$

Since the unit normal $\hat{n}$ to the ellipse at P is

$$\hat{n} = K_\theta \text{ (b cos }\theta\text{, a sin }\theta) \; ; \; K_\theta = (a^2 \sin^2\theta + b^2\cos^2\theta)^{-\frac{1}{2}};$$

it follows that

$$\Delta x = K_\theta br \cos\theta \; ; \; \Delta y = K_\theta ar \sin\theta \; ; \; r = PQ \; ,$$

and hence on substituting in equation (51) it follows that

$$\lambda = 2abr\,(a^2\sin^2\theta + b^2\cos^2\theta)^{-\frac{1}{2}} \quad (52)$$

FRACTURE COEFFICIENTS

At points Q outside, but close to, the crack front on plane z=o the singular part of any primary integral of type (34) is given by equation (38), where, since $\omega_o \to o$ as $z \to o$ and $\lambda \to o$, :

$$\bar{I} = o \text{ , if, } \alpha > o \text{ and } L > o \; ;$$
$$= \frac{2\bar{F}_1}{\sqrt{\lambda}} = \frac{2G(x,y,o)}{\sqrt{\lambda}}, \text{ if } \alpha = L = o \quad (53)$$

The contribution to the singular part, ζ_{zz}^{**}, of the normal stress ζ_{zz}' as given in equation (43), follows as :-

$$\zeta_{zz}^{**} / 2G = 2L \left\{ \int_\lambda^\infty \frac{G(s)\omega^{L-1}(s)}{s^{3/2}} ds \right\}^{**} = 4L\bar{F}_1 / \sqrt{\lambda} = \frac{4LG(o)}{\sqrt{\lambda}} \delta_1^L \quad (54)$$

on using result (53) with L replaced by L-1, δ_1^L being Kronecker delta. Hence the singular contribution from all basic integrals involving ω^L is zero except that from the member L=1, which is the first integral in equation (4) **arising from the r=1, k=1 term** in X_{ij} as given by series (8). On using equations (44) and (10) with s = λ=o it follows that

$$G(o) = Ex^{i'}y^{j'} / a^{2i''}b^{2j''} = {}^{n'}P_i \, {}^{n}P_j \, (-2)^{i+j} (\tfrac{x}{a^2})^i (\tfrac{y}{a^2})^j \; ;$$

where

$$n' = n-j = i+1 \; ; \; n = i+j+1 \; ; \; i'' = i, \; j'' = j \; ; \; i'=i \; ; \; j'=j$$

$$L' = i+j \; ; \; c(i+1),1) = 1 \; ; \; c(j+1,1) = 1 \; ; \; x = a \cos\theta, \quad y = b \sin\theta$$

$$\Rightarrow \; G(o) = \underline{|n} \; (-2)^{n-1} \left(\frac{\cos\theta}{a}\right)^i \left(\frac{\sin\theta}{b}\right)^j \tag{54}$$

On using equations (53) and (55), it follows that

$$\zeta_{zz}^{**} = \frac{8G\,G(o)}{\sqrt{2abr}} \left(a^2 \sin^2\theta + b^2 \cos^2\theta\right)^{\frac{1}{4}} . \tag{55}$$

The corresponding fracture coefficient K_1 for potential ϕ_{ij} is :

$$K_1 = \underset{r \to o}{\mathrm{Lt}} \sqrt{2\Pi r}\, \zeta_{zz}^{**} \quad ; \quad \text{set } K(i,j) = K_1 \; ;$$

$$\Rightarrow \; K(i,j) = 8G\underline{|n} \sqrt{\frac{\Pi}{ab}} \, (-2)^{n-1} \left(\frac{\cos\theta}{a}\right)^i \left(\frac{\sin\theta}{b}\right)^j \left(a^2 \sin^2\theta + b^2 \cos^2\theta\right) \tag{56}$$

Accordingly for the stress potential $\Phi = \Sigma C_{ij}\,\Phi_{ij}$, the corresponding fracture coefficient $\bar{K}_1$ follows by superposition as $\bar{K}_1 = \Sigma C_{ij} K(i,j)$. This checks with that obtained in Shah (1973) by a very complicated limiting process involving Jacobi elliptic functions.

NUMERICAL INTEGRATIONS

The typical primary integral in equations (22) is

$$V = \int_\lambda^\infty \bar{H} F_x^\alpha F_y^\beta F_z^\gamma \, ds = T(x,y,z) \int_\lambda^\infty \frac{s^{-k'}\,\omega^L}{(a^2+\lambda)^{i'}\,(b^2+\lambda)^{j'}} \, ds, \tag{57}$$

where

$$i' = i^* + \alpha \; , \; j' = j^* + \beta, \; k' = \gamma + \tfrac{1}{2} \; ,$$

and α, β, γ are integers in the range (o, 3).

The range of integration can be split into (λ, 1) and (1, ∞) and the latter range can then be inverted into (o, 1) on putting $s' = 1/s$. Accordingly the basic integral type is

$$\int_{s_o}^1 s^{k^*} f(s) \, ds, \tag{58}$$

where k^* may be negative, and some of the integrals are improper when $s_o \to o$ and $k^* = -\frac{1}{2}$.

A special subroutine was developed by Quinlan (1981) for evaluating integrals of the above type. It consists of fitting to f(s) a number, N, of parabolic arcs denoted by $\overline{f(s)}$, which have intervals in geometrical progression with ratio R. A simple subroutine produces the corresponding weight factors for any N and R. The capability to vary intervals is especially important since f(s) can often vary drastically in (o,1) when it is near singular at s = o.

NUMERICAL EXAMPLES

A Crack function capability has been inserted in computer program PQDISK Quinlan (1980) which implements "The Edge Function Method for Prismoidal Bodies" and any required number from 1 to 15 of such functions, can be included in the solution scheme for modelling an elliptic crack.

The Edge Function Method is described in Quinlan (1974, 1975, 1977, 1979), and Atluri (1978). Space prevents any further elaboration here except to point out that each boundary condition is imposed on each face in either a continuous or a discrete least squares sense. Consequently the boundary condition of zero normal stress on the crack ellipse is imposed in a least squares sense ; the zero shear condition being provided by the built in symmetries of the functions.

A cube, 2 x 2 x 2, with a centrally located elliptic plane crack, loaded normally on the faces parallel to the crack plane is considered. Four different load types - (1) point load (2) line load (3) patch load modelled by equivalent point loads using the Bousinesque solution and (4) a linear stress field producing a uniform load, can, if desired be taken together in each job ; the four loading cases processed together taking only about 15% more computer time than would be required to obtain a single solution on its own.

Three different problem sets are presented under to illustrate the use of the crack functions -

Small Elliptical Crack

The problems solved by Cruse (1975) and Kathiresan (1976) obtained K_1 for relatively small ellipses (a, b) where ab=.04 for ratios of a/b = 1, 2, 3, 4 and under a uniform stress field.

Results using PQDISK are given in Table 3 for K_1/K_1^* at 8 points on a quadrant of the crack front corresponding to values $\theta_k = k\Pi/18$ of the elliptical angle ; the residual stress Δp on the crack ellipse being appended.

As given in Kathiresan (1976) K_1^* for an ellipse of semi-axes a and b in an infinite medium is given by :-

$$K_1^* = p_o\sqrt{\Pi b}\,\{\sin^2\theta + (\tfrac{b}{a})^2\cos^2\theta\}^{\frac{1}{4}} / \int_o^{\Pi/2}\sqrt{1-k^2\sin^2\Psi}\,d\Psi \;;$$

$$k^2 = 1-b^2/a^2 \;; \quad a>b$$

a/b \ k	1	3	5	7	9	10	Δp
1	1.02	Constant at all points					$0.14 * 10^{-5}$
2	.932	1.00	1.14	1.25	1.31	1.32	$0.13 * 10^{-5}$
3	.828	.976	1.19	1.35	1.42	1.43	$0.23 * 10^{-5}$
4	.746	.961	1.22	1.39	1.48	1.49	$0.14 * 10^{-5}$

Table 3 : Ratio K_1/K^* at points θ_k for various a/b .

The results all agree to within 2% with those given by Kathiresan(1976). They were obtained using 6 crack functions, together with 60 plane functions to model the boundary effects. The computer time required for each ellipse is approximately six times that required to solve 70 simultaneous equations by ordinary Gaussian elimination and consequently is an order of magnitude less than was required by other methods. The equations took about five times as long to generate as to solve.

(b) Elongated Elliptical Crack

As a demonstration of the power of program PQDISK examples (b) and (c) are presented. The first takes a very elongated crack with a = 0.8 and b = 0.05. It uses 120 plane functions to model the boundaries and 10 crack functions. Solutions are readily obtained, in a single computer run, using N_c = 1, 2, 3 .... 10 crack functions and for four load types. Table 4 shows the variations of the residual stress Δp on the crack plane with the number N_c of crack functions used for each load type, while Table 5 gives some illustrative K_1 factors at points θ_k on the ellipse for a point load.

N_c	Point Load at (0.2,0.3)	Line Load from (0,.4 to (.5,.4)	Patch Load on 1 x 1	Uniform Stress
1	.411D 00	.172D 00	.151D-01	.275D-03
2	.407D 00	.168D 00	.150D-01	.272D-03
3	.269D 00	.171D-01	.505D-03	.184D-04
4	.266D 00	.169D-01	.505D-03	.182D-04
5	.266D 00	.169D-01	.506D-03	.181D-04
6	.112D-01	.438D-02	.362D-03	.125D-04

Table 4 : Stress Residuals on a Small elliptic Crack versus N_c

N_c \ k	1	3	5	7	9	10
1	.843	1.99	2.71	3.14	3.35	3.37
2	.842	1.98	2.68	3.08	3.26	3.29
3	.351	.994	1.92	2.98	3.71	3.81
6	.392	1.04	1.89	2.93	3.72	3.84
9	.392	1.04	1.89	2.93	3.72	3.83

Table 5 K_1 at points θ_k for point load

Table 4 shows that six crack functions are required to reduce the residual stress to 1% in the point load case, while three functions reduce the residuals to less than .01% in the case of a patch load.

The variations in the K_1 factors in Table 5 correspond to the residual stresses and no significant change occurs in the values after six crack functions. The residual stresses on the surface of the cube were less than 3% of the applied loading and could be reduced further by increasing the number of plane functions used in the modelling.

(c) Large Elliptical Crack
Program PQDISK was run for a crack with a = 0.8, b = 0.7 using the same number of functions as in (b) and results analogous to those in Table 4 are given in Table 6.

N_c	Point Load at (.2,.3)	Line Load from (0,.4) to (.4,.4)	Patch Load on 1 x 1
1	.656D 00	.223D 00	.892D-01
2	.317D 00	.169D 00	.454D-01
3	.488D 01	.170D-01	.272D-02
4	.248D-01	.160D-01	.224D-02
5	.121D-01	.135D-01	.203D-02

Table 6 Stress Residuals for large elliptic crack versus N_c

The K_1 factors were found to correspond to the residual stress pattern in a similar manner to the factors in Table 5.

Displacements and stresses are readily calculated at any point in the body, and they are consistent with the boundary residuals reported in this section. Space limitations prevents their presentation in this paper.

SURFACE FLAWS

Work is progressing rapidly in adapting PQDISK to deal with surface and corner flaws. An analysis of the singular stresses, resulting from the Crack Functions, at points where crack front meets the free surface, shows that these can be removed by setting the K_1 fracture coefficient to zero at these points. This acts as a constraint condition on the Crack Function coefficients, and the boundary stress vector is then minimised globally in a least squares sense.

Preliminary results are in good agreement with those given given by Katherisan (1976) and an extensive paper will be submitted shortly. Work is also in progress on fracture of Modes 11 and 111 and the relevant stress systems are readily generated from the Crack Functions (3) on using a scheme analogous to equations (11).

References

Atluri, S.N., Grannell, J.J., Quinlan, P.M., and Fitzgerald J.E. (1978) "Boundary-Element, Finite-Element, and Coupled Boundary-Finite-Element Methods" Internal Report GT-SA-103-78.

Cruse, T.A. (1975) "Boundary-Integral Equation Method for Three-Dimensional Elastic Fracture Mechanics Analysis" T.R. N0014-67-A-0370-0012 O.N.R.

Kathiresan, K. (1976) "Three-Dimensional Linear Elastic Fracture Mechanics Analysis by a Displacement Hybrid Finite Element Model" Ph.D. Thesis Georgia Instit. Tech.

Quinlan, P.M. (1974) "The Edge-Function in Elastostatics" paper in "Studies in Numerical Analysis" Academic Press, London (1974)

Quinlan, P.M., Fitzgerald, J.E., Atluri, S.N. (1977) "The Edge-Function Method" "Proceedings First International Symposium on Innovative Numerical Methods in Engineering Science" Versailles.

Quinlan, P.M., Grannell, J.J., Atluri, S.N., and Fitzgerald J.E., (1979) "Boundary Discretisation using The Edge-Function Method" App. Math. Modelling, 3, 1979.

Quinlan, P.M. (1980) Computer Program PQDISK for "Three Dimensional Stress Analysis in Prismoidal Bodies by The Edge-Function Method" University College Cork, Ireland.

Quinlan, P.M. (1981) "A subroutine for Weight Factors for Quadrature Formulae for Improper and Oscillatory Integrals" to appear in Journal for Computer Mathematics

Segedin, C.M. (1967) "Some Three-Dimensional Mixed Boundary Value Problems in Elasticity" Univ. of Washington R No. 67 - 3.

Shah, R.C. and Kobayashi, A.S. (1971) "Stress Intensity Factor for an Elliptical Crack under Arbitrary Normal Loading" Inter. Jour. Eng. Fracture Mechanics" Vol. 3.

A BOUNDARY ELEMENT METHOD FOR CURVED CRACK PROBLEMS IN TWO DIMENSIONS

R.N.L. Smith and J.C. Mason

Department of Mathematics and Ballistics,
Royal Military College of Science, Shrivenham, Swindon, Wilts.

ABSTRACT

A general formulation of the boundary element method is presented for applications to both straight and curved crack problems in two dimensions. A stitching technique is adopted, in which the region is subdivided along the line of the crack, and traction-singular quarter-point elements are used to give accurate crack-tip representations of displacement and traction. Values of stress intensity factors, calculated by one-point approximations, are obtained for the standard problem of a double edge-cracked plate and the new problem of an arc crack in a square plate. Good agreement is obtained with known results for the double edge-cracked plate and for the arc crack in an infinite plate.

INTRODUCTION

The importance of evaluating parameters such as stress intensity factors using linear elastic fracture mechanics has given rise to a number of strategies for crack modelling in the boundary element method (BEM) (or boundary integral equation (BIE) method). Considerable progress has taken place over recent years in extending the applicability and accuracy of the method.

An idealised crack configuration involves a surface discontinuity. However, if a BIE method is applied directly to this problem, the resulting boundary equations have a singular matrix [Cruse (1972)], since the method cannot distinguish between two surfaces in the same plane. The method must, therefore, be modified in some way. Early workers modelled the crack as two planes separated by a small finite distance and joined by some simple curve [Cruse (1973)]. This approach is inefficient, since it involves the use of a large number of elements in order to obtain a realistic approximation to a crack of small tip radius.

One special problem for which the BIE method is directly applicable is that of a crack on a plane of symmetry. Here traction can be applied to the crack surface and zero displacement and zero shear traction can be specified on the remainder of the plane. The crack is then modelled as a true slit and results compare favourably with those of other methods [Cruse (1975)].

For some special cases, usually in two dimensions, a Green's function is available, and then a formula for the BIE kernels $T_{ij}(x, y)$, $U_{ij}(x, y)$ may be determined for a crack or hole in an infinite plate. Once these modified kernels have been incorporated into the boundary integral equations, only the non-cracked boundary need be modelled. Very accurate results are possible with this approach [Snyder and Cruse (1975)].

In the "flat crack modelling" approach the displacements along the crack may be represented as relative displacements between the two surfaces. However, this has two disadvantages. Firstly both relative and total displacements remain unknown on the crack. Secondly, a non-unique boundary integral equation is generated if tractions are given on the crack but not elsewhere on the boundary [Cruse (1978)].

The more recent technique of modelling the crack by "stitching" together subregions appears to overcome the problems inherent in the various approaches discussed above. In particular the geometry of the crack may be represented accurately in either two or three dimensional problems, whether or not the crack is on a plane of symmetry. However, although the method has been applied successfully to a number of problems by Blandford et al (1981), the applications so far appear to have involved only straight cracks.

In the present paper we apply such a stitching technique, apparently for the first time, to curved cracks in two dimensional regions. In particular stress intensity factors are determined for an arc crack in a large region, and results compare favourably with those obtained from the known analytic solution for an infinite region. Stress intensity factors are also obtained for the new problem of an arc crack in a square plate.

IMPLEMENTATION OF THE BOUNDARY ELEMENT METHOD

The formulation of the system of equations for the boundary element method is well known and will not be discussed here. The reader is referred, for example, to contributions by Cruse (1969), Lachat and Watson (1975).

In the stitching method, a cracked body is regarded as consisting of (say) two subregions joined along a line which also defines the crack surfaces. Using superscripts (1) and (2) for the two subregions and a subscript I for the join between them, the tractions t and displacements u at the join are required to satisfy the continuity conditions:

$$u_I = u^{(1)} = u^{(2)}$$

$$t_I = t^{(1)} = -t^{(2)}$$

At each point of the crack surfaces and of the remainder of the boundary, one of the variables t and u is specified by the boundary conditions and the other is unknown. The boundary element equations for each subregion may therefore be combined (with u_I and t_I as unknowns) and the resulting system solved by an appropriate method.

The results given in this paper were obtained by using quadratic isoparametric elements to represent both the boundary shape and the variations in tractions and displacements.

Care must be taken in the evaluation of any integrals (in the integral equation) which involve a log (1/r) singularity. On any element in which such a singularity is present, integrals occur of the form

$$I = \int_0^1 \ell n[1/r(\xi)]\ \phi_i(\xi)\ G(\xi)\ d\xi$$

where ξ is the element parameter ($0 \leqslant \xi \leqslant 1$), $\phi_i(\xi)$ ($i = 1,2,3$) is the quadratic interpolation function, and $G(\xi)$ is the boundary Jacobian. On such elements r has the form

$$r = \xi\sqrt{a\xi^2 + b\xi + c}$$

and hence I may be written in the form

$$I = \int_0^1 \ell n(1/\xi)\ \phi_i(\xi)\ G(\xi)\ d\xi\ +$$

$$\tfrac{1}{2}\int_0^1 \ell n[1/(a\xi^2 + b\xi + c)]\ \phi_i(\xi)\ G(\xi)\ d\xi \qquad (1)$$

In our solutions, the first integral in equation (1) is approximated by a six-point Gauss quadrature formula with weight function $\ell n\ (1/r)$, while the second integral is approximated by a six-point Gauss quadrature formula with weight function 1, as are all other integrals. In the case of a straight line, r is equal to $k\xi$ (where k is a constant), $G(\xi)$ is a constant, and hence the formulation is exact if two or more Gauss points are used.

The system of linear algebraic equations produced by the BEM is singular insofar as it includes a rigid body displacement.

However, the system may be made non-singular by deleting certain equations, and replacing them by constraint equations which fix a zero displacement at certain specified points. Gauss elimination may then be used to obtain a numerical solution. This approach works well for simple problems (e.g. long straight cracks) but for less well-conditioned problems (e.g. short curved cracks) the solution can vary noticeably according to the choice of constraint equations.

An alternative approach, which appears to be more reliable and which is therefore used here, is to add extra equations to eliminate rigid body motion and then solve the resulting over-determined system by a least squares method.

CRACK TIP MODELLING

Linear elastic fracture mechanics assumes perfectly elastic regions and does not consider effects such as plasticity in the crack tip region. For a straight crack in an infinite region Williams (1957) derived an expansion for the stresses around the crack tip in terms of polar coordinates (r, θ) defined as in Figure 1.

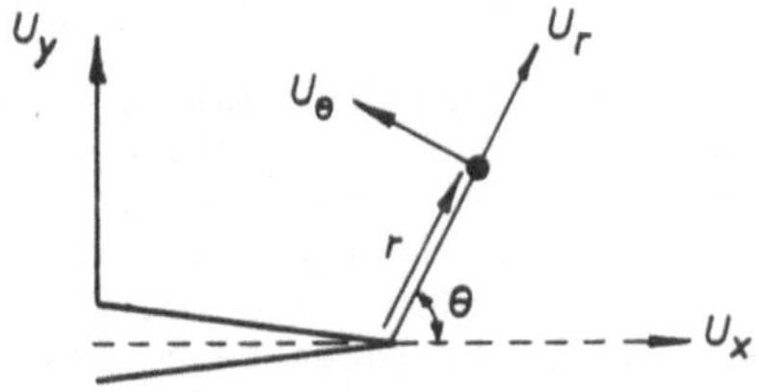

Figure 1. STRAIGHT CRACK IN AN INFINITE PLANE

When r is small, the stresses behave as $1/\sqrt{r}$, and the displacements (obtained by integrating the expansion for stresses) behave as $\sqrt{r}$. It has been shown by Henshell and Shaw (1975) that the variation of displacements and tractions with $\sqrt{r}$ can be modelled for quadratic elements by shifting the mid-point node to the quarter-point position, as in Figure 2.

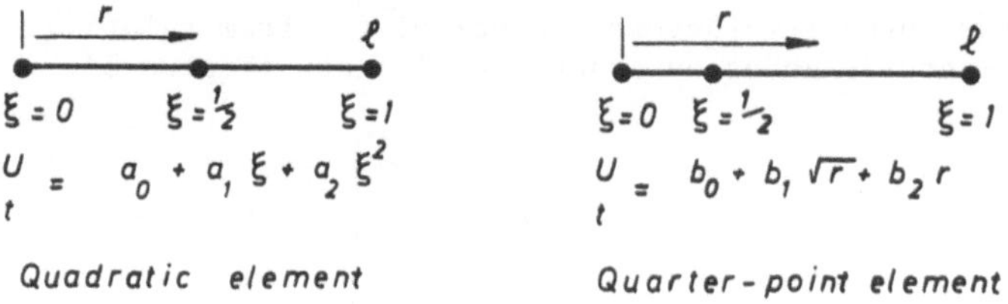

Figure 2. BOUNDARY ELEMENTS

An important advantage of quarter-point elements over other techniques for crack-tip modelling is that only very minor modifications are required to include such elements in a boundary element program.

The traction variation should, however, incorporate a $1/\sqrt{r}$ term and this may be achieved by multiplying the crack tip element coefficients by $\sqrt{\ell/r}$ [Cruse and Wilson (1977)]:

$$t = (b_0 + b_1\sqrt{r} + b_2 r)\sqrt{\ell/r} = c_0/\sqrt{r} + c_1 + c_2\sqrt{r}.$$

In the case of an arc crack an expansion has been derived by Panasyuk and Berezhnitskiy and reported by Savin (1970) for the stresses in the vicinity of the crack tip. If r is small, the series expansions for displacements may be written in the forms:

$$u_r = \frac{1}{8\mu}\sqrt{\frac{2r}{\pi}}\left\{K_I[(2\kappa - 1)\cos\frac{\theta}{2} - \cos\frac{3\theta}{2}] + K_{II}[(2\kappa - 1)\sin\frac{\theta}{2} - 3\sin\frac{3\theta}{2}]\right\} + O(r) \quad (2)$$

$$u_\theta = \frac{1}{8\mu}\sqrt{\frac{2r}{\pi}}\left\{K_I[-(2\kappa + 1)\sin\frac{\theta}{2} + \sin\frac{3\theta}{2}] + K_{II}[(2\kappa + 1)\cos\frac{\theta}{2} - 3\cos\frac{3\theta}{2}]\right\} + O(r) \quad (3)$$

with $\mu = E/[2(1+\nu)]$; E = Young's modulus; ν = Poisson's ratio. Here κ is equal to $(3 - 4\nu)$ for plane strain and to $(3-\nu)/(1+\nu)$ for plane stress. The $\sqrt{r}$ term in equations (2), (3) is, of course, identical to that for a straight crack. On the assumption that terms of order r are negligible, the angle along the crack surface may be taken to be small (since it is then a function of r), and so equations (2), (3) give the pair of equations

$$u_r = \frac{(\kappa + 1)}{4\mu} K_{II}\sqrt{\frac{2r}{\pi}} \quad (4)$$

$$u_\theta = -\frac{(\kappa + 1)}{4\mu} K_I\sqrt{\frac{2r}{\pi}}. \quad (5)$$

CALCULATION OF STRESS INTENSITY FACTORS

The stress intensity factors are calculated from relative displacements at nodes on crack tip elements (Figure 3).

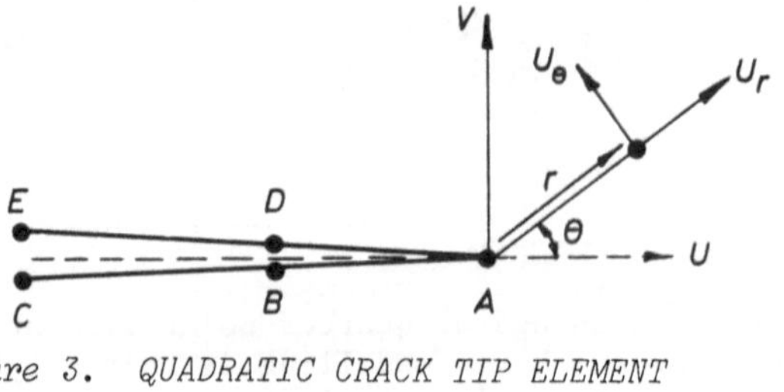

Figure 3. QUADRATIC CRACK TIP ELEMENT

Evaluating the first term in equations (2), (3) for $\theta = 180°$, we obtain the "one-point" formulae

$$K_I = \frac{2\mu}{(\kappa + 1)} \sqrt{\frac{\pi}{2r}} (v_D - v_B) \tag{6}$$

$$K_{II} = \frac{2\mu}{(\kappa + 1)} \sqrt{\frac{\pi}{2r}} (u_D - u_B) \tag{7}$$

These are the formulae which are generally used to evaluate stress intensity factors numerically for quadratic elements.

When a quarter-point element is used at the tip of a straight crack, the coefficients of $\sqrt{r}$ in this element may be equated to the parametric approximation of the coefficient of $\sqrt{r}$ in the infinite series expansion. This leads to the "two-point" formulae

$$K_I = \frac{\mu}{\kappa + 1} \sqrt{\frac{2\pi}{\ell}} [4(v_B - v_D) + v_E - v_C] \tag{8}$$

$$K_{II} = \frac{\mu}{\kappa + 1} \sqrt{\frac{2\pi}{\ell}} [4(u_B - u_D) + u_E - u_C] \tag{9}$$

which are generally used to evaluate SIF's for quarter point elements.

Equations (6), (7) may also be used to estimate stress intensity factors for straight cracks using quarter point elements, but since the $\sqrt{r}$ coefficients are not equated exactly, this may introduce errors when r is large [see Shih et al (1976)].

In the case of an arc crack, however, the approximation to u in Figure 2 is not fitted exactly and hence it is preferable to avoid using large values of r. The equations used to determine K values are therefore the one-point formulae

$$K_I = \frac{2\mu}{(\kappa + 1)} \sqrt{\frac{\pi}{2r}} (u_{\theta D} - u_{\theta B}) \tag{10}$$

$$K_{II} = \frac{2\mu}{(\kappa + 1)} \sqrt{\frac{\pi}{2r}} (u_{rD} - u_{rB}). \tag{11}$$

DOUBLE EDGE-CRACKED PLATE (Figure 4)

The example of the double edge-cracked plate under normal tension was used by Bowie (1964) to illustrate a collocation procedure for its solution. His results were estimated to be accurate to within ± 1 per cent. This same example was used as a test problem by Blandford et al (1981), who used only the two-point formulae (8), (9) to evaluate stress intensity factors for quarter-point elements.

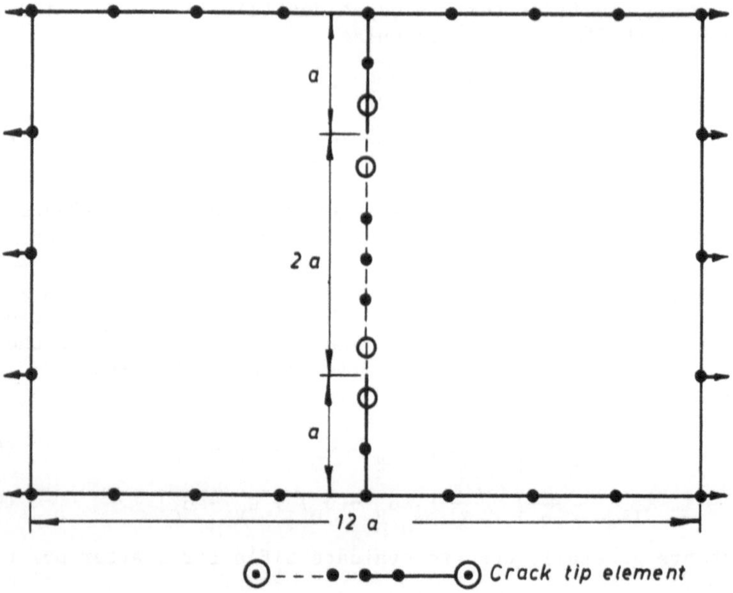

Figure 4. *BOUNDARY ELEMENT MESH FOR A DOUBLE EDGE CRACKED PLATE*

Using a fixed number of elements, we have varied the ratio (ℓ/a) of the length ℓ of the crack tip element to the length a of the crack, and the results obtained for quarter-point and traction-singular quarter-point boundary elements are shown in Figure 5.

The curves obtained by using a two point formula for K are very similar to those obtained by Blandford et al (1981), as can be seen in Figure 5.

However, the one point method appears to be advantageous for two reasons, as shown in Figure 5. Firstly, the maximum of the curve is very close to Bowie's value for the stress intensity factor, so that the error may be assumed to be negligible or negative. Secondly, variations in ℓ/a near its maximum value produce small changes in the error (for traction-singular quarter-point elements the error is less than 1 per cent for $0.06 \leqslant \ell/a \leqslant 0.45$).

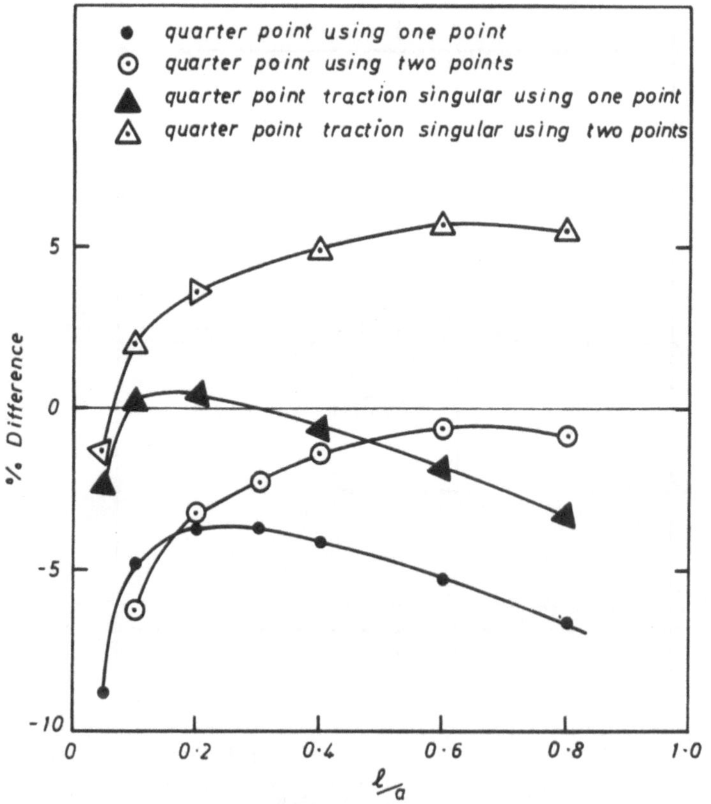

Figure 5. *PERCENTAGE DIFFERENCE IN K_I FOR DOUBLE EDGE CRACK*

ARC-CRACKED PLATE (Figure 6)

The basic mesh used for an arc crack in a square plate under constant normal tension is shown in Figure 6, which corresponds to the six element case plotted in Figure 8. In order to obtain results comparable with those for an infinite plate the ratio H/W was initially set equal to 0.1, and the effect of various parameters on the stress intensity factors was considered. In order to select an optimum tip element length c, for a crack of arc length 2a, stress intensity factors were calculated for a range of values of c/a, using a twenty four element mesh (excluding crack tip elements). The variation in the values of K_I and K_{II} is shown in Figure 7.

The convergence was examined by fixing c/a equal to 0.2 and θ equal to 30° while increasing the number of elements, as shown in Figure 8.

Values of K_I and K_{II} computed by using a mesh of 36 elements (excluding crack tip elements), a fixed ℓ/a of 0.2, and a fixed

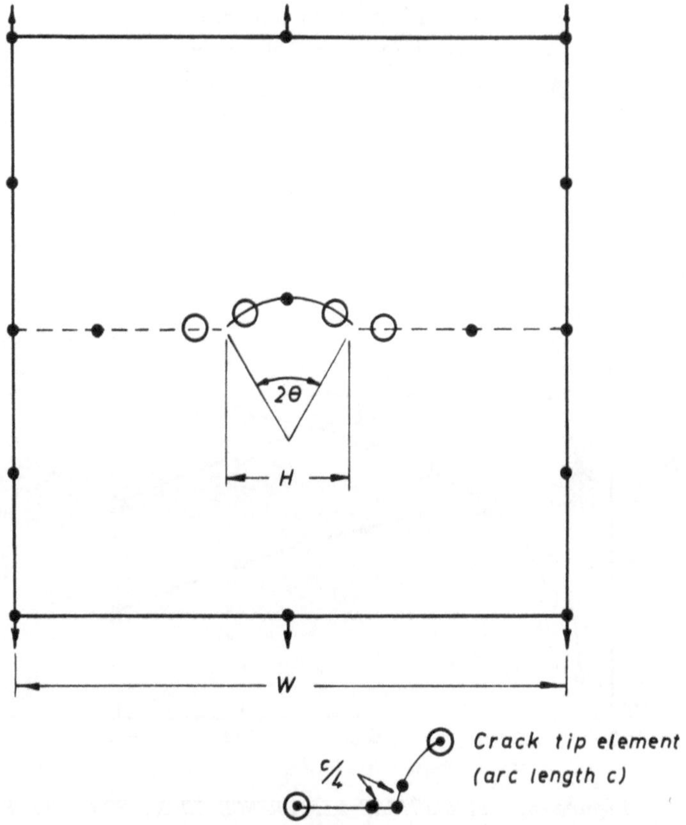

Figure 6. *SIMPLE BOUNDARY ELEMENT MESH FOR AN ARC CRACK*

chord length H, are compared with the true values for an arc crack in an infinite region in Table 1. The comparison can be seen to be favourable. The stress intensity factors have been normalized by dividing them by $\sqrt{\pi R \sin\theta}$, where R is the arc radius.

Table 1. BEM Solutions for a Large Square Plate (H/W = .1)

θ(degrees)	K_I			K_{II}		
	Infinite Plate	BEM	% error	Infinite Plate	BEM	% error
5	0.9934	0.9844	-0.9	0.0870	0.0841	-3.5
15	0.9412	0.9313	-1.1	0.2544	0.2498	-1.8
30	0.7779	0.7740	-0.5	0.4673	0.4658	-0.3
45	0.5439	0.5489	+0.9	0.6080	0.6130	+0.8
60	0.2815	0.2950	+2.8	0.6625	0.6809	+4.8
75	0.0306	0.0498	+62.7	0.6322	0.6688	+5.8

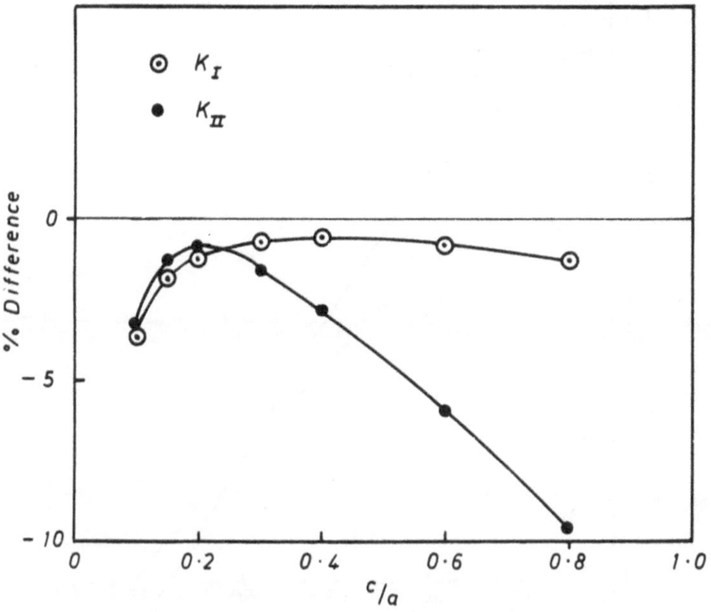

Figure 7. PERCENTAGE DIFFERENCE IN K_I AND K_{II} FOR AN ARC CRACK (θ = 30°)

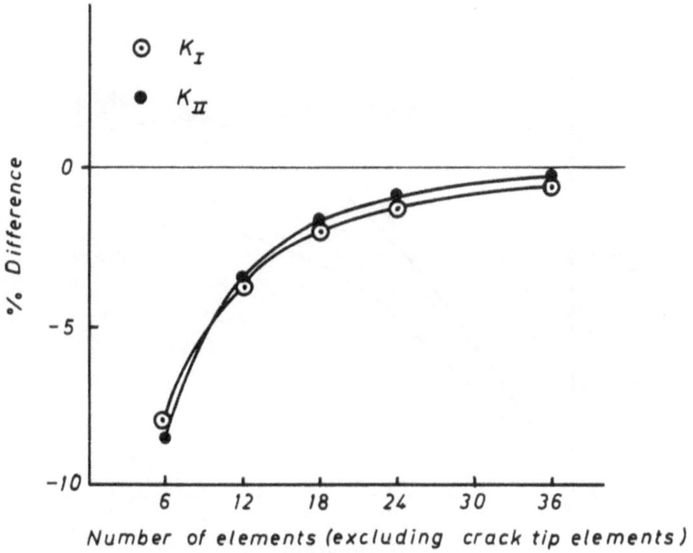

Figure 8. CONVERGENCE OF K_I AND K_{II} FOR AN ARC CRACK (θ = 30°)

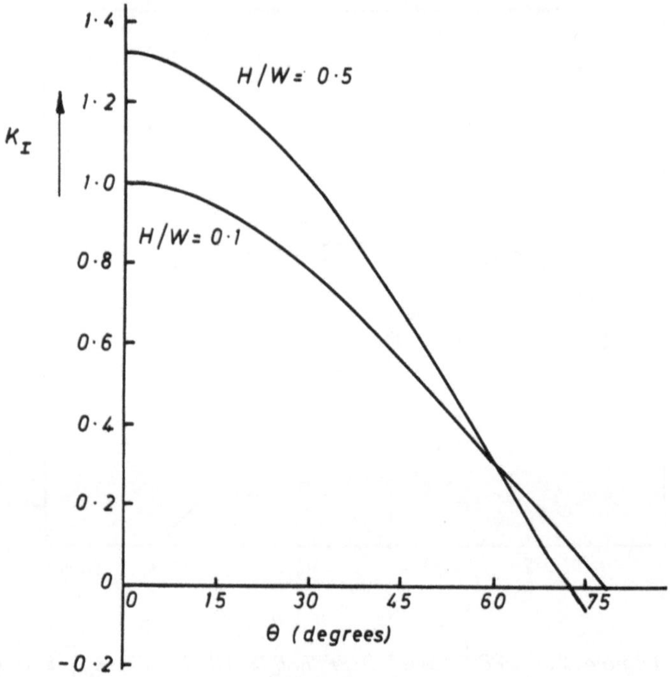

Figure 9. K_I FOR AN ARC CRACK IN A SQUARE PLATE

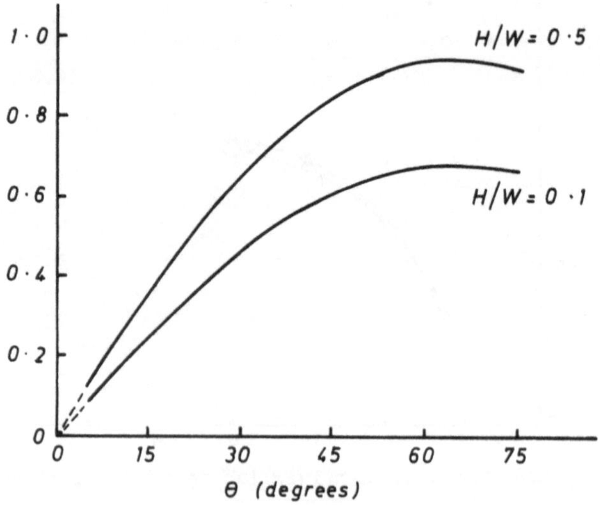

Figure 10. K_{II} FOR AN ARC CRACK IN A SQUARE PLATE

The percentage error given in Table 1 appears to increase as the angle of arc becomes larger, but excluding very small K values errors are less than 2% for arc angles of 0 to 90° and less than 6% for arc angles up to 150°.

The effect on the stress intensity factors of nearby boundaries was examined by considering an arc crack in a square plate with H/W equal to 0.5. The results for this example, as well as for the case H/W = 0.1 are plotted in Figures 9 and 10. Note that as θ tends to zero, the arc crack approaches a straight crack, and K_I and K_{II} can be seen to be approaching the appropriate values for a centre-cracked plate under constant normal tension.

CONCLUSIONS

1. The boundary element method can be applied successfully to curved crack problems. In particular, stress intensity factors have been obtained for the new problem of an arc crack in a square plate.

2. It is recommended that, in curved crack problems, traction-singular quarter-point elements should be used at crack tips, but that the one-point method should be used instead of the two-point method for the calculation of stress intensity factors.

3. The development of good curved crack techniques is an essential prerequisite for the successful solution of crack growth problems by boundary element methods. The accuracy obtained in the present work for cracks of moderate curvature should be quire adequate for most practical purposes in this context.

REFERENCES

Blandford, G.E., Ingraffea, A.R., Liggett, J.A. (1981) Two dimensional stress intensity factor computations using the boundary element method. Int. Jnl. Num. Meth. Eng., 17, 387-404.

Bowie, O.L. (1964) Recatangular Tensile sheet with Symmetric Edge Cracks. J. Appl. Mech., 31, 208-212.

Cruse, T.A. (1969) Numerical solutions in three dimensional Elastostatics, Int. J. Solids Structures 5, 1259-1274.

Cruse, T.A. (1972) Numerical evaluation of Elastic Stress intensity factors by the boundary integral equation method, in The Surface Crack: Physical Problems and Computational Solutions, (ASME New York.

Cruse, T.A. (1973) Applications of the boundary integral equation method to three dimensional stress analysis. Computers & Structures 3, 509-527.

Cruse, T.A. (1975) Boundary integral equation fracture mechanics analysis. Proceedings ASME Conference, Troy, New York. pp 31-46.

Cruse, T.A. and Wilson R.B. (1977) Boundary integral equation method for Elastic Fracture Mechanics. AFSOR-TR-78-0355, 10-11.

Cruse, T.A. (1978) Two-dimensional B.I.E. fracture mechanics analysis. Appl. Math. Modelling, 2, 287-293 (1978).

Henshell, R.D. and Shaw, K.G. (1975) Crack tip finite elements are unnecessary. Int. J. Num. Meth/ Engng., 9, 495-507.

Lachat, J.C. and Watson, J.O. (1975) A second generation boundary IE program for 3 dimensional elastic analysis Boundary Integral equation method: Computation Applications in Applied Mechanics, ASME Proc. AMD, 11, 85-100.

Savin, G.N. (1970) Stress distribution around holes. NASATT F-607, Ch VIII: 638-642.

Shih, C.F., de Lorenzi, H.H. and German, M.D. (1976) Crack extension modelling with singular quadratic isoparametric elements. Int. J. of Frac., 12: 647-651.

Snyder, M.D. and Cruse, T.A. (1975) Boundary-Integral Equation Analysis of Anisotropic cracked plates. Int. J. Fracture, 11, 2: 315-328.

Williams, M.L. (1957) On the stress distribution at the base of a stationary crack. J. Appl. Mech., 24, 109-114.

ACKNOWLEDGEMENTS

The authors wish to thank the Royal Aircraft Establishment, Farnborough for supporting this work and Mr D.P. Rooke of the Materials Department, RAE, Farnborough for his helpful advice.

Session V
Plate Bending Problems

APPLICATION OF THE DIRECT BOUNDARY ELEMENT METHOD TO REISSNER'S PLATE MODEL

Frank Van der Weeën

Department of Machine Mechanics, State University of Ghent,
Grotesteenweg Noord 2, B-9710 Ghent, Belgium

INTRODUCTION

In recent years various boundary integral methods have been applied to Kirchhoff's theory for thin elastic plate flexure : Bézine (1981), Christiansen & Hougaard (1977), Tottenham (1979), Wu & Altiero (1979) and the references included therein. However, deficiencies of this classical theory are well known: the transverse shear deformability of the plate is not taken into account, only two of three physical boundary conditions are satisfied and results are inaccurate in edge zones or around holes having a small diameter in comparison to plate thickness. The present paper describes a direct boundary integral approach to Reissner's isotropic plate model. This refined bending theory leads to a sixth-order boundary value problem in contrast to the biharmonic problem of classical theory. Consequently we must establish a system of three boundary integral equations involving generalized displacements (two rotation components plus deflection) and associated tractions (normal flexural moment, torsional moment and transverse shear force). In the numerical solution of this system we broadly follow the computational techniques of Lachat (1975) for isoparametric quadratic boundary elements. The validity of the formulation is verified on example results including comparison to analytical solutions and finite element results.

REISSNER'S PLATE EQUATIONS

Consider a homogeneous isotropic plate of uniform thickness h subjected to a transverse load q per unit area. Reissner's constitutive equations relating rotation components u_α and deflection u_3 to bending stress couples $\sigma_{\alpha\beta}$ and shear stress resultants $\sigma_{3\alpha}$ are given by

$$\sigma_{\alpha\beta} = D\,\frac{1-\nu}{2}\left(u_{\alpha,\beta}+u_{\beta,\alpha} + \frac{2\nu}{1-\nu}\,u_{\gamma,\gamma}\delta_{\alpha\beta}\right) - \frac{\nu}{1-\nu}\,\lambda^{-2} q\,\delta_{\alpha\beta}$$

$$\sigma_{3\alpha} = D\,\frac{1-\nu}{2}\,\lambda^{2}\,(u_{\alpha} + u_{3,\alpha}) \qquad (1)$$

where $\lambda = \sqrt{10}/h$, ν is Poisson's ratio and D denotes the flexural plate rigidity.
Substitution of equations (1) into the equilibrium equations

$$\sigma_{\alpha\beta,\beta} - \sigma_{3\alpha} = 0$$
$$\sigma_{3\alpha,\alpha} + q = 0 \qquad (2)$$

yields a sixth-order system of Navier equations allowing satisfaction of three boundary conditions : on the boundary $\partial\Omega$ of the midplane Ω either the displacement component u_i or the corresponding traction component $t_i = \sigma_{i\beta} n_\beta$ must be prescribed ; here, n_β are the direction cosines of the outward normal on $\partial\Omega$.

BOUNDARY INTEGRAL EQUATIONS

A straightforward application of the weighted residual technique (Brebbia, 1978) to the plate equations (1) and (2) yields the following system of integral equations :

$$C_{ij}(x)u_j(x) + \int_{\partial\Omega} [T_{ij}(x,y)u_j(y) - U_{ij}(x,y)t_j(y)]\,ds(y) = P_i(x) \qquad (3)$$

$$\text{where } P_i(x) = \int_{\Omega} [U_{i3}(x,y) - \frac{\nu}{1-\nu}\lambda^{-2} U_{i\beta,\beta}(x,y)]\,q(y)\,d\Omega(y) \qquad (4)$$

The kernel $U_{ij}(x,y)$ represents the rotation (j=1 and j=2) or the deflection (j=3) at the field point $y=(y_1,y_2)$ of an infinite plate due to a unit point couple (i=1 and i=2) or a unit point force (i=3) at the load point $x=(x_1,x_2)$. This fundamental solution was constructed using Hörmander's method (Van der Weeën, 1981) and is listed in appendix together with the corresponding traction kernel $T_{ij}(x,y)$.
The contribution $C_{ij}(x)$ of the unit point loads equals δ_{ij} for an interior point ($x \in \Omega$) and $\delta_{ij}/2$ for a boundary point ($x \in \partial\Omega$) with a smooth tangent. From a computational point of view, it is preferable to evaluate this matrix indirectly (Lachat, 1975) by expressing that the stress-free problem ($t_i \equiv 0$ and $q \equiv 0$) admits non-trivial solutions (u_1,u_2,u_3) which are arbitrary combinations of three basic rigid-body dis-

placements $(1,0,x_1-y_1)$, $(0,1,x_2-y_2)$ and $(0,0,1)$. In this way one obtains :

$$C_{i\beta}(x) = -\int_{\partial\Omega} [T_{i\beta}(x,y)+(x_\beta-y_\beta)T_{i3}(x,y)]ds(y)$$
$$C_{i3}(x) = -\int_{\partial\Omega} T_{i3}(x,y)ds(y) \tag{5}$$

The function $P_i(x)$ of equation (4), where differentiation is taken with respect to the field point y, represents the potential of the distributed transverse loading q(y). In case of a uniform loading q, this domain integral is convertible to the following boundary integral thru application of the divergence theorem :

$$P_i(x)=q\int_{\partial\Omega} [V_{i,\beta}(x,y)-\frac{\nu}{1-\nu}\lambda^{-2}U_{i\beta}(x,y)]n_\beta(y)ds(y) \tag{6}$$

where $V_i(x,y)$ is a solution of the Poisson equation

$$V_{i,\beta\beta}(x,y) = U_{i3}(x,y) \tag{7}$$

The integral equations (3) are numerically solved for unknown boundary displacement and traction components as outlined in the next section. If results at interior points are required, equations (3) and (6) with $C_{ij}(x)=\delta_{ij}$ may be used for computation of displacements $u_i(x)$. Stress resultants at interior points are obtained by differentiation of these equations with respect to the co-ordinates of the load point :

$$\sigma_{ij}(x) = -\frac{\nu}{1-\nu}\lambda^{-2}q\,\delta_{ij} + \int_{\partial\Omega} [D_{ijk}(x,y)t_k(y)-S_{ijk}(x,y)u_k(y)+q\,W_{ij}(x,y)]ds(y) \tag{8}$$

where the case i=j=3 must be excluded. The kernels $D_{ijk}(x,y)$, $S_{ijk}(x,y)$ and $W_{ij}(x,y)$ are listed in appendix.

Stress resultants at boundary points are computed from tractions and tangential derivatives of boundary displacements (Lachat, 1975).

BOUNDARY ELEMENTS

It is readily verified that all equations in the previous section remain valid if the midplane of the plate is divided in subdomains i.e. $\partial\Omega$ now represents the boundary of a subdomain Ω. In the present formulation the boundary of each subdomain is discretized into isoparametric boundary elements with quadratic variation :

$$g(\xi)= \frac{1}{2}\,\xi(\xi-1)g(-1)+(1-\xi^2)g(0)+ \frac{1}{2}\,\xi(\xi+1)g(+1) \quad (9)$$

where $\xi \in [-1,1]$ is the intrinsic element co-ordinate and g stands for cartesian co-ordinates y_α, displacements u_i or tractions t_i.
Substituting this interpolation scheme into the integral equations (3), putting the load point at each boundary node of each subdomain in turn, introducing the prescribed nodal displacements and tractions on external parts of a boundary or enforcing displacement continuity and traction equilibrium on interfaces between adjacent subdomains, one obtains a banded system of linear algebraic equations, provided that unknown values of nodal displacement and traction components have been suitably numbered. The coefficients of this system are integrals over a boundary element of kernel-interpolation function products, which are numerically integrated using Gaussian quadrature with appropriate choice of integration order. This entire procedure has been sufficiently reported in detail by Lachat (1975) but two items merit further comment :

(a) in the present case the logarithmic singularity of the kernel $U_{ij}(x,y)$, as displayed in appendix, is partially hidden in modified Bessel functions K_0 and K_1. Isolation of this singularity and subsequent Gaussian quadrature with logarithmic weight function, as proposed by Lachat, leads to akward programming. The present author found it more convenient to soften the original integrands $f(\xi)$, which are kernel - interpolation function - jacobian products, by introducing a mapping between intrinsic co-ordinate ξ and Gaussian integration variable θ such that $d\xi/d\theta$ vanishes at the value ξ^* for which the singularity occurs :

$$\int_{-1}^{1} f(\xi)d\xi = \int_{-1}^{1} [f(\xi_1(\theta))\frac{d\xi_1}{d\theta} + f(\xi_2(\theta))\,\frac{d\xi_2}{d\theta}\,]d\theta \quad (10)$$

where

$$\left.\begin{aligned} \xi_1(\theta) &= (\tfrac{\theta+1}{2})^2 - 1 \\ \xi_2(\theta) &= \tfrac{\theta+1}{2} \end{aligned}\right\} \quad \text{if } \xi^* = -1$$

$$\left.\begin{aligned} \xi_1(\theta) &= -(\tfrac{\theta-1}{2})^2 \\ \xi_2(\theta) &= (\tfrac{\theta+1}{2})^2 \end{aligned}\right\} \quad \text{if } \xi^* = 0$$

$$\left.\begin{aligned} \xi_1(\theta) &= \tfrac{\theta-1}{2} \\ \xi_2(\theta) &= 1- (\tfrac{\theta-1}{2})^2 \end{aligned}\right\} \quad \text{if } \xi^* = +1$$

In this way integration points ξ corresponding to fixed Gaussian abscissae θ are concentrated around the singularity point ξ^*.

(b) non-uniqueness of the traction vector due to a discontinuous normal at the intersection node of adjacent boundary elements requires special attention. If the total number of unknowns at the intersection node exceeds the number of independent equations obtained from the discretized versions of the integral equations and from the requirements of displacement continuity and traction equilibrium on interfaces, additional equations are introduced from the assumption of stress resultant continuity.

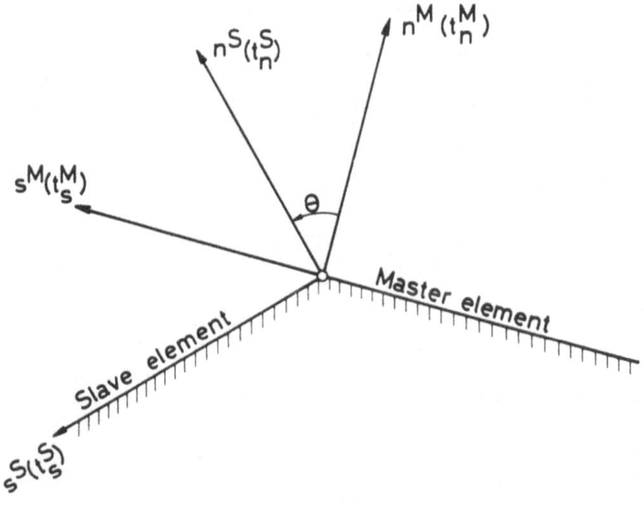

Figure 1. Intersection node with discontinuous normal

Denoting two adjacent elements by superscripts M and S (master and slave, figure 1), local traction components on the slave element are computed from stress resultants, which are in turn computed from local traction components and tangential derivatives of displacements along the master element :

$$\begin{Bmatrix} t_n \\ t_s \end{Bmatrix}^S = \begin{bmatrix} \cos^2\theta+\nu\sin^2\theta & -2\cos\theta\sin\theta \\ (1-\nu)\cos\theta\sin\theta & \cos^2\theta-\sin^2\theta \end{bmatrix} \begin{Bmatrix} t_n \\ t_s \end{Bmatrix}^M + \begin{Bmatrix} \sin^2\theta \\ -\cos\theta\sin\theta \end{Bmatrix} (D(1-\nu^2)u^M_{s,s}-\nu\lambda^{-2}q) \qquad (11)$$

$$t_3^S = \cos\theta t_3^M - \sin\theta . D \frac{1-\nu}{2} \lambda^2 (u_s^M + u_{3,s}^M)$$

where $\cos\theta = n_1^M n_1^S + n_2^M n_2^S$

$$\sin\theta = n_2^M n_1^S - n_1^M n_2^S$$

$$u_{s,s}^M = -n_2^M u_{1,s}^M + n_1^M u_{2,s}^M$$

$$u_s^M = -n_2^M u_1^M + n_1^M u_2^M$$

and derivatives with respect to the arc co-ordinate s^M are easily obtained from the interpolation scheme (9) :

$$\frac{d}{ds} = \left[\left(\frac{dy_1}{d\xi}\right)^2 + \left(\frac{dy_2}{d\xi}\right)^2 \right]^{-1/2} \frac{d}{d\xi}$$

In case of a continuous normal ($\theta=0$), equations (11) reduce to the correct result ($t_i^S=t_i^M$) in contrast to the additional equations proposed by Chaudonneret (1978).

EXAMPLE RESULTS

In the following examples boundary element results are compared to the analytical solution, if available, and to results of a finite element analysis using eight-node isoparametric plate bending elements (Hinton & Owen, 1977). Since a side of this finite element is nearly equivalent to an isoparametric quadratic boundary element, nodes in BEM-discretizations are taken coincident with boundary nodes in FE meshes, thus facilitating comparison.

Infinite plate with circular hole.
The analytical solution for a plate with a circular hole of diameter 2a and subjected to a constant stress couple $\sigma_{11}=\sigma$ at infinity was obtained by Reissner (1945). Since the proposed BEM-formulation does not allow for treatment of infinite domains, a quarter square plate has been discretized as shown in figure 2. Prescribed tractions along $x_1=b$ and $x_2=b$ are in agreement with the analytical solution and differ very little from the loading at infinity. Results for the hoop bending stress couple $\sigma_{\theta\theta}$ and shear stress resultant $\sigma_{3\theta}$ along the boundary of the hole are given in non-dimensional form in figure 3. In general BEM results are more accurate, except for the maximum value of $\sigma_{\theta\theta}/\sigma$ occuring at $\theta = 90°$. Total BEM execution time exceeds total FE execution time by 30% or 100%, depending on whether BEM results at interior nodes are required or not. These figures of merit in favour of finite elements may be explained

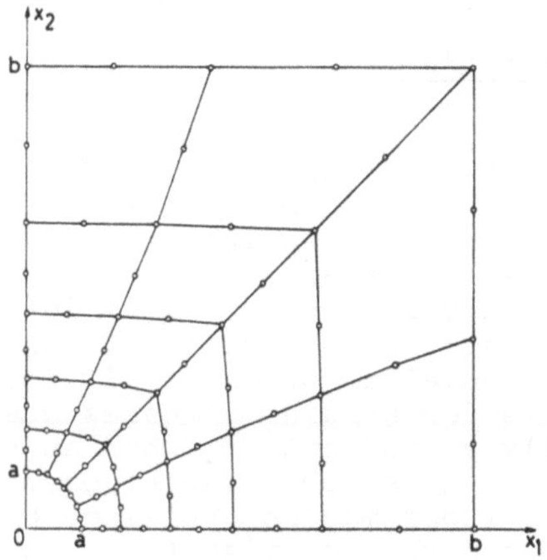

Figure 2. Square plate with central hole (b = 8a; h = 2a ; ν = 0.3)
FE mesh: 79 nodes, 20 elements
BEM mesh: 36 boundary nodes, 18 boundary elements

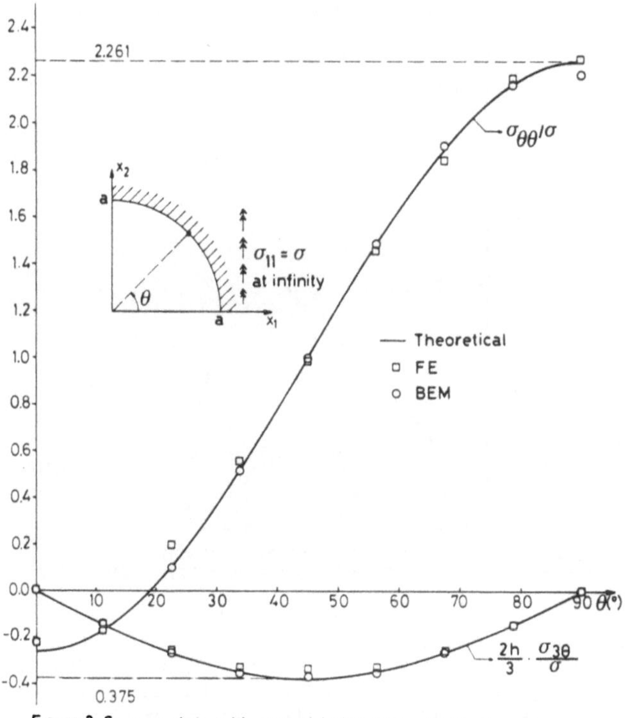

Figure 3. Square plate with central hole: stress resultants along r = a

by the use of a frontal solver (Hinton & Owen, 1977).

Clamped circular plate.

Figure 4 shows a circular plate, clamped at the outer radius r=a and subjected to a uniform loading q for $0 < r < b$. This example was choosen in order to demonstrate the use of subdomains : the quarter circle of radius r=b represents an interface in BIE-analysis. Bending stresses and shear stress are compared in figure 5 to the analytical solution, which is obtained from direct integration of the field equations with respect to the radial co-ordinate. It may be observed that BEM results for bending stresses are more accurate, especially at the center, whereas both methods give erroneous results for shear stress. This phenomenon in both methods may be explained by the small difference between the rotation (u_r) and the radial derivative of deflection ($u_{3,r}$), which effectively leads to a serious loss of précision in the computation of the shear stress resultant $\sigma_{3r} = 1/2\ D(1-\nu)\lambda^2(u_r+u_{3,r})$.

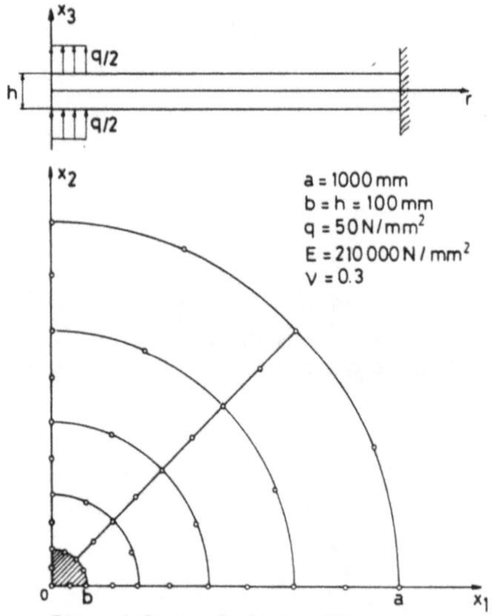

Figure 4. Clamped circular plate
FE mesh: 40 nodes, 9 elements
BEM mesh: 2 subdomains, 27 nodes, 14 elements

Plate with curved boundary.

As an example of practical interest, consider the plate depicted in figure 6. This plate, which is part of a steam vessel, is subjected to a uniform loading q and fully clamped along the boundary consisting of two straight lines and two confocal elliptical arcs.

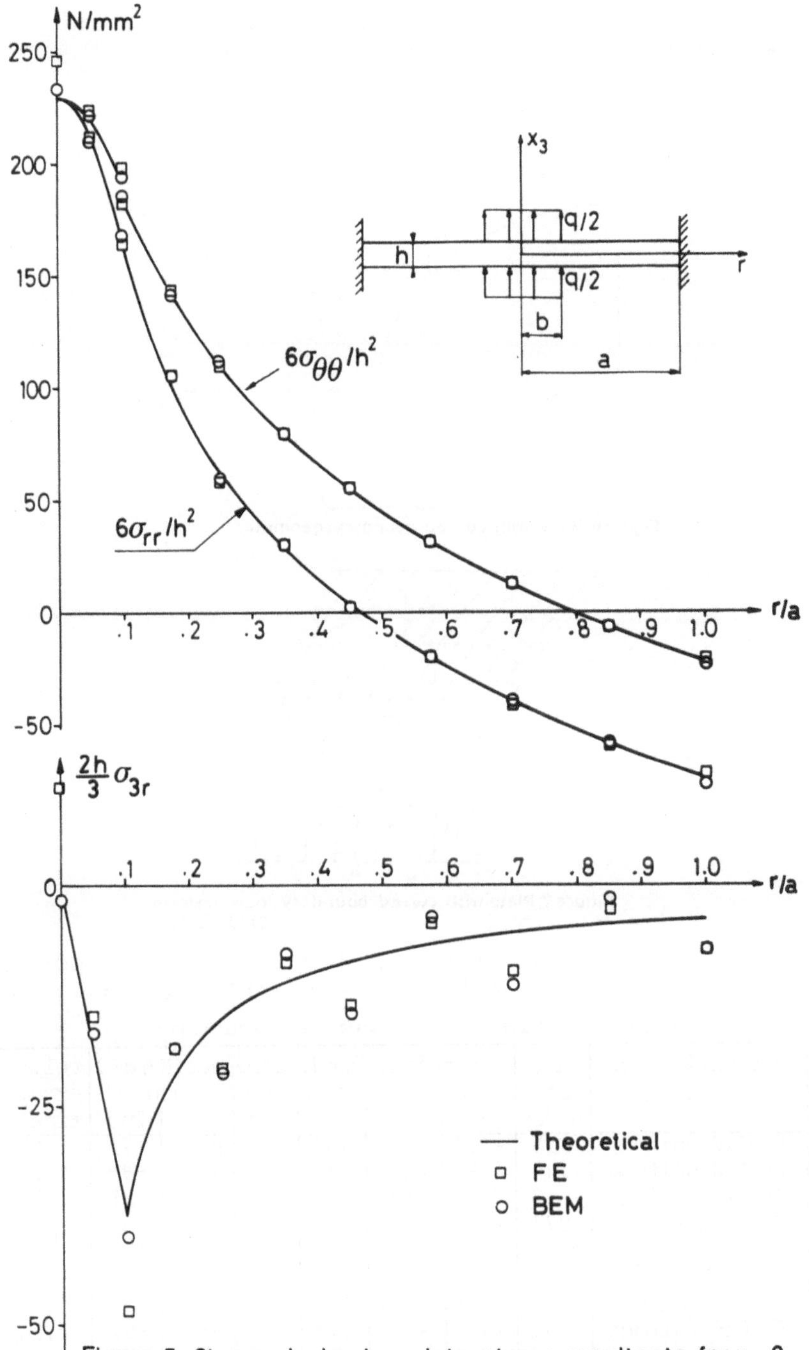

Figure 5. Clamped circular plate: stress resultants for $x_2=0$

This problem was analyzed for various discretizations, an example of which is given in figure 7.

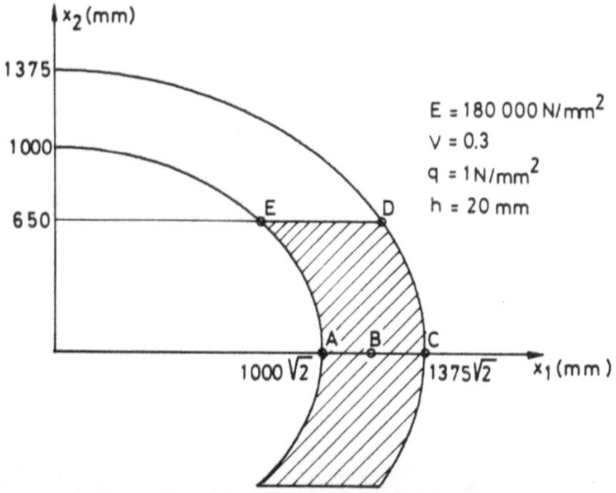

Figure 6. Plate with curved boundary: geometry

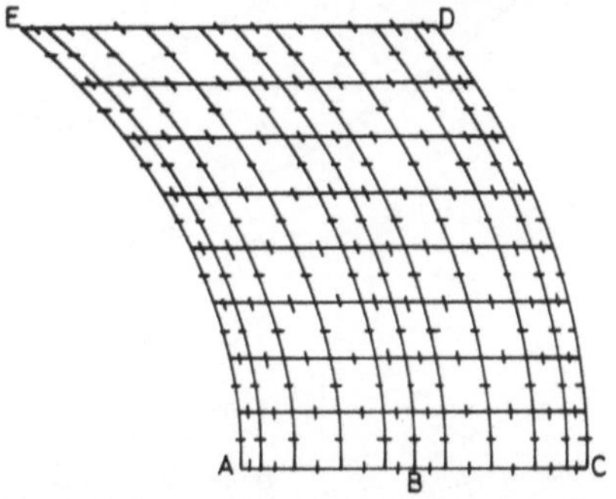

Figure 7. Plate with curved boundary: non-uniform 8x10 FE mesh

Results for maximum principal bending stresses occuring at the nodes A,B and C are listed in table 1.

Table 1. Plate with curved boundary

discretization	nbr of nodes	principal bend. stresses(N/mm^2) node A 1st	node A 2nd	node B 1st	node B 2nd	node C 1st	node C 2nd	rel. exec. time
FE 5x6 uniform	113	-287	- 86	168	60	-238	-71	1.00
FE 5x8 uniform	147	-333	-100	172	62	-269	-80	1.28
FE 8x10 uniform	277	-356	-107	173	62	-284	-85	2.70
FE 8x10 fig.7	277	-389	-116	173	62	-303	-91	2.70
BEM according to								
FE 5x6 uniform	44	-388	-119	181	63	-304	-92	1.33
FE 6x8 uniform	56	-392	-119	178	63	-307	-92	2.10

It may be observed that results for uniform FE meshes are inaccurate. Non-uniform mesh grading, as shown in figure 7, is necessary in order to obtain results comparable to those of coarse uniform BEM meshes. Comparison of relative execution time (not including BEM results for interior nodes) reveals that in this example boundary elements are far more economical for a given accuracy.

DISCUSSION

Applicability of boundary elements to a sixth-order plate bending theory has been demonstrated. Comparing accuracy and execution time, overall superiority of boundary elements over finite elements cannot be claimed. However, preparation and checking of input-data, which constitute an important cost factor in FE analysis, is very easy in case of boundary elements especially if subsequent mesh refinements are necessary.

The proposed formulation is susceptible for improvements, e.g.

- use of higher order and spline function interpolation could improve results for the tangential component of shear stress resultant along the boundary
- the technique for converting the distributed loading potential into a boundary integral is not limited to the case of uniform loading, but a simple and readily available particular solution of the Navier equations (Lachat, 1975) would require less computation time at the cost of an increase in computer storage requirements.

Extension of the reported method to non-linear problems (large plate deflections) may now be investigated.

REFERENCES

Abramowitz, M. and Stegun, I.A. (1965) Handbook of mathematical functions, Dover Publications, New York.

Bézine, G. (1981) A boundary integral equation method for plate flexure with conditions inside the domain, Int. J.Num.Meth.Engng 17, 1647-1657

Chaudonneret, M. (1978) On the discontinuity of the stress vector in the boundary integral equation method for elastic analysis, Recent advances in boundary element methods (ed. Brebbia C.A.), Pentech Press, London

Christiansen, S. and Hougaard, P. (1977) An investigation of a pair of integral equations for the biharmonic problem, DCAMM report no.121, The Technical University of Denmark.

Hinton, E. and Owen, D.R.J. (1977) Finite element programming, Academic Press, London.

Lachat, J.C. (1975) A further development of the boundary integral technique for elastostatics, Ph.D.thesis University of Southampton.

Reissner, E. (1945) The effect of transverse shear deformation on the bending of elastic plates, J.Appl.Mech. 12, A69-A77.

Tottenham, H. (1979) The boundary element method for plates and shells, Developments in boundary element methods-1 (eds. Banerjee P.K. and Butterfield R.), Appl. Science Publishers Ltd, London.

Vander Weeën, F. (1981) Randintegraalvergelijkingen voor het plaatmodel van Reissner, Ph.D.thesis, State University of Ghent.

Wu, B.C. and Altiero, N.J. (1979) A boundary integral method applied to plates of arbitrary plan form and arbitrary boundary conditions, Computers & Structures 10, 703-707.

APPENDIX

The fundamental displacement field $U_{ij}(x,y)$ for Reissner's plate model is given by

$$U_{\alpha\beta}=[8\pi D(1-\nu)]^{-1}\{[8B(z)-(1-\nu)(2\ln z-1)]\delta_{\alpha\beta}-[8A(z)+1-\nu]r_{,\alpha}r_{,\beta}\}$$

$$U_{\alpha 3} = -U_{3\alpha} = (8\pi D)^{-1}(2\ln z-1)rr_{,\alpha}$$

$$U_{33} = [8\pi D(1-\nu)\lambda^2]^{-1}[(1-\nu)z^2(\ln z-1)-8\ln z]$$

where $A(z) = K_0(z) + 2z^{-1}[K_1(z)-z^{-1}]$

$$B(z) = K_0(z) + z^{-1}[K_1(z)-z^{-1}]$$

$$z = \lambda r$$

$$r = [(y_1-x_1)^2 + (y_2-x_2)^2]^{1/2}$$

$$r_{,\alpha} = r^{-1}(y_\alpha - x_\alpha)$$

Expanding the modified Bessel functions $K_0(z)$ and $K_1(z)$ for small arguments (Abramowitz & Stegun, 1965) it is found that A(z) is continuous whereas B(z) has a singularity (- 1/2 ln z)
The corresponding traction kernel $T_{ij}(x,y)$ is given by

$$T_{\gamma\alpha} = -D_{\alpha\beta\gamma}n_\beta$$

$$T_{\gamma 3} = +D_{3\beta\gamma}n_\beta$$

$$T_{3\alpha} = +D_{\alpha\beta 3}n_\beta$$

$$T_{33} = -D_{3\beta 3}n_\beta$$

where n_β are the direction cosines of the outward normal at $y \in \partial\Omega$ and

$$D_{\alpha\beta\gamma} = (4\pi r)^{-1} [(4A+2zK_1+1-\nu)(\delta_{\alpha\gamma}r_{,\beta} + \delta_{\beta\gamma}r_{,\alpha}) + (4A+1+\nu)\delta_{\alpha\beta}r_{,\gamma} - (16A+4zK_1+2-2\nu)r_{,\alpha}r_{,\beta}r_{,\gamma}]$$

$$D_{\alpha\beta 3} = -(8\pi)^{-1}(1-\nu) [(2 \frac{1+\nu}{1-\nu} \ln z - 1)\delta_{\alpha\beta} + 2r_{,\alpha}r_{,\beta}]$$

$$D_{3\beta\gamma} = (2\pi)^{-1} \lambda^2 (B\delta_{\beta\gamma} - Ar_{,\beta}r_{,\gamma})$$

$$D_{\beta 33} = (2\pi r)^{-1} r_{,\beta}$$

The remaining kernels of the integral equations (6) and (8) are :

$$S_{\alpha\beta\gamma}=(4\pi r^2)^{-1}D(1-\nu)\{(4A+2zK_1+1-\nu)(n_\alpha\delta_{\beta\gamma}+n_\beta\delta_{\alpha\gamma}) -(16A+6zK_1+2-2\nu+z^2K_0)[(n_\alpha r_{,\beta}+n_\beta r_{,\alpha})r_{,\gamma}+(\delta_{\alpha\gamma}r_{,\beta}+\delta_{\beta\gamma}r_{,\alpha})r_{,n}] -(16A+4zK_1+2+2\nu)(n_\gamma r_{,\alpha}r_{,\beta}+\delta_{\alpha\beta}r_{,\gamma}r_{,n})\}$$

$$S_{\alpha\beta 3}=(4\pi r)^{-1} D(1-\nu)\lambda^2 [2A\, \delta_{\alpha\beta}r_{,n}-(8A+2zK_1)r_{,\alpha}r_{,\beta}r_{,n} + (2A+zK_1)(n_\alpha r_{,\beta}+ n_\beta r_{,\alpha})]$$

$$S_{3\beta\gamma} = -(4\pi r)^{-1} D(1-\nu)\lambda^2 [2An_\gamma r_{,\beta}- (8A+2zK_1)r_{,\beta}r_{,\gamma}r_{,n} + (2A+zK_1) (\delta_{\beta\gamma}r_{,n} + n_\beta r_{,\gamma})]$$

$$S_{3\beta 3} = (4\pi r^2)^{-1} D(1-\nu)\lambda^2 [(1+z^2B)n_\beta - (2+z^2A)r_{,\beta}r_{,n}]$$

$$V_{\alpha,\beta} = (128\pi D)^{-1} r^2 [(4\ln z - 5)\delta_{\alpha\beta} + 2(4\ln z - 3)r_{,\alpha}r_{,\beta}]$$

$$V_{3,\beta} = [128\pi D(1-\nu)\lambda^2]^{-1} rr_{,\beta} [32(2\ln z - 1)-z^2(4\ln z - 5)]$$

$$W_{\alpha\beta} = -(64\pi)^{-1}r\{(4\ln z - 3)[(1+3\nu)\delta_{\alpha\beta}r_{,n}+(1-\nu)(n_\alpha r_{,\beta}+n_\beta r_{,\alpha})] +[4\nu\delta_{\alpha\beta}+4(1-\nu)r_{,\alpha}r_{,\beta}]r_{,n}\}-\nu(1-\nu)^{-1} \lambda^{-2} D_{\alpha\beta\gamma}n_\gamma$$

$$W_{3\beta} = (8\pi)^{-1} [(2\ln z - 1)n_\beta + 2r_{,\beta}r_{,n}] - \nu(1-\nu)^{-1} \lambda^{-2}D_{3\beta\gamma}n_\gamma$$

where $r_{,n} = r_{,\zeta}n_\zeta$

APPLICATION OF COMPLEX VARIABLE THEORY TO BOUNDARY ELEMENT METHOD

Bulent A. Ovunc

University of Southwestern Louisiana

ABSTRACT

The application of the complex variable theory to the plates subjected to externally applied in-plane or transversal loads provides a unique solution if the boundaries of the plates can be expressed in terms of a single parameter, or if the boundaries can be transformed by conformal mapping into one which can be expressed in terms of a single parameter. Herein, an approximate solution is given for plates whose boundaries can neither be expressed in terms of a single parameter nor be transformed by conformal mapping. The plates whose boundaries can not be expressed by a single parameter are divided into elements. The only condition on an element is that a single parameter or a conformal mapping function must describe its boundaries. The two types of boundaries that an element may have are: free boundaries and common boundaries which are the boundaries between two adjacent elements or the boundaries between an element and the external supports.

In an element the stresses are given on its free boundaries. On its common boundaries, the stresses are defined by complex fourier series with unknown coefficients. Since the boundaries are expressed terms of a single parameter and since the stresses are described along the boundaries, the stress boundary condition can be written in terms of two complex potential functions. The analytical functions of the two potentials are determined by means of the Cauchy integrals applied to the stress boundary conditions. The two analytical potential functions, obtained in the way just indicated, are in series form and contain the unknown coefficients of the series which describe the stresses along the common boundaries. The displacements and the stresses at a point within the element or on its boundaries are resolved by means of the two potential functions. The unknown coefficients stemming from the stress conditions along a common boundary are determined by considering the continuity of the displacements at as many discrete points as the number of unknown coefficients introduced to describe the stress distribution along this common boundary. For better accuracy, the number of common boundaries, thus the number of elements must be minimum, since the approximation in the solution stem from the stress distribution along the common boundaries. The sizes of elements do not need to be reduced around the critical points of the plates. Moreover, only the common boundaries of the elements need to be divided into discrete points, reducing the number of unknowns tremendously.

INTRODUCTION

The boundary element method has been overshadowed by finite element method for many years. Although the early formulations of the boundary element method date from the beginning of this century (9),(11). The simplicity in the application of the boundary element method to fluid mechanics has been recognized and the mathematical difficulty stemming from the nonlinear boundary condition has been overcome (12). The weighted residual techniques have been introduced to the formulation to gain in clarity (2). The early applications of the methods have been related to the problems of the classical elastostatics (17), and to the general transient elastodynamics (4). The constraint equation which relates the values of the surface displacements to the surface tractions has been considered as the basis. The continuous boundary has been replaced by piecewise flat segments. The unknown boundary parameters have been determined by numerical techniques applied to the constraint equation along the boundary. The stress state at a point in the domain has been obtained by Somigliana identity. A variety of problems including two and three dimensional cracked bodies has been investigated (4),(5). The two basic items of the boundary element method: the fundamental solution of the field equations and the reciprocity theorem, which is the Green formula in potential theory, have been singled out. The Green formula has been particularized, then discretized following the isoparametric finite element idea. The variables are expressed in terms of the assumed shape functions. The Dirichlet condition has been well managed to circumvent the difficulties at the sharp corners along the boundaries (1). The Somigliana identity has been used as the base equation with resulting kernels of the fundamental solution of Navier's equation. For the points along the boundary, the discretization of the base equation has been performed by using approximate shape functions. The stress state in the domain has been computed by applying the Lamé's operator to the base equation (8). Based on Boundary element method, general purpose computer programs have been generated (8), (6). The advantages of the boundary element method in comparison to the finite element method have been the reduction in dimension of the domain to be discretized which further reduces the data preparation and CPU time. For problems with cracks where good resolutions of stresses near the cracks are needed or when the domain is extended to infinity, the boundary element method is superior. But the determination of information at internal points requires lengthy calculations and their accuracy are comparatively low. Whereas, the finite element method is more powerful and has a vast field of application (3). The complex variable method, used herein, assumes that the boundary conditions be given so that the two potential functions be determined by the contour integrals (13). Once the two potentials are determined, the displacements and the stress state on the boundaries and within the domain are obtained with same accuracy by the existing expressions related to the two potentials without recurring any intermediate derivations.

PROCEDURE OF ANALYSIS

The analysis of plates by the complex variable theory consists of the determination of the two analytical potential functions through the application of Cauchy integrals to the boundary conditions. The displacements and the stresses within the plate are given by simple expressions in terms of these two potentials. For a given plate, the determination of the two potential functions depends on the characteristics of its boundary. If the boundary of the plate is a simple smooth line, in a sense that if the function which describes the boundary and its first derivative are continuous along the boundary, the two potentials are determined directly. In the case when the boundary is formed by simple smooth lines with first order discontinuities on their first derivatives at finite number of points, if possible, the entire boundary is mapped into a smooth curve. Otherwise, the plate is divided into elements such that the boundary of each element is transformable to a smooth curve by a corresponding conformal mapping function. The two potentials are determined for each element separately, by means of the conformal mapping function which corresponds to the element in consideration. A deep cantilever beam with variable moment of inertia is shown in Fig. 1. The beam is divided to three elements only. The unknowns are the normal and shearing stresses at the discrete points of the common boundaries AB, DE and FG. The rectangular and triangular elements are used in the solution of the above cantilever problem.

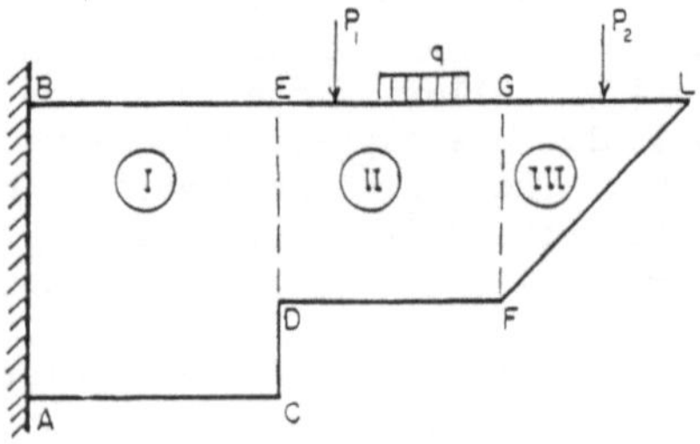

Figure 1. Cantilever beam divided in elements.

In the analysis, the given domain is divided into elements. The distribution of the normal stress X_n and the shearing stress Y_n are approximated by unknown parametric values at selected discrete points (Fig. 2). The normal and shearing stresses at a discrete point are expressed in terms of a single complex parameter

$$Z = X_n + i\, Y_n \tag{1}$$

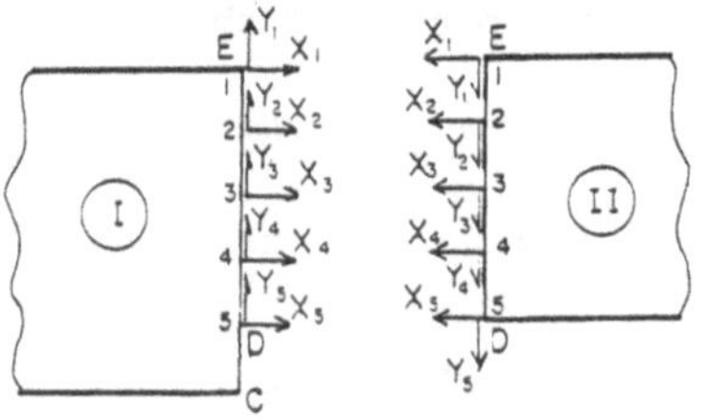

Figure 2. Stresses at discrete points along a common boundary.

Types of elements.
Various types of elements may be used as long as their boundaries are either formed by a single smooth curve or can be transformed to a single smooth curve by conformal mapping. At the present time, only circular, rectangular, triangular and half plane elements are considered in the analysis.

Conformal mapping. The rectangular and triangular elements are mapped to the circular ones. The boundaries of rectangular or triangular elements are mapped into a semi infinite region by means of Schwarz-Christoffel transformation, then into a circle by means of bilinear transformation. (Fig. 3). The expressions of the conformal mapping are obtained in elliptic integral form. In order to eliminate the multivaluedness of the elliptic integral the integrand is expanded to a series then integrated (10), (7), (14). The conformal mapping functions are: for rectangular elements:

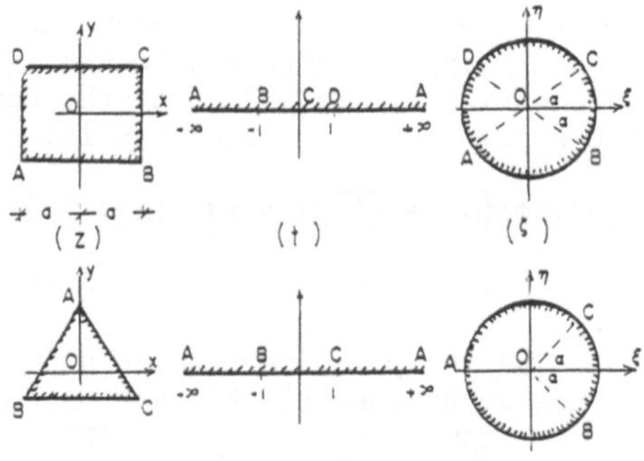

Figure 3. Rectangle and triangle mapped to circles.

$$z = \omega(\zeta) = \frac{2a}{K'} \sum \frac{1}{2n+1}\, P_n\,(\cos 2\alpha)\, \zeta^{2n+1} \tag{2}$$

where:

$\sin\alpha = k$ = modulus of the Jacobian elliptic function, sinu

P_n = Legendre coefficients

$b/a = K/K'$

$4K + 2iK'$ = period of the function,

for equilateral triangle:

$$\omega(\zeta) = K(\zeta - \frac{2}{3}\frac{\zeta^4}{4.1!} + \frac{2.5}{3^2}\frac{\zeta^7}{7.2!} - \quad (3)$$

The accuracy of the conformal mapping function depends on the number of terms in the series describing the function. By considering 26 first terms of the mapping function, the edges of a square become fairly accurate straight lines. But the number of terms required for the series to represent sharp corners is too large (Fig. 4).

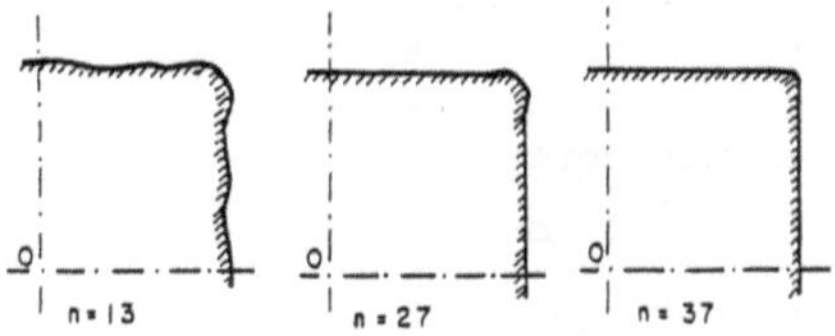

Figure 4. Quarter of a square mapped to circle.

Potential functions of elements.
The two potential functions of an element are determined from the stress boundary condition. The stress boundary condition for an element can be written in terms of two potential functions $\Phi(z)$ and $\chi(z)$, where: $z=x+iy$ is the coordinates of a point within the domain S occupied by the element. If the externally applied stresses X_n and Y_n are given numerically or parametrically (Eq. 1) along the boundary L of the domain S, the stress boundary condition is:

$$\Phi(\sigma)+\sigma\overline{\Phi'(\sigma)} + \overline{\chi'(\sigma)} = i\int_L (X_n+iY_n)\, dL = f_1+if_2 = F(\sigma) \quad (4)$$

where σ is the value of Z along the boundary L.

The two potentials which appear at the left side of the stress condition (Eq. 4) are expressed by two series:

$$\Phi(z) = \sum_{n=} a_n z^n \text{ and } \chi'(z) = \sum_{n=} a'_n z^n \quad (5)$$

Their values at the boundary $z = \sigma$, are

$$\Phi(\sigma) = \sum_{n=} a_n \sigma^n \text{ and } \chi'(A) = \sum_{n=} a'_n \sigma^n \quad (6)$$

where:

$\sigma = R\, e^{i\Theta}$

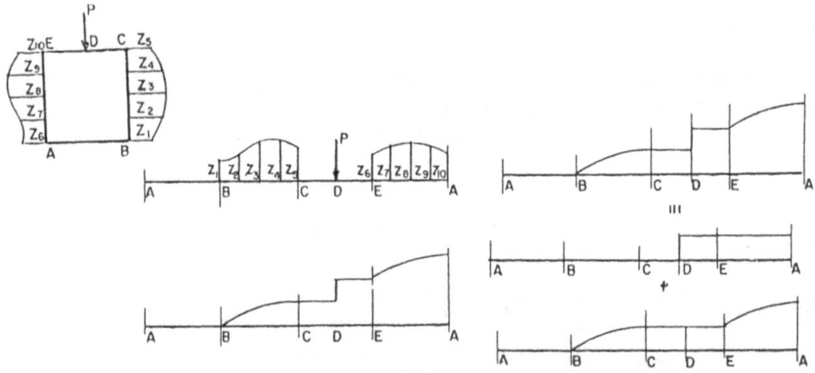

Figure 5. Stresses and summation of stresses along the boundary.

The values of $f_1 + if_2$ in terms of the external stresses are (Fig. 5):

at a discrete point j,

$$F_j = (f_1+if_2)_j = F_{j-1}+.5\ (Z_j+Z_{j-1})\ ds + \text{function of P}$$

at a point σ between two discrete points, j and j+1,

$$F(\sigma) = F_j+.5(\sigma-jd\sigma)\ (Z_{j+1}(\frac{\sigma}{d\sigma} - j) + Z_j(2- \frac{\sigma}{d\sigma} + j))$$

$$+ \text{function of P} \tag{7}$$

If the domain is simply connected, the sum of the externally applied stresses acting along the boundary must be equal to zero:

$$\int_L (X_n + iY_n)\ dL = 0 \tag{8}$$

which leads to revert $f_1 + if_2$ to its original value as σ executes a complete circuit, and vice versa. Moreover, the sum of the moments of the externally applied stresses must be zero, which yields to:

$$[x\ f_1 + \ yf_2]_L = 0 \tag{9}$$

By expanding the function $F(\sigma)$ in odd Fourier series, one has,

$$F(\sigma) = \sum_{n=1} (A_n + B_n)\ \sigma^n \tag{10}$$

where:

$$A_n = \sum_{k=1}^{N} g_{nk}\ F_k = \{g_n\}^T \{F\}$$

and B_n is a known coefficient in terms of the given externally applied stresses.

If the boundary of the element is a smooth curve, the functions $\Phi(\sigma)$, $\chi(\sigma)$ and $F(\sigma)$ where $\sigma = R\, e^{i\Theta}$ (Eqs. 6 and 10) are substituted into the stress boundary condition (Eq. 4) to have:

$$\sum_{k=1} a_k R^k e^{ik\Theta} + \bar{a}_1 R\, e^{i\Theta} + \sum_{k=0} (k+2)\, \bar{a}_{k+2} R^{k+2} e^{-ik\Theta} +$$

$$\sum_{k=0} \bar{a}'_k R^k e^{-ik\Theta} = \sum_{-\infty}^{+\infty} (A_k + B_k)\, e^{ik\Theta}$$

Comparing the coefficients of $e^{ik\Theta}$, one obtains, for n=1

$$a_1 = \mathrm{Re}\left((\{g_1\}^T \{F\} + B_1)/R \right)$$

for $n > 1$

$$a_n = (\{g_n\}^T \{F\} + B_n)/R^n$$

and for $n \geq 0$

$$a'_n = (\{\bar{g}_n\}^T \{\bar{F}\} - (n+2)\{g_{n+2}\}^T \{F\} + \bar{A}_{-n} - (n+2) A_{n+2})/R^n$$

Thus, the potentials become

$$\Phi(z) = \sum a_n z^n \quad \text{and} \quad \chi(z) = \sum a'_n z^n \tag{11}$$

If the boundary of the element is converted to a smooth curve by a conformal mapping function,

$$z = \omega(\zeta) = \sum c_n \zeta^n \tag{12}$$

The stress boundary condition (Eq. 4) is transformed to:

$$\Phi(\sigma) + \omega(\sigma) \frac{\overline{\Phi'(\sigma)}}{\overline{\omega'(\sigma)}} + \overline{\chi(\sigma)} = F(\sigma) \tag{13}$$

By taking the Cauchy integrals of both sides one has:

$$\Phi(\zeta) + \frac{\omega(\zeta)}{\bar{\omega}'(\frac{1}{\zeta})} \bar{\Phi}'(\frac{1}{\zeta}) + \bar{a} = F(\zeta) \tag{14}$$

The series expansion of the functions which appear in the above expression are:

$$\Phi(\zeta) = \sum a_k \zeta^k \tag{15}$$

$$\frac{\omega(\zeta)}{\overline{\omega}'\left(\frac{1}{\zeta}\right)} = \zeta^n \frac{\sum c_k \zeta^{k-1}}{\sum k \, \overline{c}_k \, \zeta^{n-k}} = \sum b_k \zeta^k \tag{16}$$

$$\frac{\omega(\zeta)}{\overline{\omega}'(\zeta)} \overline{\Phi}' \left(\frac{1}{\zeta}\right) = \sum_{k=1}^{n} K_k \zeta^k = \sum_{k=1}^{n} \left(\sum_{j=k}^{n} (j-k+1)\right.$$

$$\left. \overline{a}_{(j-k+1)} b_j \right) \zeta^k \tag{17}$$

$$F(\zeta) = \sum_n (\{g_n\}^T \{F\} + B_n) \zeta^n \tag{18}$$

Substituted in Equation 14, one has

$$\sum_{k=1}^{n} (a_k + \sum_{j=k}^{n} (j-k+1) \overline{a}_{(j-k+1)} b_j) \zeta^n = \sum_k (\{g_k\}^T$$

$$\{F\} + B_k) \zeta^k \tag{19}$$

Comparing the coefficients of ζ^n, one obtains,

$$a_k + \sum_{j=k}^{n} (j-k+1) \overline{a}_{(j-k+1)} b_j = \{g_k\}^T \{F\} + B_k \tag{20}$$

If a_k coefficients are real, the above equations yield to a set of n equations which can be written in matrix form, as follows

$$[H] \{a\} = [G] \{F\} + \{B\} \tag{21}$$

If a_k coefficients are complex, a similar matrix equation can be written for imaginary part of the a_k coefficients. Solving for $\{a\}$ one has:

$$\{a\} = [H]^{-1} [G]' \{F\} + [H]^{-1} \{B\} \tag{22}$$

Thus, the potential function $\Phi(\zeta)$ (Eq. 15) is determined in terms of the common boundary parameters $\{F\}$.

The conjugate of the stress boundary condition (Eq. 13) can be written as follows:

$$\overline{\Phi(\sigma)} + \overline{\omega(\sigma)} \frac{\Phi'(\sigma)}{\omega'(\sigma)} + \chi'(\sigma) = F(\sigma) \tag{23}$$

By taking the Cauchy integrals of both sides, one has:

$$\overline{a}_o - \sum \left(\frac{K_k}{\zeta^k}\right) + \frac{\overline{\omega}\left(\frac{1}{\zeta}\right)}{\omega'(\zeta)} \Phi'(\zeta) + \chi'(\zeta) = \frac{1}{2\pi i} \int \frac{F(\sigma)}{\sigma - \zeta} d\sigma \tag{24}$$

or

$$\Psi(\zeta) = \frac{1}{2\pi i}\int \frac{F(\sigma)}{\sigma-\zeta}\, d\sigma - \frac{\omega(\frac{1}{\zeta})}{\omega'(\zeta)}\ \Phi'(\zeta) +$$

$$\sum(\frac{K_n}{\zeta^n}) + \bar{a}_o \tag{25}$$

where,

$$\Psi(\zeta) = \chi'(\zeta) \tag{26}$$

The two potential functions $\Phi(\zeta)$ and $\Psi(\zeta)$ can be written in matrix form. For the first one (Eqs. 15 and 22) one has,

$$\Phi(\zeta) = \Phi_o(\zeta) + \{\alpha(\zeta)\}^T ([K]\{F\} + \{D\}) \tag{27}$$

where

$\Phi_o(\zeta)$ = a holomorphic function within the domain

$\{\alpha(\zeta)\}^T = (\zeta \quad \zeta^2 \quad \zeta^3 \cdot \cdot \quad \zeta^n)$

$[K] = [H]^{-1}\ [G]$

$\{\alpha(\zeta)\}^T\ \{D\} = \{\alpha(\zeta)\}^T\ \{B\} - \Phi_o(\zeta)$

Similarly, the second potential $\Psi(\zeta)$ (Eq. 25) can be also written in matrix form as follows:

$$\Psi(\zeta) = \Psi_o(\zeta) + \{\beta(\zeta)\}^T\ \lfloor L\rfloor\ \{F\} + \{\gamma(\zeta)\}^T\ \{E\} \tag{28}$$

Characteristic matrix and load vector of an element.
The characteristic matrices of elements are obtained from the displacement boundary condition with respect to the coordinates of the original domain,

$$\times\Phi(\sigma) - \omega(\sigma)\,\overline{\frac{\Phi'(\sigma)}{\omega'(\sigma)}} - \overline{\Psi(\sigma)} = 2\mu(g_1(\sigma) + ig_2(\sigma)) =$$

$$2\,\mu g(\sigma) \tag{29}$$

where $\times$ and μ are the constants characterizing the elastic behavior of the element, and $g_1(\sigma)$ and $g_2(\sigma)$ are the values of the displacements along the boundaries of the element. The value of the displacements at a discrete point j, located along the common boundary is:

$$2\mu g(\sigma_j) = \times\ \Phi(\sigma_j) - \frac{\omega(\zeta_j)}{\omega'(\zeta_j)}\ \overline{\Phi'(\sigma_j)} - \overline{\Psi(\sigma_j)} \tag{30}$$

By substituting the expressions of the potential functions

(Eqs. 27 and 28) in the above expression, (Eq. 30), one has,

$$g(\sigma_j) = \Gamma_o(\sigma_j) + \{\Gamma(\sigma_j)\}^T \{F\} \tag{31}$$

where:

$$\Gamma_o(\sigma_j) = \frac{1}{2\mu}\left(\times\Phi_o(\sigma_j) - \frac{\omega(\sigma_j)}{\overline{\omega'(\sigma_j)}}\overline{\Phi_o'(\sigma_j)} - \overline{\Psi_o(\sigma_j)} + (\times\{\alpha(\sigma_j)\}^T - \frac{\omega(\sigma_j)}{\overline{\omega'(\sigma_j)}}\{\overline{\alpha'(\sigma_j)}\}^T)\{\overline{D}\} - \{\overline{\gamma\ (\sigma_j)}\}^T\{\overline{E}\}\right)$$

and

$$\{\Gamma(\sigma_j)\}^T = \frac{1}{2\mu}\left((\times\{\alpha(\sigma_j)\}^T - \frac{\omega(\sigma_j)}{\overline{\omega'(\sigma_j)}}\{\overline{\alpha'(\sigma_j)}\}^T\ [\overline{K}] - \{\overline{\beta(\sigma_j)}\}^T[\overline{L}]\right)\{\overline{F}\} \tag{32}$$

The characteristic matrices of element are obtained by writing in matrix form, the displacements (Eq. 32), at all the selected discrete points along the common boundaries of an element, as follows,

$$\{g\} = \{\Gamma_o\} + [\Gamma]\{F\} \tag{33}$$

where

$$\{g\} = \begin{Bmatrix} g(\sigma_1) \\ \cdots\cdot \\ g(\sigma_n) \end{Bmatrix} \tag{34}$$

$$\{\Gamma_o\} = \begin{Bmatrix} \Gamma_o(\sigma_1) \\ \cdots\cdot \\ \Gamma_o(\sigma_n) \end{Bmatrix} \tag{35}$$

$$[\Gamma] = \begin{Bmatrix} \{\Gamma(\sigma_1)\}^T \\ \cdots\cdot\cdot \\ \{\Gamma'(\sigma_n)\}^T \end{Bmatrix} \tag{36}$$

and n is the total number of discrete points on the common boundaries of the element. The vector $\{\Gamma_o\}$ and the matrix $[\Gamma]$ are called the load vector and the characteristic matrix of an element, respectively.

Characteristic matrix and load vector of the system.
The characteristic matrix $[\Lambda]$ and the load vector $\{\lambda\}$ of the system are obtained from those, $[\Gamma]$ and $\{\Gamma_o\}$, of the elements by using the code number (16), (18) and the relationship stating that the sum of the displacements $g_p(\sigma_j)$ and $g_s(\sigma_j)$ of the p'th and s'th elements at a discrete point σ_j along the common boundary of the p'th and s'th elements. Recalling

that the code numbers are the numbers assigned to the freedoms of the system, the general terms Λ_{ij}, λ_i and T_i of the characteristic matrix load vector and the boundary parameter of the system can be written, respectively, as follows,

$$\Lambda_{ij} = \sum_{r=1}^{s} \Gamma^{r}_{lm}$$

$$\lambda_i = \sum_{r=1}^{s} \Gamma^{r}_{o l}$$

$$T_i = F^{r}_{l}$$

where the freedom numbers l,m of the r'th element correspond to the code numbers i,j of the system, respectively, s is the total number elements having such properties. The relationship for the sum of the displacements at each discrete points are written for the whole system to form the system equation,

$$[\Lambda]\,\{T\} + \{\lambda\} = \{0\} \tag{37}$$

The vector of parameters {T}, at discrete points of the system is determined from the system equation (Eq. 37) as follows,

$$\{T\} = -\,[\lambda]^{-1}\,\{\lambda\} \tag{38}$$

The vector of parameter {F} of an element is obtained from {T} (Eq. 38) by means of code numbers then substituted in the expression of the potentials $\Phi(\zeta)$ and $\Psi(\zeta)$ (Eqs. 27 and 28) in order to have these two functions be completely determined.

Displacements and stresses.
The displacements and stresses at an arbitrary point of an element are obtained by means of the two potential functions which are already completely determined.

The displacements are: in cartesian coordinates,

$$2u(u+iv) = \times\Phi(\zeta) - \frac{\omega(\zeta)}{\overline{\omega'(\zeta)}}\,\overline{\Phi'(\zeta)} - \overline{\Psi(\zeta)} \tag{39}$$

or in curvilinear coordinates,

$$2u(v_\theta + iv_\rho) = \frac{\bar{\zeta}}{\rho}\,\frac{\overline{\omega'(\zeta)}}{|\omega'(\zeta)|}\left[\times\Phi(\zeta) - \frac{\omega(\zeta)}{\overline{\omega'(\zeta)}}\,\overline{\Phi'(\zeta)} - \overline{\Psi(\zeta)}\right] \tag{40}$$

and the stresses are: in cartesian coordinates,

$$\sigma_x + \sigma_y = 2[\Phi'(\zeta) + \overline{\Phi'(\zeta)}]$$

$$\sigma_y - \sigma_x + 2i\tau_{xy} = \frac{2}{\omega'(\zeta)} [\overline{\omega(\zeta)}\Phi'(\zeta) + \omega(\zeta)\, \Psi'(\zeta)] \qquad (41)$$

or in curvilinear coordinates,

$$\sigma_\rho + \sigma_\theta = \sigma_x + \sigma_y$$

$$\sigma_\theta - \sigma_\rho + 2i\tau_{\rho\theta} = \frac{2\zeta^2}{\rho^2\overline{\omega'(\zeta)}} [\overline{\omega(\zeta)}\Phi'(\zeta) + \omega'(\zeta)\Psi'(\zeta)] \qquad (42)$$

The displacements and the stresses obtained from the above analysis are of the same accuracy whether they are calculated at a point on the boundary or in the domain.

APPLICATION

The analysis described above has a wide range of applications including plates which have part of their boundaries at infinity. The procedure of analysis can be extended very easily to include the bending of plates.

The analysis is applied to a disk subjected to external loads (15). The polar coordinates (r, Θ) are used to locate the position elements: the outer ring, element 1 and and the inner disk element 2, along a common boundary r = mR (Fig. 6).

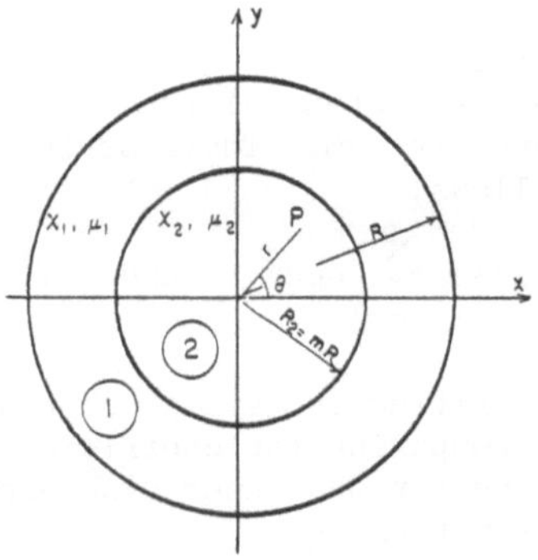

Figure 6. Disk made with two different materials.

Characteristics of the elements.
For element 1. The series expansions of the two potentials are (Eq. 11)

$$\Phi(z) = \sum_{-\infty}^{+\infty} a_n\, z^n \quad \text{and}$$

$$\Psi(z) = \sum_{-\infty}^{+\infty} a_n'\, z^n$$

The stress boundary conditions are written as follows (Eq. 4): for the outer boundary

$$\sum_{k=2} (a_k R^k - (k-2)\bar{a}_{-k+2} R^{-k+2} + \bar{a}'_{-k} R^{-k})\ \sigma^k + \sum_{k=1} (a_{-k} R^{-k} + (k+2)\ \bar{a}_{k+2} R^{k+2}$$

$$+\bar{a}'_k R^k)\ \sigma^k = \sum_{k=1} B^o_k \sigma^k \qquad (44)$$

and for the inner boundary

$$\sum_{k=2} (a_k m^k R^k - (k-2)\bar{a}_{-k+2}\ m^{-k+2} R^{-k+2} + \bar{a}'_{-k} m^{-k} R^{-k})\ \ \sigma^k + \sum_{k=1}$$

$$(a_{-k} m^{-k} R^{-k} + (k+2)\ \bar{a}_{k+2} m^{k+2} R^{k+2} + \bar{a}'_k\ m^k R^k)\sigma^k =$$

$$\sum_{-\infty}^{+\infty} (A^i_k + B^i_k)\ \sigma^k \qquad (45)$$

In the example, the outer circle is considered as unit circle, R=1, and the external load is assumed to be self equilibrating forces acting along the outer boundary (Fig. 6). Thus, the coefficients of the series of the summation of the stresses become

$$B^i_k \neq 0 \qquad \text{and} \qquad B^i_n = 0$$

The coefficients a_n and a'_n of the series for the two potentials are determined by comparing the coefficients of σ^n. Therefore, the two potential functions are determined in terms of the parameters A^i_k, related to the summation of the contact stresses along the common boundary.

For element 2.
The two potential functions are holomorphic inside of the boundary r=m. Their series expansions can be written as follows,

$$\Phi(z) = a_o + \sum_{k=1}^{\infty} a_k\, z^k \qquad \text{and} \qquad \Psi(z) = a'_o + \sum_{k=1}^{\infty} a'_k\, z^k$$

By writing the stress boundary conditions along the r=m and by comparing the coefficients of σ^k, the two potential functions for the inner disk are also determined in terms of the parameters A^i_k.

Characteristics of the system.
The characteristics of the system is obtained from the displacement boundary condition which can be written in polar coordinates as follows,

$$(v_r + iv_\Theta)\Big|_1 + (v_r + iv_\Theta)\Big|_2 = 0$$

The two potentials for individual elements are completely determined by the resolution of the unknown parameters A^i_k, from the above equation. If the radial stresses becomes tension along a part of the common boundary, the two elements separate from each other along that part. In such cases, the problem is analyzed by an iterative process which takes into account that the contact stresses along the separated part of the boundary are zero.

CONCLUSION

The application of the complex variable theory to the plates provides a unique solution. Some approximations are introduced if the plate has to be divided into elements. In the division, the size of the elements must be the largest possible. Only the common boundaries are divided into discrete points where the unknown parameters are introduced. Therefore, the number of elements, the number of unknowns, the required storage area in the memory and the CPU time are minimum. The magnitudes of the displacements and the stresses are of the same accuracy at the points on the boundaries or within the domain. The number of the terms in the series which describes the conformal mapping function must be large enough to represent the boundaries properly. Still the corners are somewhat rounded.

ACKNOWLEDGEMENTS

Gratitudes are expressed to the Computing Center of the University of Southwestern Louisiana for making their computer HONEYWELL 68/80 (Multics) available to him and to Miss Debra Boudreaux for her conscientious typing of the manuscript.

REFERENCES

1. Alarcon E., Martin A. and Paris F., (1979). Boundary Elements in Potential and Elasticity Theory. Compt. and Struct., 10, 351-362.

2. Brebbia C. and Dominguez J., (1978). Boundary Element Methods versus Finite Elements - Applied Numerical Modelling, Ed. C.A. Brebbia, John Wiley & Sons, 571-586.

3. Brebbia C. and Dominguez J., (1978). Discussion of "Boundary Element Methods versus Finite Elements, Applied Numerical Modelling," Ed. C.A. Brebbia, John Wiley & Sons, 709-711.

4. Cruse T.A. and Van Buren W., (1971). Three Dimensional Elastic Stress Analysis of a Fracture Specimen with an Edge Crack. Int. J. Fract. Mech., 7, 1-16.

5. Cruse, T.A., (1973). Application of Boundary Integral Equation Method to Three Dimensional Stress Analysis. Comp. and Struct. 3, 509-527.

6. Danson D., Brebbia C.A. and Adey R.A., (1981). The BEASY System. Handbook of Finite Element Systems, Ed. C.A. Brebbia, CML Publ., 77-91.

7. Deverall L.J., (1957). Solution of Some Problems in Bending of Thin Clamped Plates by Means of the Method of Muskhelishvili. J. Appl. Mech. 24, 295.

8. Doblare M. and Alarcon E. (1981). A Three Dimensional B.I.E.M. Program. Handbook of Finite Element Systems, Ed. C.A. Brebbia, CML Publ., 318-338.

9. Fredholm J. (1906). Solution d'un Problem Fondamental de la Theorie d'Elasticite', Ark. Mat. Ast. Fys. 2, No. 28, 3-7.

10. Gray, C.A.M., (1951). Polynomial Approximations in Plane Elastic Problems. J. Mech. and Appl. Math. 4, Part 4, 444.

11. Lauricella G. (1906). Atti Reale. Accad. Lincer, 15, 426-432.

12. Liu P.L-F. and Liggett, (1978). Boundary Integral Solutions to Groundwater Problems, Applied Numerical Modelling, Ed. C.A. Brebbia, John Wiley and Sons, 559-570.

13. Muskhelishvili N.J., (1963). Some Basic Problems of the Mathematical Theory of Elasticity. Trans. J.R.M. Radok, P. Noordhoff, Ltd.

14. Ovunc B.A.,,(1964)., De la Solution Elastic d'une Plaque Mince, Carree Chargee Suivant ses Cotes. Bull. Tech. Univ. of Istanbul. 17, 37-62.

15. Ovunc B.A., (1969). On the Splitting Tests of the Hollow Cylinder Specimens, IABSE 29/I, 49-55.

16. Ovunc, B.A., (1981). STDYNL Computer Program from Structures. Handbook of Finite Element Systems, Ed. C.A. Brebbia, CML Publ. 434-445.

17. Rizzo F.J.,(1967). An Integral Equation Approach to Boundary Value Problems of Classical Elastostatics. Q. Appl. Math. 25, 83-95.

18. Tezcan S.S. (1966). Computer Analysis of Plane and Space Structures. ASCE ST2. 143-173.

BOUNDARY ELEMENT NONLINEAR BENDING ANALYSES OF CLAMPED SANDWICH PLATES AND SHELLS

N. Kamiya, Y. Sawaki and Y. Nakamura

Department of Mechanical Engineering, Mie University
Kamihamacho, Tsu 514 Japan

INTRODUCTION

Present study concerns numerical analysis of flat plates and shallow shells with sandwich-type layered structure in the case of finite deflection by boundary element method (BEM), which are fundamental lightweight structural elements. The structures are assumed to be made of thin face layers and soft core, which sustain bending and shear deformations, respectively. Conventional governing field equations in such large deformation problems are nonlinear and coupled, and thus are not easily solvable. Therefore, we derive another governing field equations with the aids of an extended version of so-called Berger approach or "hypothesis" which was originally proposed and applied to nonlinear bending problem with solid flat plates [1]. For sandwich plates and shallow shells with isotropic faces and core, the field equations may be reduced to a simplified single differential equation in terms of deflection alone with a supplementary constrained condition; i.e., invariance of a particular parameter [2, 3].

Boundary integral equation is formulated for the above-derived field equation for flat sandwich plates and shallow sandwich shells subjected to mechanical as well as thermal loadings. Similar integral formulation for nonlinear bending problems with solid flat plates has been reported by the present authors [4, 5]. Two-dimensional region occupied by plate or shell and its boundary are discretized into respective elements; domain elements and boundary elements, and corresponding algebraic equations are solved numerically. Iteration numerical procedures are repeated until convergence of solution is satisfied. Results are obtained and shown for clamped sandwich plates and shallow shells bent by uniform lateral load and by temperature difference between upper and lower faces. This study partially demonstrates a potentiality of the boundary element method to a complicated nonlinear deflection analysis of sandwich structural elements.

GOVERNING FIELD EQUATIONS

With reference to the orthogonally curvilinear coordinate system x, y and z as shown in Fig. 1, we consider a symmetrically laminated shallow sandwich shell of double curvature k_x and k_y. The axes x and y are directed to the principal axes of curvature in the middle plane of the shell and the z axis is directed to downward. The region occupied by the shell and its boundary are denoted by S and C, respectively. We introduce the following assumptions for the geometries and kinematics of the sandwich shell:

(i) the thicknesses t of the upper and lower faces are identical and are sufficiently thin in comparison with the core thickness h ($h \gg t$);

(ii) the bending rigidity of the faces is neglected;

(iii) the core withstands shear deformation, whereas the faces withstand bending deformation;

(iv) the faces are thin enough to ignore variations of inplane displacements and stresses in their thickness direction;

(v) the deflection w does not vary through the entire thickness of the shell; and

(vi) the faces and the core are made of each isotropic elastic material.

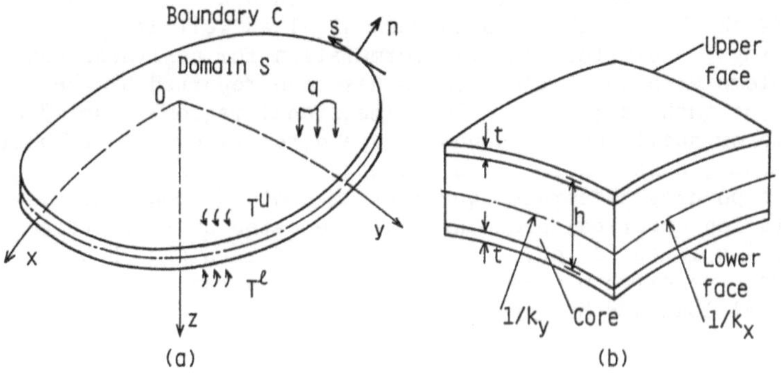

Fig. 1. Shallow sandwich shell (a) and its element (b).

According to the above-stated assumptions and classical uncoupled thermoelasticity, the total potential energy of a whole system of sandwich shell subjected to distributed lateral load q and surface temperature fields T^u and T^ℓ is expressed as follows:

$$\Pi = \int_S \Big\{ \frac{E^f t}{1-\nu^{f2}} \{ I_1^{m2} - 2(1-\nu^f) I_2^m + \frac{1}{4} [(\frac{\partial r}{\partial x})^2 + (\frac{\partial s}{\partial y})^2$$

$$+ 2\nu^f \frac{\partial r}{\partial x}\frac{\partial s}{\partial y}] + \frac{1}{8}(1-\nu^f)(\frac{\partial r}{\partial y} + \frac{\partial s}{\partial x})^2\} + \frac{1}{2} G^c h[(\frac{r}{h} - \frac{\partial w}{\partial x})^2$$

$$+ (\frac{s}{h} - \frac{\partial w}{\partial y})^2] - \frac{2E^f \alpha^f t}{1-\nu^f}[T^m I_1^m + \frac{1}{4} f(\frac{\partial r}{\partial x} + \frac{\partial s}{\partial y})]$$

$$- qw \Big\} dS \qquad (1)$$

In the above equation the following notations are employed:

E^f, ν^f, α^f : Young's modulus, Poisson's ratio and linear thermal expansion coefficient of the faces, respectively

G^c : shear modulus of the core

$r = u^u - u^\ell$, $s = v^u - v^\ell$: differences of inplane displacements

$T^m = \frac{1}{2}(T^u + T^\ell)$, $f = T^u - T^\ell$: average and difference temperatures of the faces

u, v : inplane displacements

T : temperature measured from the undeformed reference state

I_1^m, I_2^m : first and second invariants of the average strains of the faces

Superfixes f, u, ℓ and m refer to the face, upper face, lower face and average values on the both faces, respectively.

Using the variational calculus to minimize the total potential energy Π, Eq. (1), of the elastic system of the shallow shell, we can derive the conventional governing field equations expressed by the deflection w, inplane displacements u^u, u^ℓ, v^u, v^ℓ and their differences r, s, which are fully coupled and are not easily solvable.

Now, following one of the present authors' line of thought, we neglect the term I_2^m appearing in Eq. (1) as an extended version of the Berger original idea for sandwich structures. Varying the four inplane displacements and the deflection to minimize Π, we obtain, after lengthly calculation, three simultaneous differential equations in terms of w, r and s and the identity

$$I_1^m - (1+\nu^f)\alpha^f T^m = \text{const.} \tag{2}$$

The boundary conditions which complete the system of field equations are derived as usual.

For sandwich shallow shells or plates made of isotropic faces and core, we can eliminate the differences of the inplane displacements r and s in virtue of the above-derived differential equations and finally arrive at the following equation expressed in terms of the deflection alone:

$$\begin{aligned}\nabla^4 w - \frac{\varkappa^2}{A\varkappa^2+1}(\nabla^2 w + k_x + k_y)\\ - \frac{1}{A\varkappa^2+1}\frac{1}{D_s}[q - A\nabla^2 q + B\nabla^2 f - AD_s\varkappa^2\nabla^2(k_x + k_y)] = 0\end{aligned} \tag{3}$$

where the constant in Eq. (2) is represented as

$$I_1^m - (1+\nu^f)\alpha^f T^m = \frac{h^2\varkappa^2}{4} \tag{4}$$

and

$$I_1^m = \frac{\partial u^m}{\partial x} + \frac{\partial v^m}{\partial y} + \frac{1}{2}\left(\frac{\partial w}{\partial x}\right)^2 + \frac{1}{2}\left(\frac{\partial w}{\partial y}\right)^2 - (k_x + k_y)w \tag{5}$$

$$D_s = \frac{E^f th^2}{2(1-\nu^{f2})}, \quad A = \frac{D_s}{G^c h}, \quad B = \frac{D_s(1+\nu^f)\alpha^f}{h}$$

$$\nabla^4 = \nabla^2\nabla^2 = \left(\frac{\partial^2}{\partial x^2} + \frac{\partial^2}{\partial y^2}\right)^2$$

In consequence, the field equation governing finite deflection of the sandwich plate and shell is reduced to a single equation, (3), in terms of the deflection with a subsidiary condition (4). The nonlinearity of the deformation rests on only the one parameter $\varkappa^2$ appearing in Eq. (4). Such formulation facilitates subsequent integral formulation and calculations in the extreme.

INTEGRAL EQUATION FORMULATION

For simplicity, let the boundary curve C surrounding the sandwich shell be smooth. It has been well-known that the following generalized Green identities hold for arbitrary differentiable functions w and w* with respect to harmonic and biharmonic differential operators [6]:

$$\int_S (w^*\nabla^2 w - w\nabla^2 w^*)\,dS = \int_C (w^*\frac{\partial w}{\partial n} - w\frac{\partial w^*}{\partial n})\,ds \tag{6}$$

and

$$\int_S (w\nabla^4 w^* - w^*\nabla^4 w)\,dS$$
$$= \int_C [w^*K'(w) - \frac{\partial w^*}{\partial n}M'(w) + \frac{\partial w}{\partial n}M'^*(w^*) - wK'^*(w^*)]\,ds \tag{7}$$

where the operators M' and K' respectively stand for

$$M' = -[(1-\nu^f)(\ell^2\frac{\partial^2}{\partial x^2} + 2m\ell\frac{\partial^2}{\partial x\partial y} + m^2\frac{\partial^2}{\partial y^2}) + \nu^f\nabla^2] \tag{8}$$

and

$$K' = (1-\nu^f)\frac{\partial}{\partial s}[\ell m(\frac{\partial^2}{\partial x^2} - \frac{\partial^2}{\partial y^2}) - (\ell^2 - m^2)\frac{\partial^2}{\partial x\partial y}] - \nabla^2\frac{\partial}{\partial n} \tag{9}$$

Direction cosines of the outward normal to the boundary curve C are denoted by ℓ and m, and normal and tangential derivatives on the boundary are $\partial/\partial n$ and $\partial/\partial s$, respectively.

As a function w*, we take a fundamental solution to the biharmonic equation

$$\nabla^4 w^*(P, Q) + \delta(P, Q) = 0 \tag{10}$$

i.e.,

$$w^* = -\frac{1}{8\pi}r^2\log r \qquad (r = |\overline{PQ}|) \tag{11}$$

where P and Q are arbitrary points on the x, y plane.

Making use of Eqs. (6), (7) and (10), we can formulate the following integral equation:

$$c(P)w(P) = - \int_S \left\{ \frac{\varkappa^2}{A\varkappa^2+1} \{[\nabla^2 w^*(P, Q)]w(Q) + w^*(P, Q)[k_x(Q)+k_y(Q)]\} \right.$$

$$+ \frac{1}{D_s(A\varkappa^2+1)} \{q(Q) - A\nabla^2 q(Q) + B\nabla^2 f(Q)$$

$$\left. - AD_s\varkappa^2\nabla^2[k_x(Q) + k_y(Q)]\}w^*(P, Q) \right\} dS$$

$$- \int_C \{K(\hat{Q})w^*(P, \hat{Q}) - M(\hat{Q}) \frac{\partial w^*}{\partial n}(P, \hat{Q})$$

$$+ M'^*[w^*(P, \hat{Q})] \frac{\partial w}{\partial n}(\hat{Q}) - K'^*[w^*(P, \hat{Q})] w(\hat{Q})\}ds \quad (12)$$

$$(Q \in S, \hat{Q} \in C)$$

where

$$K(\hat{Q}) = K'[w(\hat{Q})] + \frac{\varkappa^2}{A\varkappa^2+1} \frac{\partial w}{\partial n}(\hat{Q}) \quad (13)$$

$$M(\hat{Q}) = M'[w(\hat{Q})] + \frac{\varkappa^2}{A\varkappa^2+1} w(\hat{Q}) \quad (14)$$

and

$$c(P) = \begin{cases} 1 & (P \in S) \\ 1/2 & (P \in C) \end{cases} \quad (15)$$

And further, differentiating Eq. (12) in the normal direction on the boundary, the second equation can be obtained:

$$c(P)\frac{\partial w}{\partial n_0}(P) = - \int_S \left\{ \frac{\varkappa^2}{A\varkappa^2+1} \{\frac{\partial}{\partial n_0} [\nabla^2 w^*(P, Q)]w(Q) \right.$$

$$+ \frac{\partial w^*}{\partial n_0}(P, Q)[k_x(Q) + k_y(Q)]\} + \frac{1}{D_s(A\varkappa^2+1)} \{q(Q)$$

$$- A\nabla^2 q(Q) + B\nabla^2 f(Q) - AD_s\varkappa^2\nabla^2[k_x(Q) + k_y(Q)]\}$$

$$\left. \times \frac{\partial w^*}{\partial n_0}(P, Q) \right\} dS - \int_C \{K(\hat{Q})\frac{\partial w^*}{\partial n_0}(P, \hat{Q}) - M(\hat{Q})\frac{\partial w^*}{\partial n_0 \partial n}(P, \hat{Q})$$

$$+ \frac{\partial M'^*}{\partial n_0}[w^*(P, \hat{Q})]\frac{\partial w}{\partial n}(\hat{Q}) - \frac{\partial K'^*}{\partial n_0}[w^*(P, \hat{Q})]w(\hat{Q})\}ds \quad (16)$$

In order to analyze nonlinear bending problems with solid elastic plates, the authors derived and employed efficiently integral equation formulations similar to Eqs. (12) and (16) [4, 5].

CLAMPED SANDWICH PLATES AND SHELLS

The governing field equation after the Berger hypothesis is known to give fairly correct estimation to the rigorous solution under the restricted boundary conditions where the inplane displacements are constrained on the boundary. In what follows, we consider clamped sandwich plates and shallow sandwich shells as applications of the above-derived integral formulations.

If the domain integral terms appearing in Eqs. (12) and (16) are known, equations (12) and (16) construct a system of simultaneous integral equations for the four boundary values w, $\partial w/\partial n$, K and M. In the case of clamped plates or shells, both w and $\partial w/\partial n$ vanish on the boundary and the equations are thought for the remaining two.

The magnitude of the Berger constant $\varkappa^2$ can be estimated as follows. Integrating both side of Eq. (4) over the whole domain S occupied by the shell, we get the expression

$$\int_S [I_1^m - (1+\nu^f)\alpha^f T^m] dS = \frac{h^2\varkappa^2}{4} S \qquad (17)$$

where the area of the domain S is denoted by S. When the inplane displacements are constrained on the boundary, Eq. (17) can be reduced to

$$\varkappa^2 = \frac{2}{Sh^2} \int_S [(\frac{\partial w}{\partial x})^2 + (\frac{\partial w}{\partial y})^2 - 2(k_x + k_y)w - 2(1+\nu^f)\alpha^f T^m] dS \qquad (18)$$

which is numerically estimated by virtue of distributions of the deflection alone.

The boundary C and the domain S are dicretized into one- and two-dimensional elements (domain elements and boundary elements), respectively, and the boundary values constructing integral equations (12) and (16) are assumed invariant on each boundary element (piecewise constant element). Equations (12) and (16) are reduced to a system of discretized algebraic equations on the boundary with additional domain integral terms representing nonlinear effect of the deformation. Calculation of the last terms which also relate to the magnitude of the Berger constant may be performed numerically by an appropriately assumed distributions of the deflection (for the most case, linear solution can be used effectively as an initial estimation).

Such numerical iteration calculation must be repeated until a convergence condition is satisfied.

Figures 2 to 6 depict the results of calculation for circular flat sandwich plates and square shallow sandwich shells and plates bent by uniformly distributed lateral load and by specified temperature difference between the upper and lower faces. The following geometrical and material parameter are used in the calculation:

$$t = 0.635 \text{ mm}, \qquad h = 17.13 \text{ mm}, \qquad \nu^f = 0.3$$

$$E^f = 0.72 \times 10^5 \text{ MPa}, \qquad G_0^c = 0.414 \times 10^2 \text{ MPa}$$

For circular sandwich plates:

$$R = 254 \text{ mm} \quad : \quad \text{radius}$$

$$T^u = \frac{2}{3} T_0\{1 + [1 - (\frac{r}{R})^2]\}$$

$$T^\ell = \frac{4}{3} T_0\{1 + [1 - (\frac{r}{R})^2]\} \qquad (r^2 = x^2 + y^2)$$

For square shallow sandwich shells:

$$2a = 508 \text{ mm} \quad : \text{ edge length}$$

$$T^u = \frac{2}{3} T_0\{1 + [1 - (\frac{x}{a})^2][1 - (\frac{y}{a})^2]\}$$

$$T^\ell = \frac{4}{3} T_0\{1 + [1 - (\frac{x}{a})^2][1 - (\frac{y}{a})^2]\}$$

Circular plates are discretized into 24 piecewise constant boundary elements and 72 triangular domain elements whereas square sandwich plates and shells are divided into 40 piecewise constant elements and 100 square domain elements. The present BEM solutions marked by small circles in Figs. 2, 3 and 4, maximum deflections and distributions of deflection in the radial direction, coincide well with the corresponding numerical solutions (solid lines) obtained by the well-populated Runge-Kutta-Gill method with a sufficiently fine integration interval, which could be carried out only for the axisymmetric problem. Figures 5 and 6 are results for square sandwich plates and shallow shells whose mean curvatures are inserted in the figures. Various results for the sandwich shells with positive and negative initial curvatures are obtained analogously within reduced computational cost. The present simplified method of nonlinear bending analysis based on the decoupled field equation due to the Berger hypothesis and integral equation formulation can be applied to sandwich plates and shallow shells with arbitrary contours.

REFERENCES

1. H. M. Berger, A New Approach to the Analysis of Large Deflection of Plates, J. Appl. Mech., 22, 1955, 465-472.
2. N. Kamiya, Governing Equations for Large Deflections of Sandwich Plates, AIAA J., 14, 1976, 250-253.
3. N. Kamiya, Analysis of the Large Thermal Bending of Sandwich Plates by a Modified Berger Method, J. Strain Anal., 13, 1978, 17-22.
4. N. Kamiya and Y. Sawaki, An Integral Equation Approach to Finite Deflection of Elastic Plates, Int. J. Non-Linear Mech., 1982 (in Press).
5. N. Kamiya, Y. Sawaki, Y. Nakamura and A. Fukui, An Approximate Finite Deflection Analysis of a Heated Plate by the Boundary Element Method, Appl. Math. Modell., 6, 1982, 23-27.
6. S. Bergman and M. Schiffer, Kernel Functions and Elliptic Differential Equations in Mathematical Physics, Academic Press Inc., New York, 1953.

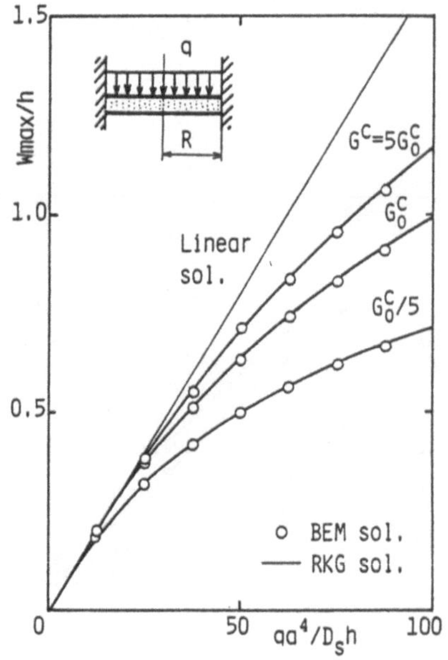

Fig. 2. Maximum deflections of flat circular sandwich plates (Mechanical load).

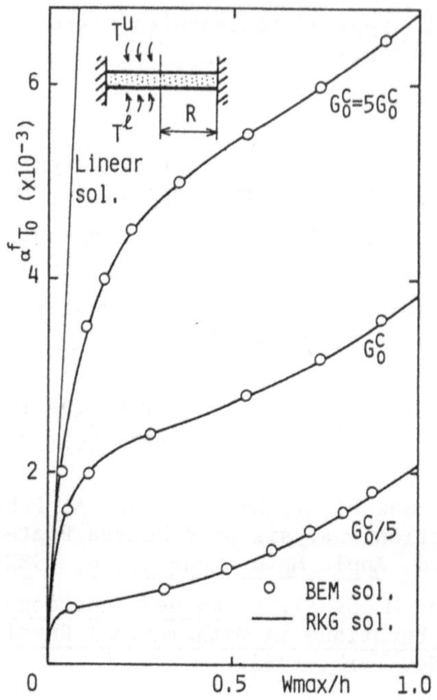

Fig. 3. Maximum deflections of flat circular sandwich plates (Thermal load).

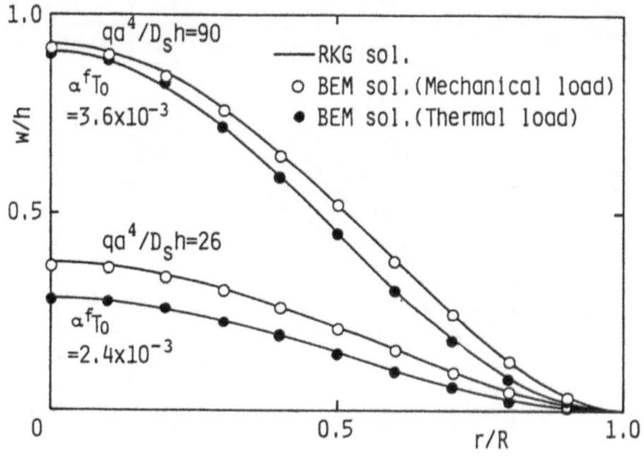

Fig. 4. Distributions of deflection in flat circular sandwich plates.

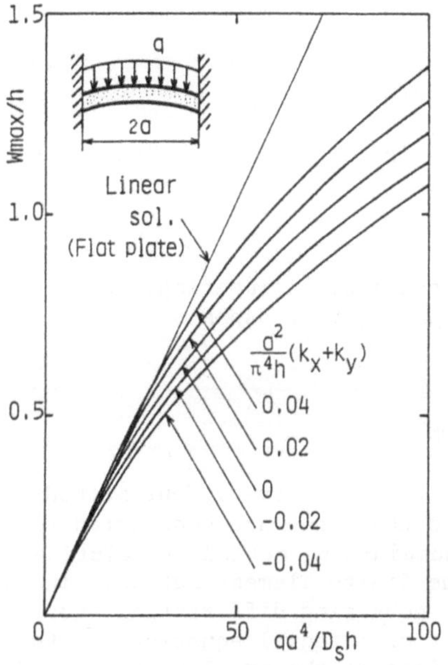

Fig. 5. Maximum deflections of shallow square sandwich shells (Mechanical load, $G^c = G^c_o$).

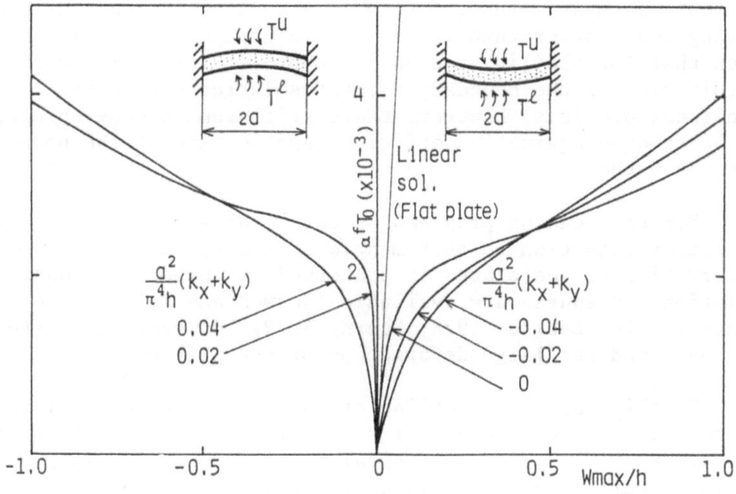

Fig. 6. Maximum deflections of shallow square sandwich shells (Thermal load, $G^c = G^c_o$).

INTEGRAL EQUATION APPROACH TO SMALL AND LARGE DISPLACEMENTS OF THIN ELASTIC PLATES

Masataka TANAKA

Department of Mechanical Engineering, Osaka University
Yamada Oka 2-1, Suita 565 / Japan

1. INTRODUCTION

In recent years, the boundary element methods (BEM) have been so well developed that they are considered to be a powerful alternative to a domain-type method of solution such as the finite difference or the finite element method. In the boundary element methods the governing differential equations are transformed into a set of integral equations on the boundary and hence the dimensions of the problem considered can be reduced by one. The resulting integral equations are discretized by introducing a finite number of elements located on the boundary. The final set of simultaneous equations to be solved include only the nodal unknowns on the boundary alone.

Most of linear problems can be computed by means of the boundary element method, with good accuracy at a smaller computing cost rather than in the domain-type methods. It can be seen that for the linear problems there is left only some standardization of the boundary element solution procedures. For nonlinear problems, however, there still remain many obstacles to be removed against an efficient application of the boundary element methods.

For the bending problems of thin elastic plates, we have a similar situation to that mentioned above. As far as small deformation of the plates is concerned, a number of boundary solution procedures are available (Jaswon and Maiti, 1968; Hansen, 1976; Bezine, 1978; Stern, 1979). However, few attempts can be found for large deformation of the plates.

In this paper, an outline of the direct boundary element method is first presented for small displacements of thin elastic plates. Then, a new integral equation approach is proposed to large deflection analysis of the plates. The proposed integral equation method of solution is based on the von Kármán theory and formulated in terms of incremental variables.

2. FORMULATION FOR SMALL DISPLACEMENTS

Let us consider an elastic plate subjected to the distributed pressure q . The boundary and the inner domain of the plate are denoted by Γ and Ω , respectively. We use a cartesian coordinate system $O\text{-}x_1x_2x_3$ as shown in Fig.1. The governing differential equation of the plate can be expressed in terms of the lateral displacement of the plate middle plane as

$$\nabla^4 w = q / D \tag{1}$$

where

$$\nabla^4 = (\nabla^2)^2 \equiv \left(\frac{\partial^2}{\partial x_1^2} + \frac{\partial^2}{\partial x_2^2}\right)^2 \tag{2}$$

and D denotes the flexural rigidity of the plate which is assumed in this study to be constant.

We now introduce a deformation function w^* which is produced by another loading system. Then, we consider the bilinear form which is symmetric with respect to w and w^* defined as

$$B(w, w^*) = \int_\Omega \left\{ \nabla^2 w \nabla^2 w^* + (1-\nu) L(w, w^*) \right\} d\Omega \tag{3}$$

where ν denotes Poisson's ratio and $L(w,(\ ))$ a differential operator which is given by

$$L(w,(\ )) \equiv 2 \frac{\partial^2 w}{\partial x_1 \partial x_2} \frac{\partial^2(\ )}{\partial x_1 \partial x_2} - \frac{\partial^2 w}{\partial x_1^2} \frac{\partial^2(\ )}{\partial x_2^2} - \frac{\partial^2 w}{\partial x_2^2} \frac{\partial^2(\ )}{\partial x_1^2} \tag{4}$$

It is well known (Bergman and Schiffer, 1953) that Green's second identity for the biharmonic equation such as equation (1) can be expressed as

$$\int_\Omega \nabla^2 w \, \nabla^2 w^* \, d\Omega = \int_\Omega w^* \nabla^4 w \, d\Omega + \int_\Gamma \left(\nabla^2 w \frac{\partial w^*}{\partial n} - w^* \frac{\partial \nabla^2 w}{\partial n} \right) d\Gamma \tag{5}$$

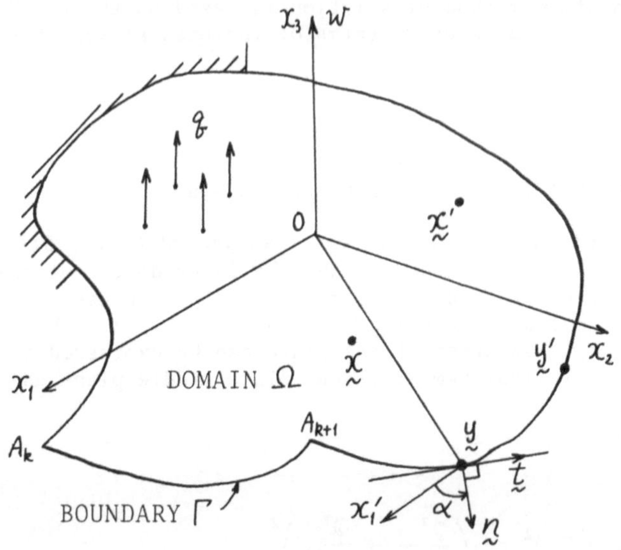

Figure 1 Coordinate system and notations

where $\partial(\ )/\partial n$ denotes the normal derivative on the boundary.

Integrating by parts equation (3) repeatedly with respect to w^* and using the Green's identity (5), we can obtain (Bezine, 1978; Stern, 1979; Bergman and Schiffer, 1953)

$$
\begin{aligned}
B(w, w^*) = \int_\Omega w^* \nabla^4 w \, d\Omega + \int_\Gamma \Big[& w^* \frac{\partial \hat{M}_{nt}(w)}{\partial t} \\
& - \frac{\partial w^*}{\partial n} \hat{M}_n(w) - w^* \frac{\partial (\nabla^2 w)}{\partial n} \Big] d\Gamma \\
& + \int_\Gamma \frac{\partial}{\partial t} \left[w^* \hat{M}_{nt}(w) \right] d\Gamma
\end{aligned}
\tag{6}
$$

where $\partial(\ )/\partial t$ denotes the tangential derivative on the plate boundary. In equation (6) $\hat{M}_n$ and $\hat{M}_{nt}$ are the differential operators defined on the plate boundary as follows:

$$
\hat{M}_n \equiv -\nabla^2 + (1-\nu)\left\{ n_2^2 \frac{\partial^2}{\partial x_1^2} + n_1^2 \frac{\partial^2}{\partial x_2^2} - 2 n_1 n_2 \frac{\partial^2}{\partial x_1 \partial x_2} \right\} \tag{7}
$$

$$
\hat{M}_{nt} \equiv -(1-\nu)\left\{ n_1 n_2 \left(\frac{\partial^2}{\partial x_1^2} - \frac{\partial^2}{\partial x_2^2} \right) + (n_1^2 - n_2^2) \frac{\partial^2}{\partial x_1 \partial x_2} \right\} \tag{8}
$$

where $n_1 = \cos\alpha$ and $n_2 = \sin\alpha$ (Fig.1).

It is interesting to note that when the boundary is smooth the last integral in equation (6) vanishes and this equation results in the so called Rayleigh-Green identity. For simplicity's sake, we restrict ourselves only to the case of a smooth boundary. If the boundary has K corners as shown in Fig.1, we must add the following term:

$$\int_\Gamma \frac{\partial}{\partial t}[w^* \hat{M}_{nt}(w)]\, d\Gamma = \sum_{k=1}^{K} [\![w^* \hat{M}_{nt}(w)]\!] \tag{9}$$

to the right hand side of equation (6) and calculate the discontinuity jump denoted by the symbol $[\![\ ]\!]$.

Using the symmetry property of the bilinear form $B(w,w^*)$ with respect to the arguments, we can derive

$$\int_\Omega (w \nabla^4 w^* - w^* \nabla^4 w)\, d\Omega$$
$$= \int_\Gamma (w^* V_n - T_n^* M_n + M_n^* T_n - V_n^* w)\, d\Gamma \tag{10}$$

where the following notations are used:

$$\left.\begin{aligned} T_n^* &= \hat{T}_n(w^*),\ M_n^* = \hat{M}_n(w^*),\ V_n^* = \hat{V}_n(w^*) \\ T_n &= \hat{T}_n(w),\ M_n = \hat{M}_n(w),\ V_n = \hat{V}_n(w) \end{aligned}\right\} \tag{11}$$

in which the differential operators $\hat{V}_n$ and $\hat{T}_n$ are defined by

$$\hat{V}_n \equiv \frac{\partial \hat{M}_{nt}}{\partial t} - \frac{\partial \nabla^2}{\partial n},\quad \hat{T}_n \equiv \frac{\partial}{\partial n} \tag{12}$$

As the deflection function w^* we shall use the fundamental solution to equation (1), which is governed by

$$\nabla^4 w^*(\underset{\sim}{x}, \underset{\sim}{y}) = \delta(\underset{\sim}{x}, \underset{\sim}{y}) \tag{13}$$

for an infinite plate made of the same material as that of the plate to be analyzed. Here we denote by $\mathbf{x}$ and y arbitrary points in the infinite domain and by $\delta(\ )^{\sim}$ the Dirac delta function. It is well known that the fundamental solution w^* is given by

$$w^*(\underset{\sim}{x}, \underset{\sim}{y}) = \frac{1}{8\pi} r^2 \ln r \tag{14}$$

where r denotes the distance between the two points $\underset{\sim}{x}$ and $\underset{\sim}{y}$

defined as $r \equiv |\underset{\sim}{x}-\underset{\sim}{y}|$.

Substitution of equations (1) and (13) into equation (10) and use of the property of the Dirac delta function lead to

$$w(\underset{\sim}{x}) + \int_\Gamma [V_n^*(\underset{\sim}{x},\underset{\sim}{y}')w(\underset{\sim}{y}') - M_n^*(\underset{\sim}{x},\underset{\sim}{y}')T_n(\underset{\sim}{y}') + T_n^*(\underset{\sim}{x},\underset{\sim}{y}')M_n(\underset{\sim}{y}') - w^*(\underset{\sim}{x},\underset{\sim}{y}')V_n(\underset{\sim}{y}')]d\Gamma(\underset{\sim}{y}') = \frac{1}{D}\int_\Omega w^*(\underset{\sim}{x},\underset{\sim}{x}')q(\underset{\sim}{x}')d\Omega(\underset{\sim}{x}') \tag{15}$$

Taking the limiting process in which the point $\underset{\sim}{x}$ approaches from inside the domain Ω to an arbitrary point $\underset{\sim}{y}$ on the boundary, we can derive the following integral equation:

$$c(\underset{\sim}{y})w(\underset{\sim}{y}) + \int_\Gamma [V_n^*(\underset{\sim}{y},\underset{\sim}{y}')w(\underset{\sim}{y}') - M_n^*(\underset{\sim}{y},\underset{\sim}{y}')T_n(\underset{\sim}{y}') + T_n^*(\underset{\sim}{y},\underset{\sim}{y}')M_n(\underset{\sim}{y}') - w^*(\underset{\sim}{y},\underset{\sim}{y}')V_n(\underset{\sim}{y}')]d\Gamma(\underset{\sim}{y}') = \frac{1}{D}\int_\Omega w^*(\underset{\sim}{y},\underset{\sim}{x}')q(\underset{\sim}{x}')d\Omega(\underset{\sim}{x}') \tag{16}$$

The coefficient $c(\underset{\sim}{y})$ depends only on the geometrical property of the boundary at the point $\underset{\sim}{y}$, and is obtainable from

$$c(\underset{\sim}{y}) = 1 + \lim_{\varepsilon\to 0}\int_{\Gamma_\varepsilon} V_n^*(\underset{\sim}{y},\underset{\sim}{y}')d\Gamma(\underset{\sim}{y}') \tag{17}$$

where Γ_ε denotes a boundary portion of a semicircle whose radius and centerpoint are ε and $\underset{\sim}{y}$, respectively. It is noted that $c = 1/2$ when the boundary is smooth.

If the normal derivative of equation (16) is considered, the following integral equation is obtained in a similar manner:

$$c'(\underset{\sim}{y})T_n(\underset{\sim}{y}) + \int_\Gamma [\tilde{V}_n^*(\underset{\sim}{y},\underset{\sim}{y}')w(\underset{\sim}{y}') - \tilde{M}_n^*(\underset{\sim}{y},\underset{\sim}{y}')T_n(\underset{\sim}{y}') + \tilde{T}_n^*(\underset{\sim}{y},\underset{\sim}{y}')M_n(\underset{\sim}{y}') - \tilde{w}^*(\underset{\sim}{y},\underset{\sim}{y}')V_n(\underset{\sim}{y}')]d\Gamma(\underset{\sim}{y}') = \frac{1}{D}\int_\Omega \tilde{w}^*(\underset{\sim}{y},\underset{\sim}{x}')q(\underset{\sim}{x}')d\Omega(\underset{\sim}{x}') \tag{18}$$

where $c'(\underset{\sim}{y})=1/2$ for the smooth boundary and in general

$$c'(\underset{\sim}{y}) = 1 - \lim_{\varepsilon\to 0}\int_{\Gamma_\varepsilon} \tilde{M}_n^*(\underset{\sim}{y},\underset{\sim}{y}')d\Gamma(\underset{\sim}{y}') \tag{19}$$

In equation (18) we use the following notations:

$$\tilde{w}^* = \hat{T}_n(w^*) \quad , \quad \tilde{T}_n^* = \hat{T}_n(T_n^*) \qquad (20)$$
$$\tilde{M}_n^* = \hat{T}_n(M_n^*) \quad , \quad \tilde{V}_n^* = \hat{T}_n(V_n^*)$$

Equations (16) and (18) are a set of integral equations on the boundary for the linear bending problems of elastic plates. Since two of the four boundary variables w, T_n, M_n and V_n are prescribed by the boundary conditions, we can determine the remaining unknowns from which the whole behavior of the plate is governed within the small deformation range.

3. FORMULATION FOR LARGE DISPLACEMENTS

The large deformation theory for the bending problems of thin elastic plates proposed by von Kármán (see Timoshenko and W.-Krieger, 1959) can be expressed as

$$\nabla^4 F = E\left[(\partial^2 w/\partial x_1 \partial x_2)^2 - (\partial^2 w/\partial x_1^2)(\partial^2 w/\partial x_2^2) \right] \qquad (21)$$

$$\nabla^4 w = q/D + (h/D)\left[(\partial^2 F/\partial x_2^2)(\partial^2 w/\partial x_1^2) + (\partial^2 F/\partial x_1^2)(\partial^2 w/\partial x_2^2) - 2(\partial^2 F/\partial x_1 \partial x_2)(\partial^2 w/\partial x_1 \partial x_2) \right] \qquad (22)$$

where E and h denote Young's modulus and the plate thickness, respectively. The function F in the above equations is a stress function which identically satisfies the equilibrium equations of the membrane forces:

$$\frac{\partial N_{11}}{\partial x_1} + \frac{\partial N_{12}}{\partial x_2} = 0 \ , \quad \frac{\partial N_{12}}{\partial x_1} + \frac{\partial N_{22}}{\partial x_2} = 0 \qquad (23)$$

if the membrane forces N_{11}, N_{12} and N_{22} are assumed to be derivable from

$$N_{11} = h\frac{\partial^2 F}{\partial x_1^2} \ , \quad N_{22} = h\frac{\partial^2 F}{\partial x_2^2} \ , \quad N_{12} = -h\frac{\partial^2 F}{\partial x_1 \partial x_2} \qquad (24)$$

Since an incremental approach may have a wider applicability to highly complex problems which include both the geometrical and physical nonlinearities, we express equations (21) and (22) in the incremental form. Denoting an incremental variable by the superimposed dot, we obtain

$$\nabla^4 \dot{F} = E\,L(w, \dot{w}) \qquad (25)$$

$$\nabla^4 \dot{w} = \frac{\dot{q}}{D} - \frac{h}{D}L(F, \dot{w}) - \frac{h}{D}L(w, \dot{F}) \qquad (26)$$

where the differential operator $L(F,(\ ))$ is defined in a similar manner to equation (4). The right hand sides of equations (25) and (26) include the unknown increments $\dot{w}$ and $\dot{F}$ as well as their total values at the current deformation state under consideration which are considered to be known in the incremental approach. The terms of $L(w,\dot{w})$, $L(w,\dot{F})$ and $L(F,\dot{w})$ in these equations are linear with respect to the incremental variables, and they result from the nonlinearities involved in the von Kármán's large deformation theory.

Since it is almost impossible to find the exact fundamental solution to the set of coupled differential equations (25) and (26), we shall use instead the fundamental solution to the biharmonic equation. This fundamental solution is defined by equation (13) and given by equation (14). Use of equations (13) and (26) for equation (10) leads to the following integral equation:

$$\begin{aligned}\dot{w}(\underset{\sim}{x}) &+ \int_\Gamma \Big[V_n^*(\underset{\sim}{x},\underset{\sim}{y}')\,\dot{w}(\underset{\sim}{y}') - M_n^*(\underset{\sim}{x},\underset{\sim}{y}')\,\dot{T}_n(\underset{\sim}{y}') \\ &+ T_n^*(\underset{\sim}{x},\underset{\sim}{y}')\,\dot{M}_n(\underset{\sim}{y}') - w^*(\underset{\sim}{x},\underset{\sim}{y}')\,\dot{V}_n(\underset{\sim}{y}')\Big]\,d\Gamma(\underset{\sim}{y}') \\ &= \frac{1}{D}\int_\Omega w^*(\underset{\sim}{x},\underset{\sim}{x}')\Big[\dot{q}(\underset{\sim}{x}') - hL(F,\dot{w}) - hL(w,\dot{F})\Big]\,d\Omega(\underset{\sim}{x}')\end{aligned} \tag{27}$$

Combination of equation (13) with (25) yields

$$\begin{aligned}\dot{F}(\underset{\sim}{x}) &+ \int_\Gamma \Big[V_n^*(\underset{\sim}{x},\underset{\sim}{y}')\,\dot{F}(\underset{\sim}{y}') - M_n^*(\underset{\sim}{x},\underset{\sim}{y}')\,\dot{S}_n(\underset{\sim}{y}') \\ &+ T_n^*(\underset{\sim}{x},\underset{\sim}{y}')\,\dot{N}_n(\underset{\sim}{y}') - w^*(\underset{\sim}{x},\underset{\sim}{y}')\,\dot{Q}_n(\underset{\sim}{y}')\Big]\,d\Gamma(\underset{\sim}{y}') \\ &= E\int_\Omega w^*(\underset{\sim}{x},\underset{\sim}{x}')\,L(w,\dot{w})\,d\Omega(\underset{\sim}{x}')\end{aligned} \tag{28}$$

In the above integral equations the following notations are introduced:

$$\left.\begin{aligned}\dot{T}_n = \hat{T}_n(\dot{w}),\quad \dot{M}_n = \hat{M}_n(\dot{w}),\quad \dot{V}_n = \hat{V}_n(\dot{w}) \\ \dot{S}_n = \hat{T}_n(\dot{F}),\quad \dot{N}_n = \hat{M}_n(\dot{F}),\quad \dot{Q}_n = \hat{V}_n(\dot{F})\end{aligned}\right\} \tag{29}$$

If we evaluate the singular integrals appearing in equations (27) and (28) when the point $\underset{\sim}{x}$ approaches from inside the domain Ω, we can obtain

$$\begin{aligned}C(\underset{\sim}{y})\,\dot{w}(\underset{\sim}{y}) &+ \int_\Gamma \Big[V_n^*(\underset{\sim}{y},\underset{\sim}{y}')\,w^*(\underset{\sim}{y}') - M_n^*(\underset{\sim}{y},\underset{\sim}{y}')\,\dot{T}_n(\underset{\sim}{y}') \\ &+ T_n^*(\underset{\sim}{y},\underset{\sim}{y}')\,\dot{T}_n(\underset{\sim}{y}') - w^*(\underset{\sim}{y},\underset{\sim}{y}')\,\dot{V}_n(\underset{\sim}{y}')\Big]\,d\Gamma(\underset{\sim}{y}')\end{aligned}$$

$$= \frac{1}{D}\int_{\Omega} w^*(\underset{\sim}{y},\underset{\sim}{x}')\left[\dot{q}(\underset{\sim}{x}') - h\,L(F,\dot{w}) - h\,L(w,\dot{F})\right] d\Omega(\underset{\sim}{x}') \tag{30}$$

and

$$c(\underset{\sim}{y})\,\dot{F}(\underset{\sim}{y}) + \int_{\Gamma} [V_n^*(\underset{\sim}{y},\underset{\sim}{y}')\,\dot{F}(\underset{\sim}{y}') - M_n^*(\underset{\sim}{y},\underset{\sim}{y}')\,\dot{S}_n(\underset{\sim}{y}')$$
$$+ T_n^*(\underset{\sim}{y},\underset{\sim}{y}')\,\dot{N}_n(\underset{\sim}{y}') - w^*(\underset{\sim}{y},\underset{\sim}{y}')\,\dot{Q}_n(\underset{\sim}{y}')\,]\,d\Gamma(\underset{\sim}{y}')$$
$$= E\int_{\Omega} w^*(\underset{\sim}{y},\underset{\sim}{x}')\,L(w,\dot{w})\,d\Omega(\underset{\sim}{x}') \tag{31}$$

where the coefficient $c(\underset{\sim}{y})$ is given by equation (17).

In a similar manner to that in the small deformation theory, the following integral equations can be derived with respect to the normal derivatives of $\dot{w}$ and $\dot{F}$ on the boundary, i.e.

$$c'(\underset{\sim}{y})\,\dot{t}_n(\underset{\sim}{y}) + \int_{\Gamma} [\tilde{V}_n^*(\underset{\sim}{y},\underset{\sim}{y}')\,\dot{w}(\underset{\sim}{y}') - \tilde{M}_n^*(\underset{\sim}{y},\underset{\sim}{y}')\,\dot{t}_n(\underset{\sim}{y}')$$
$$+ \tilde{T}_n^*(\underset{\sim}{y},\underset{\sim}{y}')\,\dot{M}_n(\underset{\sim}{y}') - \tilde{w}^*(\underset{\sim}{y},\underset{\sim}{y}')\,\dot{V}_n(\underset{\sim}{y}')\,]\,d\Gamma(\underset{\sim}{y}')$$
$$= \frac{1}{D}\int_{\Omega} \tilde{w}^*(\underset{\sim}{y},\underset{\sim}{x}')\left[\dot{q}(\underset{\sim}{x}') - h\,L(F,\dot{w}) - h\,L(w,\dot{F})\right] d\Omega(\underset{\sim}{x}') \quad \ldots\ldots(32)$$

and

$$c'(\underset{\sim}{y})\,\dot{S}_n(\underset{\sim}{y}) + \int_{\Gamma} [\tilde{V}_n^*(\underset{\sim}{y},\underset{\sim}{y}')\,\dot{F}(\underset{\sim}{y}') - \tilde{M}_n^*(\underset{\sim}{y},\underset{\sim}{y}')\,\dot{S}_n(\underset{\sim}{y}')$$
$$+ \tilde{T}_n^*(\underset{\sim}{y},\underset{\sim}{y}')\,\dot{N}_n(\underset{\sim}{y}') - \tilde{w}^*(\underset{\sim}{y},\underset{\sim}{y}')\,\dot{Q}_n(\underset{\sim}{y}')\,]\,d\Gamma(\underset{\sim}{y}')$$
$$= E\int_{\Omega} \tilde{w}^*(\underset{\sim}{y},\underset{\sim}{x}')\,L(w,\dot{w})\,d\Omega(\underset{\sim}{x}') \tag{33}$$

The coefficient $c'(\underset{\sim}{y})$ is the same as given by equation (19).

The four integral equations (30) through (33) on the boundary together with equations (27) and (28) in the inner domain govern the incremental behavior of the large deformation problems of elastic plates. They are solved under appropriate boundary conditions.

4. BOUNDARY CONDITIONS

We can imagine various boundary conditions of the plates such as free edge, simple support, clamped edge, etc. For the deflection function w, the boundary conditions can be easily expressed as follows:

(a) Free edge

$$\hat{M}_n(\dot{w}) \equiv \dot{M}_n = 0 \;, \quad \hat{V}_n(\dot{w}) \equiv \dot{V}_n = 0 \tag{34a}$$

(b) Simple support

$$\dot{w} = 0 \;, \quad \dot{M}_n = 0 \tag{34b}$$

(c) Clamped edge

$$\dot{w} = 0 \;, \quad \hat{T}_n(\dot{w}) \equiv \dot{T}_n = 0 \tag{34c}$$

These conditions are related directly to the boundary variables. On the other hand, the boundary conditions with respect to the stress function F are somewhat complicated. Since the stress function is connected directly with the membrane forces as in equation (24), the so called stress boundary conditions are rather easily expressible than the displacement boundary conditions. In the following, considerations are restricted to the stress boundary conditions for the in-plane deformation of the plates.

As is well known, the in-plane tractions N_1 and N_2 are related to the membrane force components in the following manner:

$$\left.\begin{aligned} \dot{N}_1 &= \dot{N}_{11}\, n_1 + \dot{N}_{12}\, n_2 \\ \dot{N}_2 &= \dot{N}_{12}\, n_1 + \dot{N}_{22}\, n_2 \end{aligned}\right\} \tag{35}$$

For the direction cosines of the normal to the plate boundary there are the following relationships(Fig.2):

$$\left.\begin{aligned} n_1 &\equiv \cos(\underset{\sim}{n}, x_1) = \partial x_1/\partial n = \partial x_2/\partial t \\ n_2 &\equiv \cos(\underset{\sim}{n}, x_2) = \partial x_2/\partial n = -\partial x_1/\partial t \end{aligned}\right\} \tag{36}$$

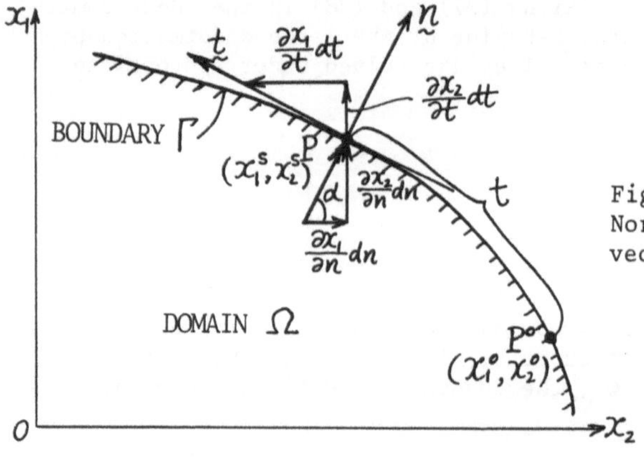

Figure 2
Normal and tangential vectors to boundary

Substitution of equations (24) and (36) into (35) leads to

$$\left.\begin{aligned} \dot{N}_1 dt &= \frac{\partial(\partial\dot{F}/\partial x_2)}{\partial x_1} dx_2 + \frac{\partial(\partial\dot{F}/\partial x_2)}{\partial x_2} dx_2 = d\left(\frac{\partial\dot{F}}{\partial x_2}\right) \\ -\dot{N}_2 dt &= \frac{\partial(\partial\dot{F}/\partial x_1)}{\partial x_1} dx_1 + \frac{\partial(\partial\dot{F}/\partial x_1)}{\partial x_2} dx_2 = d\left(\frac{\partial\dot{F}}{\partial x_1}\right) \end{aligned}\right\} \tag{37}$$

Integration of equation (37) yields

$$\begin{aligned} \dot{F} = &-\int_0^t \dot{F}_1 dx_1 + \int_0^t \dot{F}_2 dx_2 \\ &+ C_1(x_1 - x_1^\circ) + C_2(x_2 - x_2^\circ) + C_3 \end{aligned} \tag{38}$$

where

$$\dot{F}_1 = \int_0^t \dot{N}_2 dt \quad , \quad \dot{F}_2 = \int_0^t \dot{N}_1 dt \tag{39}$$

C_1, C_2 and C_3 are integration constants which can be put equal to zero for a simply connected domain. It can be seen in this case that equation (38) leads to

$$\begin{aligned} \dot{F} &= \int_0^t (-\dot{F}_1 dx_1 + \dot{F}_2 dx_2) = \left[-\dot{F}_1 x_1 + \dot{F}_2 x_2\right]_0^t + \int_0^t (x_1 d\dot{F}_1 - x_2 d\dot{F}_2) \\ &= \int_0^t (x_1 - x_1^s)\dot{N}_2 dt - \int_0^t (x_2 - x_2^s)\dot{N}_1 dt \end{aligned} \tag{40}$$

Equation (40) implies that the value of $\dot{F}$ can be interpreted as the moment of the boundary tractions with respect to the point P under consideration located on the boundary (Fig.2).

The normal derivative $\partial\dot{F}/\partial n$ can be related to the prescribed boundary tractions in the following manner. That is, the chain rule of differentiation is combined with equation (36) to yield

$$\frac{\partial\dot{F}}{\partial n} = \frac{\partial\dot{F}}{\partial x_1}\frac{\partial x_1}{\partial n} + \frac{\partial\dot{F}}{\partial x_2}\frac{\partial x_2}{\partial n} = \frac{\partial\dot{F}}{\partial x_1}\frac{\partial x_2}{\partial t} - \frac{\partial\dot{F}}{\partial x_2}\frac{\partial x_1}{\partial t} \tag{41}$$

Substitution of equation (37) leads to

$$\frac{\partial\dot{F}}{\partial n} = (-\dot{F}_1 + C_1)n_1 + (\dot{F}_2 + C_2)n_2 \tag{42}$$

For a simply connected domain the integration constants C_1 and C_2 are again equal to zero. Then we have

$$\frac{\partial\dot{F}}{\partial n} = -n_1\int_0^t \dot{N}_2 dt + n_2\int_0^t \dot{N}_1 dt \tag{43}$$

This relation implies that $\partial\dot{F}/\partial n$ is the projection of the resulting forces acting along the boundary portion t onto the tangential direction of the boundary.

5. SOLUTION SCHEME AND DISCUSSION

5.1 Iterative Method

We shall first consider an iterative solution scheme for the boundary integral equations (30) through (33). These equations include the incremental variables $\dot{w}$, $\dot{T}_n$, $\dot{M}_n$, $\dot{V}_n$, $\dot{F}$, $\dot{S}_n$, $\dot{N}_n$ and $\dot{Q}_n$. Since four of them are prescribed by the boundary conditions, the remaining four boundary variables are the real unknowns. It can be seen that the number of the integral equations coincides with that of the boundary unknowns and hence the boundary integral equations may be solved by means of an appropriate iterative procedure.

As in the usual boundary element method, the plate boundary Γ is divided into a series of boundary elements. It is then assumed that the boundary variables are interpolated with the corresponding nodal values by using appropriate interpolation functions as in the finite element method. Under this discretization the set of integral equations are approximated to the following discretized system:

$$[A]\{\dot{w}\} + [B]\{\dot{T}_n\} + [C]\{\dot{M}_n\} + [D]\{\dot{V}_n\} = \{\dot{G}\} \tag{44}$$

$$[a]\{\dot{F}\} + [b]\{\dot{S}_n\} + [c]\{\dot{N}_n\} + [d]\{\dot{Q}_n\} = \{\dot{g}\} \tag{45}$$

where, for example, $\{\dot{w}\}$ denotes the nodal values of $\dot{w}$ on the boundary. The coefficient matrices in equations (44) and (45) are all known. It is noted that the column vectors $\{\dot{G}\}$ and $\{\dot{g}\}$ can be calculated from the increment of the distributed pressure $\dot{p}$ as well as $\dot{w}$ and $\dot{F}$ in the inner domain Ω.

For the solution of the system of linear equations (44) and (45), we may use the iterative solution scheme which consists of the following steps:

(1) Assume the initial values $\dot{F}$ in Ω and compute $\{\dot{G}\}$ of equation (44).
(2) Compute two sets of the nodal variables $\{\dot{w}\}$, $\{\dot{T}_n\}$, $\{\dot{M}_n\}$ and $\{\dot{V}_n\}$ on Γ by use of equation (44).
(3) Calculate $\dot{w}$ in Ω by means of equation (27), then use its values for computation of $\{\dot{g}\}$.
(4) Solve equation (45) for the nodal unknowns of $\{\dot{F}\}$, $\{\dot{S}_n\}$, $\{\dot{N}_n\}$ and $\{\dot{Q}_n\}$.
(5) Compute $\{\dot{G}\}$ of equation (44).
(6) If an error norm between the assumed values and the newly obtained ones for $\{\dot{G}\}$ falls within a tolerable limit, proceed to the next loading step. Otherwise, iterate steps (2) through (5) until convergence is achieved.

5.2 Direct Method

To the integro-differential equations under consideration, we can also apply a rather simple solution scheme called the boundary-volume element method, which has been proposed for inhomogeneous thermoelasticity and elastoplasticity (Tanaka and Tanaka, 1980, 1981). In this method the whole domain $\Omega+\Gamma$ is divided into a finite number of elements. The elements in the inner domain Ω play a similar role as in the finite element method. Under this discretization scheme the nodal unknowns of $\dot{w}$ and $\dot{F}$ in Ω are taken out from the surface integrals involved in the right hand sides of the integral equations. Denoting these nodal unknowns by $\{\dot{w}\}_i$ and $\{\dot{F}\}_i$, we have instead of equations (44) and (45) the following set:

$$[A_1]\{\dot{w}\} + \underline{[A_2]\{\dot{w}\}_i + [\tilde{A}_1]\{\dot{F}\}_i} + [B]\{\dot{T}_n\} + [C]\{\dot{M}_n\} + [D]\{\dot{V}_n\} = \{\dot{G}_0\} \tag{46}$$

$$[a_1]\{\dot{F}\} + \underline{[a_2]\{\dot{F}\}_i + [\tilde{a}_1]\{\dot{w}\}_i} + [b]\{\dot{S}_n\} + [c]\{\dot{N}_n\} + [d]\{\dot{Q}_n\} = \{\dot{g}_0\} \tag{47}$$

Furthermore, the integral equations (27) and (28) provide the following discretized system:

$$[A_{1i}]\{\dot{w}\} + \underline{[A_{2i}]\{\dot{w}\}_i + [\tilde{A}_{1i}]\{\dot{F}\}_i} + [B_i]\{\dot{T}_n\} + [C_i]\{\dot{M}_n\} + [D_i]\{\dot{V}_n\} = \{\dot{G}_i\} \tag{48}$$

$$[a_{1i}]\{\dot{F}\} + \underline{[a_{2i}]\{\dot{F}\}_i + [\tilde{a}_{1i}]\{\dot{w}\}_i} + [b_i]\{\dot{S}_n\} + [c_i]\{\dot{N}_n\} + [d_i]\{\dot{Q}_n\} = \{\dot{g}_i\} \tag{49}$$

The underlined terms in equations (46) through (49) result from the surface integrals of the integral equations. It is again noted that all the coefficient matrices are known in the set of equations (46) through (49).

If the total number of nodal points in Ω and that of boundary nodes are assumed to be $\tilde{N}$ and N, respectively, we have $2(\tilde{N}+2N)$ nodal unknowns to be determined by the set of equations (46) through (49). There are $4N$ equations in equations (46) and (47), while $2\tilde{N}$ equations are included in equations (48) and (49). Therefore, the number of the resulting equations is $2(\tilde{N}+2N)$, and hence they can be solved simultaneously.

Although additional nodal unknowns are added to the final set of discretized equations to be solved, there is no need for iterative computations within each loading step, if the direct method of solution is applied.

6. CONCLUSIONS

The boundary element methods have been studied for small and large displacements of thin elastic plates. The integral equation formulation was presented in terms of incremental variables for the large displacements of the plates, and solution procedures were investigated in some detail. The boundary-volume element method which is a combined boundary and finite element method was proposed as an alternative solution procedure to an iterative one for the coupled integral equations between the in-plane and the out-of-plane displacements. The proposed methods of solution could also be applied to the problems with higher nonlinearities due to large deformation as well as nonlinear constitutive relation.

ACKNOWLEDGEMENTS

The author wishes to express his cordial thanks to Professors M. Hamada, K. Ohji, H. Fukuoka and M. Naruoka for their helpful support. Parts of this work were sponsored by the Ministry of Education of the Japanese Government.

REFERENCES

Bergman, S. and Schiffer, M. (1953) Kernel Functions and Elliptic Differential Equations in Mathematical Physics. Academic Press, New York.

Bezine, G. (1978) Boundary Integral Formulation for Plate Flexure with Arbitrary Boundary Conditions. Mech. Res. Comm., 5, 197-206.

Hansen, E.B. (1976) Numerical Solution of Integro-Differential and Singular Integral Equations for Plate Bending Problems. J. Elasticity, 6, 39-56.

Jaswon, M.A. and Maiti, M. (1968) An Integral Equation Formulation of Plate Bending Problems. J. Eng. Math., 2, 83-93.

Stern, M. (1979) A General Boundary Integral Formulation for the Numerical Solution of Plate Bending Problems. Int. J. Solids & Struct., 15, 769-782.

Tanaka, M. and Tanaka, K. (1980) On Numerical Solution Scheme for Thermoelastic Problems in Inhomogeneous Media by Means of Boundary-Volume Element. Z. Angew. Math. Mech., 60, 719-723.

Tanaka, M. and Tanaka, K.(1981) On a New Boundary Element Solution Scheme for Elastoplasticity. Ing.-Arch., 50, 289-295.

Timoshenko, S.P. and Woinowsky-Krieger, S. (1959) Theory of Plates and Shells. 2nd ed. McGraw-Hill, London-New York-Tokyo.

ON THE COMPUTATION OF THE STRESS INTENSITY FACTORS IN ELASTIC PLATE FLEXURE VIA BOUNDARY INTEGRAL EQUATIONS

J. W. KIM

Korea Military Academy, Korea

INTRODUCTION

A fairly general boundary integral equation (BIE) formulation for elastic plate flexure has been given earlier by STERN(1979) in terms of a pair of coupled singular integral equations involving the natural variables of deflection, normal slope, bending moment, and equivalent shear on the plate boundary. While this formulation allows for discontinuities in the boundary variables as might naturally occur at a corner of the plate boundary, or where the boundary support conditions undergo a sudden change in type as from clamped to free, the calss of admissible problems is still required to have bounded moment and shear resultants which produce bounded stresses everywhere in the plate. However, there are important classes of problems for which the moment and shear resultants as calculated within the framework of linear theory are not bounded, for example at sharp notches or cracks as shown by WILLIAMS(1951). At such singular points a knowledge of the so-called stress intensity factors governing the growth rate of the stresses has proven useful in linear fracture mechanics.

In this chapter we indicate how to adapt the BIE formulation so that the stress intensity factors also become natural variables to be determined by the solution of coupled singular integral equations. While the basic method follows closely in spirit the ideas outlined for plane elastostatic calculations by BARONE and ROBINSON (1972) and by STERN (1975), the complexity of the problem is magnified by the higher-order equations of plate theory, and some new difficulties not present in the plane elastostatic case arise.

BOUNDARY INTEGRAL REPRESENTATION

We first briefly summarize the development of the BIE representation in STERN(1979). The plate is modeled by a bounded region Ω with total boundary $\partial\Omega$ containing a finite number of corner points $\ell_1 \cdot\cdot\ \ell_k$ as indicated in Fig.1 where other relevant notation is shown. The plate deflection w is governed by the differential equation

$$\nabla^4 w = q/D \qquad \text{in } \Omega \tag{1}$$

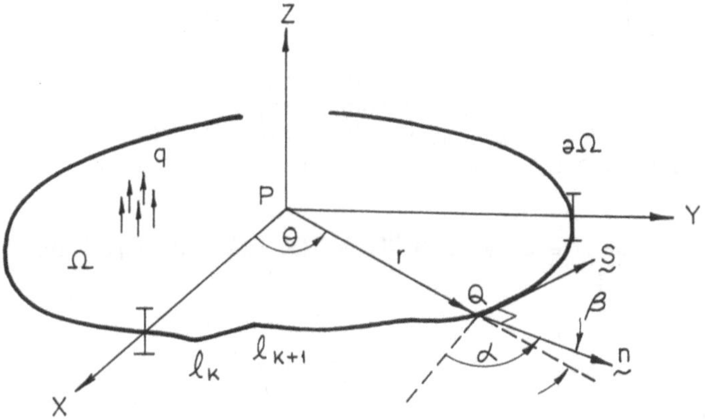

FIGURE 1. Coordinates and Notation.

where q is the transverse load on the plate, and suitable boundary conditions, left unspecified at the moment, are imposed on $\partial\Omega$. Also $D = Eh^3/12(1-\nu^2)$ denotes the flexural rigidity of the plate and ∇^4 is the iterated laplacian

For u any other sufficiently smooth function on Ω, interpretable as a possible deflection for suitable loading and support conditions, we have a natural reciprocal work identity in the form

$$\int_{\partial\Omega} \{V_n(u)w - M_n(u)\frac{dw}{dn} + \frac{du}{dn}M_n(w) - uV_n(W)\}\, ds \tag{2}$$

$$+ \sum_{k=1}^{K} \Big[[\![M_t(u)]\!]\ w - [\![M_t(w)]\!]\ u \Big]_{\ell_k} = D\int_{\Omega} (u\nabla^4 w - w\nabla^4 u)\, da$$

The notation and development was detailed in STERN(1979) ; briefly $V_n(\cdot)$, $M_n(\cdot)$, $\frac{d\cdot}{dn}$ are the equivalent shear, bending moment and normal slope on the boundary $\partial\Omega$, and $[\![M_t(\cdot)]\!]_{\ell_k}$ is the discontinuity jump in twisting moment (interpertable as a concentrated force reaction) at the corner ℓ_k.

Now let $p \in \Omega$ be an interior point of the plate (the origin of coordinates in Fig.1) and introduce the special "singluar solution"

$$w(p) = \frac{1}{8\pi D} r^2 \ln r \tag{3}$$

with N, V, M and T the corresponding normal slope, equivalent shear, bending moment and twisting moment on $\partial\Omega$. Deleting a small circular region centered at P, and applying the identity Eq. (2) to Ω so modified, produces two contributions to the boundary integral. The first integral is still over the entire boundary $\partial\Omega$ just as in Eq. (2). The second however is over the small circle surrounding P. With the particular choice of the function W defined in Eq. (3) for the auxiliary function u, this last integral may be evaluated in the limit as the circle surrounding P shrinks to P.
This produces, from Eq. (2), the representation

$$w\Big|_P = -\int_{\partial\Omega}\left[Vw - M\frac{dw}{dn} + NM_n(w) - WV_n(w)\right] ds$$

$$-\sum_{k=1}^{K}\left[\left[\!\left[T\right]\!\right] w - \left[\!\left[M_t(w)\right]\!\right] W\right]_{\ell_k} + \int_{\Omega} qW\, da \tag{4}$$

where we have replaced $D\nabla^4 w$ with the distributed load intensity q. Since only the functions derived from W depend on the point P, Eq. (4) may be differentiated (in the interior of Ω) to whatever extent is permitted by the regularity of the load q. Thus the plate deflection and quantities derivable from it are determined by a knowledge of the deflection, normal slope, bending moment, and equivalent shear on the plate boundary, and the discontinuity jumps in the twisting moment (concentrated support forces) which might occur at corners. These quantities in turn may be determined using the following considerations.

The same type of argument used in obtaining Eq. (4) is repeated for the point P on the boundary $\partial\Omega$. Now only a semi-circular region is deleted from Ω and two new corners are introduced into the boundary of the modified region as indicated in Fig. 2. In addition to the special singular function W defined in Eq. (2), we define another

$$W_\gamma = \frac{1}{2\pi D} r \ln r \cos(\theta + \gamma) \tag{5}$$

with γ the angle from the outer normal at P to the line $\theta = 0$, and N_γ, V_γ, M_γ, T_γ the corresponding normal slope, equivalent shear, bending moment and twisting moment. This leads to the pair of coupled boundary integral representations

$$\frac{1}{2} w\Big|_P + \oint_{\partial\Omega}\left[Vw - M\frac{dw}{dn} + NM_n(w) - WV_n(w)\right] ds$$

$$+\sum_{k=1}^{K}\left[\left[\!\left[T \right]\!\right] w - \left[\!\left[M_t(w) \right]\!\right] W\right]_{\ell_k} = \int_{\Omega} q \ W \ da \tag{6}$$

$$\left.\frac{dw}{dn}\right|_p + \oint_{\partial\Omega}\left[V\left(w-\left.w\right|_p\right) - M_\gamma\frac{dw}{dn} + N_\gamma M_n(w) - W_\gamma V_n(w)\right] ds$$

$$+\sum_{k=1}^{K}\left[\left[\!\left[T_\gamma \right]\!\right]\left(w-\left.w\right|_p\right) - \left[\!\left[M_t(w) \right]\!\right] W_\gamma\right]_{\ell_k} = \int_{\Omega} q \ W_\gamma \ da \tag{7}$$

where $\oint$ denotes a Cauchy principal value integral. If the point p should lie at a corner of the plate boundary additional considerations are required; these are developed in detail in STERN(1979).

Finally, two boundary conditions involving the boundary variables are available from the particualr nature of the supports of lack of them; for example, on a clamped portion of the boundary the deflection w and normal slope $\frac{dw}{dn}$ are required to vanish, whereas if the boundary is free of support then the bending moment $M_n(w)$ and equivalent shear $V_n(w)$ are zero. The boundary conditions, together with Eqs. (6) and (7), are sufficient to determine the boundary variables everywhere on $\partial\Omega$.

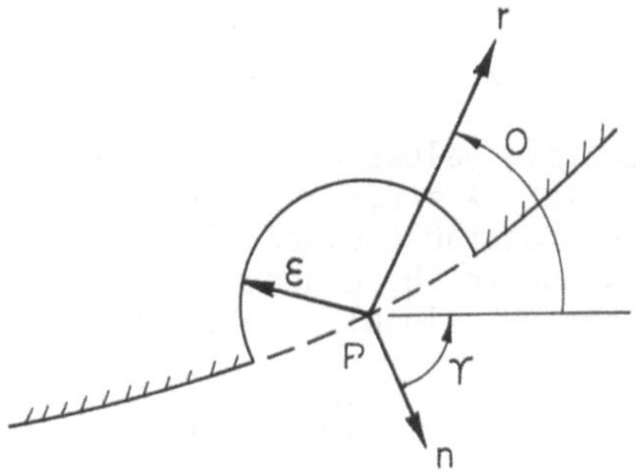

FIGURE 2. Origin on the boundary.

SINGULARITY EQUATIONS

The equations obtained in the preceeding section are based on the presumption that the moment and shear resultants remain bounded near the origin point P. We now consider how to obtain appropriate integral equations for other cases of interest at boundary points where the stresses become unbounded. For brevity we outline the major ideas only for the special case of a through crack with free edges.

Following the ideas of Williams (1951), we infer the asymptotic singular behavior of the plate deflection at the base of a crack from the nontrivial solutions of the homogeneous boundary value problem corresponding to an unloaded plate containing a straight semi-infinite crack with free edges:

$$\begin{aligned} \nabla^4 w(r,\theta) &= 0 \quad &&\text{for } r > 0,\ 0 \leq \theta \leq 2\pi \\ M_n(w) = V_n(w) &= 0 \quad &&\text{for } \theta = 0\ (\beta = \tfrac{3\pi}{2}) \\ M_n(w) = V_n(w) &= 0 \quad &&\text{for } \theta = 2\pi\ (\beta = \tfrac{\pi}{2}) \end{aligned} \tag{8}$$

A separation of variables solution results in an eigenvalue problem; with

$$w(r,\theta;\lambda) = r^{\lambda+1}\Big[b_1^{(\lambda)} \sin(\lambda+1)\theta + b_2^{(\lambda)} \cos(\lambda+1)\theta + b_3^{(\lambda)} \sin(\lambda-1)\theta + b_4^{(\lambda)} \cos(\lambda-1)\theta\Big] \tag{9}$$

the eigenvalues are given by

$$\sin 2\pi\lambda = 0 \tag{10}$$

We reject negative values of λ as leading to unbounded strain energy, while for $\lambda = 0$, $\lambda = 1$ or $\lambda \geq 2$ the stresses remain bounded. The acceptable eigensolutions with singular stresses thus correspond to the eigenvalues $\lambda=1/2$ and $\lambda=3/2$ which leads to the general singular solution

$$w^s(r,\theta) = r^{3/2}\left\{ b_1\left[\sin\frac{3\theta}{2} + \frac{3(1-\nu)}{7+\nu}\sin\frac{\theta}{2}\right] + b_2\left[\cos\frac{3\theta}{2} + \frac{3(1-\nu)}{5+3\nu}\cos\frac{\theta}{2}\right]\right\} \tag{11}$$

$$+ r^{5/2}\left\{ c_1\left[\sin\frac{5\theta}{2} - \frac{5(1-\nu)}{9-\nu}\sin\frac{\theta}{2}\right]\right.$$

$$\left. + c_2\left[\cos\frac{5\theta}{2} + \frac{5(1-\nu)}{3+5\nu}\cos\frac{\theta}{2}\right]\right\}$$

Equation (11) defines the asymptotic singular behavior of the plate deflection near the base of a crack in terms of the four parameters b_1, b_2, c_1, c_2. (This result is of course well known, for example WILLIAMS(1961).

The idea now is to substitute for u in the reciprocal work identity Eq. (2) particular biharmonic functions with the right order of singular behavior at the base of the crack, so that when the argument leading to the representations Eqs. (6) and (7) is repeated for P at the crack tip we obtain representations for b_1, b_2, c_1, c_2. Furthermore, if these functions are also solutions of the boundary value problem Eq. (8) then no contribution to the integrals will result from the crack flanks.

Without attempting to furnish details we merely list main results. The particular solution of Eq. (8) required (called the complementary solution) is of the form

$$\begin{aligned} u(r,\theta) = {} & r^{1/2}\left\{B_1\left[\sin\frac{\theta}{2} + \frac{1-\nu}{5+3\nu}\sin\frac{3\theta}{2}\right]\right. \\ & \left. + B_2\left[\cos\frac{\theta}{2} + \frac{1-\nu}{7+\nu}\cos\frac{3\theta}{2}\right]\right\} \\ & + \bar{r}^{1/2}\left\{C_1\left[\sin\frac{\theta}{2} + \frac{1-\nu}{3+5\nu}\sin\frac{5\theta}{2}\right]\right. \\ & \left. + C_2\left[-\cos\frac{\theta}{2} + \frac{1-\nu}{9-\nu}\cos\frac{5\theta}{2}\right]\right\} \end{aligned} \tag{12}$$

where B_1, B_2, C_1, C_2 are arbitrary constants. By deleting a small circular region surrounding the crack tip at P as indicated in Fig. 3 we add to Eq. (2) a contour integral (on C_ε) and two dorner jump terms (at ℓ^+ and ℓ^-) which are evaluated in the limit as $\varepsilon \to 0$ to produce, after routine but tedious calculation,

$$J_{tip} = \lim_{\varepsilon\to 0}\left\{ \int_{C_\varepsilon}\left[V_n(u)w - M_n(u)\frac{dw}{dn} + \frac{du}{dn}M_n(w) - uV_n(w)\right]ds\right.$$

$$+ \left\{ \left[[\![M_t(u)]\!] \, w - [\![M_t(w)]\!] \, u \right]_{\ell^+} + \left[[\![M_t(u)]\!] \, w - [\![M_t(w)]\!] \, u \right]_{\ell^-} \right\} \Bigg\}$$

$$= \frac{24(1-\nu)\,(3+\nu)\pi D}{(7+\nu)\,(5+3\nu)} \left[B_1 b_1 + B_2 b_2 \right]$$

$$- \frac{120(1-\nu)(3+\nu)\pi D}{(9-\nu)(5+3\nu)} \left[C_1 c_1 + C_2 c_2 \right] \qquad (13)$$

By suitable choice of the arbitrary constants B_1, B_2, C_1, C_2 in the complementary solution Eq. (12) we can obtain representations analogous to Eqs. (6) and (7) for the parameters b_1, b_2, c_1, c_2. Alternatively, more physically meaningful parameters might be introduced in place of these; for example a symmetric moment intensity factor $K_1^{(1)}$ which governs the rate of growth of the bending moment:

$$K_1^{(1)} = \lim_{r \to 0} r^{1/2} \, M_n(w) \Big|_{\substack{\theta=\pi \\ (\beta=\pi/2)}} = - \frac{3D(1-\nu)(3+\nu)}{7+\nu} \, b_1 \qquad (14)$$

An antisymmetric moment intensity factor may also be defined and satisfies

$$K_2^{(1)} = \lim_{r \to 0} r^{1/2} \, M_t(w) \Big|_{\substack{\theta=\pi \\ (\beta=\pi/2)}} = - \frac{3D(1-\nu)(1+\nu)}{5+3\nu} \, b_2 \qquad (15)$$

Then, for example, if we take

$$B_1 = - \frac{5+3\nu}{8\pi} \quad , \quad B_2 = C_1 = C_2 = 0 \qquad (16)$$

in Eq. (12) (denoting this special function u_1) we find Eq. (13) reduces to $J_{tip} = K_1^{(1)}$ and the representation becomes

$$K_1^{(1)} + \oint_{\partial\Omega} \left[V_n(u_1) w - M\,(u_1) \frac{dw}{dn} + \frac{du_1}{dn} M_n(w) - u_1 V_n(w) \right] ds$$

$$+\sum_{k=1}^{K} \left[\llbracket M_t(u_1) \rrbracket \; w - \llbracket M_t(w) \rrbracket \; u_1 \right]_{\ell_k} = D \int_{\Omega} u_1 \; q \; da \tag{17}$$

with similar equations for $K_2^{(1)}$, c_1 and c_2

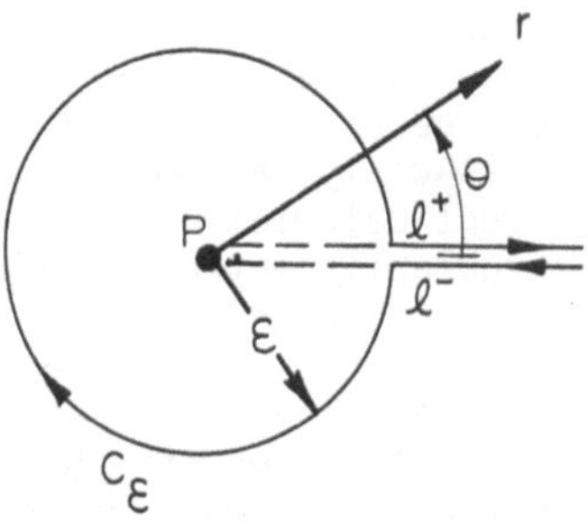

FIGURE 3. Integratinon near crack tip.

NUMERICAL IMPLEMENTATION

The incorporation of the additional singularity intensity variables in the numerical BIE scheme described in STERN (1979) involves seral new considerations. Briefly the original scheme consists of discretizing the boundary by the introduction of a finite number of nodal points, care being taken to place a nodal point at corners and other exceptional points. The primary variables are then interpolated between nodal points which reduces the problem to the determination of nodal values of each of the four primary variables (deflection, normal slope, bending moment and equivalent shear) at the "regular nodes", with corners requiring some added considerations as detailed in STERN(1979). We have generally at each node two boundary conditions to be imposed as well as the pair of integral equations (6) and (7) which are easily discretized in terms of the nodal variables by any convenient quadrature rule applied to the intervals between adjacent nodes. The solution of this system of linear equations (including additional corner variables and equations) then completes the process.

Now we have introduced singularity intensity factors as additional unknown variables, and with each an additional integral equation such as Eq. (17) is also furnished so that the system of equations is still determinate. However, some as yet untreated questions arise in the limiting process leading to the integral equations for origin at the singularity.

The evaluation of various limits in obtaining Eqs. (6) and (7) was accomplished with the presumption that the deflection function behaved smoothly (bounded third derivatives) near the boundary of the plate. This of course is not the case near a singular point of the type under consideration. It is not much more difficult to verigy these limits even in the singular case if one notes the form of the asymptotic behavior of the plate deflection near the singularity, for example Eq. (11) at the base of a through crack.

A similar question arises in deriving representations such as Eq. (17) since the deflection and rotation of the plate at the crack tip were (tacitly) assumed to be zero. Again it can be verified by direct calculation (as well as by an energy argument) that the singularity intensity equations are unaltered by any superposed rigid body displacement of the plate.

Finally, the boundary segments adjacent to a singular node should receive special treatment since the eigensolutions furnish additional information concerning the behavior of the primary variables. For example consider the segment from the crack tip to the adjacent node on the crack edge a distance d away. Normally we would interpolate the displacement on this interval linearly writing

$$w(r) = w_0 + (w_1 - w_0)r/d \tag{18}$$

where w_0 is the deflection at the crack tip node and w_1 is at the adjacent node. However, from Eq. (11), with whatever rigid body displacement of the plate is needed at the crack tip, the deflection along this segment must be of the form

$$w(r) = w_0 + w_0'r + pr^{3/2} + qr^{5/2} + \text{remainder} \tag{19}$$

where w_0' is the slope along the crack edge at the tip, with a corresponding representation similar to Eq. (7), and p, q are particular linear combinations of the intensity parameters K_1, K_2, c_1, c_2. The remainder term may be approximated to the order of r^2 so that the deflection is continuous at the adjacent node yielding the interpolation

$$w(r) = w_0 + w_0'r + pr^{3/2} + qr^{5/2} + \left\{ w_1 - w_0 + w_0'd + pd^{3/2} + qd^{5/2} \right\} (r/d)^2 \tag{20}$$

Similar results are obtained to interpolate the other variables.

As an example of the entire procedure consider the symmetric bending of a centrally cracked square plate as illustrated in Fig. 4. Symmetric considerations permit us to analyze only one quarter of the plate (the second quadrant, for example

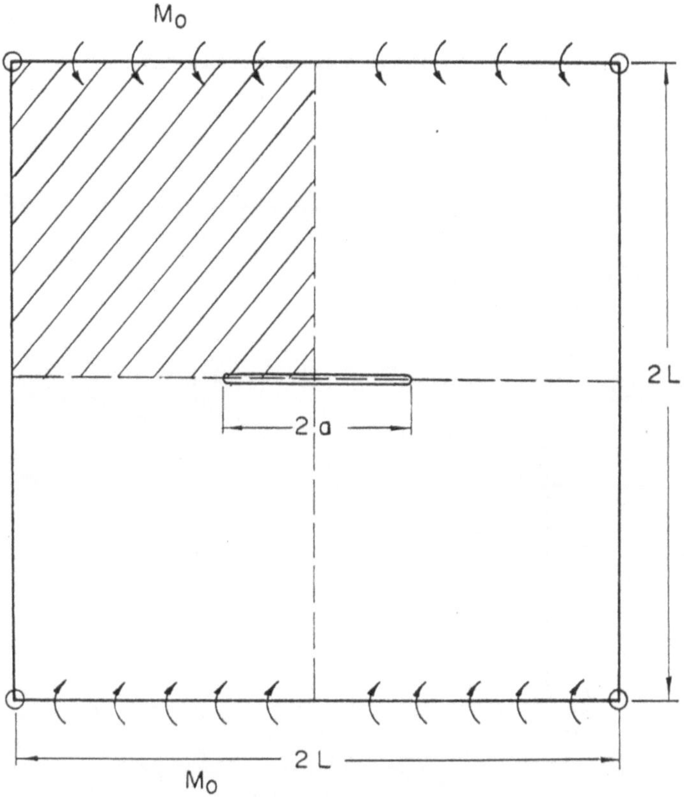

FIGURE 4. Symmetric Bending of a Centrally Cracked Square Plate.

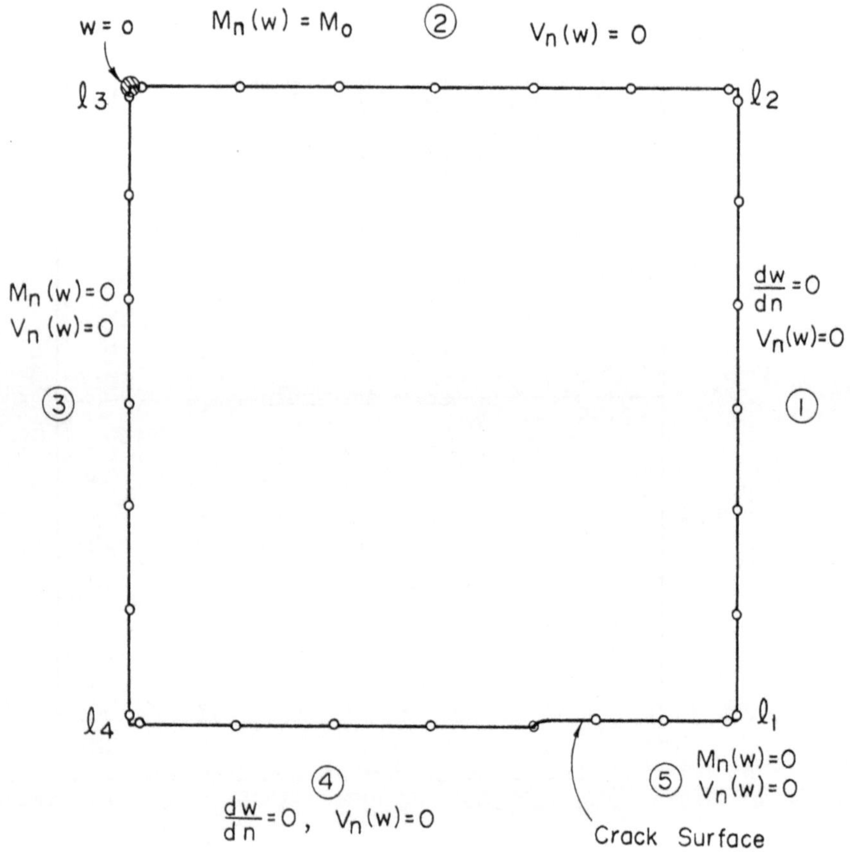

FIGURE 5. Nodal Distribution and Boundary Conditions for One Quarter of the Plate.

which is isolated in Fig. 5. Boundary conditions are also shown in Fig. 5.

Results for the symmetric moment intensity factor $K_1^{(1)}$ are shown in Table 1. for five ratios of a/L ranging from..5 to.1. An eight node mesh on every side was used in all five cases. The results are nondimensionalized using the solution for an infinite plate (a/L → 0) obtained by Sih, et al(1962). These nondimensional intensity factors for a square plate are compared with those of Wilson's finite element solution (1971) for an infinite strip of finite width with symmetric uniform bending. As indicated in Table 1. there is reasonable agreement but the BIE solutions exhibit rather poor behavior over the range of crack size. This may be due to inherent poor conditioning of the BIE equations and the delicate nature of the CPV evaluations.

While the numerical results appear to be generally correct, they are far from satisfactory. In the numerical solution there were indications that the system of equations was poorly conditioned. It was also found that the results were sensitive to the form of interpolation used on the elements adjacent to the crack tip and in the CPV evaluations. Further study is needed to identify and correct, or at least to minimize the effects of the major error mechanisms. Numerical experiments with different geometries, mesh patterns and interpolation forms could be useful.

Additional areas also merit investigation. The treatment of curved boundaries should be investigated to determine whether significant errors are introduced by crude approximation (say piecewise linear) of its shape. In principle, the shape of the boundary may be treated as accurately as desired, but with a significant increase in computation cost. Another area of investigation which might prove fruitful is the possible gain in accuracy for a given mesh using a smoother than piecewise linear representation of the boundary variables, for example cubic splines. There are no theoretical analyses presently available to suggest how the accuracy of the solution depends on the smoothness of the boundary approximations, so numerical experiments will probably be the major investigative tool here as well.

TABLE 1. Nondimensional Symmetric Moment Intensity Factors

$\frac{a}{L}$	.1	.2	.3	.4	.5
$K_1^{(1)}/K^{\infty}$ (BIE)	1.0130	1.0480	1.0578	1.1011	1.2117
$K_1^{(1)}/K^{\infty}$ (FEM)	1.006	1.024	1.058	1.105	1.181

REFERENCES

BARONE, M. R. and ROBINSON, A.R. (1972) Int. J. Solids Struct. 8, 1319-1338.
STERN, M. (1975) Recent Advances in Engr. Sc. 7, 359-368.
STERN, M. (1979) Int. J. Solids Struct. 15, 769-782.
SIH, G. C. PARIS, P. C. and ERDOGAN, F. (1962) J. Appl. Mech. 29, 78-82.
WILLIAMS, M. L. (1951) Proc. 1st U. S. Nat. Cong. of Appl. Mech. 352-329.
WILLIAMS, M. L. (1961) J. Appl. Mech. 28, 78-82.
WILSON, W. K. and THOMPSON, D. G. (1971) Engr. Frac. Mech. 3, 97.

Session VI
Applications

BEASY A BOUNDARY ELEMENT ANALYSIS SYSTEM

D.J. Danson

C.M. Consultants, Southampton, U.K.

1. INTRODUCTION

The theoretical basis for the Boundary Element Method (BEM) has been understood for some time [1]-[3] and for the case of linear potential and stress analysis, it can fairly be said that the method no longer lies in the realms of research. Nevertheless, there is a distinct lack of commercially available software making use of the method, particularly when compared with the large number of available Finite Element Systems [4]. Good commercial software must exhibit the following criteria.

i) Accuracy and reliability. Programs must provide accurate results for all the types of problem they are intended to solve. It is no use producing results in which the designer cannot have confidence.
ii) Ease of use. This not only refers to the necessity of adequate pre- and post-processing but also to the fact that Boundary Element software will be used by engineers who do not have a thorough understanding of the method.
iii) Ease of maintenance and updating. This may seem of more importance to the software house supplying the program than to the end user. However, it is important for the user to have any bugs corrected quickly and this is very difficult if the program is written using convoluted and badly documented code. It should be remembered that the person rectifying a bug or introducing a new feature is often not the same person who wrote the original program.
iv) Efficient use of memory. This is important in so far as clients often wish to mount the program on relatively small machines and efficient use of memory enables larger problems to be solved. However, they often use a machine which is already in use running Finite Element programs and since the BEM tends to use considerably less memory size is rarely a problem.
v) Efficient code to enable fast program execution. This,

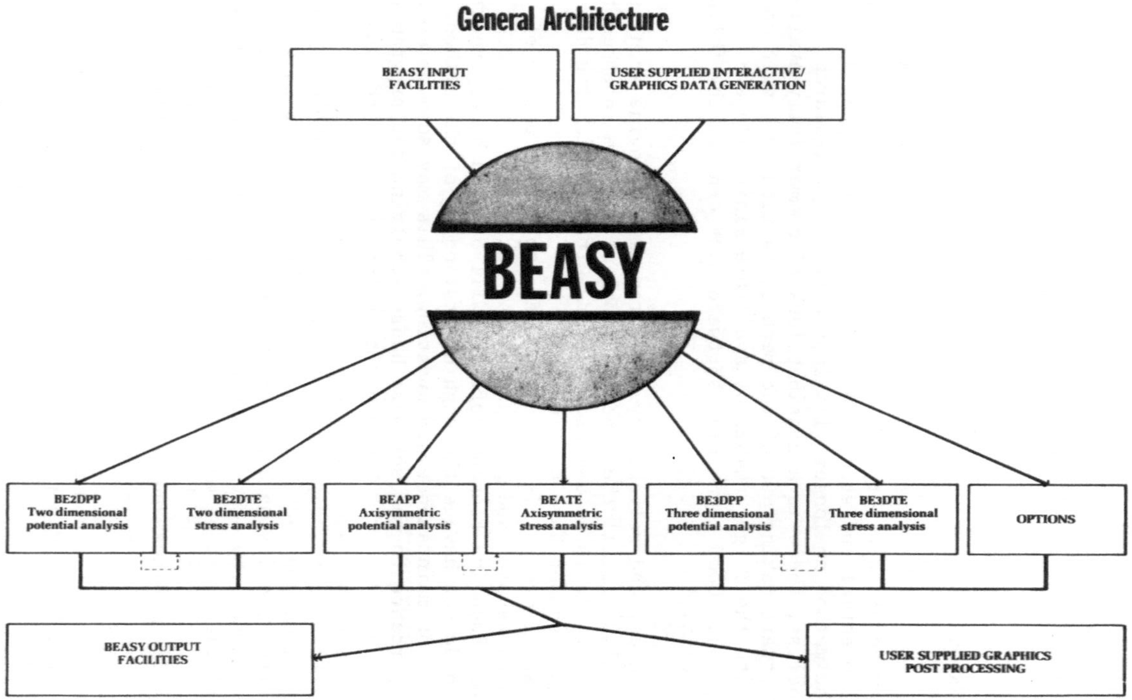
General Architecture
BEASY INPUT FACILITIES
USER SUPPLIED INTERACTIVE/ GRAPHICS DATA GENERATION
BEASY
BE2DPP Two dimensional potential analysis
BE2DTE Two dimensional stress analysis
BEAPP Axisymmetric potential analysis
BEATE Axisymmetric stress analysis
BE3DPP Three dimensional potential analysis
BE3DTE Three dimensional stress analysis
OPTIONS
BEASY OUTPUT FACILITIES
USER SUPPLIED GRAPHICS POST PROCESSING

Figure 1

in the author's opinion, is becoming much less important with the decline in the costs of CPU time. Above all, clarity of coding should not be sacrificed to execution speed as this seriously compromises ease of maintenance. Moreover, only those parts of the code which are heavily used by the program should be optimized. It is a complete waste of time optimizing parts of the code only used once during a run.

2. SPECIFICATION

The general architecture of BEASY is shown in Fig. 1.

As can be seen from Fig. 1 BEASY consists of six independent modules for the solution of problems in potential theory and linear isotropic stress analysis. The box labelled OPTIONS covers various more advanced applications of the BEM.

Element Types

BEASY uses constant, linear and quadratic non-conforming boundary elements. These are illustrated in Fig. 2.

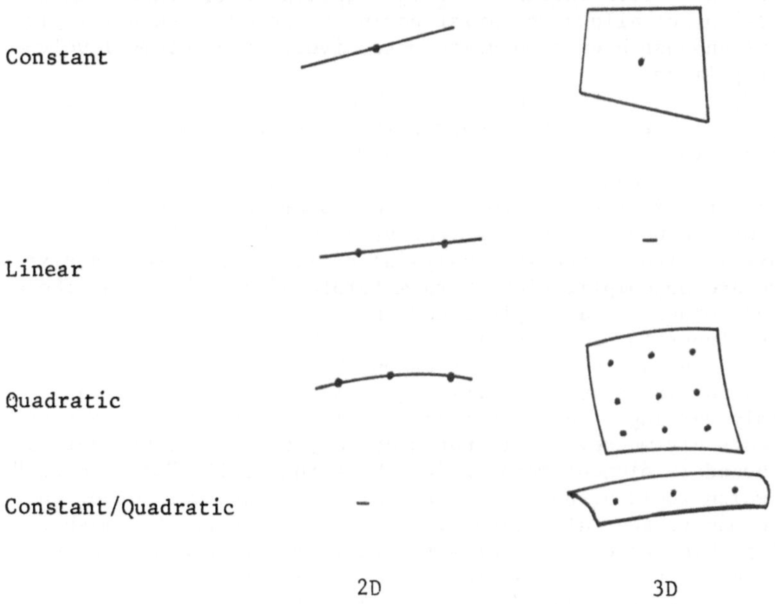

Figure 2 BEASY's Elements

The constant and linear elements have linear geometry described by linear shape functions and the quadratic and constant/quadratic elements have fully quadratic geometry described by quadratic shape functions. Actually, the author can see no theoretical objection for describing the geometry of linear elements with quadratic shape functions and vice-versa and it may be decided to allow the user complete flexibility in the choice of geometry and order of element in the next release of BEASY. At the moment the 3D modules do not contain a linear element since most of the time users are content to use either constant or quadratic elements. This linear element will be added, together with triangular elements, in the next release of BEASY.

The 3D quadratic element is a complete bi-quadratic element containing nine nodes as opposed to the more common incomplete 8-node quadrilateral. The reason for choosing a 9-node element is not only because it contains all the terms for a full bi-quadratic expansion but because, remembering that the classical BEM is a collocation technique with the nodes as collocation points, the use of a 9-node elements gives rise to an even pattern of collocation points. The necessity for this is borne out by the fact that the 9-node element behaves much better than the 8-node one particularly in problems where a change in the type of boundary condition produces a singularity in the flux (for potential problems) or in the stresses (in elasticity problems).

The constant/quadratic element which has three nodes and allows the system variables to vary quadratically in one direction but allows no variation in the other is surprisingly useful and has been used quite extensively on problems involving pipelines.

Non-conforming elements have been chosen not because the author believes they are inherently superior to conforming elements (he doesn't), but for the practical reason that they are very much easier to handle in such a way as to provide maximum flexibility to the user. Nodes are always on a smooth part of the boundary so that the value of the C matrix is always 1/2 (for potential analysis) or $\frac{1}{2}\delta_{ij}$ (for stress analysis). There are no complications with multiple nodes. Handling these on the corners of a single zoned problem is bad enough but the coding tends to become horrendous when handling problems containing several zones or sub-regions. It is very easy to mix non-conforming elements putting quadratic elements in areas of rapidly varying stresses (or potential) and linear or constant elements elsewhere. It is not necessary for the points defining the boundary element mesh, which the author calls "mesh points", and which would be nodes if conforming elements were used, to be common to several elements. Fig. 3 shows part of a mesh for a 3D problem which is a perfectly valid mesh for BEASY but would be difficult to handle if conforming elements were used. There are no problems of "fanning out" elements from areas of high element density.

Applied Loads and Boundary Conditions

The boundary conditions allowed for in the potential modules are potential or flux density prescribed or a linear relation between the two (generally called heat transfer boundary con-

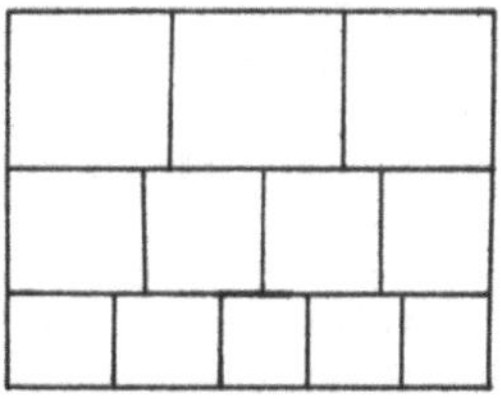

Figure 3 Part of a typical BE mesh

dition as it is a very common boundary condition in thermal analyses). Non-linear boundary conditions as specified by clients for a particular application are easily modelled and this has been done successfully in the field of cathodic protection where the potential and flux (current) density on the cathode are related by the polarization curve. The 2D and 3D modules can also model concentrated point and line sources which is a useful feature for many applications.

The stress analysis modules allow for prescribed displacement or traction or spring boundary conditions. The boundary conditions may be entered either in the global coordinate system or in a local system, one of whose axes always coincides with the normal to the boundary surface. This local system is not only useful when applying boundary conditions of a single type (e.g. when specifying an internal pressure in a spherical pressure vessel) but is absolutely essential when specifying a mixed type of boundary condition such as sliding, where the displacement is prescribed normal to the boundary and the traction is prescribed tangential to the boundary. Problems with gravitational or rotational loading and problems where the stresses are due to steady state thermal loading may also be analysed. These problems involving body forces may be analysed without having to divide the domain into integration cells [5]. In the case of thermal analyses it is necessary to first solve the potential problem to obtain the temperatures and fluxes at the boundary nodes and the temperatures at any internal points. This information is then fed to the stress analysis module which then calculates the thermal stresses. Exactly the same data file may be used for both analyses.

Analysis Steps

BEASY carries out a typical Boundary Element analysis in six distinct steps. (see Table 1). The analysis may be started or stopped at each step.

Step	Input Required	Computation	Output Generated
1	Data file	Check on data	
2	Data file	Formation of influence matrices	Influence matrices
3	Data file Influence matrices	Application of the type of boundary conditions to form the system matrix	System matrix $\underset{\sim}{A}$ (Part of system of equations $A \underset{\sim}{x} = \underset{\sim}{d}$)
4	Data file System matrix	Reduction of the left hand side of the system of equations	Reduced left hand side
5	Date file Reduced left hand side	Application of the magnitude of the boundary conditions to form the right hand vector $\underset{\sim}{d}$ Reduction and back-substitution to obtain the boundary solution	Boundary solution
6	Data file Boundary solution	Computation of results at internal points	Results at internal points

Table 1

BEASY carries out some fairly simple checks on the data. If clients request the addition of more sophisticated checks then these can be provided.

The influence matrices $\underset{\sim}{H}$ & $\underset{\sim}{G}$ are the matrices in the equation

$$\underset{\sim}{H}\,\underset{\sim}{u} = \underset{\sim}{G}\,\underset{\sim}{p} + \underset{\sim}{b}$$

where $\underset{\sim}{u}$ is the vector of nodal potentials/displacements
$\underset{\sim}{p}$ is the vector of nodal fluxes/tractions
and $\underset{\sim}{b}$ is a vector resulting from the influence of sources/body forces

In a typical problem the formation of the influence matrices H and G takes a large proportion of the total run time. These matrices $\underset{\sim}{H}$ & $\underset{\sim}{G}$ are stored on disc.

Application of the type of boundary conditions enable the elements of the matrices to be rearranged so that we may write

$$\underset{\sim}{A}\,\underset{\sim}{x} = \underset{\sim}{B}\,\underset{\sim}{f} + \underset{\sim}{b}$$

where $\underset{\sim}{A}$ is the system matrix
$\underset{\sim}{x}$ is the vector of unknowns
$\underset{\sim}{B}$ is the complementary matrix
and $\underset{\sim}{f}$ is a vector of prescribed boundary values.

The system matrix $\underset{\sim}{A}$ and the complementary matrix $\underset{\sim}{B}$ are stored on disc.

Gauss elimination is now applied to the system matrix $\underset{\sim}{A}$ so that it may be expressed as the product of an upper and lower triangular matrix

$$\underset{\sim}{L}\,\underset{\sim}{U}\,\underset{\sim}{x} = \underset{\sim}{B}\,\underset{\sim}{f} + \underset{\sim}{b}$$

This reduction of the left hand side also consumes a large proportion of the total run time for a typical problem. The matrices $\underset{\sim}{L}$ and $\underset{\sim}{U}$ are stored on disc.

Not until this stage is the vector $\underset{\sim}{b}$ due to sources/body forces evaluated and the right hand side vector $\underset{\sim}{d}$ calculated from the equation

$$\underset{\sim}{d} = \underset{\sim}{B}\,\underset{\sim}{f} + \underset{\sim}{b}$$

We now have the equation

$$\underset{\sim}{L}\,\underset{\sim}{U}\,\underset{\sim}{x} = \underset{\sim}{d}$$

which may be solved for the vector of unknowns $\underset{\sim}{x}$ by the normal right hand side reduction/backsubstitution process.

BEASY uses an out-of-core block solver based on the one published by Das [6]. This enables large problems to be solved efficiently on quite small machines.

The final step calculates the values at internal points using the boundary solution just obtained.

The reason for carrying out the analysis in the manner described above is to minimize the run times required for repeated analyses. The influence matrices are dependent only on the mesh geometry. Once the user has got that right he should never need to form the influence matrices more than once. The analysis for subsequent boundary conditions need only be restarted at step 3 if the type of boundary condition has been altered. e.g. a prescribed displacement is specified where previously the traction was specified. If only the magnitude of the boundary condition is altered, or new sources/ body forces added then it is not necessary to repeat steps 3 and 4 but the run may be restarted at step 5. Thus both the time consuming steps of forming the influence matrices and of decomposing the system matrix are avoided.

Once the boundary solution is obtained the solution at internal points is calculated by a fairly simple procedure. No further equation solution is required. If after looking at the results the user decides results are required at some more internal points then only step 6 needs to be repeated.

Symmetry

Symmetry is handled by the simple expedient of reflecting the boundary about the plane of symmetry and continuing the boundary integration around the reflected part of the structure. By making use of the fact that not only the geometry but also the potentials fluxes, displacements and tractions are symmetric the number of equations is not increased. Indeed it is reduced by the fact that no elements are required on the plane of symmetry. The time taken to compute the influence matrices is increased but the extra cost is nearly always outweighed by the reduced number of man-hours taken up in data preparation.

Output

The boundary values output by the potential modules are potential and the component of flux density crossing the boundary. These may be output at the nodes and/or at the mesh points depending on what the user has requested. The flux density crossing the boundary is integrated over each element to give the total flux flowing through the element. Elements may be given a label so that the flux flowing through elements with the same label is added together but keeping positive and negative fluxes separate so that the total flux flowing through a particular area of boundary may be automatically calculated. It is often convenient to use the boundary condition type code as an element label and BEASY enables the user to do this. Finally the total flux flowing out of each zone and the total flux flowing into the zone are printed out. The difference between these two figures should be the total strength of the sources within the zone. This provides the user with a quick and simple check on the accuracy of the calculation.

The boundary values output by the stress analysis modules are displacements and tractions. As with the potential modules, these may be printed out at the nodes and/or at the mesh points. The tractions are integrated over each element to provide the total force on each element. These forces may be added together in a manner analogous to the fluxes in the potential modules thus providing the user with a check on equilibrium. Clearly the net force acting on the boundary of a body must be equal and opposite to the total body forces acting on that body.

When results are output at mesh points this is done on an element by element basis. The user may also request the calculation of mesh point averages. These are calculated by adding together the values of the variables in question for each element meeting at that mesh point and dividing by the number of elements meeting there. If the contribution from a particular element differs from the average by more than 20% then an asterisk is printed alongside the average to warn the user that something may be wrong. The reason for the discrepancy is usually either

i) The numbers concerned are all very close to zero. Thus the error is still small when compared with a typical problem value.

or

ii) There is a discontinuity in either the fluxes or tractions due to the geometry or boundary conditions so the values should not have been averaged in the first place. This problem can be avoided by defining two coincident mesh points at such a discontinuity. Notice that if conforming elements had been used it would have been necessary, not merely desirable, to put two nodes at such a point.

or

iii) The mesh needs refining.

After solution by the BEM the boundary tractions are known. Unfortunately, most users require the complete stress tensor on the boundary. There are two approaches to this problem. One is to take the Somigliana Identity for stresses to the boundary. This has two disadvantages.

i) The kernels in the boundary integral expression have a high order of singularity which has to be coped with. This is a particular problem with axisymmetric analysis where the kernels are extremely messy.

ii) It is quite expensive on CPU time.

The alternative method is to find the missing components of the stress tensor by differentiating the displacements in the plane of the element in the same way as it is done with finite elements. This is the method adopted by BEASY. This has the effect, as with finite elements, of reducing the order of variation on an element by one. e.g. a quadratic element can only represent linear variation of boundary stresses. Hence boundary stresses are not obtainable with constant elements, only centroidal (or element) stresses are obtainable with linear elements, and the nodal stresses on quadratic elements will display a linear variation. As with finite elements the boundary stresses will be less accurate than the displacements, although the tractions show the same degree of accuracy as the displacements.

BEASY also outputs the potentials and fluxes/displacements and stresses at specified internal points.

BEASY options

The options in BEASY cover a number of more advanced BEM applications. They each have only one element type and cannot handle zoned problems. They are:-

BETA2D Solves the diffusion equation in two dimensions. Its main application is in time-dependent thermal calculations. Uses a conforming linear element.

BETAAX As BETA2D but for axisymmetric geometries.

BEPLAS Solves problems in elasto-plasticity in two dimensions. Uses a conforming linear element.

BEREPOT Solves the scalar/potential wave equation in three dimensions. Has been used to analyse the transient response of liquids due to an explosion at a point [7]. Uses a non-conforming complete bi-quadratic quadrilateral element.

3. ACCURACY

The BEM is extremely accurate. However, there are three possible sources of error.

i) Round-off error.

Machines cannot handle an infinite number of decimal places. Round off error is only a real problem when the system of equations is ill-conditioned. An ill-conditioned system of equations usually results from physical ill-conditioning inherent in the problem. Fortunately most problems engineers have to deal with tend to be well conditioned.

ii) Modelling error.

If a particular problem is being modelled using quadratic elements (say), then the boundary potentials/displacements, and fluxes/tractions are constrained to vary in a piecewise quadratic manner. If the true values cannot be described in this manner then some inaccuracy will result. This error reduces rapidly with refinement of the mesh. In other words, convergence is rapid. Another source of modelling error is geometrical. For example, parabolic elements are normally used to model circular boundaries and the difference will lead to some inaccuracy. Errors due to geometrical inaccuracy will also reduce rapidly with refinement of the mesh.

iii) Integration errors.

In the BEM one has to perform integrations of the type

$$\int_{\Gamma e} f(r)\ln\frac{1}{r}\,\underset{\sim}{\phi}\,d\Gamma\ ,\quad \int_{\Gamma_e} g(r)\,\frac{1}{r}\,\underset{\sim}{\phi}\,d\Gamma\ ,\quad \int_{\Gamma_e} h(r)\,\frac{1}{r^2}\,\underset{\sim}{\phi}\,d\Gamma$$

where Γ_e is the part of the boundary modelled by element e
r is the distance between the source point and the field point (see Fig. 4). The field point is on the element e.
$f(r), g(r), h(r)$ are various well behaved functions of r
$\underset{\sim}{\phi}$ are the element interpolation functions

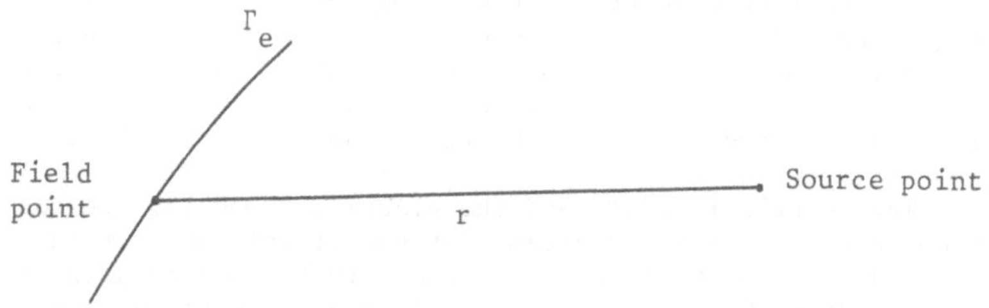

Figure 4

These integrations are normally performed using Gaussian Quadrature. An n-point Gauss scheme will exactly integrate a polynomial of crder 2n-1. However, the integrands encountered in the BEM are not, as they are in the Finite Element Method (FEM) finite polynomials. Moreover, the author's experience indicates that it is essential to perform these integrations accurately if the method is to produce consistently accurate results. The problem of performing these integrations accurately has been recognized previously [8]. If the integrations are not performed accurately then mesh refinement will have little effect on the results. It is the author's opinion that convergence studies showing poor convergence using the BEM are almost certainly due to insufficient care being taken over the integrations. Various techniques have been used in BEASY to ensure that the integrations are evaluated accurately.

i) When the source point is not on the field element

BEASY compares the distance from the source point to the field element with the size of the field element and uses this information to select an appropriate integration scheme. BEASY has a selection of Gauss schemes to choose from ranging from a one point to a 32 point scheme. Fortunately the latter scheme is rarely called upon! In BE2DPP it was found possible to do all the integrations analytically, even for a quadratic element, provided the element was flat. If the element is curved the integration may be performed analytically over a flat element tangential to the actual element with an expression for the difference between the actual and flat elements integrated numerically. When the element is a long way from the source point analytical integration was found to break down as it involves computing the difference between two very similar numbers and numerical integration was used. This technique was used both for the boundary solution and for the solution for potentials and fluxes at internal points. The result is an extremely accurate and fast module.

BE3DPP employs straightforward Gaussian Quadrature unless the source point is quite close to the field element. It was found possible to integrate in a radial direction analytically over a constant element and then numerically in the other direction. Thus it was found worthwhile to use this semi-analytical scheme over a constant element tangential to the actual element and to then integrate the expression representing the difference between the two elements numerically.

The kernels in BEAPP and the elasticity modules are too complicated to make any attempt at analytical integration worthwhile. For field elements close to the source point a high order of Gaussian Quadrature is used (the sledgehammer technique).

ii) When the source point is on the field element

For modules BE2DPP and BE3DPP techniques similar to the ones described above were used. These analytical and semi-analytical techniques are actually easier for this case than for the case when the source point is not on the field element. For the remaining modules special numerical integration schemes for integrating functions containing singularities were used [9].

4. EXAMPLES

1. Uniaxial Compression (2D)

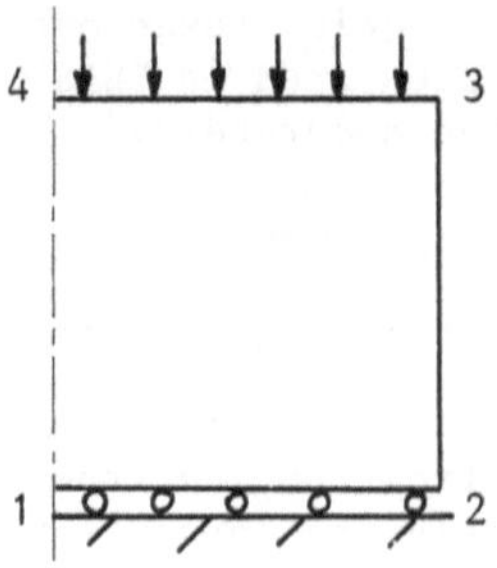

$E = 2.1 \times 10^8$

$\nu = 0.35$

Plane stress analysis

Figure 5

This is a very trivial example using three linear elements. The purpose of including it is to demonstrate that since the the exact solution is linear the BEM ought to give an

ZONE 1

ELEMENT	NODE	COORDINATES		DISPLACEMENTS		TRACTIONS	
		X	Y	U	V	T(X)	T(Y)
1	1	250.0	0.0000E+00	4.1667E-06	0.0000E+00	0.0000E+00	10.00
	2	750.0	0.0000E+00	1.2500E-05	0.0000E+00	0.0000E+00	10.00
2	3	10000.	250.0	1.6667E-05	-1.1905E-05	0.0000E+00	0.0000E+00
	4	10000.	750.0	1.6667E-05	-3.5714E-05	0.0000E+00	0.0000E+00
3	5	750.0	1000.	1.2500E-05	-4.7619E-05	0.0000E+00	-10.00
	6	250.0	1000.	4.1664E-05	-4.7619E-05	0.0000E+00	-10.00

ELEMENT	FORCES	
	F(X)	F(Y)
1	0.00000E+00	10000.
2	0.00000E+00	0.00000E+00
3	0.00000E+00	-10000.

Figure 6 Actual output for Example 1

exact solution with a minimum number of linear elements. Fig. 6 shows the relevant part of the computer output which demonstrates that the exact solution is indeed obtained.

2. Hollow cylinder subjected to a temperature difference between its inside and outside faces

This problem was run on BE2DPP using 4 elements, on BEAPP using 6 elements and on BE3DPP using 3 elements. Details of the cylinder are as follows.

Inner radius	$a = 30$ mm
Outer radius	$b = 80$ mm
Temperature on inside face	$T_i = 100°C$
Temperature on outer face	$T_o = 0°C$
Conductivity	$k = 60\ W\ m^{-1}\ °K^{-1}$

The boundary element meshes for this problem are shown in Fig. 7 and the results for a line of internal points at 5 mm centres are tabulated in Table 2.

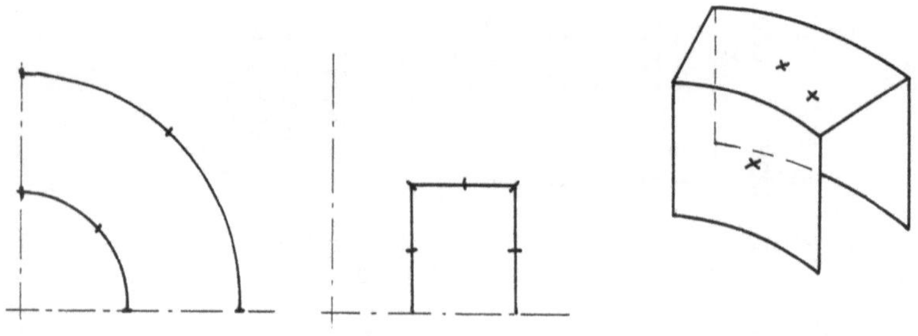

BE2DPP BEAPP BE3DPP

Figure 7 Boundary Element Meshes for Example 2

Radius [mm]	Exact	Temperature [°C] BE2DPP 4 Elements	BEAPP 6 Elements	BE3DPP 3 Elements
35.0	84.2840	84.235	84.284	83.861
40.0	70.6698	70.633	70.670	70.083
45.0	58.6613	58.632	58.662	57.976
50.0	47.9192	47.895	47.920	47.228
55.0	38.2019	38.180	38.203	37.567
60.0	29.3306	29.310	29.332	28.755
65.0	21.1699	21.147	21.171	20.704
70.0	13.6142	13.549	13.615	13.262
75.0	6.5800	6.5043	6.5804	6.3950

Radius [mm]	Exact	Flux [kW/m²] BE2DPP 4 Elements	BEAPP 6 Elements	BE3DPP 3 Elements
35.0	0.17478	0.17448	0.17477	0.17242
40.0	0.15293	0.15283	0.15293	0.15447
45.0	0.13594	0.13587	0.13594	0.13602
50.0	0.12235	0.12230	0.12234	0.12204
55.0	0.11122	0.11119	0.11122	0.11055
60.0	0.10195	0.10196	0.10196	0.10094
65.0	9.4118E-02	9.41509E-02	9.41148E-02	9.2784E-02
70.0	8.7389E-02	8.74867E-02	8.73930E-02	8.5554E-02
75.0	8.1564E-02	8.17049E-02	8.15672E-02	7.9317E-02

Table 2 Temperature and fluxes in a hollow cylinder

The results from the thermal analysis of BE2DPP were used to calculate the thermal stresses at the same internal points using BE2DTE. Additional constants required for the stress analysis were:

Young's Modulus	210,000 N/mm^2
Poisson's Ratio	0.33
Coefficient of linear expansion	11×10^{-6} /°C

The results for the radial and hoop stresses obtained in the boundary element analyses are given in table 3.

Radius	Exact Solution [N/mm²] Radial Stress	Hoop Stress	BE2DTE with 5 elements Radial Stress	Hoop Stress
35	-26.1264	-145.127	-25.736	-144.90
40	-37.1991	-87.1149	-36.995	-86.998
45	-40.1798	-42.7314	-40.046	-42.661
50	-38.6011	-7.27359	-38.499	-7.2396
55	-34.3827	22.0115	-34.292	22.012
60	-28.6223	46.8373	-28.534	46.803
65	-21.9727	68.3243	-21.872	68.257
70	-14.8339	87.236	-14.696	87.149
75	-7.45664	104.111	-7.2005	104.04

Table 3 Radial and Hoop stresses under Thermal Loading

Figure 8 shows the boundary element mesh required for the analysis of the impressed current corrosion protection system for an oil platform to be operated in the North Sea. The requirement is to solve Laplace's equation in the infinite domain of the seawater and the designer wishes to know the current density on the cathode, which is the hull of the platform. Since the boundary of the seawater is the platform itself this example is ideally suited to the BEM. There are three planes of symmetry in the problem. Two of them are genuine planes of symmetry. The mesh shown in Figure 8 represents only one quarter of the total problem. The third plane of symmetry is put on the sea surface to enforce the boundary condition $\partial u/\partial n = 0$ on the sea surface, which saves having to put elements there. This is the reverse of the normal procedure used in the FEM where a plane of symmetry is represented by the boundary condition $\partial u / \partial n = 0$. There are also some elements not shown in Figure 8 a large distance from the platform. These were put there to enforce the boundary condition required by the designer of $\partial u/\partial n = 0$ at infinity. Without these elements the BEM would automatically enforce a boundary condition of $u = 0$ at infinity which is not what the designer wanted. It should be noted that this mesh at "infinity" is not connected to the mesh on the structure itself. The smallest elements used in the problem had dimensions of 460 × 45 mm (on the anodes) and the largest had dimenions of 30 × 30 m (on the "infinite" boundary). The problem was run on an earlier version of BE3DPP which only had constant elements, and was analysed using 653 elements. If it were to be re-analysed today the number of elements could be considerably reduced by using quadratic elements. Several analyses were performed as the designer not only wanted results for the complete system but also wanted to know what would happen if various combinations of anodes were switched off. The ability to restart the analysis at various intermediate points was very useful here. There is little point in presenting any results

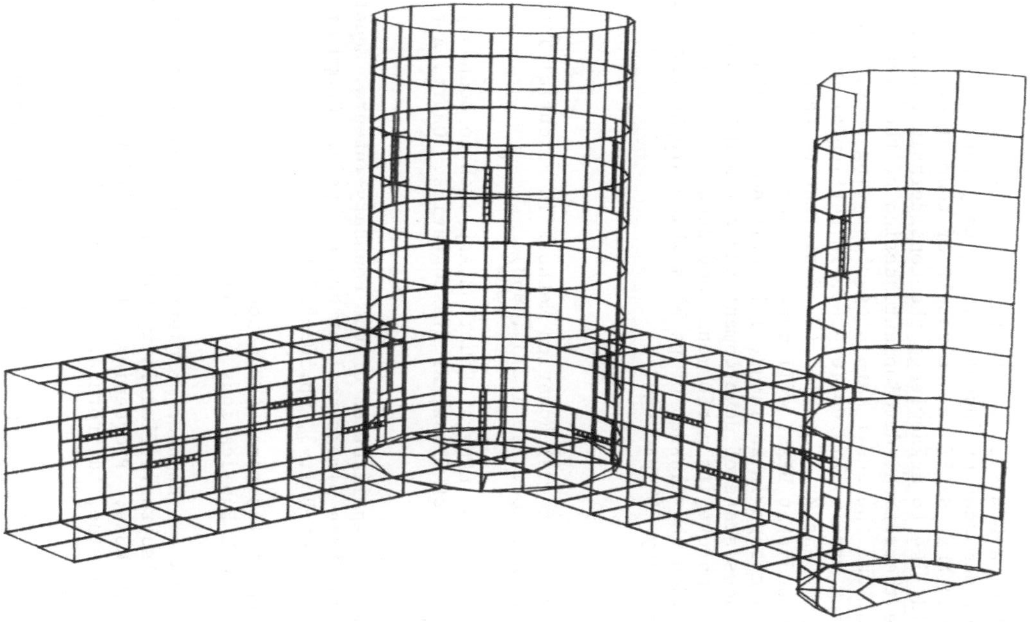

Figure 8 BE Mesh for an Oil Platform

in this paper as there is nothing with which to compare them. However, the designer assured us that the results were very much as he expected and saw no reason to doubt them. The analyses were run on a PDP11/23 with a total physical memory of 200 KBytes which demonstrates how large 3D BEM analyses may be performed on quite small machines.

4. CONCLUSIONS

Development of a BEM system with the reliability, flexibility and accuracy demanded by industry is no easy task. It should be emphasized that BEASY is the result of team effort. The author would like to acknowledge the considerable contributions made to the theoretical development, design of the structure of the modules, and the sheer hard slog of writing code, testing and debugging made by his colleagues, in particular Dr. Adey, Dr. Collington and Mr. Tugwell. We feel that we have succeeded in our aims and that BEASY will make a major contribution to analysis and design in an industrial environment. BEASY is in a continuous state of development and our aim is to continually to improve its performance and accuracy and to gradually widen the classes of problem which are better suited to BEASY than any FEM package.

REFERENCES

1. G.T. SYMM. Integral Equation Methods in Potential Theory, I, Proc. Roy. Soc. Ser A 273 (1963).

2. F.J. RIZZO. An Integral Equation Approach to Boundary Value Problems of Classical Elastostatics. Quart. Appl. Math. 25 (1967).

3. T.A. CRUSE. Application of the Boundary-Integral Equation Method to Three Dimensional Stress Analysis. Computers and Structures Vol. 3, pp.509-527 (1973).

4. C.A. BREBBIA (Editor) A Handbook of Finite Element Systems. CML Publications (1981).

5. D.J. DANSON. A Boundary Element Formulation of Problems in Linear Isotropic Elasticity with Body Forces. Boundary Element Methods. Proceedings of the Third International Seminar, Irvine, California, July 1981. C.A. Brebbia (Editor).

6. P.C. DAS. A Disc Based Block Elimination Technique used for the Solution of Non-Symmetrical Fully Populated Matrix Systems Encountered in the Boundary Element Method. Recent Advances in Boundary Element Methods. C.A. Brebbia (Editor) Pentech Press 1978.

7. P.H.L. GROENENBOOM. The Application of Boundary Elements to Steady and Unsteady Potential Fluid Flow Problems in Two and Three Dimensions. Boundary Element Methods. Proceedings of the Third International Seminar, Irvine, California, July 1981. C.A. Brebbia (Editor)

8. T. ANDERSSON. Boundary Element Integration Technique. A Survey on Recent Developments. Report LiTH-IKP-R-190 Linköping University Institute of Technology, Dept. of Mech. Eng. Sweden, 1981.

9. A.H. STROUD and D. SECREST. Gaussian Quadrature Formulas. Prentice Hall 1966.

THE BOUNDARY ELEMENT METHOD IN AN INDUSTRIAL ENVIRONMENT

Gero Kuich, Consulting Engineer

Brown, Boveri & Cie., Baden/Switzerland

ABSTRACT

The use of the Boundary Element Method (BEM) in an environment that was before oriented mainly towards the Finite Element Method is shown on the example of introducing the BEASY-System at Brown Boveri in Switzerland.

The CAD-programs and the pre- and postprocessors used in connection with the Finite-Element-programs could be adapted with little effort to the generation of geometry and the plotting of results for the Boundary Element Method.

The great advantages of the BEM are shown on several practical applications, especially in connection with an existing program for contact calculations. Also comparisons with results of Finite Element calculations are presented.

THE BOUNDARY ELEMENT METHOD

The application of the Boundary Element Method (BEM) for temperature and elasticity problems (1) has only in the very last years come out of the state of research into a form usable by industry.

The main difference to the Finite Element Method (FEM), which is already well established for many years (2), is the fact, that only the boundary has to be divided into elements and the boundary values are assumed to vary according to the shape function of the elements.

In steady-state temperature analysis the boundary values are temperature and heat flux, in stress analysis they are displacements and tractions (pressure) and always one of both must be prescribed and the other is unknown.

Starting from the conversion of volume integrals into a boundary integral formulation and a known 'Fundamental Solution' (influence of a source point on a field point), the influence matrices for all the boundary values can be calculated.

Then these matrices are rearranged, so all prescribed values are on one side and the unknowns on the other, leading to a set of equations in terms of unknown nodal values, with the nodes only on the boundary.

These equations can be solved by any standard procedure to get the unknown boundary values. In addition the results can be calculated at any internal points, even in restarting the calculation only with the known boundary solution and the geometry.

It is quite obvious that in this formulation loads may only act on the boundary, but it is also possible to convert volume integral formulations of body forces to surface integrals. So gravitational forces, centrifugal loads and steady-state temperature distributions can be considered as load for stress analysis.

In spite of the advantages of the BEM many engineers are a bit reluctant to use such a new method, especially if they are used to the Finite Element Method for many years.

This lecture shows that the introduction of the BEM can be done very easily in an industrial environment oriented mainly towards the FEM and all the advantages of the BEM can be used beside the FEM programs.

ADVANTAGES AND DISADVANTAGES OF THE BEM COMPARED TO FEM

As seen from the user's point of view one can summarize the following advantages and disadvantages of the Boundary Element Method compared to the Finite Element Method.

Advantages:

- Only boundary to discretize (1 dim less than FEM)
- Not so much experience necessary as in FEM to decide for the optimium element size
- Coupling to Computer Aided Design programs much easier
- Accuracy much better with the same computer time, especially for stress concentrations
- Infinite problems easy to solve
- Numbering of nodes and elements not important
- Simple interpretation of results
- Results at internal points after the main calculation as requested by the user

Disadvantages:

- For nonlinear problems the inner region must be discretized
- Not economic if surface/volume large
- Material properties constant in zones
- No internal forces except gravity, centrifugal forces and steady-state temperature
- No advantages for beams, plates and shells

Conclusions:

The BEM will certainly not replace the FEM for structural analysis, but it is a valuable supplement for certain types of problems:

- 2D, axisymmetric and 3D, if ratio of volume/surface is large
- stress concentrations
- contact problems
- infinite problems

INTRODUCTION OF BEASY

It is not so easy to introduce a new method in an industrial environment which is already oriented towards the FEM.

But it is not necessary to repeat all work which had to be done to have a good pre- and postprocessing system for FEM.

As example we can see how the BEASY-System (3), developed by Computational Mechanics in Southampton was introduced at Brown Boveri in Switzerland, where a lot of FE-calculations are made using the FE-programs ASKA and ADINA.

Generation programs can be used to generate boundary elements instead of finite elements (Fig. 1). The shape of the elements will be the same only the order one less, so plane problems can be solved by generating 'FE-beams' around the region. Threedimensional problems can be generated by using twodimensional mesh generators like for plate and shell problems.

Also plotting of results can be done using the FE-postprocessors (Fig. 2). At least the most important geometry check and some result plotting is possible, but not all BEM-specific types of result plotting.

But it is not necessary to have all this pre- and postprocessors because also the BEASY-System has means of mesh-generation and results plotting (Fig. 3). It is only a great help in the first phase of introduction, because the FEM-user's are accustomed to their programs and like to use them in the usual way. Also in the first phase of introducing the BEASY-System not all of the possibilities were already contained in the BEASYP-programm.

A typical job-sequence can be seen in Fig. 4. Usually one starts with generating the boundary elements and checking the grid with the plot-program. Then the main job is running, yielding the boundary solution and perhaps results at some internal points.

If one wants to know results at more internal points afterwards one can make an easy restart, specifying only the coordinates of the desired internal points.

The interpretation of results can than be aided by plotting with the postprocessor program.

This job sequence is the same for temperature or stress calculations, only a different file is used. The temp-file or stress-file can either be the formatted BEASY-file or the unformatted file as used in the ASKA-database.

For a thermoelastic analysis the temperatures and heat fluxes from the temperature calculation are automatically read from the temp-file and used as input for the stress analysis.

EXAMPLES

Three typical examples for the application of the BEM are shown and comparisons to FEM-calculations are presented:

Temperature calculation for a part of a turbine rotor

The temperature calculation with heat transfer boundary condition from node 8 to 65 (Fig. 5) showed very good coincidence with a FE-calculation. Linear boundary elements were used and the temperature was also calculated at 14 internal points.

Fig. 6 shows the temperature distribution from node 8 to 30.

Rotorshaft with different mesh sizes

The stresses in the notch of a rotorshaft were investigated using finite elements and boundary elements, both with different mesh sizes. This shaft was only used for these test calculations and does not represent a real design.

Using finite elements a very fine mesh (Fig. 7) was necessary to get exact values for all stress components (Fig. 8).

The much simpler boundary element discretization of Fig. 9 gave about the same results (Fig. 10) with about half of the computer costs and much more savings in mantime for data preparation.

Contact calculation

Fig.11 shows the FE-discretization for a three-dimensional contact-calculation of a blade root. This very coarse mesh could not give the right results at the notches, so additional calculations for the notches were necessary.

The same contact-program (4) as used with the finite elements can be used for boundary element calculations, because this program was already changed before from using contact forces to using contact pressures as unknowns for the contact calculation.

So contact calculations can be made in a much simpler way with boundary elements (Fig. 12), because the pressures are already the unknowns in the BEM-formulation.

Also coupling of FEM-meshes with BEM-meshes can be done with help of this contact program.

CONCLUSIONS

The introduction of a Boundary Element program like BEASY in an industrial environment is most easily done, if the boundary elements are treated like simple finite elements and the existing FE pre- and postprocessors can be used.

After the FEM-user's have tried examples and made comparisons with FEM-results, they soon will be convinced of the advantages of the BEM for certain types of problems. Especially the easy preparation of input data and interpretation of results will help the spreading of the BEM in the future.

With the steadily increasing use of Computer Aided Design programs another great advantage of the BEM will get more and more importance. One can use the basic geometry, generated by CAD-programs, directly as input to the BEASY-System and has only to add the boundary conditions for completing the input data.

In general one can say that the BEM will certainly not replace the FEM for temperature and stress analysis, but it will be a valuable supplement for certain types of problems, like stress concentrations, infinite problems and contact problems.

REFERENCES

1. Brebbia, C.A. The Boundary Element Method for Engineers, Pentech Press, 1978

2. Zienkiewicz, O.C., The Finite Element Method in Engineering Science, McGraw Hill, 1971

3. BEASY User's Manual, Computational Mechanics, 1982 German version by FEMCAD AG, Oberrohrdorf, Switzerland

4. Kuich G., Berechnung allgemeiner Schrumpf- und Kontaktprobleme, FEM-Kongress Baden-Baden, 1975

FINITE-ELEMENT GENERIERUNG

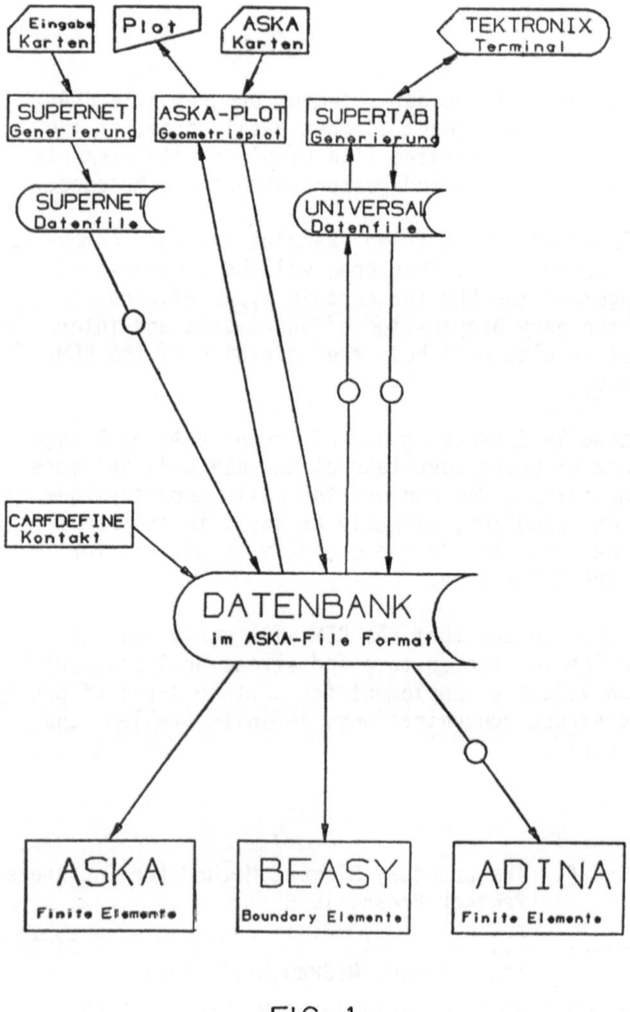

FIG. 1

FINITE-ELEMENT AUSWERTUNG

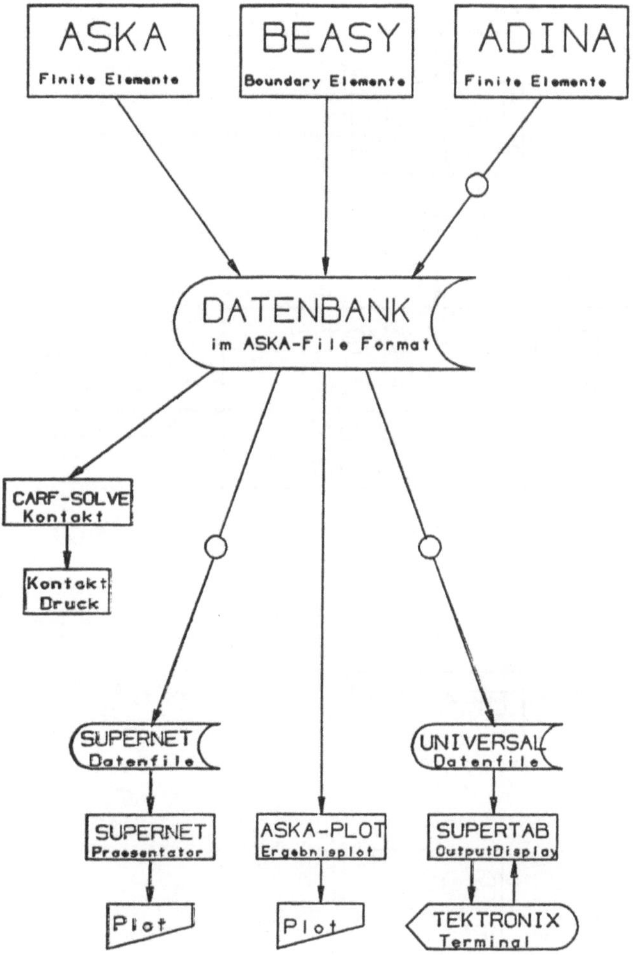

FIG. 2

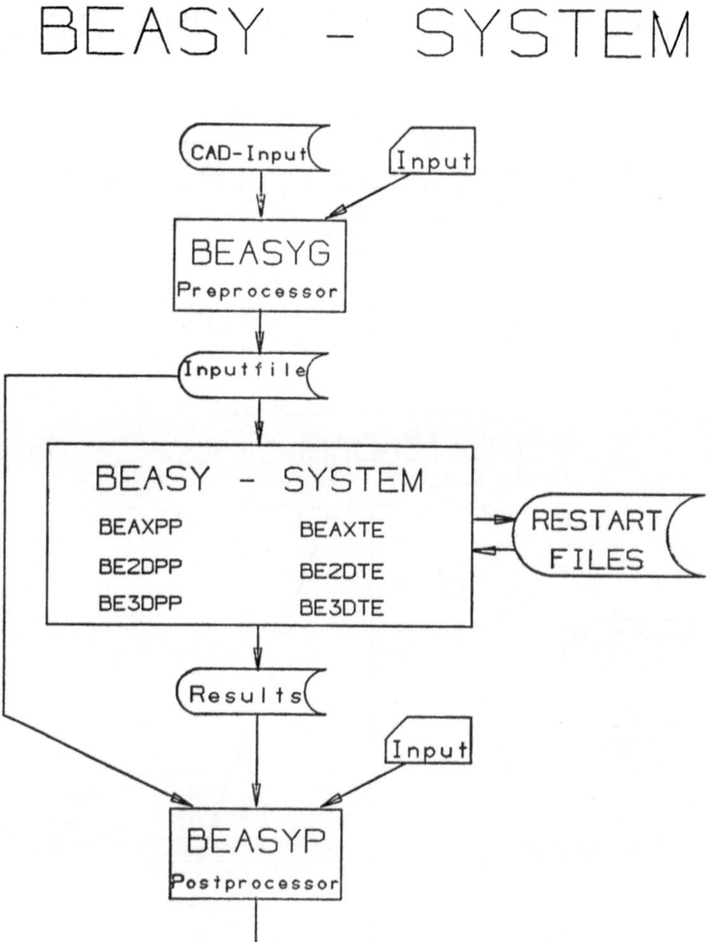
BEASY - SYSTEM
CAD-Input
Input
BEASYG
Preprocessor
Inputfile
BEASY - SYSTEM
BEAXPP
BE2DPP
BE3DPP
BEAXTE
BE2DTE
BE3DTE
RESTART
FILES
Results
Input
BEASYP
Postprocessor
PLOT

FIG. 3

TYPICAL JOB SEQUENCE

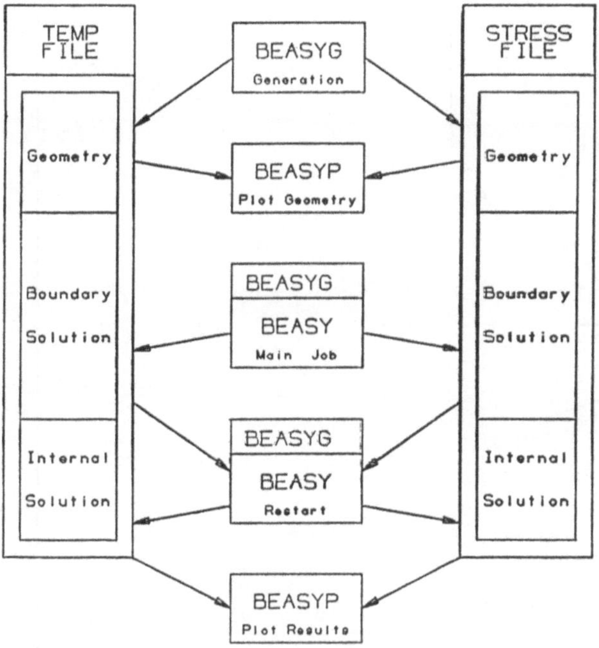

FIG. 4

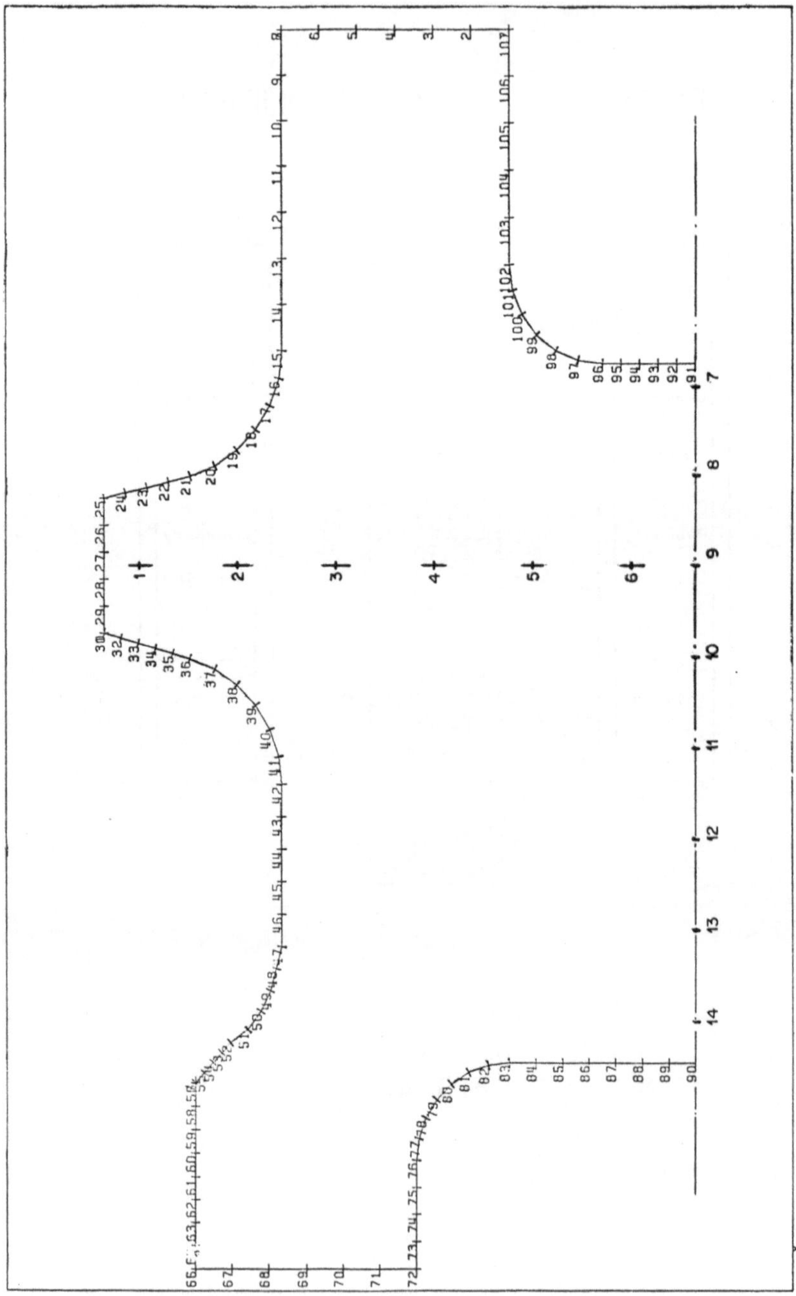

FIG. 5

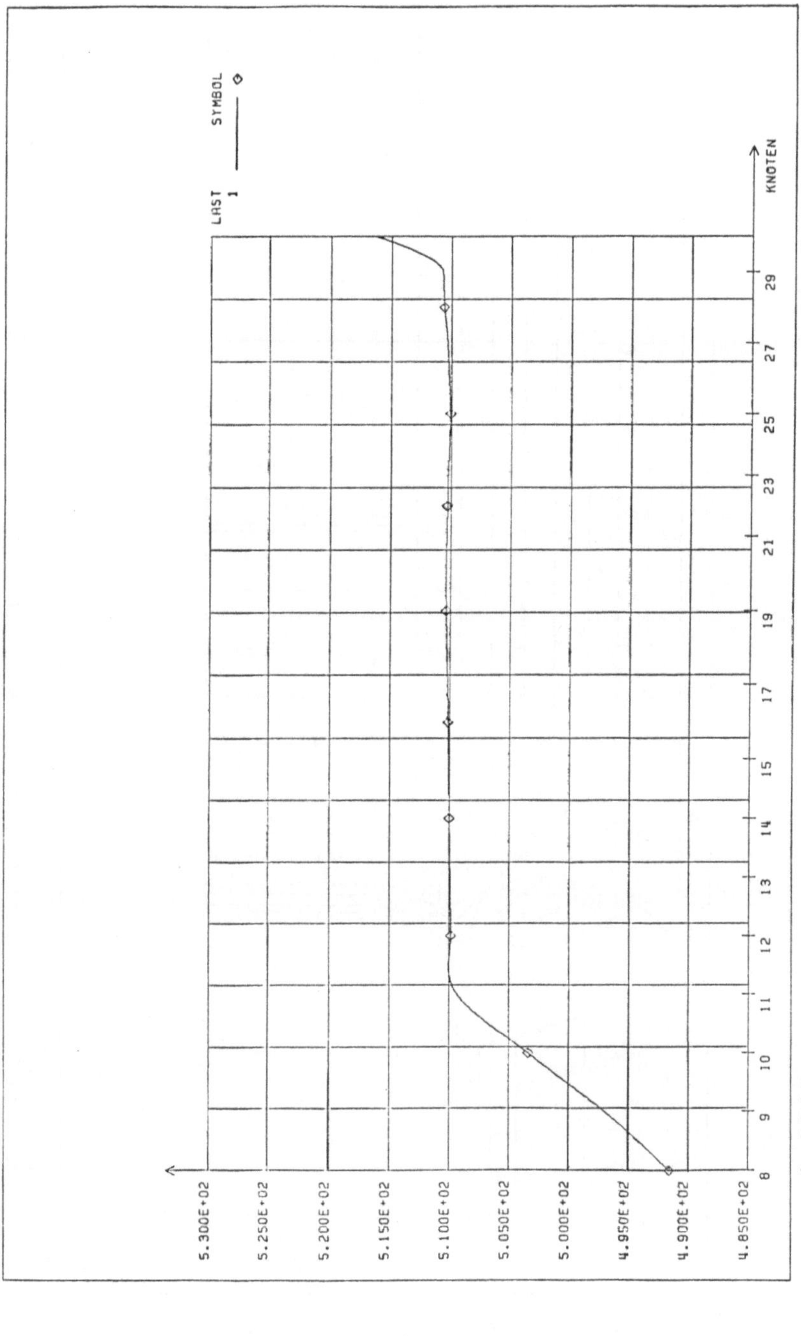

HD-ARAD

TEMP RK BESCH

FIG. 6

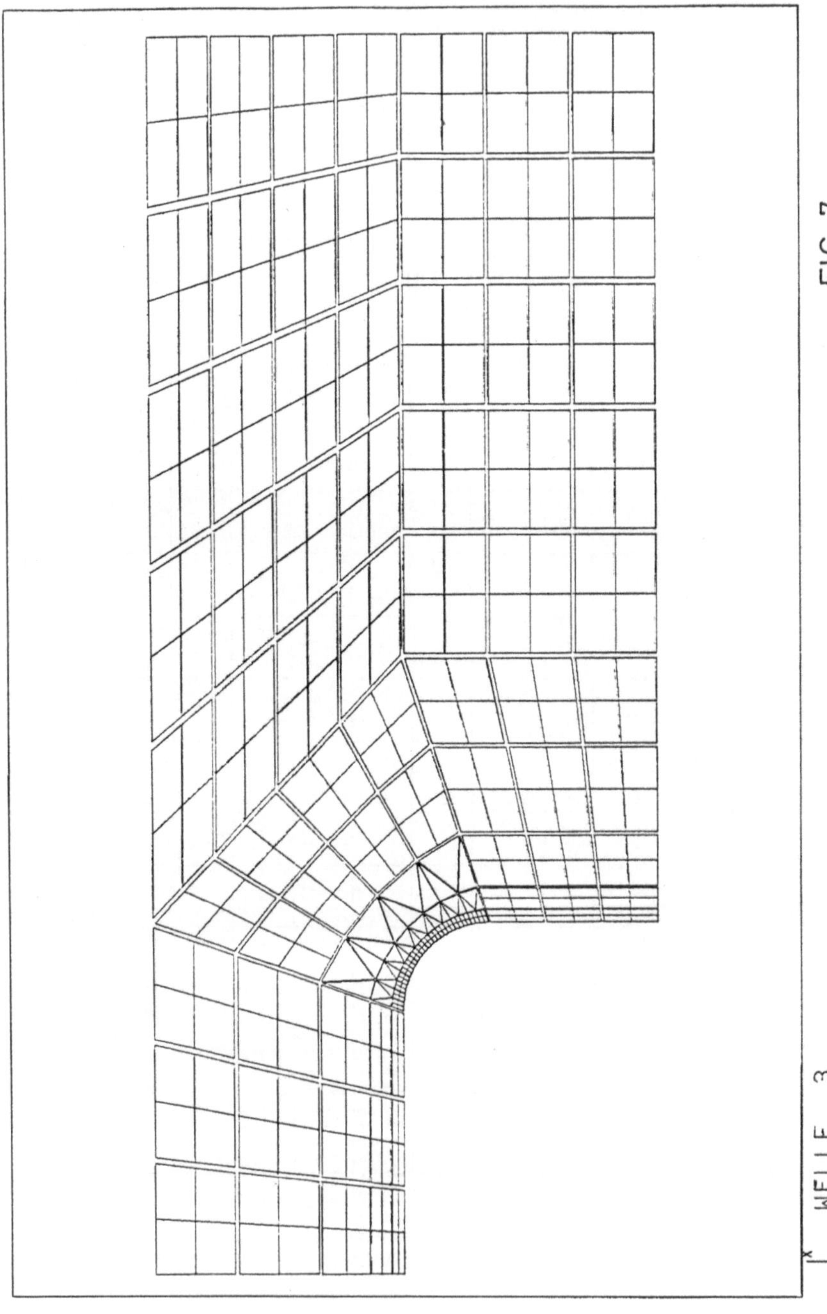

FIG. 7

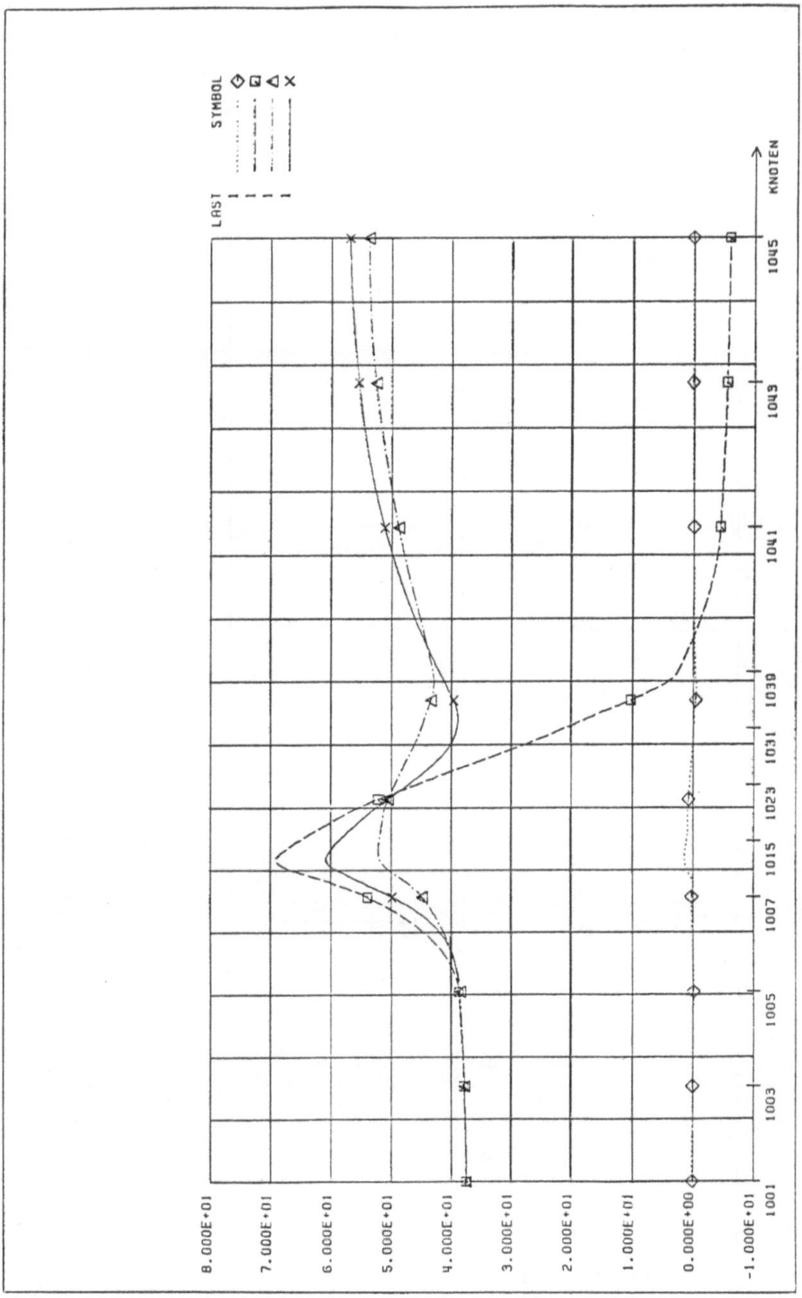

WELLE 3 ASKA-RESULTATE

SIGN SIGT SIGU SIGM

FIG. 8

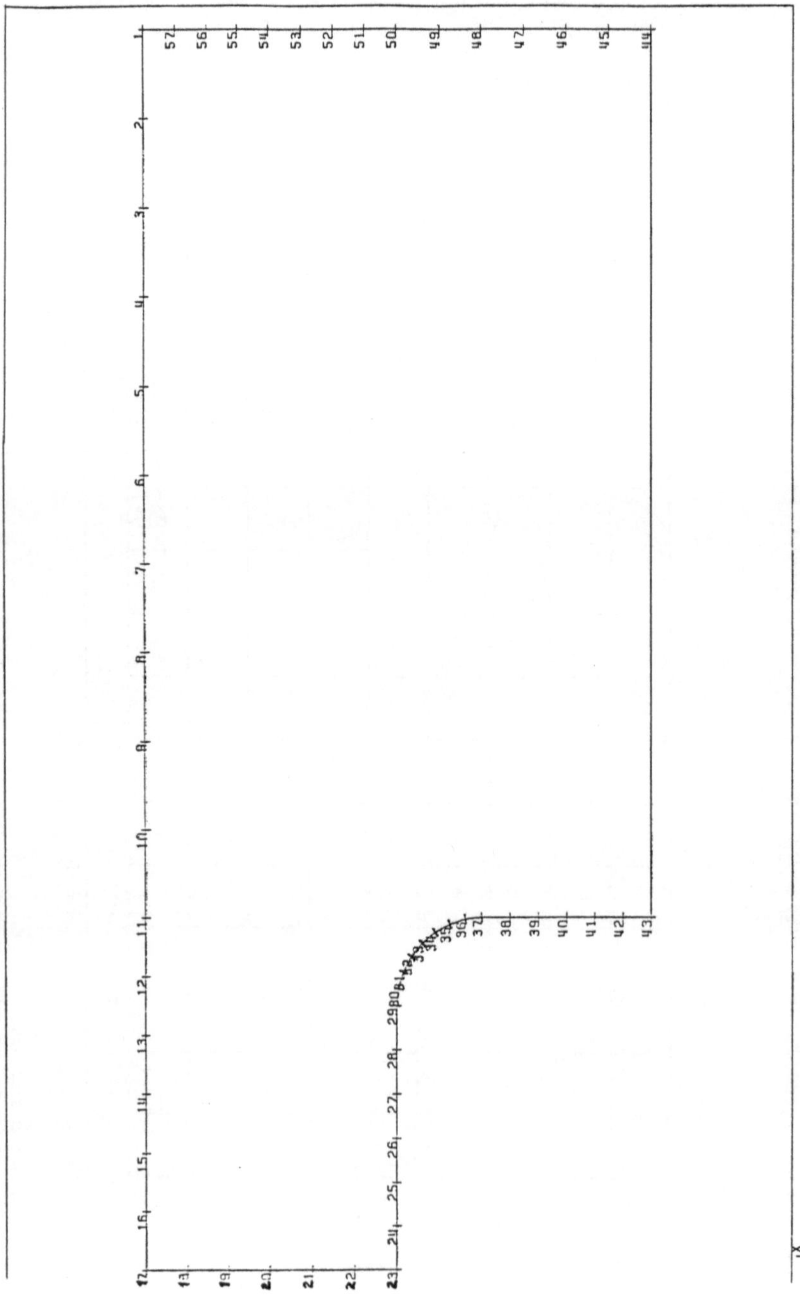
IGM-RØTSYM FEINHEIT 2 QUADRATISCH - QUADRATISCH
WINKEL 90.0 180.0 90.0 PROPORTIONEN 1.000 : 1.000 :0
x
y

FIG. 9

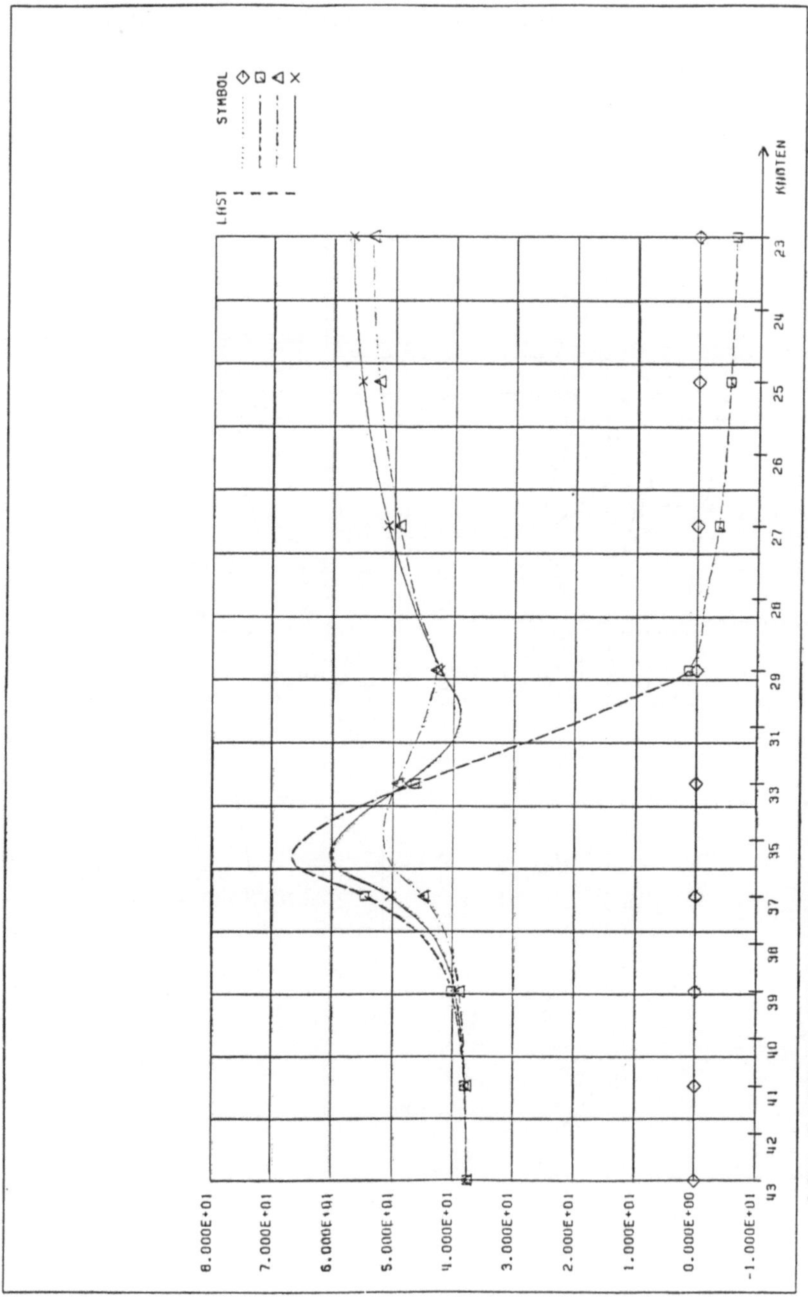

IGM-ROTSYM FEINHEIT 2 QUADRATISCH - QUADRATISCH

SIGN SIGT SIGU SIGM

FIG. 10

Finite-Element-Idealisierung des Schaufelfusses und der Wellennut

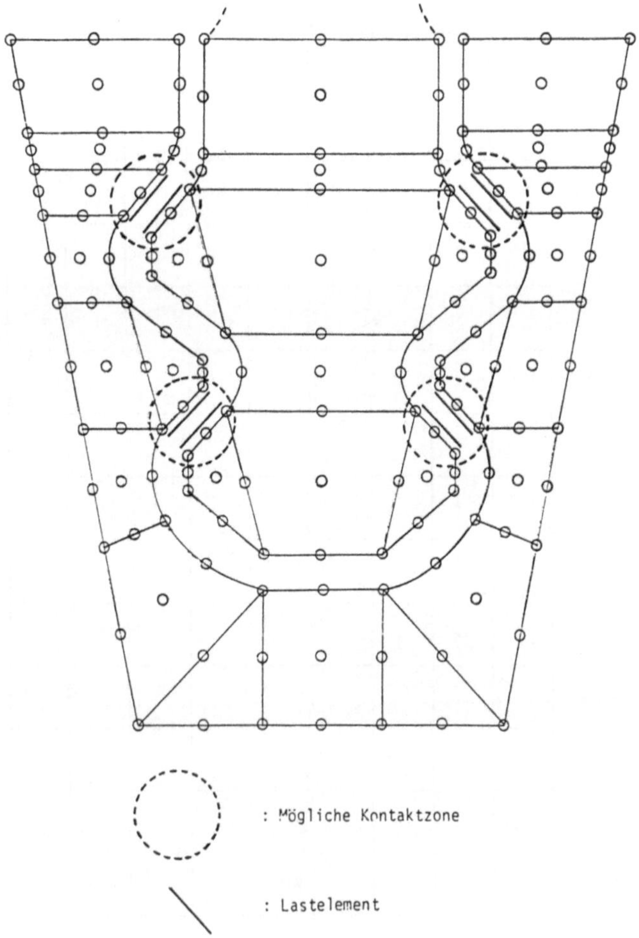

FIG. 11

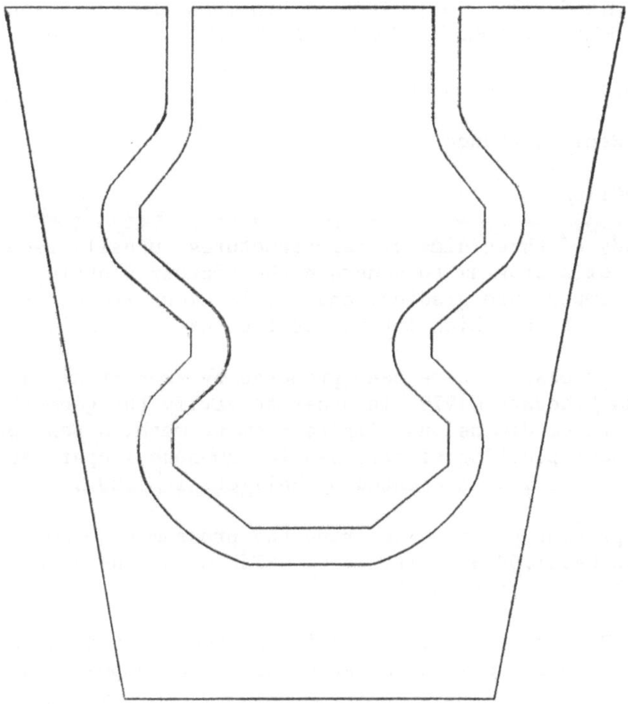

2-dim. Boundary

FIG. 12

PLANE INTERSECTIONS OF A THREE DIMENSIONAL BOUNDARY ELEMENTS MESH AND STRESS, DISPLACEMENT CONTOUR PLOTTING

M. Afzali and A. Chaudouet

CETIM, Senlis, France

INTRODUCTION

The study of three dimensional structures, usually requires the use of a program to generate the contour plotting of stress/temperature gradient and displacements/temperatures on a given plane which intersects the three dimensional mesh.

Two pre-processors have been proposed by Frey et al | 1979 |, and Akin and Stoddart | 1979 | in order to verify the geometric data of three dimensional finite element mesh. A post-processor for contour plotting of stresses for 20-node isoparametric brick elements, is presented by Hall et al | 1980 |.

In the present paper we describe the program PP3D-COUPE available in CA.ST.OR 3D software of CETIM using the boundary elements. This program can be used as :

(i) Pre-processor : to verify the geometric data, detecting voids or anomalies when the structure is decomposed into elements,

(ii) Post-processor : to plot the iso-values of stresses/temperature gradient and displacements/temperatures on the plane of intersection in elastic/thermal analysis.

When the pre-processor is used, only the contours on the plane are generated.

The main input data are :

(i) The definition of the three dimensional boundary elements (subregions, elements, node coordinates, etc ...) available on a geometry file. If the iso-values of stresses and displacements or temperature gradients and temperatures are requested, a 3D stress and displacement or temperature file obtained by standard 3D elastic or thermal analysis is also needed.

(ii) The plane definition, and the subregions numbers to be intersected.

With the informations given above, the program generates the contours for the subregions of interest. The coordinates of a groupe of points, defining the contours, are saved on "contour file". In order to decompose the surfaces bounded by the contours, into two dimensional elements the automatic mesh generator M2D (Boissenot |1975|) is used. This program is available in CETIM structure analysis software CA.ST.OR (CAlcul de STructures sur ORdinateur). However, a program ensuring the PP3D-COUPE and M2D interface is required. The interface program transfers the points coordinates to M2D using contour file. It is possible to discretize only a part of surfaces bounded by the contours. M2D generates a two dimensional mesh and corresponding file which is saved by the user. The contours and/or discretized surfaces may be plotted by using the option D2D of CA.ST.OR and 2D mesh file and with the aid of a Benson plotter.

The stresses/temperature gradients and displacements/temperatures at any point on the surface or inside the structure are computed by the aid of elastic (E3D)/thermal (TH3D) three dimensional structure analysis (CA.ST.OR |1981| using Boundary Element (BIE) method (Boissenot et al |1978|, and Lachat and Watson |1977|).

A program ensures the interface between PP3D-COUPE and E3D/TH3D in order to compute the solutions at any point of two dimensional mesh on the intersection plane. In PP3D-COUPE by defining the plane and using the solutions file of boundary elements nodes, one may obtain the 2D mesh file and the solution files on the plane of intersection in one job.

When a given point corresponds to two or more boundary elements, one may obtained different stress tensors and displacement vectors. The choice of unique stress tensor and displacement vector is discussed.

PLANE - ELEMENTS INTERSECTION

The elements considered in 3D structure analysis are the isoparametric 8-node quadrilateral or 6-node triangular elements shown in the figure 1. The global coordinates of a point on the element may be written as follows :

$$\begin{Bmatrix} x \\ y \\ z \end{Bmatrix} = \sum_{i=1}^{n} N_i \, (\xi, \eta) \begin{Bmatrix} x_i \\ y_i \\ z_i \end{Bmatrix} \qquad (1)$$

Where x_i are the global coordinates of elements nodes and n is the number of nodes. The shape functions N_i (ξ, η) are such that N_i gives unity at node i and zero at all other nodes

(Zienkiewicz |1977|).

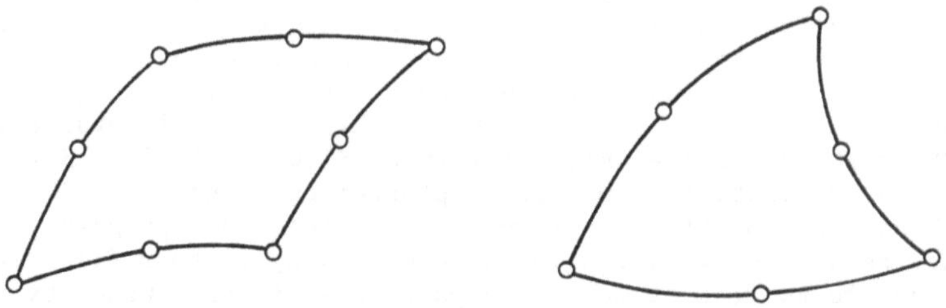

Quadrilateral Element Triangular Element

Figure 1.

Let now consider a plane defined as follows :

$$ax + by + cz = d \tag{2}$$

The point P(x, y, z) is an intersection point of the plane-element if the following nonlinear equation is satisfied :

$$\phi\ (\xi,\ \eta) \equiv ax\ (\xi,\eta) + by\ (\xi,\eta) + cz\ (\xi,\eta) - d = 0 \tag{3}$$

The function ϕ is a second degree polynomial of ξ for η = constant. By subtituting the equation (1) into equation (3), we have :

$$\phi\ (\xi,\eta) \equiv A_1\xi^2\eta + A_2\xi\eta^2 + A_3\xi^2 + A_4\eta^2 + A_5\xi\eta + A_6\xi + A_7\eta + A_8 = 0. \tag{4}$$

Where A_1, ..., A_8 are parameters related to the global coordinates of element nodes and plane coefficients. These parameters are constant for a given element and a plane. A similar equation may be obtained for other type of shape functions.

The local coordinates of the points of plane-element intersection may be obtained using equation 4. Subtituting the local coordinates into equation (1), global coordinates of the intersection points are calculated.

When the plane intersects the element in only one curve, for a given value of $\xi = \xi_1, \ldots, \xi_n$, equation (4) gives $\eta = \eta_1, \ldots, \eta_n$ $(-1<\xi,\ \eta<1)$; then the local coordinates of intersection points are calculated. This method may not be adequate if the plane intersects the element in two or more curves. In this case equation (4) gives two solutions η_1' and η_1'' for $\xi = \xi_1$. Another method called "Dichotomic Circles" is used. The basic steps of this method are :

(1) determination of the intersection points of element

sides and plane by solving equation (4) for $\xi = \pm 1$ and $\eta = \pm 1$,

(2) construction of a circle with one of the intersection points (e.g.P_1) as center with radius $r \leqslant 0.1$, see figure 2,

(3) determination of the plane-circle intersection points (M_i, i = 1, 2, ...) by iteration and using equation (4). In iteration process the position of any point on the circle is defined by angle Θ varying between Θ_{min} and Θ_{max}. As shown in figure 2, only the intersection points with $\Theta_{min} \leqslant \Theta \leqslant \Theta_{max}$ are considered.

The radius r of any circle should be less than the distance between point M_i and other points P_j (j = 2, 3, 4 in figure 2(a)). When equation (5) gives two acceptable solutions ($-1 \leqslant \xi$, $\eta \leqslant 1$), in order not to take into account the points on the other curve, the circle radius r is reduced to r/2. This decreases the number of solutions to one.

At arriving point (P_2), if $|M_i P_2| \leqslant 0.1$, it is considered that the curve is completed,

(4) determination of the middle point P_o of the curve P_1P_2 by integrating the distance between the points on the curve such that $\widehat{P_1Po} = \widehat{PoP_2}$. Each intersection curve is represented by the points P_1, Po and P_2,

(5) determination of other curve if it remains other not connected points on elements sides (P_3, P_4).

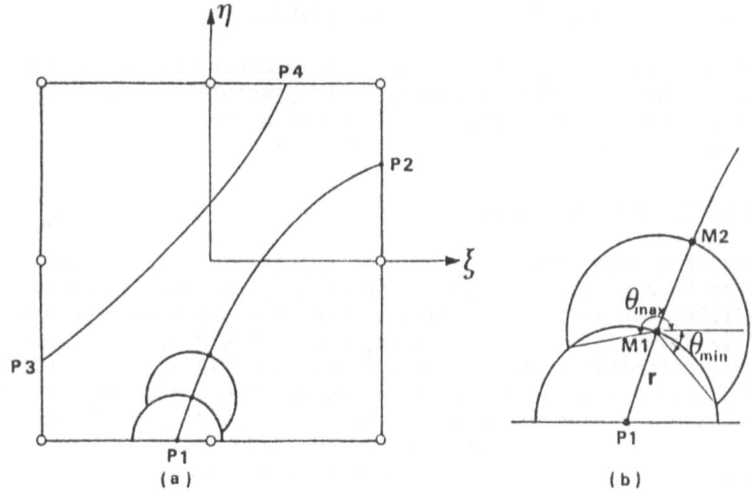

Figure 2. Dichotomic Circles method

The global coordinates of the points P_1, Po and P_2 in 3D space are calculated by subsituting the local coordinates of the points into equation (1). In special case when the plane intersects an element without intersecting the sides a closed contour is obtained by the first method.

2D Coordinates system

The 2D coordinate system related to the plane may be defined by the user by giving three points defining 2 directions, or will be chosen by the program. Consider the unit normal vector to the plane $\underset{\sim}{n}$, and points P and Q on the plane. The unit vectors $\underset{\sim}{u}$, $\underset{\sim}{v}$ and $\underset{\sim}{w}$ of 2D coordinates system with origine P, are defined as follows :

$$\underset{\sim}{u} = \underset{\sim}{PQ}/|\underset{\sim}{PQ}|$$

$$\underset{\sim}{w} = \underset{\sim}{n}$$

$$\underset{\sim}{v} = \underset{\sim}{w} \times \underset{\sim}{u}$$

The coordinates of the intersection points (P_1, Po, P_2) of each curve are calculated in the new system by a simple coordinate transformation. The new coordinates of the intersection points are stored in the tables for each element.

Contour construction

Once all curves of plane-elements intersection are obtained in the subregions of interest, they are connected by identification of the extreme points (P_1, P_2) coordinates of each curve. When a contour is obtained, the points coordinates are stored on the contour file. The other contours in the subregions are obtained in the same manner. The elements whose four sides are on the intersection plane are not taken into account. For the detail of the algorithm see Afzali |1981|.

In determination of stresses and displacements on an limited area on the intersection plane, it is possible for the user to decrease the computation time by giving the limits X_{min}, X_{max}, Y_{min} and Y_{max} of region of interest.

AUTOMATIC MESH GENERATOR

In order to plot the iso-values of stresses and displacements/ temperature gradients and temperatures by the aid of D2D of CA.ST.OR, the surfaces bounded by the contours should be discretized into elements. For this, the option M2D, two dimensional automatic mesh generator (Boissenot |1975|) available in CA.ST.OR is used. The data in M2D are the coordinates of the points on the contours and a factor corresponding to each point and an elementary distance that define the mesh refinement. The elements in this program are the 6-node isoparametric triangles. Between the points defining the contour other points are added depending upon the elementary distance and the factors of mesh refinement between two points on the contour.

The algorithm consists of successive decomposition of the surface into the triangles. The triangles in the first decomposition are constructed by connecting the points on the contour. In the other decompositions the number of the triangles is increased until the suitable mesh refinement is achieved.

In PP3D-COUPE, a program is provided to ensure the interface between PP3D-COUPE and M2D. The coordinates of the points on the contours are transfered by the contour file. It is possible to define a refinement factor for each boundary element intersected by the plane. If this factor is not given, it is considered to be equal to 1. In PP3D-COUPE, the refinement factor corresponds to the points on the contour. The factors values are stored on a file, at the same time that the points coordinates are stored on the contour file. The elementary distance is given for each subregion. Other information that M2D needs is the curvature of the curves on the contour. The curvature is calculated in the interface program by using the points (P_1, Po, P_2) coordinates of the contours.

The two dimensional mesh file generated by M2D may be saved by the user. By using 2D mesh file and the option D2D of CA.ST.OR and with the aid of Benson plotter, the contours and/or discretized surfaces may be plotted.

STRESS AND DISPLACEMENT COMPUTATION

Once the 2D mesh of the plane intersection with the 3D geometry is performed, and in order to be able to plot iso-values curves on this plane, the solutions (displacement and stress for elastic analyses or temperature and gradient of temperature for thermal analyses) must be computed at each node of the 2D mesh.

For points inside the volume, the displacements are calculated by using Somigliana identity and the stresses are calculated by using a formula obtained by applying Hooke's Law to the Somigliana identity (Boissenot, Chaudouet, Dubois |1978|)

$$u_i = \int_S (T_{ij}\, u_j + U_{ij}\, t_j) ds$$

$$\sigma_{ij} = \int_S (D_{ijk}\, u_k + S_{ijk}\, t_k) ds$$

For points on the surface, the displacements are determined by interpolating the displacements obtained at each node of the 3D mesh in the standard 3D computation. As far as stresses on each element are concerned, the computation is performed by using the tensions (t) and the derivatives of the displacements (ε), these quantities are obtained by interpolating the solutions (t and u) at each node of the element. Let (3) be the normal direction on an element and (1 and 2) two orthogonal directions in the tangential plane of this element, then

$\sigma_{33} = t_3 \qquad \sigma_{32} = t_2 \qquad \sigma_{31} = t_1$

$\sigma_{12} = f(\varepsilon_{12})$

$\sigma_{11} = g(t_3, \varepsilon_{11}, \varepsilon_{22})$

$\sigma_{22} = h(t_3, \varepsilon_{11}, \varepsilon_{22})$

It can be seen on the above expressions that σ_{33}, σ_{32} and σ_{31} are calculated with a good precision (t being a solution of the linear system generated by the BIE method) where as the precision on σ_{12}, σ_{11} and σ_{22} is not so good since ε is obtained by using the derivative of the interpolation functions with respect to the intrinsic coordinates.

For points on a smooth surface, the stress tensor can be calculated as the average of the stress tensors components obtained on each element including this point.

For points situated on a sharp edge (figure 3), by taking into account the above considerations, it is obvious that :
- on element A, σ_{11} and σ_{12} will have a good precision
- on element B, σ_{12} and σ_{22} will have a good precision

Thus by writing the stress tensors in a local basis composed of the normal directions on element A (direction 1) and on element B (direction 2) and of a direction (3) orthogonal to directions 1 and 2, a better result than a simple average can be obtained by writing :

$\sigma_{11} = \sigma_{11}\ (A) \qquad \sigma_{22} = \sigma_{22}\ (B)$

$\sigma_{ij} = (\sigma_{ij}\ (A) + \sigma_{ij}\ (B))/2. \qquad (ij = 12 - 13 - 23 - 33)$

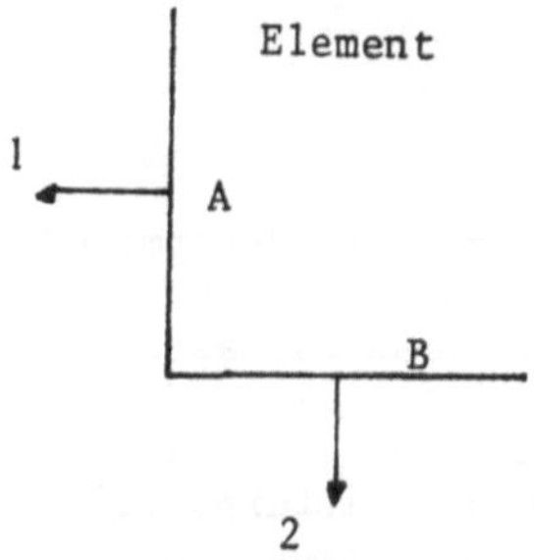

Figure 3

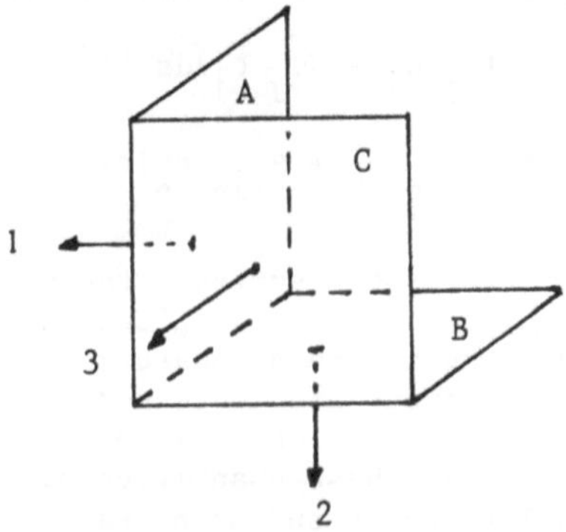

Figure 4

For points situated at a corner, the same procedure can be applied. Let direction 1 be the normal direction of element A, direction 2 the one of element B and direction 3 the one of element C (figure 4), then :

$\sigma_{11} = \sigma_{11}$ (A) $\quad \sigma_{22} = \sigma_{22}$ (B) $\quad \sigma_{33} = \sigma_{33}$ (C)

$\sigma_{12} = (\sigma_{12}$ (A) $+ \sigma_{12}$ (B))/2

$\sigma_{13} = (\sigma_{13}$ (A) $+ \sigma_{13}$ (C))/2.

$\sigma_{23} = (\sigma_{23}$ (B) $+ \sigma_{23}$ (C))/2.

For points where four or more normals may exist, this procedure can no longer be applied. A simple average of the stress tensors obtained on each element is performed.

EXAMPLES

Stress and displacements contour plotting

Figure 5 shows a clamp with a uniform internal pressure of $P_i = 0.45$ daN/mm^2 at R = 194.5 mm and a uniform pressure of 0.84 daN/mm^2 around the half-hole A due to tightening. A uniform traction distributed force of 2.08 daN/mm^2 is applied on face B of the clamp. The boundary conditions are : zero displacement on direction normal to the face of θ = 15° and zero displacement on direction z on ligne L.

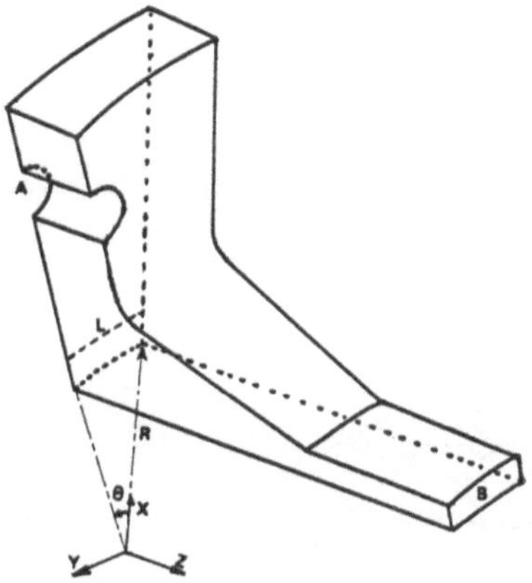

Figure 5

The three dimensional boundary elements are shown in the figure 6 with four subregions. By using the boundary elements file and solutions file one can obtain the iso-values of stresses and distored structure on a given plane intersecting the structure. The plane may be defined by its coefficients or by the coordinates of three points. The files corresponding to 2D mesh and stresses and displacements of 2D elements nodes on the plane should be saved by the user.

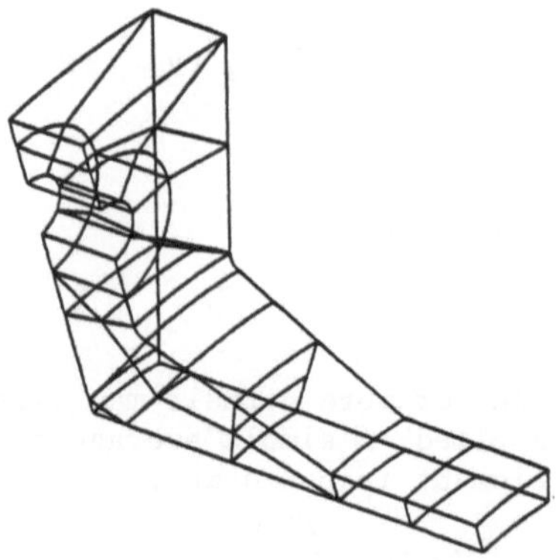

Figure 6. Three dimensional boundary elements

Figure 7 shows the discretized surfaces bounded by the contours obtained by intersecting the structure with plane y = 0.27 x. This figure is plotted by using the 2D mesh file and with the aid of Benson plotter. The 2D elements are 6 node-triangular element.

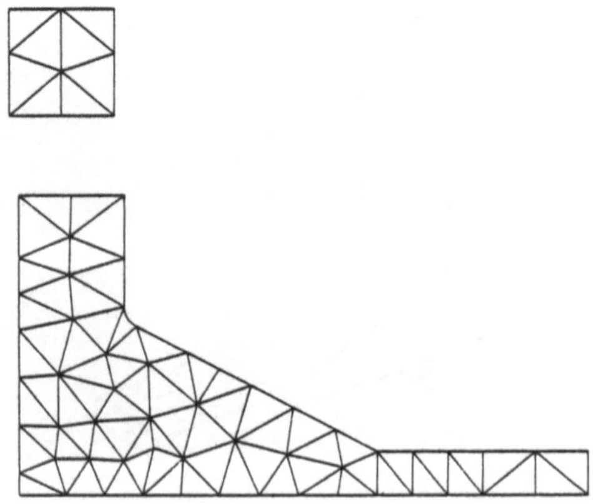

Figure 7. Discretized surfaces of cross section on the plane of intersection y = 0.27x

PP3D-COUPE calculates the stresses and displacements at any node using the 3D solution file of boundary elements obtained after a standard 3D elastic analysis. Using the option D2D of CA.ST.OR and 2D stress file, the iso-values of Von Mises stresses are plotted in figure 8. As it can be seen the stress concentration around the fillet is very important. If the solutions are requested only on a part of cross section, by defining the regions with the limits X_{min}, X_{max}, Y_{min} and Y_{max}, the computation time may be reduced considerably.

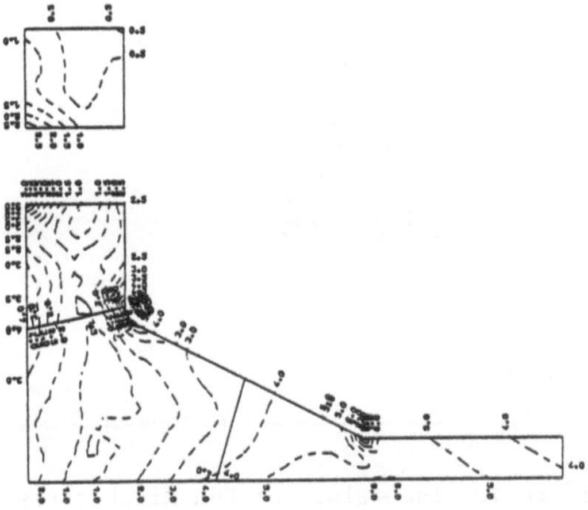

Figure 8. Iso-values of Von Mises stresses on the plane of intersection y = 0.27x

In figure 9 the stresses are requested on the shaded area of the intersection of the structure with plane Y = 0.13 x. The iso-values of Von Mises stresses are plotted in figure 10 on this plane.

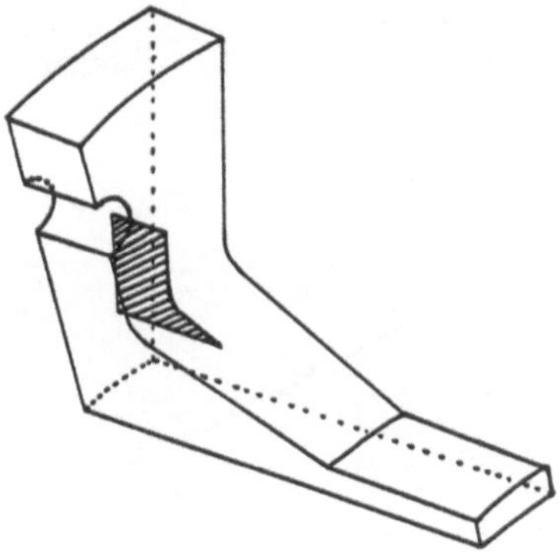

Figure 9. Area of intersection to be discretized

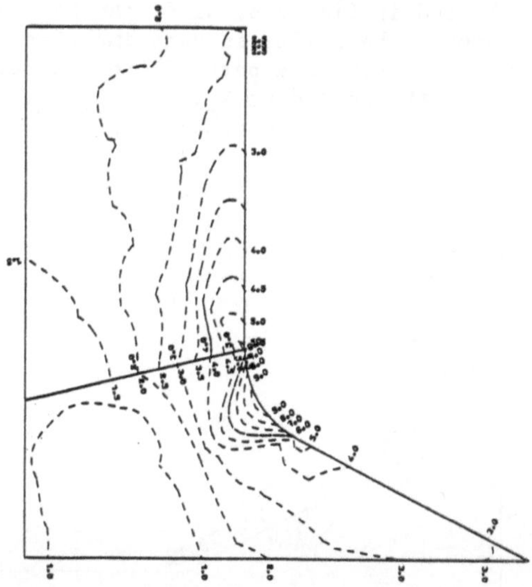

Figure 10. Iso-values of Von Mises stresses
for fillet area on plane y = 0.13x

Figure 11 shows the distored structure after applied forces. The complete and dotted lines represente respectively the initial geometry and distored structure on the plane y = 0.27x.

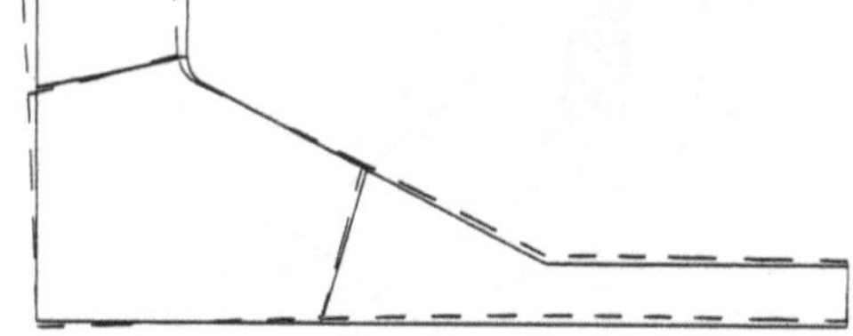

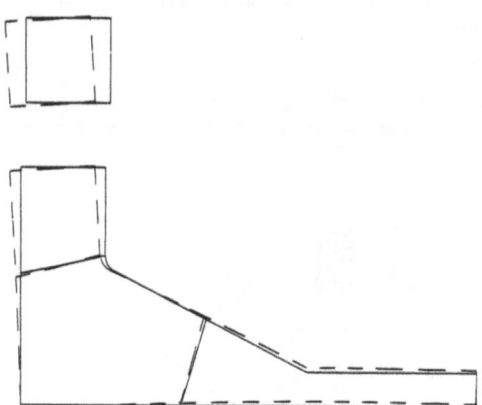

Figure 11. Initial geometry and distored structure
on the plane y = 0.27x

Temperature contour plotting

An example of three dimensional thermal analysis is given in figure 12. The boundary conditions are shown on six faces of the structure in the figure. The faces perpendicular to the axes y and z are isolated (flux ϕ = o). The temperature at faces x = 0 and x = 8 are respectively T = 1 and T = 0. The iso-values of temperatures on the plane of intersection y = 1. are plotted in figure 13.

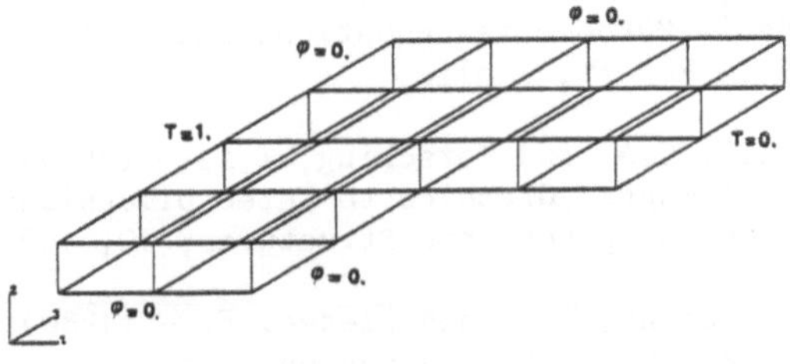

Figure 12. Three dimensional boundary elements for thermal analysis

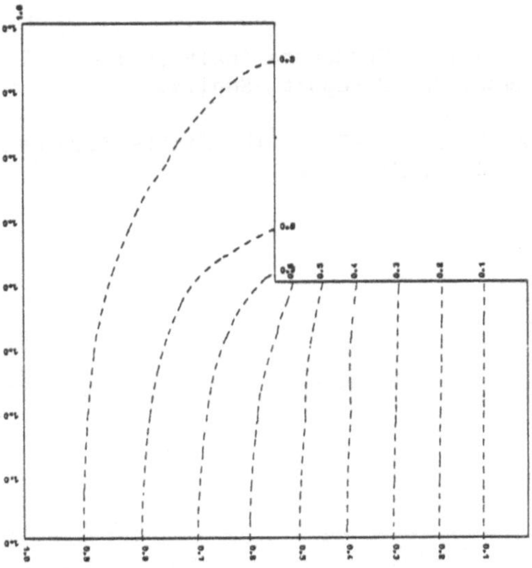

Figure 13. Isothermal lines on the plan y = 1 which intersects the structure in figure 12.

REFERENCES

Afzali, M. (1981) - Coupe de Structures Tridimensionnelles. CETIM Report, Senlis.

Akin, J.E. and Stoddart, W.C.T. - (1979) - Plane Intersections and Contours for General Isoparametric Solids. Computers and

Structures, 10, 155-157.

Boissenot, J.M. - (1975) - Méthode de Maillage Automatique des Pièces Planes ou Axisymétriques. CETIM Report, Senlis.

Boissenot, J.M., Chaudouet, A. and Dubois, M. - (1978) - Application de la Méthode des Equations Intégrales à la Mécanique. CETIM Report, Senlis.

Frey, A.E., Hall, C.A. and Porsching, T.A. - (1979) - An Application of Computer Graphics to Three Dimensional Finite Element Analyses. Computers and Structures, 10, 149-154.

Hall, C.A., Porsching, T.A. and Sledge, F. - (1980) - STRS IS : - Contour Plotting of Stresses on Planes of Intersections. Computers and Structures, 12, 221-224.

Lachat, J.C. and Watson, J.O. - (1977) - Progress in the Use of Boundary Integral Equations Illustrated by Examples. Computer Methods in Applied Mechanics and Engineering, 10, N° 3.

CA.ST.OR - (1981) - Structure Analysis ; CAlcul de STructures sur ORdinateur. CETIM Report, Senlis.

Zienkiewicz, O.C. - (1977) - The Finite Element Method 3rd edn, Mc Graw-Hill, New York.

A NEW BOUNDARY CONDITION SOLVED WITH B.I.E.M.

J.J.Anza, E.Ahedo,I.Da Riva, E.Alarcón.

Polytechnical University Madrid

1. INTRODUCTION

Among the classical operators of mathematical physics the Laplacian plays an important role due to the number of different situations that can be modelled by it. Because of this a great effort has been made by mathematicians as well as by engineers to master its properties till the point that nearly everything has been said about them from a qualitative viewpoint.

Quantitative results have also been obtained through the use of the new numerical techniques sustained by the computer. Finite element methods and boundary techniques have been successfully applied to engineering problems as can be seen in the technical literature (for instance [1], [2], [3]).

Boundary techniques are especially advantageous in those cases in which the main interest is concentrated on what is happening at the boundary. This situation is very usual in potential problems due to the properties of harmonic functions.

In this paper we intend to show how a boundary condition different from the classical, but physically sound, is introduced without any violence in the discretization frame of the Boundary Integral Equation Method.

The idea will be developed in the context of heat conduction in axisymmetric problems but it is hoped that its extension to other situations is straightforward. After the presentation of the method several examples will show the capabilities of modelling a physical problem.

2. AXISYMMETRIC PROBLEMS

It is well known that problems with axial symmetry can be treated by reducing their dimensionality, for instance the discretization with Finite Elements can be done only in a transverse section. It is clear that this possibility in connection with the reduction, intrinsic to BIEM, will transform an axisymmetric problem into a series of equations on the line defining the boundaries of a typical cross-section.

This has been fully exploited by several researchers (4 , 5) and can be summarised as follows.

The general basic equation is

$$c\,\phi + \int_{\partial\Omega} \phi \frac{\partial \psi}{\partial n}\, ds = \int_{\partial\Omega} \frac{\partial \phi}{\partial n}\, ds \qquad (1)$$

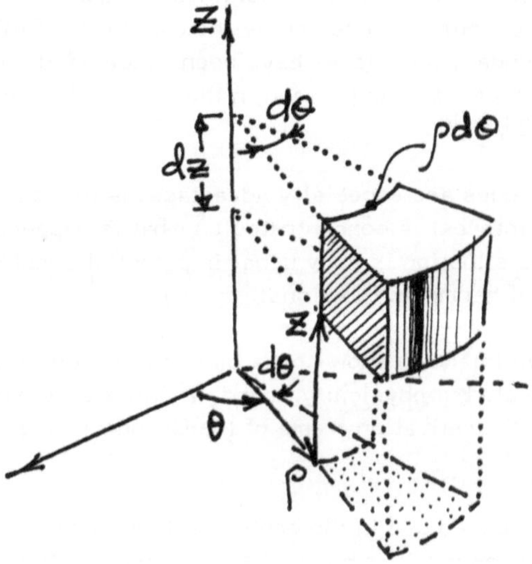

Figure 1

where

$$\psi = \frac{1}{4\pi\rho}$$

is the fundamental solution of Laplace's equation and ρ the distance between the observation point and that in which ψ is "applied".

In polar coordinates (Figure 1)

$$ds = \rho \; d\theta \; dz \qquad (2)$$

and taking into account the symmetry of the problem equation (1) can be transformed into

$$c\,\phi + \int_{z_1}^{z_2} \rho \; \phi \; dz \; \frac{\partial \psi_{as}}{\partial n} = \int_{z_1}^{z_2} \rho \; \frac{\partial \phi}{\partial n} \; dz \; \psi_{as} \qquad (3)$$

where

$$\psi_{as} = \int_0^{2\pi} \psi \; d\theta = \int_0^{2\pi} \frac{1}{4\pi\rho} \; d\theta = \frac{Q_{-\frac{1}{2}}(\gamma)}{2\pi\,(\rho\rho_i)^{\frac{1}{2}}}$$

$$\frac{\partial \psi_{as}}{\partial n} = \int_0^{2\pi} \frac{\partial \psi}{\partial n} \; d\theta = \frac{1}{2\pi\,(\rho\rho_i)^{\frac{1}{2}}} \; \Big\{ \Big(-\frac{Q_{-\frac{1}{2}}(\gamma)}{2} + \frac{\rho^2-\rho_i^2-(z-z_i)^2}{2\,\rho\rho_i} \; \frac{dQ_{-\frac{1}{2}}(\gamma)}{d\gamma}\Big) \; \rho,n + \Big(\frac{z-z_i}{\rho_i} \; \frac{dQ_{-\frac{1}{2}}(\gamma)}{d\gamma}\Big) \; z,n \Big\} \qquad (4)$$

where

$$\gamma = 1 + \frac{(\rho-\rho_i)^2 + (z-z_i)^2}{2\,\rho\rho_i}$$

and

$Q_{-\frac{1}{2}}$ is the Legendre function of the second kind.

Discretization of equation (3) is done in the usual way (see for instance ref. [6]) by defining the evolution of inside "boundary elements" as function of ϕ and its derivatives at the "nodes".

It is possible to define as many equations of type (3) as nodes and the imposition of the boundary conditions allows the solution of the problem.

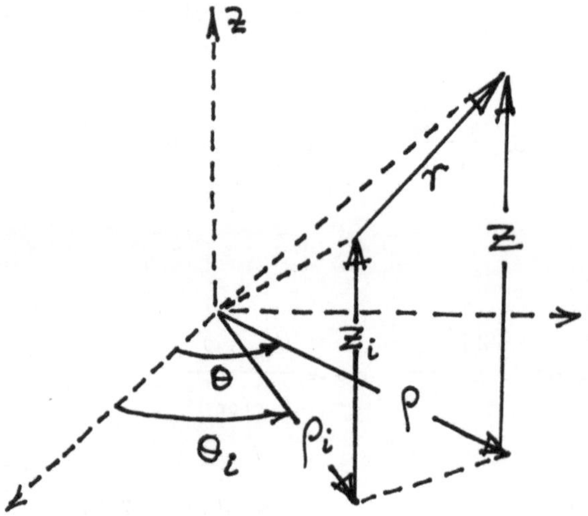

Figure 2

As is well known the boundary conditions for a classical properly posed problems are of the following types

a) Dirichlet condition

$$\phi = \phi_o$$

b) Newmann condition

$$\frac{\partial \phi}{\partial n} = q_o$$

c) Newton (or Robin) condition

$$\alpha \phi + \beta \frac{\partial \phi}{\partial n} = 0 \qquad (6)$$

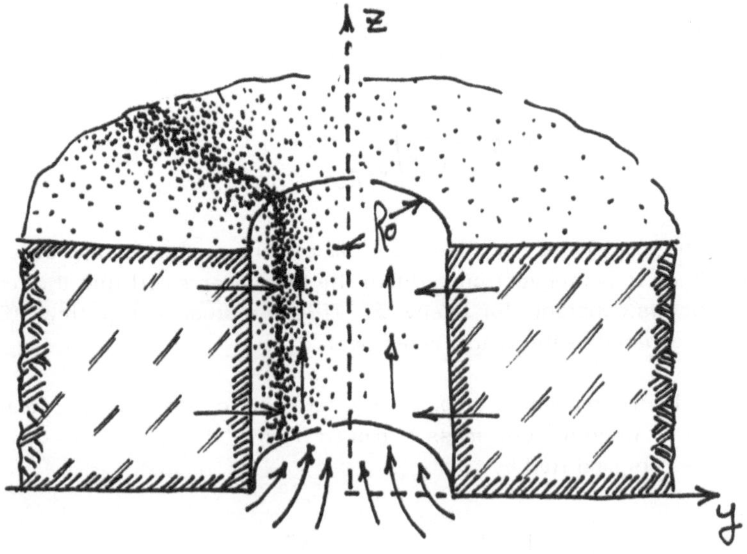

Figure 3

The discussion of the different possibilities for imposing these conditions can be seen for instance in [4], [6], [7].

In what follows we describe a different condition and how to treat it in the BIEM context.

3. SPECIAL BOUNDARY CONDITIONS

Imagine an axisymmetric situation as that sketched in Figure 3.

A fluid receives heat through the walls of a circular channel and eliminates it by mass-flow. The problem can be imagined in the heat propagation field with a field equation of the type

$$\frac{k_z}{k_y} \frac{\partial^2 T}{\partial z^2} + \frac{1}{y} \frac{\partial}{\partial y} (y \frac{\partial T}{\partial y}) = 0 \qquad (7)$$

where k_x and k_y are the conductivities of the solid body in orthogonal directions. As is well known [8], this equation can be reduced to an isotropic situation by a simple change of scale in the direction x.

The flux of heat in a tube element is

$$k_y \, q(R_o) \; 2\pi R_o \, dz$$

$$q(R_o) = \left(\frac{\partial T}{\partial y}\right)_{y=R_o} \qquad (8)$$

T is the temperature in the tube-fluid interface, that is it is assumed that there is no convection and that the tube is so small that the temperature is constante for every z. If it is a steady-state flow, (8) can be equated to the transported heat.

Let m be the transport velocity (kg/seg) and c_p the specific heat of the fluid. When the unit mass is moved from z to $z + dz$ its quantity of heat is modified by

$$c_p \left(T + \frac{\partial T}{\partial z} dz\right) - c_p T = c_p \frac{\partial T}{\partial z} dz \qquad (9)$$

so that

$$k_y \, q(R_o) \, 2\pi R_o \, dz = m \, c_p \frac{\partial T}{\partial z} dz \qquad (10)$$

The last equation can be written as a condition on the boundary

$$\frac{\partial T}{\partial y} = \mu \frac{\partial T}{\partial z} \qquad (11)$$

with

$$\mu = \frac{m \, c_p}{2\pi k_y R_o}$$

Equation (11), which is the new boundary condition, relates the normal derivative to the tangential one.

In order to develop its numerical treatment let us remember the right hand side of equation (3)

$$\text{RHS} = \int_{z_1}^{z_2} \rho \frac{\partial \phi}{\partial n} dz \; \psi_{as} \qquad (12)$$

The parts of the boundary with the classical conditions are treated as usual. Along the tube and after substitution of (11) in (12), and integrating by parts

$$RHS = \int_{z_1}^{z_2} \psi_{as}\, \mu \frac{\partial T}{\partial z} \rho\, dz = \mu \left(\left(\psi_{as}\, \rho\, \phi \right)_{z_1}^{z_2} - \int_{z_1}^{z_2} \phi \frac{\partial}{\partial z} (\psi_{as}\, \rho)\, dz \right) \tag{13}$$

At this moment the discretization used to model ϕ can be introduced

$$\phi = \underset{\sim}{N}\, \underset{\sim}{\phi}^{e} \tag{14}$$

and the last integral in (13) can be expressed as

$$\left(\int \underset{\sim}{N} \frac{\partial}{\partial z} (\psi_{as}\, \rho)\, dz \right) \underset{\sim}{\phi}^{e} \tag{15}$$

It is clearly advantageous to use constant elements in which case

$$RHS = \mu \left(\left(\psi_{as}\, \rho\, \phi \right)_{z_1}^{z_2} - \sum_{j=1}^{M} \phi_j \int \frac{\partial}{\partial z} (\psi_{as}\, \rho)\, dz \right) =$$

$$= \mu \left(\left(\psi_{as}\, \rho\, \phi \right)_{z_1}^{z_2} - \sum_{j=1}^{M} \phi_j \left(\psi_{as}\, \rho \right)_{j_1}^{j_2} \right) \tag{16}$$

Clearly z_1 and z_2 indicate the limits of the zone where the new conditions are applied and (j_1, j_2) to the points defining the $\bar{j}$ element.

As the form of the terms is the same it is possible to include the first two into the sum without problem.

4. EXAMPLES

As an example of the application of the previous method imagine the problem of Figure (4a). A thick-wall tube with internal diameter $_o$ and external R is formed with an orthotropic material with conductivities k_y, k_z while the extreme bases of the cylinder are maintained at temperatures T_o and T_H.

Equation (7) can be transformed into

$$\frac{k_x R^2}{k_y H^2} \frac{\partial^2 \theta}{\partial \xi^2} + \frac{1}{\eta} \frac{\partial}{\partial \eta} \left(\eta \frac{\partial \theta}{\partial \eta} \right) = 0 \tag{17}$$

by the rule

$$\theta = \frac{T - T_o}{T_H - T_o}$$

$$\xi = \frac{z}{H} \tag{18}$$

$$\eta = \frac{y}{R}$$

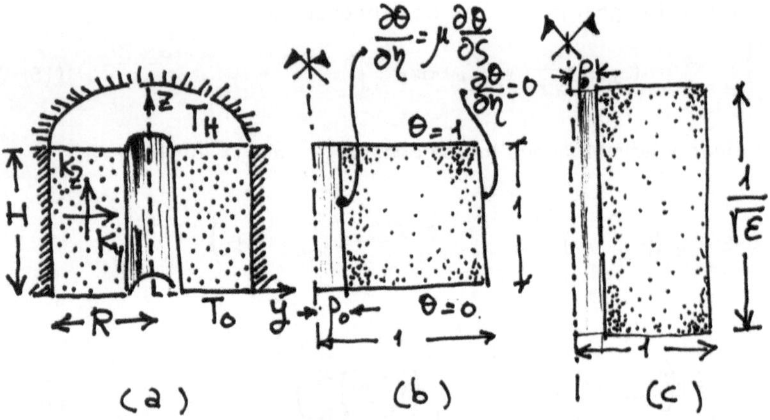

Figure 4

over the domain of Figure (4b).

Condition (11) is also modified accordingly. The BIEM can be applied directly to the problem but it is also possible to change again the geometry by doing

$$\xi' = \frac{1}{\sqrt{\frac{k_x R^2}{k_y H^2}}} \qquad \xi = \frac{H}{R}\sqrt{\frac{k_y}{k_x}} \; \xi \tag{19}$$

obtaining finally

$$\frac{\partial^2 \theta}{\partial \xi'} + \frac{1}{\eta} \frac{\partial}{\partial \eta} \left(\eta \frac{\partial \theta}{\partial \eta} \right) = 0 \tag{20}$$

which is the differential equation of an axisymmetric potential problem

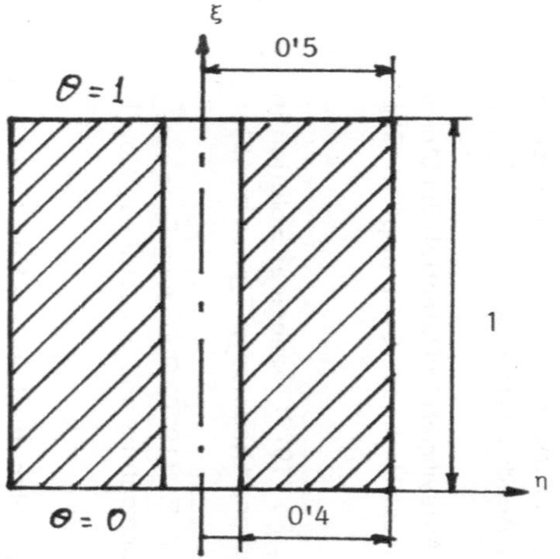

ξ	Analytical μ' =0	μ'=0	μ' =1	μ'=5
0'975	0'975	0'9765	0'921	0'578
0'775	0'775	0'7753	0'603	0'250
0'575	0'575	0'5751	0'393	0'125
0'375	0'375	0'3748	0'218	0'053
0'175	0'175	0'1746	0'077	0'013

Temperature at $\eta = 0'1$

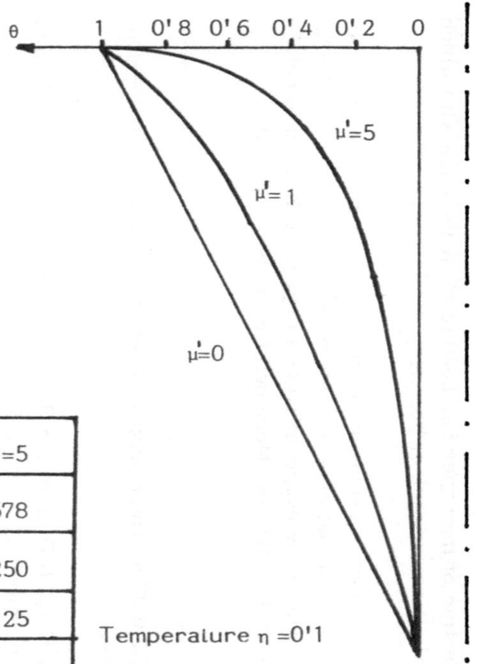

Temperature $\eta = 0'1$

$$\left(\frac{\partial \theta}{\partial \eta} = \mu' \frac{\partial \theta}{\partial \xi'}\right)$$

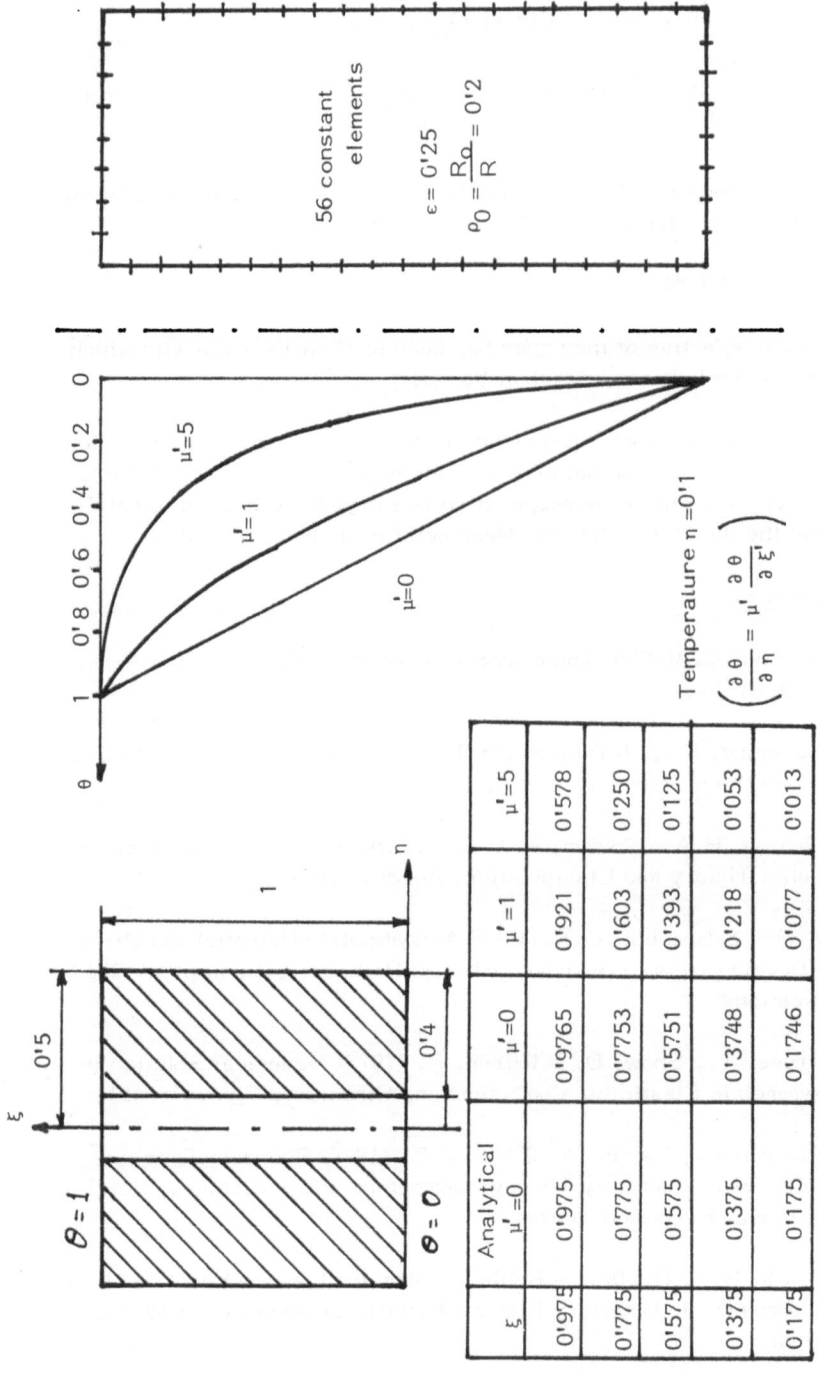

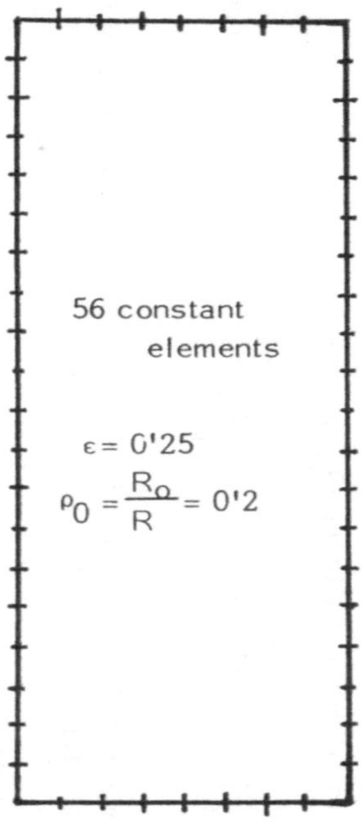

defined over the domain of Figure (4c) where

$$\sqrt{\varepsilon} = \frac{R}{H}\sqrt{\frac{k_x}{k_y}} \qquad \mu' = \frac{\mu}{\sqrt{\varepsilon}} \tag{21}$$

Figure 5 shows the discretization for a particular case as well as the evolution of the temperature in the channel.

5. CONCLUSIONS

The main objective of the paper has been to show the ease with which BIEM can include non-classical boundary conditions.

The problem posed was described elsewhere [9] and corresponds to the physcial situation that arises in the neck of vapour cooled shield (VCS) dewars where the evaporation of a cryogenic is used to refrigerate the thermal isolation. More details can be seen in (9).

References

[1] Brebbia, C.A(1978) The Boundary Element Method for Engineers. Pentech Press.

[2] Banerjee, P.K. & Butterfield R. (1981) Boundary Element Methods in Engineering Science. McGraw-Hill.

[3] Jaswon, H.A. & Symm, G.T. (1977) Integral Equation Methods in Potential Theory and Elastostatics. Academic Press.

[4] Wrobel & Brebbia, C.A. (1980) Axisymmetric Potential Problems. New Developments in Boundary Element Methods, Southampton. CML Publications.

[5] Cruse, T., Snow, D. & Wilson, P. (1977) Numerical Solution in Axisymmetric Elasticity. Computers and Structures 7, pp.445-451.

[6] Alarcón, E., Martín, A. & París, F. (1979) Boundary Elements in Potential in Elasticity Theory. Computers and Structures, Vol.10, pp.351-362, Pergamon Press.

[7] Symm, G.T. (1980) The Robin Problem for Laplace Equation. New Developments in Boundary Element Method, Southampton. CML Publications.

[8]París, F, Martin, A. & Alarcon, E. (1981) Potential Problems. Progress in Boundary Elements I, Chap.8 . Ed. C. Brebbia. Pentech Press.

[9]Da Riva, I. (1979) Spacecraft Thermal Control Design Data. ESA (TST-02) Issue No.1, Vol.3. ETS Aeronáuticos de Madrid.

Session VII
Combination with Other Techniques

A CRITICAL STUDY OF DIFFERENT BOUNDARY ELEMENT STIFFNESS MATRICES

O. Tullberg and L. Bolteus,
Dept of Structural Mechanics and Dept of Structural Design,
Chalmers University of Technology,
SWEDEN

ABSTRACT

Seven different boundary element stiffness matrices, for two-dimensional elasticity, are developed and critically studied in order to be able to choose the best one for the coupling of FEM and BEM.

First, a non-symmetric direct stiffness matrix is deduced with use of the direct boundary element method. This matrix does not fulfill the equilibrium requirements for stiffness matrices. To satisfy this requirement and to make the matrix symmetric six different methods are developed. To the authors knowledge, one of these are novel but the others have been proposed elsewhere. The methods are compared by studying the rate of convergence of the results in two numerical examples. Special attention has been paid to the discontinuous tractions and a general method to take these into account, in the different formulations, is presented.

INTRODUCTION

Although the conventional finite element method (FEM) is the most general technique for numerical analysis of partial differential equations, in engineering areas, it is not always the one that is best suited or gives the most accurate results. For two- and three-dimensional problems in elasticity and potential theory the boundary element method is better suited than FEM. This because of the small input data needed for this method and the accurate results of both primary and secondary variables. However, for non-linear problems FEM is still the most efficient method to use. The non-linear regions of a structure is often of limited size and the main region can approximately be treated as linear. In such problems a coupling of FEM and BEM should be fruitful. A coupling of FEM and BEM, or some other numerical method, is also needed when the domain extends to infinity in some or all directions. This is a necessity in dynamical problems. Different methods for such problems have been proposed in the literature. Hybrid methods have been developed by Lysmer and Kuhlemeyer (1969), Waas (1972), Chen and Mei (1976), and Gupta et al. (1982). Another way to treat the problem is to use infinite elements, Bettes (1977) and Bettes and Zienkiewicz (1977). Murukami et al. (1981) proposed a method which makes use of Green functions. The method is close to the conventional BEM. In Taylor and Zietsman (1981) a comparison is made between hybrid methods and BE-methods. They concluded that BEM was more flexible and needed less computations than the hybrid method. The coupling of FEM and BEM for static problems was first proposed by Zienkiewicz et al. (1977) and Shaw (1978). In these papers, the integral equation was used in connection with an energy functional. The method gave rise to symmetric stiffness matrices. A more direct method to generate a stiffness matrix was shown by Brebbia (1978). This matrix was, however, not symmetric. It is possible to show that the symmetric part of that matrix is the same as that derived with the energy approach. This will be shown here, with discontinuous tractions taken into account. There are other ways to couple BEM and FEM then by forming a stiffness matrix. This was discussed in Brebbia (1978) and Bolteus and Tullberg (1981). However, that method could not be used in situations with non-linear FE-regions. The extension of that method to such problems will be discusssed by the authors in forthcoming paper.

The first combination of FEM and BEM for elastostatics appears to be by Osias (1977). After that the paper by Brebbia and Gergiou (1979) was presented, in which they compared the symmetrized direct stiffness matrix with conventional BEM. The boundary elements were of constant function type. No differences between the methods could be observed in the tested examples. Since the constant boundary element only can be used in a limited class of problems, as was shown in Bolteus and Tullberg (1981), the conclusions from that investigation do not hold in a general case. In Kelly et al. (1979) the non-symmetric and the symmetric stiffness formulations for potential problems were compared. In the symmetric version also an equilibrium equation was added to ensure equilibrium of the boundary fluxes. Quadratic shape functions were used and discontinuous tractions were taken into account by placing the collocation points inside the elements. In that special example the symmetric matrix gave better results than the non-symmetric. This contradicts the results from the present investigation. Mustoe and Volait (1980) tested the syme-metric matrix with the equilibrium constraint for elastostatic problems. The results were in good agreement with analytical solutions. Hartman (1981) gave a theoretical discussion on the two main problems with the stiffness matrix generated from the direct boundary element method. These are,

1) it is non-symmetric
2) the columns do not sum to zero

In the article he gives some advices on how to fulfil the requirements for a sound stiffness matrix. These will be tested here.

In this paper seven different boundary element stiffness matrices, for two-dimensional elasticity, will be developed and compared. The first is the direct non-symmetric matrix from the direct boundary element method. The second is the symmetric part of this one. Then three variations of the first matrix are deduced with the ideas from Hartman (1981) in mind. The next method follows the idea of Mustoe and Volait (1980). The last method rests upon a modified functional with an equilibrium constraint. Discontinuous tractions are taken into account in all methods. The critical study is done on two numerical examples in which we measure the rate of convergence for the nodal values of the different methods.

The program used is a modified version of BEMSTAT, presented in Bolteus and Tullberg (1981). In the study quadratic elements are used . The singular integrals are calculated analytically and the others with Gauss quadrature.

DIRECT FORMULATION

Problem formulation

The boundary value problem of elastostatics in a finite or infinite domain D with the boundary S, can be defined by a boundary integral equation and boundary conditions as;

(1) $(E+L)u(x) - Kt(x) = 0 \qquad x \text{ on } S$

(2) $u(x) = \bar{u} \qquad x \text{ on } S_1$

(3) $t(x) = \bar{t} \qquad x \text{ on } S_2$

The integral operators working on the displacement vector, u, and the boundary traction, t, can be defined as (if v is a general vector function written as a column matrix)

(4) $$Ev(x) = \int_S D(s)v(s)\delta(x-s)ds = D(x)v(x)$$

(5) $$Lv(x) = \int_S T^t(x,s)v(s)ds$$

(6) $$Kv(x) = \int_S U(x,s)v(s)ds$$

In equation (4) the sifting property of the Dirac function has been used. The kernel functions D,T and U are, in a n-dimensional case, n by n matrices. D depends on the geometrical shape of the boundary and for a smooth boundary we have

(7) $$D = 0,5\delta_{ij} \qquad i,j = 1,n$$

where the Kronecker delta is understood. The kernel functions U and T are the fundamental matrices for the Navier differential operator. If a concentrated force, with the direction vector e, acts in a point X in an infinite and elastic space the displacement and traction (on a boundary in this space) can be measured in an arbitrary point x with aid of the fundamental matrices as

(8) $$u(x) = U(x,X)e$$

(9) $$t(x) = T(x,X)e$$

where u, t, and e are column matrices, $n*1$, and U and T are $n*n$ matrices. U is symmetric and depends only on the distance between the points, but T is non-symmetric and depends both on the distance and the angle between the normal and the vector e.

In this paper we will concentrate on the two-dimensional case and special attention will be paid to the points of discontinuous tractions. These appears in sharp corners and where load is increased stepwise. We denote these points with z and the idea is shown in Figure 1.

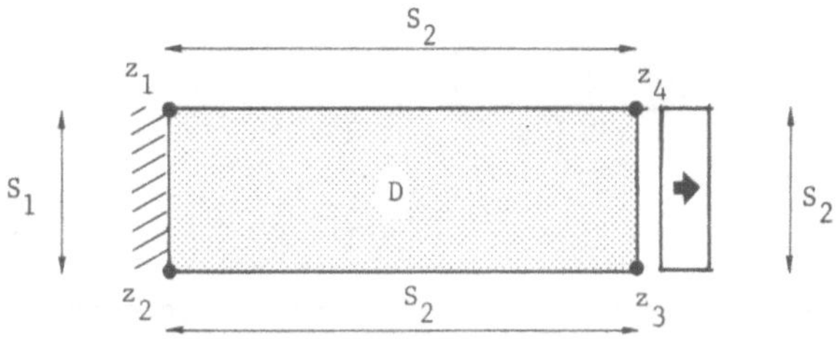

Fig. 1 Definition of the boundary conditions and the additional points in which the tractions is discontinuous.

Numerical formulation

In the numerical formulation the boundary is divided into nodes and elements on which local basis functions of polynomial type are defined for the function approximation. Further, the collocation method is used with the collocation points placed in the nodes on the boundary. The integration method used is described in Bolteus and Tullberg (1981). Here, we only mention that the singular integrals are integrated analytically and the others with Gauss quadrature.

The displacement field on the boundary is approximated with continuous basis functions;

$$(10) \qquad u(x) = \Phi_1^t(x)\tilde{u}_1 + \Phi_2^t(x)\tilde{u}_2 = \Phi^t(x)\tilde{u}$$

where u is a 2*1 column matrix, u tilde a 2N*1 column matrix with nodal displacements (N is the number of nodes). In the first expression in Equation (10) the submatrices includes the nodal variables belonging to boundary type 1 or 2. The matrix of basis functions can be expressed in block matrix form as

$$(11) \qquad \Phi(x) = [\Phi_1(x) \quad \Phi_2(x)]^t =$$

$$= [\underline{\phi}_1(x) \quad \underline{\phi}_2(x) \ \dots \ \underline{\phi}_N(x)]^t \qquad [2N\times 2]$$

In the last expression the subscripts denote node numbers and the shape function matrix for a node can be written as

$$(12) \qquad \underline{\phi}_I(x) = \phi_I(x)\delta_{ij} \qquad\qquad i,j = 1,2$$

A set of linear basis functions are shown in Figure 2.

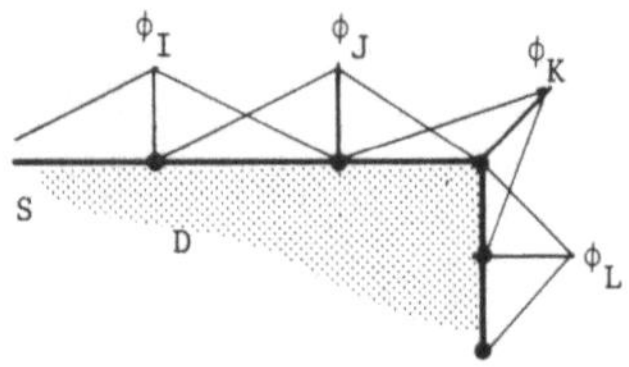

Fig. 2 Linear basis functions on boundary elements

The displacement will be continuous in the nodes. However, the traction have jumps in the z-points defined in Figure 1. To be able to take this into account we approximate the traction as

$$(13) \quad t(x) = \Psi_1^t \tilde{t}_1 + \Psi_2^t \tilde{t}_2 + \Psi_s^t \tilde{t}_s$$

$$= \Psi^t \tilde{t} + \Psi_s^t \tilde{t}_s$$

where t is a 2*1 column matrix and t tilde a 2N*1 column matrix with nodal tractions. The matrix of basis functions are defined in the same manner as for the displacements

$$(14) \quad \Psi(x) = [\Psi_1(x) \quad \Psi_2(x)]^t =$$

$$= [\underline{\psi}_1(x) \quad \underline{\psi}_2(x) \ \dots \ \underline{\psi}_N(x)]^t \qquad [2N\times 2]$$

$$(15) \quad \underline{\psi}_I(x) = \psi_I(x)\delta_{ij} \qquad i,j = 1,2$$

In the first definition in Equation (14) the subscripts denote on which boundary part the node is located. In the z-points where the traction experience a jump we place two nodes, with no element between. The nodes are called master and slave. The reason is that we collocate only in the master node and not in the slave. The traction must be prescribed in the slave node. Thus, no unknowns will be coupled to this node and it will only contribute to the load vector. The basis functions and nodal values coupled to the slaves are separeted from the others and denoted with subscript s in Equation (13). The idea is shown in Figure 3.

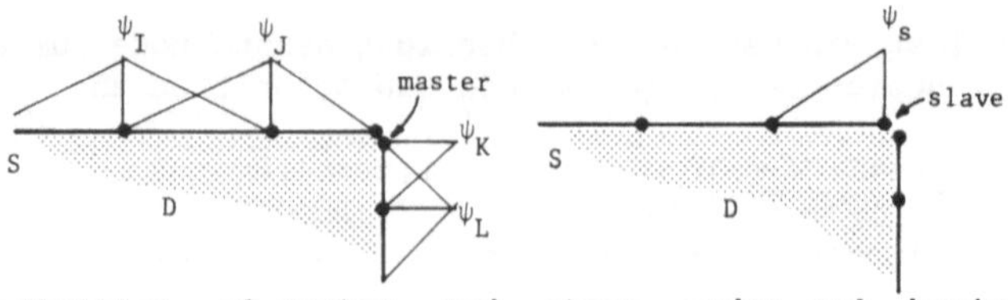

Fig. 3 Definition of master and slave nodes and basis functions

The number of slave nodes are as many as the z-points.

Other ways of treating this problem have been reported by Chaudonneret (1978), Alarcon et al. (1979), and by Mustoe and Volait (1980).

Introducing the function approximations in the boundary integral equation (1) and collocating in the nodes, the following system of matrices arise

$$(16) \quad H\tilde{u} = G\tilde{t} + \tilde{g}$$

where

$$(17) \quad H_{IJ} = \int_S T^t(\tilde{x}_I, s)\Phi_J(s)ds + D(\tilde{x}_I)\delta_{IJ} \quad [2N\times 2N]$$

no sum on I

$$(18) \quad G_{IJ} = \int_S U(\tilde{x}_I, s)\Psi_J(s)ds \quad [2N\times 2N]$$

$$(19) \quad \tilde{g}_I = \int_S U(\tilde{x}_I, s)\Psi_s(s)ds\tilde{t}_s \quad [2N\times 1]$$

In the conventional BEM the system in Equation (16) is rearranged with respect to the boundary conditions to a solvable system of equations with both displacements and tractions as unknowns.

It is possible to form a stiffness relation directly from Equation (16) which was shown by Brebbia (1978). If the equation first is pre-multiplied with the inverse of G, the tractions can be expressed as

$$(20) \quad \tilde{t} = C\tilde{u} - \tilde{s}$$

where

$$(21) \quad C = G^{-1}H$$

$$(22) \quad \tilde{s} = G^{-1}\tilde{g}$$

Then we have to find a matrix that converts the nodal tractions to nodal point forces which gives us the final system. However, in the case with discontinuous tractions some care must be taken in this step. If we form equivalent nodal point forces from the tractions, in the same manner as in FEM,

$$(23) \quad \tilde{P} = \int_S \Phi t ds$$

where we have used the displacement basis functions as weight-functions, and insert the approximate expression of the traction the desired matrix is found in the following equation

(24) $\tilde{P} = M\tilde{t} + \tilde{T}$

where

(25) $M = \int_S \Phi\Psi^t ds$

(26) $\tilde{T} = \int_S \Phi\Psi_s ds\tilde{t}_s$

Notice that the M-matrix in Equation (25) is non-symmetric and can be partitioned in block form as,

(27) $M = \left[\begin{array}{c|c} M_{11} & M_{12} \\ \hline 0 & M_{22} \end{array}\right] \begin{array}{l} 2N-2L \\ 2L \end{array}$ (2L is the number of nodes on S_2)

The expression in Equation (20) can now be inserted in (24) and reordering yields the stiffnes relationship for the direct boundary element method, here called M1;

(28) $S_1\tilde{u} = \tilde{P} + (\tilde{S}-\tilde{T})$

where

(29) $S_1 = CM$

(30) $\tilde{S} = M\tilde{s}$

The stiffness matrix in Equation (28) is not symmetric and the columns do not sum to zero, which implies that equilibrium is not satisfied. The rows sum to zero due to the exact integrations. Hence rigid body motions are possible without that non-zero forces are introduced. The mentioned problems will now be treated in different ways in the succeding.

SYMMETRIC FORMULATIONS

A stiffness matrix should from basic principles be symmetric. The easiest way to achieve this, Brebbia (1978), is to take the symmetric part of the non-symmetric matrix;

(31) $S_2 = (1/2)(MC + C^tM^t)$

This matrix can be deduced in a more rigorous manner by using an energy approach, Zienkiewicz et al. (1977). Here we show it once again but with discontinuous tractions taken into account, since they were not included in that paper.

The potential energy in an elastic body, with no body forces, can be written as;

(32) $$\pi(u,t) = (1/2)\int_S u^t t ds - P_i^t u_i$$

where we have used matrix notation. Introducing the approximating expressions for the traction and the displacement yields,

(33) $$\tilde{\pi}(\tilde{u},\tilde{t}) = (1/2)(\tilde{u}^t M\tilde{t} + \tilde{u}^t\tilde{T}) - \tilde{u}^t\tilde{P}$$

The formulae for the traction, Equation (20), is now introduced in this expression to give a functional which depends only on the displacement variables,

(34) $$\tilde{\pi}(\tilde{u}) = (1/2)(\tilde{u}^t MC\tilde{u} - \tilde{u}^t\tilde{S} + \tilde{u}^t\tilde{T}) - \tilde{u}^t\tilde{P}$$

Minimizing and equating to zero yields the symmetric stiffness matrix in Equation (31) but with a different r.h.s. than in Equation (28);

(35) $$S_2\tilde{u} = \tilde{P} + (1/2)(\tilde{T}-\tilde{S})$$

We call this method M2. A comment on the different r.h.s. is on its place. Already in Equation (27), the parenthesis looked a bit odd, and now we have derived the same equations but came up with another value of the parenthesis it is clear that has to vanish. However, in a normal discretization it takes a finite value. From several tests it was made clear that the r.h.s. in (35) gave the best results.

In M2 the stiffness matrix is symmetric but the sum of rows or columns do not equal zero. Hartman (1981) discussed the reasons why, and pointed out that the error is in the C-matrix. However, we cannot make any corrections in this so he advices us to make them on CM. Thus, let us sum the x- and y-terms separately and for each column in CM. Then we are able to form different methods. In M3 the error is divided by N and added to each term x- and y- respectively. Then it is symmetrized as in M2. In M4 the error is placed in the diagonal and then the matrix is symmetrized. The stiffness matrix will be symmetric in M3 and M4 but the sum of rows and columns will not equal zero due to the symmetrization. To force the sums to zero we first symmetrize CM and then calculate the error in the row or column and add it to the diagonal. This method is called M5.

A different approach to assure the equilibrium of the matrix has been proposed by Mustoe and Volait (1980) and Kelly et al. (1979). Here, they add the constraint

(36) $$\int_S t ds = 0$$

i.e. the sum of the traction should equal zero. This equation was added to the system in Equation (16). In the discontinuous approach presented in this paper this take the form;

$$(37)\quad \begin{bmatrix} H & 0 \\ 0 & 0 \end{bmatrix} \begin{bmatrix} \tilde{u} \\ 0 \end{bmatrix} = \begin{bmatrix} G & Q \\ Q^t & 0 \end{bmatrix} \begin{bmatrix} \tilde{t} \\ \lambda \end{bmatrix} + \begin{bmatrix} \tilde{g} \\ r \end{bmatrix}$$

or

$$(38)\quad H^*\tilde{u}^* = G^*\tilde{t}^* + \tilde{g}^*$$

where

$$(39)\quad Q = \int_S \Psi ds$$

$$(40)\quad r = \int_S \Psi_s ds \tilde{t}_s$$

in which λ takes the place of a Lagrangian multiplyer (2*1 matrix) which introduces a controlled pertubation in each equation. After inversion, multiplication and partitioning, the traction can be written in the same way as in Equation (20),

$$(41)\quad \tilde{t} = C^*\tilde{u} - \tilde{s}^*$$

with

$$(42)\quad C^* = G^{*-1}H^*$$

$$(43)\quad \tilde{s}^* = G^{*-1}\tilde{g}^*$$

Then symmetrizing as in M2 gives us the sixth method, M6. The next method is proposed by the authors and is a variation of M6. In the method we introduce a modified functional with the constraint in Equation (36). Thus, we may write

$$(44)\quad \pi^*(u,t) = \pi(u,t) + \lambda^t \int t ds$$

where we have used the functional in Equation (32). Introducing the approximating expressions for the functions and making use of the boundary integral formulation of the tractions in Equation (20), the modified functional will only depend on the nodal displacements and the Lagrangian multiplier,

$$(45)\quad \tilde{\pi}^*(\tilde{u},\lambda) = \tilde{\pi}(\tilde{u}) + \lambda^t(\int \Psi^t ds\tilde{t} + \int \Psi_s ds \tilde{t}_s)$$

$$= \tilde{\pi}(\tilde{u}) + \lambda^t(Q^t C\tilde{u} - Q^t\tilde{s}) + \lambda^t r$$

Minimizing and equating to zero yields

$$(46)\quad \begin{bmatrix} S_2 & C^tQ \\ Q^tC & 0 \end{bmatrix} \begin{bmatrix} \tilde{u} \\ \lambda \end{bmatrix} = \begin{bmatrix} \tilde{P} \\ r \end{bmatrix}$$

This method is called M7 and is the last one to be studied. We notice that M7 needs more computations than M6.

CRITICAL STUDY

The critical study of the seven methods is done with aid of two numerical examples. The plate in Figure 4 is in the first example loaded with a stretching force. In the other example the load bends the plate and it then works as a high cantilever beam. Three different boundary meshes are used with 8, 16, and 24 elements respectively.

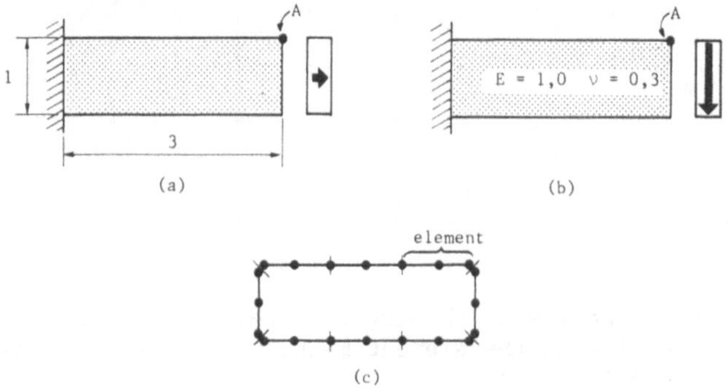

Fig. 4 a) stretching problem
b) bending problem
c) 8-element mesh

The displacement in point A will be studied since the error there is about the largest. We now assume that the rate of convergence of the displacements in the nodal points can be written in the following form

$$(47)\quad |u - \tilde{u}| \leq ch^{\alpha}||u||_{H^r}$$

where u is the exact solution and u-tilde the numerical result, c is a generic constant, h is the mean element length and on the r.h.s. some kind of norm is defined. Error bounds and rate of convergence for BE-formulations have been presented in Hsiao and Wendland (1981). Results for nodal quantities are also given.

The results from this critical study is presented in Figure 5. Two things can be read from the figures; 1) The relative error for the different methods can be compared for same discretizations, 2) The slope on the curves measures the rate of convergence.

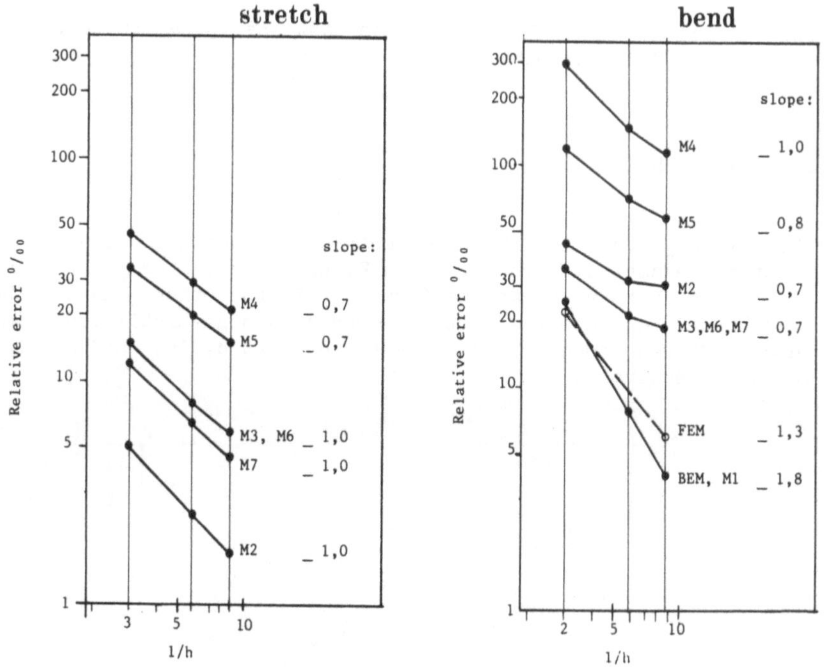

Fig. 5 Error analysis showing the relative error, as a function of the element mesh, and rate of convergence (slope.)

In the first example, the results for the conventional BEM and M1 (non-symmetric matrix) was correct to 6 figures already for the 8-element mesh. All methods are in error less than 5%. The M2-method (only symmetrized) gives very good results in this simple example and the rate of convergence is approximately h. M3, M6, and M7 shows a similar behaviour. M4 and M5 have the lowest accuracy.

The second example, which is more critical, shows the very different behaviour between M1 and the other methods. First we notice that the relative error is only 2% or lower for the conventional BEM and M1, and the rate of convergence is almost h^2. For the other methods the rate of convergence is only about h or less. Moreover, the manipulated methods M2-M7 shows a decreasing convergence as the number of elements are increased. The three methods with equilibrium control, M3, M6, and M7 are in this case better than M2 which is the symmetrized version without equilibrium correction. M4 and M5 give bad results.

A test was also made in this case with a 8-node isoparametric FEM. The test showed that, with the same number of boundary nodes, the conventional BEM and M1 gives better results and a higher rate of convergence.

The error of the column sum in M1 is studied in Figure 6. This shows that the error is about 1 or 2% of the diagonal element, in a normal mesh. These figures were also obtained by Mustoe and Volait (1980).

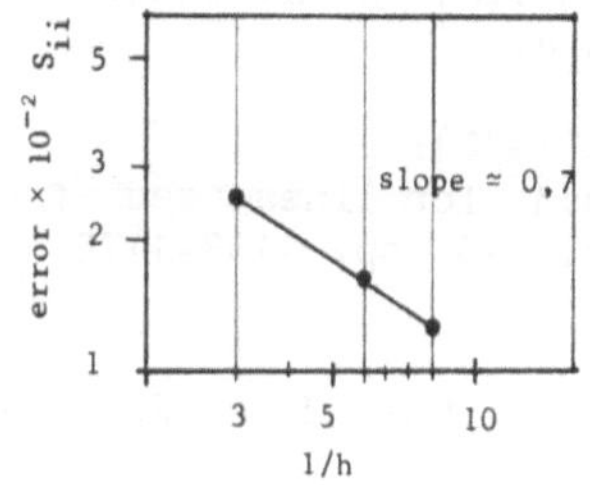

Fig. 6 The sum of the column terms in M1 in percent of the diagonal element

CONCLUSIONS

From the critical study it is possible to draw the following important conclusions,

(i) The direct non-symmetric stiffness matrix is the best one.

(ii) The direct non-symmetric stiffness matrix is as good as, or better, than a FE-matrix.

(iii) If one wants a symmetric matrix M3, M6 or M7 should be used. M6 needs less computations.

It should be noted that quadratic elements have been used. If linear elements were used, e.g. in the bending problem, about 100 elements would be needed to reproduce the results given here. The result should not converge at all if constant elements were used, as was shown in Bolteus and Tullberg (1981).

At last we point out that the CPU-times have not been measured since the program was not implemented in an optimal manner. However, it is easy to understand that the CPU-time for these methods are much larger than for an equivalent FEM since an inversion of a full non-symmetric matrix is needed.

REFERENCES

Lysmer, J. and Kuhlemeyer, R. (1969):
Finite dynamic model for infinite media, J. Eng. Meth. Div., ASCE, 95.

Waas, G. (1972):
Linear two-dimensional analysis of soil dynamic problems in semi-infinite layered media, Ph. D. Thesis, Univ. of California, Berkely, California.

Mei, C.C. and Chen, H.S. (1976):
A hybrid element method for linearized free surface flows, Int. J. Num. Meth. Eng., 10, pp. 1153-1175

Bettes, P. (1977):
Infinite elements, Int. J. Num. Meth. Eng., 11, pp. 53-64.

Bettes, P. and Zienkiewicz, O.C. (1977):
Diffraction and refration of surface waves using finite and infinite elements, Int. J. Num. Meth. Eng., 11, pp. 1271-1290.

Osias, J.R. et al. (1977):
Proc. First Symp. Num. Anal. Appl. Eng. Sci., Versailles, France 23-27 May.

Zienkiewicz, O.C., Kelly, D.W. and Bettes, P. (1977): The coupling of finite element and boundary solution procedures, Int. J. Num. Meth. Engng., Vol. 11, pp. 355-376.

Brebbia C.A. (1978): The Boundary Element Method for engineers. Pentech Press, London.

Chaudonneret, M. (1978):
On the discontinuity of the stress vector in the boundary integral equation method for elastic analysis, Int. Symp. Recent Adv. BEM, Southampton, England, July.

Shaw, R.P. (1978):
Coupling boundary integral method to othe numerical techniques, Int. Symp. Recent Adv. BEM. Southampton, England, July.

Alarcon, E., Martin, A., and Paris, F. (1979):
Boundary elements in potential and elasticity theory, Comp. Struc., 10, pp. 351-362.

Brebbia, C.A. and Gergiou, P. (1979):
Combimation of boundary and finite elements in elastostatics, Appl. Math. Modelling, 3.

Kelly D.W., Mustoe G. and Zienkiewicz O.C. (1979): Coupling boundary element methods with other numerical methods. In Developments in Boundary Element Methods -1, Applied Science Publishers Ltd., London.

Mustoe, G. and Volait, F. (1980): A symmetric direct boundary integral equation method for two-dimensional elastostatics. Paper presented at 2nd Int. Sem. on Boundary Element Methods, Southampton.

Bolteus, L. and Tullberg, O. (1981): BEMSTAT - a new type of boundary element program for two-dimensional eleasticity problems. In Boundary Element Methods (Edt. C.A. Brebbia), Springer-Verlag, Berlin.

Hartman, F. (1981): The derivation of stiffness matrices from integral equations. In Boundary Element Methods (Edt. C.A. Brebbia), Springer-Verlag, Berlin.

Hsiao G. and Wendland, W.L. (1981): The Aubin-Nitsche lemma for integral equations, J. Int. Eq., 3, pp. 299-315.

Murakami, H., Shioya, S., Yamada, R., and Luco, E. (1981): Transmitting boundaries for time harmonic elastodynamics on infinite domains, Int. J. Num. Meth. Eng., 17, pp. 1697-1716.

Taylor, E. and Zietsman, J. (1981): A comparison of localized finite element formulations for two-dimensional wave diffraction and radiation problems, Int. J. Num. Meth. Eng., 17, pp. 1355-1384.

Gupta, S., Penzien, J., Lin, T. and Yeh, C. (1982): Three-dimensional hybrid modelling of soil-structure interaction, Earthquake Eng. Struc. Dyn., 10, pp. 69-87.

ANALYTICAL COMBINATION OF BOUNDARY ELEMENT METHOD AND THIN-WALLED SEGMENT METHOD AND IT'S APPLICATION TO BOX GIRDER BRIDGES

S. Komatsu* and M. Nagai**

* Department of Civil Engineering, Osaka University, Osaka, Japan
** Steel Structure and Industrial Equipment Division,
Kawasaki Heavy Industries, Ltd., Tokyo, Japan

INTRODUCTION

In large thin-walled structures such as large-span box girder bridges and cable-stayed bridges, the stress distribution in the plate elements are most likely to deviate from those predicted by the primary bending theory or the torsion-bending theory applicable to only the frame work consisting of slender bar elements. That is to say, the shear-lag phenomena and the distortion of the cross-section often occurs in the box girder so that the reactions on the supports are sometimes different from the values calculated by the conventional grid analysis. It may be considered as a main cause of such phenomena that the reactions are strongly affected by the local deformation of diaphragms on the end supports. This is serious problem for both skew girder and curved girder. These stress deviation as induced mainly due to thinness of plate element may often causes miserable accident.[1] Hence, the sophisticated method for estimating more accurately the stress deviation in large thin-walled structures should be developed. For analysing such structures including simultaneously both the primary and deviating stress distribution, it is well-known fact that the finite element method is a powerful tool. However, unfortunately, necessity of large CPU time and large memory size become serious defect for practical application. Furthermore, for the case of large stress gradient, the accuracy of numerical results tends to decrease because of element discretisation. Nevertheless this characteristics is almost inevitable, such tendency injures substantially reliability of this method.

The boundary element method[2],[3] is suitable for evaluating the stress distribution with large gradient and also is expected to reduce the required data rather than FEM. Especially, time-consuming for data preparation is an important factor from practical point of view in structual design.

This paper proposes the happy combination of the boundary element method[4] and the thin-walled segment method[5] developed by the authors for three-dimensional problems utilizing the characteristics of the both approaches. In the combination method, the former can be applied to the regions including the large stress gradient, while the latter is used in the regions with comparatively small stress variation. Then the proposed method can be extended to solve the thin-walled box girder bridges.

Several realistic problems encountered in both numerical analysis of BEM and the combination technique such as the singular integration at corner points on the boundary, the discontinuity of surface tractions and the inaccuracy of numerical solution in the close vicinity of the boundary are discussed first. Then, some numerical examples are given to verify the validity of this approach. Next, the combining technique is described for the three-dimensional thin-walled structures such as the box girder bridges. As a numerical example, a box girder with right supports as well as a skew box girder are analysed for making a comparison between the analytical solution obtained by the combination method and by the conventional FEM as well as the experimental results.

NUMERICAL ANALYSIS

A number of problems encountered in numerical computation are discussed. The concrete problems caused by introduction of some higher order interpolation functions and improvement of the numerical solutions in the vicinity of the boundary are discussed in detail.

Singular integration at a corner point on the boundary

For particular node i on the boundary, a foundamental equation of BEM is written as follows:

$$[C]\{U\}_i+\sum_j\int_{S_j}[P^*]\{U\}dS_j=\sum_j\int_{S_j}[U^*]\{P\}dS_j \tag{1}$$

writing eq. (1) for all nodes on the boundary

$$[H]\{U\}=[G]\{P\} \tag{2}$$

is obtained. Now, considering the case where rigid body motion takes place, the following equation with respect to matrix [C] is obtained.

$$[C]=-\int_{S-\varepsilon}[P^*]ds \tag{3}$$

Eq. (3) gives nothing else but the indirect evaluation of the singular integration.

Discontinuity of surface traction

The discontinuity of surface tractions as shown in Figure 1 is easily treated by rearranging matrix [G] as follows:

$$[H]\{U\}=[\cdots\cdots|\overset{p_{i(-1)}}{g_{i(-1)}}|\overset{p_{i(+)}}{g_{i(+)}}|\cdots\cdots]\{P\} \tag{4}$$

where $[g_{i(-1)}]$ and $[g_{i(+)}]$ are the matrix associated with element $(i-1)\sim i$ and $i\sim(i+1)$, respectively.

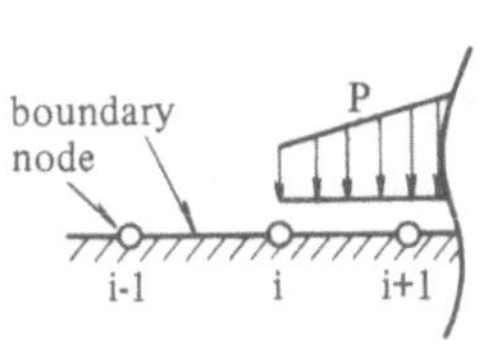

Figure 1 Discontinuity of surface tractions

Improvement of numerical solution at inner point near the boundary
It was seen that the error included in numerical solution at inner point becomes seriously large as the distance of that point to the boundary tends to zero. Generally the displacements in the region $\{U_i\}$ are determined as

$$\{U_i\}=[-\int_S[P^*]\{U\}ds+\int_S[U^*]\{P\}ds] \tag{5}$$

If, the distance between the inner point and the boundary tends to zero, however, eq. (5) is not valid. So, let us now consider again the rigid body motion, the alternative equation giving the displacements within the region is given as follows:

$$\{U_i\}=[C]^{-1}[-\int_S[P^*]\{U\}ds+\int_S[U^*]\{P\}ds] \tag{6}$$

Then, the strain $\{U_i\}_{,K}$ becomes

$$\{U_i\}_{,K}=[C]^{-1}[\int_S[P^*_{,K}]ds\{U_i\}-\int_S[P^*_{,K}]\{U\}ds+\int_S[U^*_{,K}]\{P\}ds] \tag{7}$$

where ( ),k is the differentiation with respect to $k(k=x,y)$. Eq. (7) also satisfies the condition of the rigid body motion.

NUMERICAL EXAMPLE

Uni-compressed square plate
A square plate with the width 10 cm is uniformly compressed. Only eight nodes and four quadratic elements are arranged for whole boundary. A complete agreement with analytical solutions is recognized at any points in the region.

Uni-compressed circular plate
A circular plate with the radius 10 cm is compressed uniformly in the radial direction. Linear and quadratic elements are adopted and various number of nodes are arranged for whole boundary and the radial displacement on the boundary is compared with each other. Superiority of high-order interpolation function regarded as the approximation of the geometry can be clearly recognized.

Stress concentration of plate with circular hole
A square plate with a circular hole as shown in Figure 2 is analysed. From symmetry of the problem, only a quarter of the model is treated. Two types of boundary element discretization are used and the numerical results are shown in Figure 3. The stress concentration can be sufficiently evaluated by the model with 36 nodes.

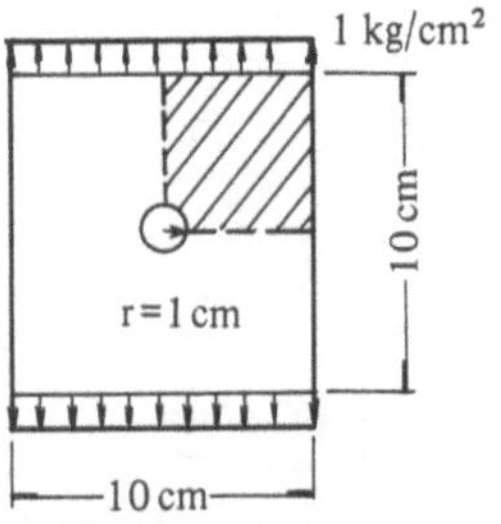

Figure 2 **Square plate with a circular hole**

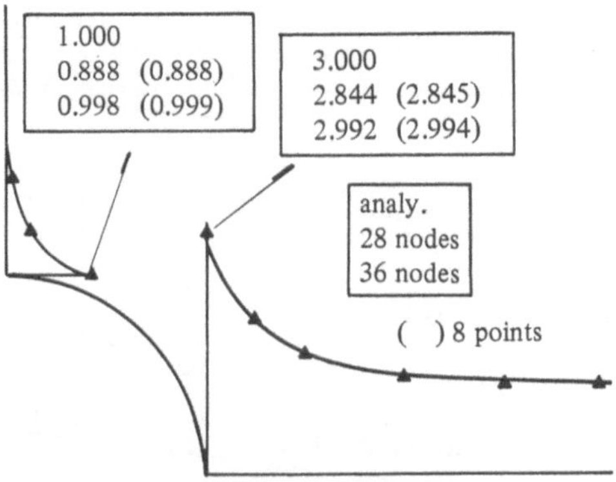

Figure 3 **Stress concentration**

Improvement of numerical solutions near the boundary

Figure 4 shows the displacement within a circular plate submitted to uniform pressure. The boundary of the circular plate is divided into eight quadratic elements and the numerical integration is executed by one-dimensional Gaussian quadrature with 8 integration points. The white circle ○ and black one ● show the numerical results obtained by eq. (5) and eq. (6), respectively. Table 1 shows the values of r/R for which the error falls within 3%. A fine improvement has been attained.

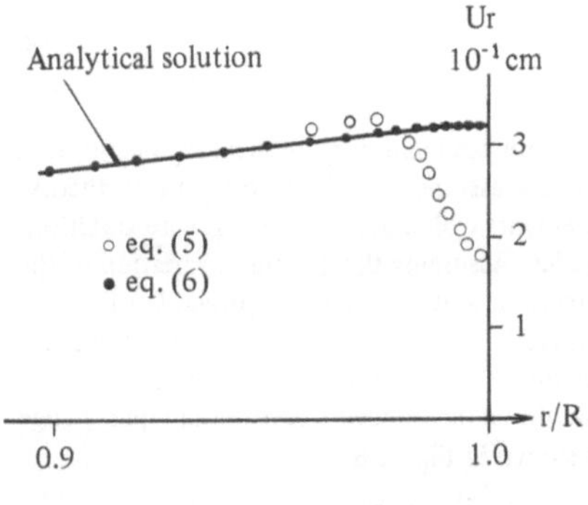

Figure 4 **Improvement of the displacement near the boundary**

Table 1 **r/R with Error less than 3% of Radial Normal Stress of Circular Plate**

Points of Gauss quad.	r/R	(Eq. 7)
4	0.80	(0.920)
6	0.92	(0.960)
8	0.93	(0.980)
12	0.96	(0.990)
16	0.98	(0.994)

ANALYTICAL COMBINATION OF BEM AND THIN-WALLED SEGMENT METHOD (TSM) FOR THIN-WALLED BOX GIRDER

The box girder is divided into three parts. Then, the boundary element and the thin-walled segment are used in both end parts and an intermediate part, respectively, as shown in Figure 5. The variation of both the displacements and the tractions can be expressed by a quadratic function on the borderline between two kinds of parts. Therefore, eight nodes only have to be arranged at the borderline as shown in Figure 6.

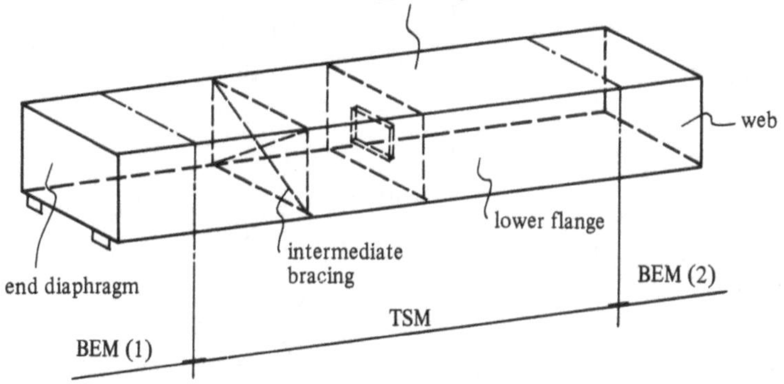

Figure 5 Box girder bridge

Since the boundary element formulation for three-dimensional thin-walled structures has been reported in Ref. 4, the thin-walled segment formulation for the box girder analysis is presented here briefly.

The displacement fields of the box girder are assumed as the following:

$$u(s,\ z)=\sum_{i}^{8}\phi^{(i)}(s)\cdot U_i(z) \tag{8}$$

$$v(s,\ z)=\sum_{j}^{4}\psi^{(j)}(s)\cdot V_j(z) \tag{9}$$

where $\phi^{(i)}(s)$ and $\psi^{(j)}(s)$ are the generalized coordinates including the shear lag phenomena and the distortional modes together with the modes assumed by the primary beam theory, while $U_i(z)$ and $V_j(z)$ are the generalized displacements which denote the spanwise variations of the shear lag, the distortion and the other modes. Assuming the quadratic variation of the generalized displacements in the z (spanwise) direction, and applying the principle of virtual displacement, the stiffness equation for segment is easily derived. Figure 7 shows the freedoms of thin-walled segment. For the case where the thin-walled segment is connected with the boundary element, however, a specific element should be introduced to avoid the rapid change of the stiffness between two kinds of elements as shown in Figure 8.

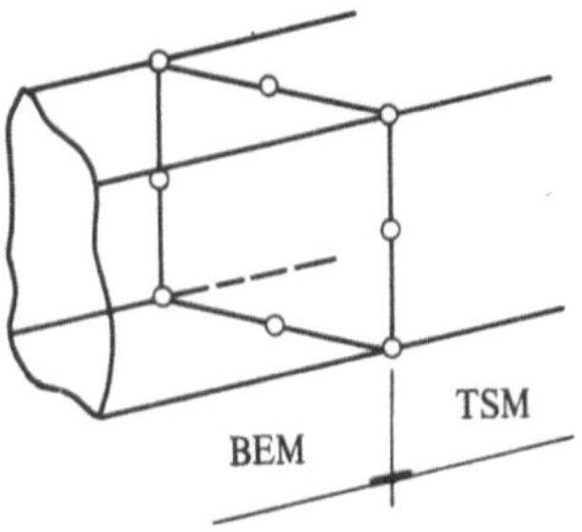

Figure 6 Nodes arrangement at connecting borderline

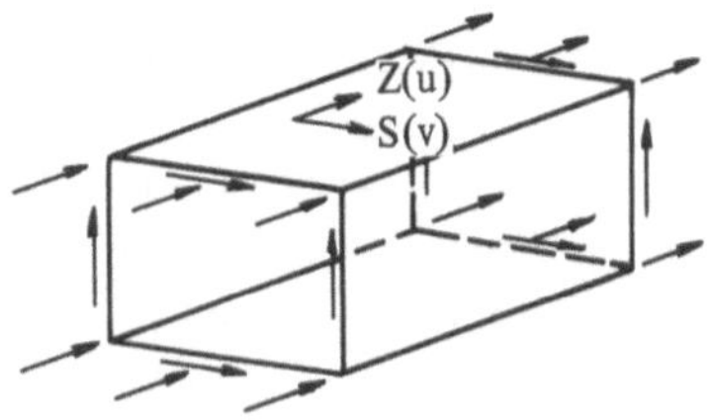

Figure 7 Freedoms of TSM

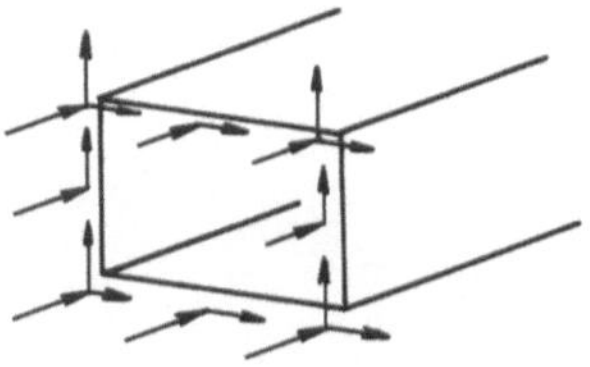

Figure 8 Freedoms of TSM at connecting borderline

Now the elastic equations for the region of the boundary element, together with segment stiffness equation are written as follows:[6)]

$$[H_O^{(1)} \vdots H_i^{(1)}]\begin{Bmatrix}U_O^{B(1)}\\ U_i^{B(1)}\end{Bmatrix} = [G_O^{(1)} \vdots G_i^{(1)}]\begin{Bmatrix}P_O^{B(1)}\\ P_i^{B(1)}\end{Bmatrix} \tag{10}$$

$$[K_i^{(1)} \vdots K_O \vdots K_i^{(2)}]\begin{Bmatrix}U_i^{F(1)}\\ U_O^{F}\\ U_i^{F(2)}\end{Bmatrix} = [M_i^{(1)} \vdots L \vdots M_i^{(2)}]\begin{Bmatrix}P_i^{F(1)}\\ F_O\\ P_i^{F(2)}\end{Bmatrix} \tag{11}$$

$$[H_i^{(2)} \vdots H_O^{(2)}]\begin{Bmatrix}U_i^{B(2)}\\ U_O^{B(2)}\end{Bmatrix} = [G_i^{(2)} \vdots G_O^{(2)}]\begin{Bmatrix}P_i^{B(2)}\\ P_O^{B(2)}\end{Bmatrix} \tag{12}$$

where (ℓ), $\ell=1$ and 2 corresponds to boundary element regions (1) and (2), respectively. $\{F_O\}$ denotes the nodal force vector which is not linked to the boundary elements and $[M_i]$ is defined for borderline S_i of element as follows:

$$[M] = \int_{S_i}\{N_1\}^T\{N_2\}ds_i = \int_{-1}^{1}\{N_1\}^T\{N_2\}|J|d\eta \tag{13}$$

where $\{N_1\}$ is the vector including interpolation functions associated with displacements, $\{N_2\}$ is the vector associated with traction and $|J|$ the Jacobian. The vector $\{N_1\}$ and $\{N_2\}$ are given as follows:
For linear element

$$\{N_1\}\text{or}\{N_2\} = \left[\frac{1}{2}(1-\eta) \vdots \frac{1}{2}(1+\eta)\right], \quad |J| = \frac{1}{2}\sqrt{(x_j-x_i)^2+(y_j-y_i)^2} \tag{14}$$

For quadratic element

$$\{N_1\}\text{or}\{N_2\}=\left[\frac{1}{2}\eta(\eta-1) \vdots 1-\eta^2 \vdots \frac{1}{2}\eta(\eta+1)\right]$$

$$|J|=\sqrt{\{(x_i-2x_j+x_K)\eta+(-\frac{1}{2}x_i+\frac{1}{2}x_K)\}^2 + \{(y_i-2y_j+y_K)\eta+(-\frac{1}{2}y_i+\frac{1}{2}y_K)\}^2} \tag{15}$$

where x_m and y_m (m=i, j, k) are the nodal coordinates and η the dimensionless coordinate. And [L] is defined as follows:

$$[L]=\begin{bmatrix} & 0 & \\ 1 & & \\ & \ddots & \\ & & 1 \\ & 0 & \end{bmatrix} \text{segment region} \tag{16}$$

Using the equilibrium and compatibility conditions on the borderline, the resulting system of equations is obtained as follows:

$$\begin{bmatrix} H_o^{(1)}-G_i^{(1)}H_i^{(1)} & & & & \\ M_i^{(1)}K_i^{(1)} & & & & \\ & \ddots & & & \\ & & K_o & & \\ & & & \ddots & \\ & & & & K_i^{(2)} \quad M_i^{(2)} \\ & & & & H_i^{(2)}-G_i^{(2)}H_o^{(2)} \end{bmatrix}\{\delta\}=[B_o^{(1)} \vdots L \vdots B_o^{(2)}]\begin{Bmatrix} P_o^{B(1)} \\ F_o \\ P_o^{B(2)} \end{Bmatrix} \tag{17}$$

where $\{\delta\}^T=\{U_o^{B(1)}, P_i^{(1)}, U_i^{(1)}, U_o^F, U_i^{(2)}, P_i^{(2)}, U_o^{B(2)}\}$

The stress concentration occuring in the end diaphragms, the complex stress distribution of main box girder near the supports, the ordinary shear lag phenomena of the main girder as well as the distortional behaviour closely related to a rigidity of both the end diaphragms and intermediate bracings can be simultaneously analysed by using eq. (17). In addition to those above mentioned, the skew girder bridges can be easily treated by the appropriate coordinate transformation in the boundary element regions.

NUMERICAL EXAMPLE FOR BOX GIRDER

Comparison by ordinary finite element analysis

The analytical combination of BEM and TSM is applied to the model girder shown in Figure 9. From symmetry, only a half of the span is analysed. The boundary element method is applied to the region ranging from 0 m to 2 m from the end supports and the remainder is analysed by TSM (Figure 10). The numerical results are compared with those by the ordinary finite element analysis. (981 nodes and 284 elements for half model)

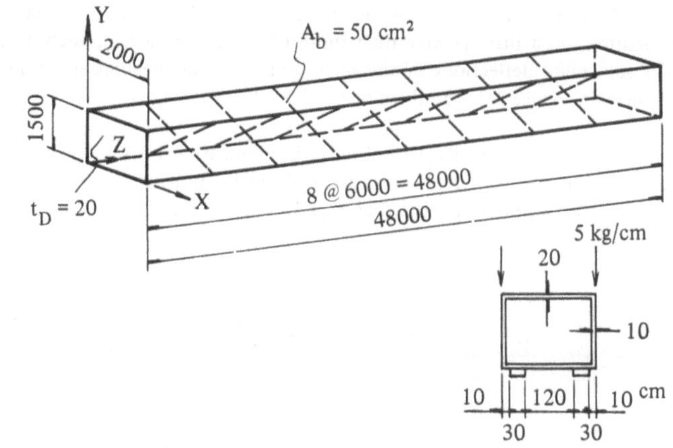

Figure 9 Model girder

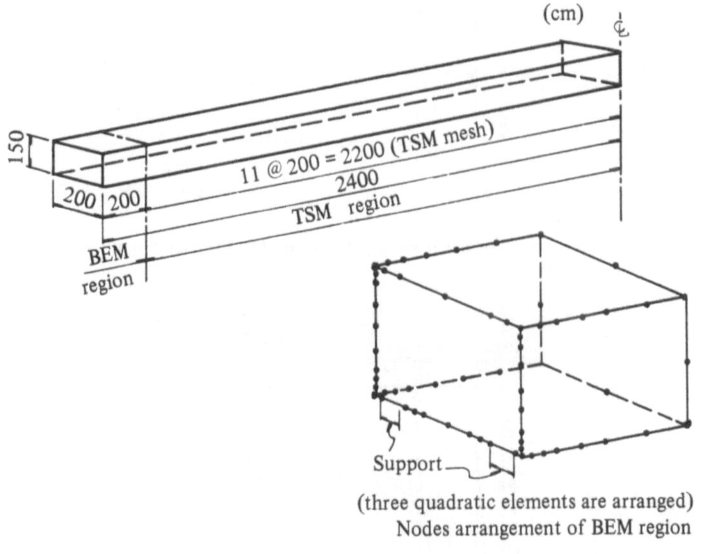

Figure 10 Mesh division of the combination analysis

The normal stresses σ_x and σ_z induced in the end diaphragm due to uniform loads acting along both webs are illustrated in Figures 11 and 12. In these figures, the white circles ○ show the results obtained by the ordinary finite element analysis. A fine agreement between two approaches can be recognized expect for the regions near the supports. The continuous curves expressing the stress concentrations produced near the supports are explicity obtained by the method of analytical combination. It appears that the ordinary finite element approach needs much more fine element discretization to pursue exactly the large stress gradient. This is said for the case of uniform loads acting along only one web.

Table 2 shows the vertical deflections of the end diaphragm. The values obtained by the analytical combination are a little greater than those of the finite element analysis. And with regard to the maximum deflections of the main girder and the bending stress distribution, there exists fine coincidence between the numerical results of two approaches.

Table 3 shows the CPU times really needed until the eliminating process has been carried out by using IBM3033 computer for both approaches. The CPU times in analytical combination method can be seen to be about 30% less than that in ordinary finite element method.

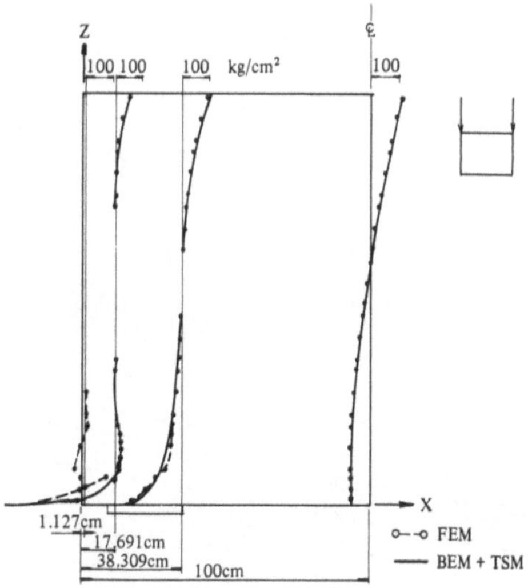

Figure 11 Normal stress σ_x distribution in the diaphragm

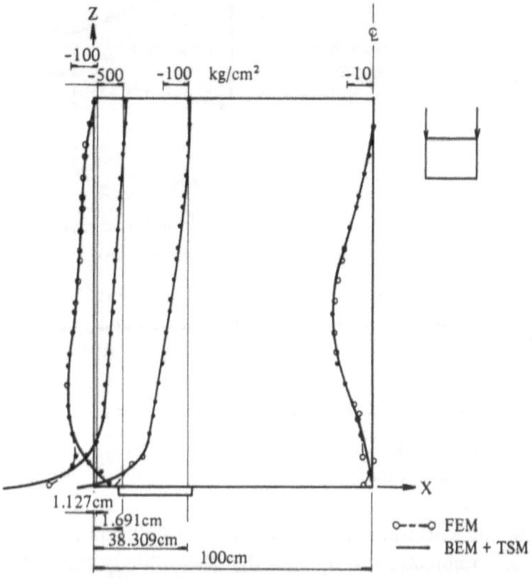

Figure 12 Normal stress σ_z distribution in the diaphragm

Table 2 Vertical deflections of the end diaphragm

(cm)

	Case 1	Case 2
A	−1.0303 (−1.0315)	−0.2054 (−0.2025)
B	−0.5671 (−0.5503)	−0.0890 (−0.0846)
C	−0.3407 (−0.3352)	−0.1703 (−0.1676)
D	−0.3504 (−0.3003)	−0.1527 (−0.1501)
E	−1.0303 (−1.0315)	−1.2357 (−1.2340)
F	−0.5671 (−0.5503)	−0.6560 (−0.6349)

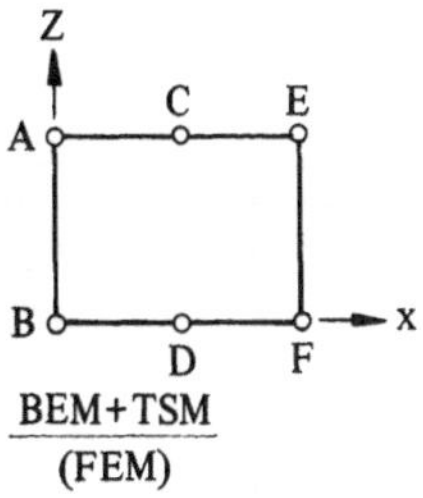

Table 3 Comparison of the CPU times

(sec.)

FEM	BEM + TSM
101	72

Box girder with wide flang

Box girder with wide flang and with comparatively wide distance of intermediate bracings as shown in Figure 13 is analysed by author's combination method. Figure 14 shows the normal stresses (σ_x, σ_y) and the shearing stress (τ_{xy}) in the end diaphragm under uniform loads (25 kg/cm) acting along on both webs. Figure 15 is the normal stress (σ_z) of upper flang of box girder. The intermittent line is the results by uniform loads acting along only on web. It is clearly seen that the complex stress distribution in the end diaphragm and also the non-uniform stress distribution in the main girder due to shear-lag and distortion can be simultaneously obtained by combination analysis.

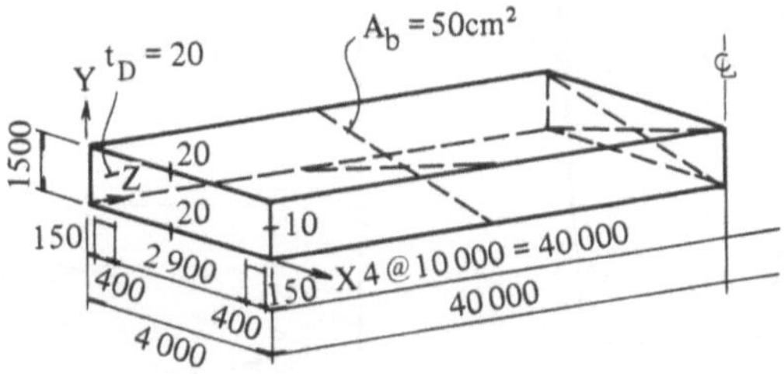

Figure 13 Box Girder with wide flang

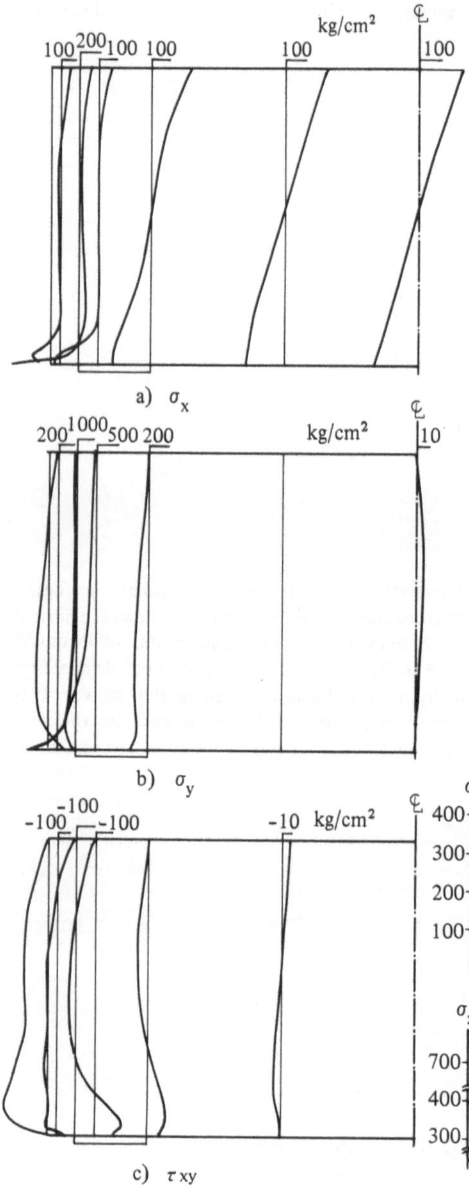

Figure 14 Stress distribution in the diaphragm

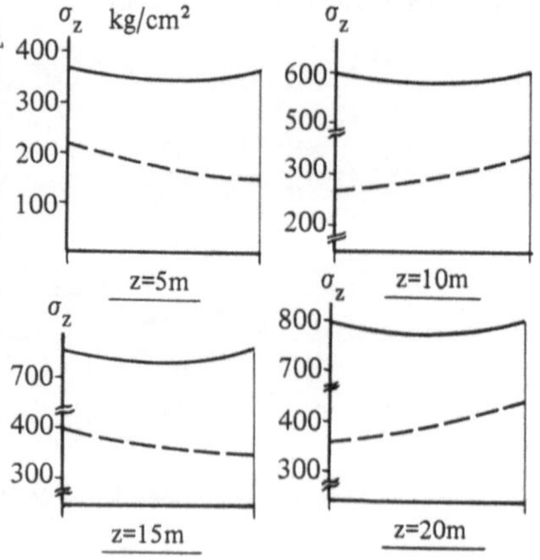

Figure 15 Normal stress σ_z distribution in upper flang

Skew box girder

An experimental research has been performed in order to examine the validity of the proposed combination method. A model of skew girder used in the experiments is shown in Figure 16. The model girder is made of acrylite, of which young's modulus is 2.9 x 10^4 kg/cm^2 at temperature of 28°C. The vertical deflection of the both webs of the girder are shown in Figure 17. The stresses in the end diaphragm are shown in Figure 18. In all cases, good agreement between analytical values and experimental ones can be seen.

Table 4 shows the reactions obtained by two theoretical approaches and those measured in experiments. In the table, BEAM means a grid analysis by means of the displacement method for frame structure based on the beam theory. The great difference in the reactions between the beam theory and the experiments can be apparently seen. While fair coincidence between the theoretical results and the experimental ones can be seen except for small uplift reactions.

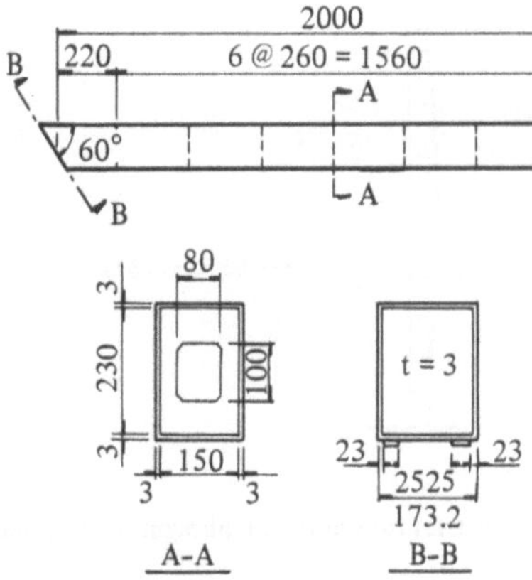

Figure 16 Test girder

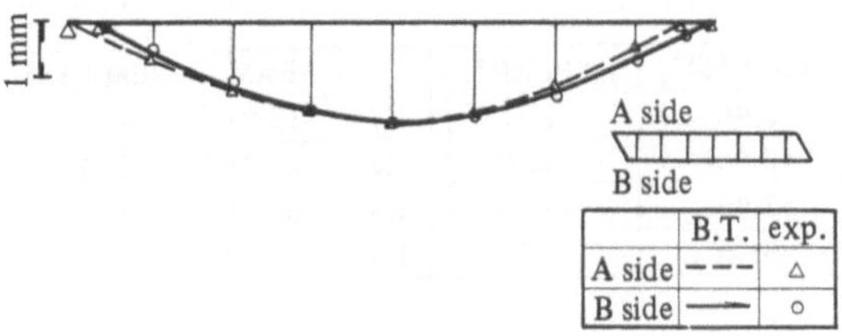

Figure 17 Deflection of the girder

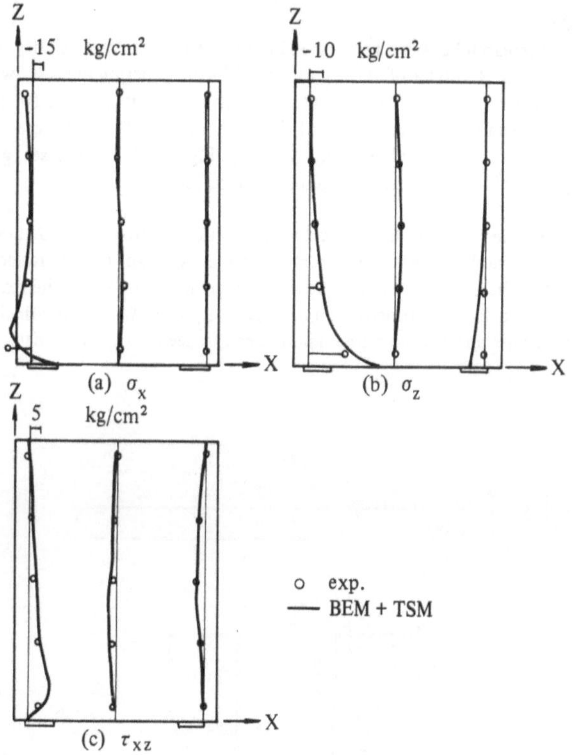

Figure 18 Stress distribution in the diaphragm for L/2 loading

Table 4 Reactions

unit : kg

(a) 1/2·L LOADING

	BEAM	BEM + TSM	EXPERIM.
R1	−17.36	−3.46	1.77
R2	67.36	53.46	48.20
R3	67.36	53.46	48.20
R4	−17.36	−3.46	1.77

(b) 1/4·L LOADING

	BEAM	BEM + TSM	EXPERIM.
R1	−18.91	−9.06	−5.59
R2	42.91	33.12	29.11
R3	68.91	59.01	56.60
R4	7.09	16.95	18.80

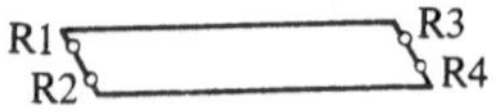

CONCLUSION

The usefulness of the approaches developed by the authors was verified through various kinds of numerical examples.

The analytical combination of the boundary element method and the thin-walled segment method has been developed into three-dimensional analysis of thin-walled structure. In the process of the combining technique, the possibility of using the higher-order interpolation functions for both the displacements and the tractions at the borderline also has been discussed.

The validity of the proposed analytical combination is examined by comparison with the ordinary finite element analysis. The main features of the former as compared with the latter are as follows:
The number of data required can be considerably reduced.
The continuous variations of the stress distribution can be accurately found and the stress concentration phenomena can be pursured effectively.
The time consumption for numerical calculations is very much small as compared with the finite element analysis.

Skew box girder bridges can be easily treated and the validity of the program developed here is verified through some experimental results.

The proposed approach is the most suitable one for making clear the local deformation and stress irregularity together with the overall behaviour of the thin-walled structure such as box girder bridges.

References

1) Loading in Collapse Range, The Consulting Engineering, October, 1970
2) Brebbia, C. A. and Walker, S.: Introduction to boundary element method, Recent Advances in Boundary Element Method, Pentech Press, 1978
3) Brebbia, C. A. and Nakaguma, R.: Boundary elements in stress analysis, Proc. of ASCE, Vol.105, No.EM1, February, 1979
4) S. Komatsu and M. Nagai: Application of Boundary Element Method to Thin-Walled Structures, the 15th JSSC Matrix Structual Analysis Symposium, 1981, Tokyo, Japan
5) F. Sakai and M. Nagai: Three-Dimensional Analysis of Thin-Walled Box Girder Briges by Block Finite Element Method, Proc. of JSCE, No.255, November, 1976 (in Japanese)
6) S. Komatsu, M. Nagai and T. Nishimaki: Analytical Combination of Boundary Element Method and Finite Element Method and it's Application to Thin-Walled Box Girder Bridges, the 15th JSSC Matrix Structual Analysis Symposium, 1981, Tokyo, Japan